ENERGY

Gordon J. Aubrecht II
The Ohio State University
at Marion

Merrill Publishing Company
A Bell & Howell Information Company
Columbus Toronto London Melbourne

**For my children, and
my children's children.
May the Earth abide.**

Published by Merrill Publishing Company
A Bell & Howell Information Company
Columbus, Ohio 43216

This book was set in Garamond

Administrative Editor: David Gordon
Developmental Editor: Wendy Jones
Production Coordinator: Rex Davidson
Art Coordinator: James H. Hubbard
Cover Designer: Cathy Watterson
Text Designer: Cynthia Brunk

Library of Congress Catalog Card Number: 88–61047
ISBN: 0–675–20426–7
Printed in the United States of America
1 2 3 4 5 6 7 8 9—92 91 90 89

Cover Photo: Wind generators at Altamont,
California. Copyright George Elich.

Photos not otherwise credited were supplied by
author.

Photo credits: Dixon, *The Dynamic World of Physics,*
Merrill: 1984, pp. 243, 253; Lennox Industries,
p. 323; Ed Linton, U.S. Windpower, p. 242; KFA Jülich,
p. 220; Merrill/Kevin Fitzsimmons, pp. 30, 33, 48, 53,
54, 58, 237, 331, 350, 352; NASA, pp. 15, 19, 20; New
York State Energy Research and Development
Administration, pp. 161, 162; U.S. Department of
Energy, pp. 243, 299.

PREFACE

"Then idiots talk," said Eugene, leaning back, folding his arms, smoking with his eyes shut, and speaking slightly through his nose, "of energy. If there is a word under any letter from A to Z that I abominate, it is energy. It is such a conventional superstition, such parrot gabble!"

Charles Dickens
Our Mutual Friend (1865)

This book originally grew out of a course I taught at the University of Oregon in 1974 and 1975, just after the first of the "energy crises" wakened interest in the topic. The one-quarter course was one of a series of "minicourses" instituted in the physics department because of student desire for relevance and because of a decline in enrollment in other physics courses.

Preparation for the course was difficult due to the diversity of the students who enrolled. I decided at the outset that I would not be apologetic about introducing the physics needed for the ensuing discussions. The ability to sift through data, winnow the salient features, and act upon that basis was something I tried to emphasize throughout the course. A few of the problems were discussed in some detail in response to specific student interest, e.g., the material on insulation and lighting in Chapter 19 was incorporated because of interest expressed by a group of architecture students.

The course focused in part on the *moral* dilemmas facing our technological society in its quest for some control of its own destiny. The lack of sensitivity to moral issues is embarrassingly present all around us. When moral issues are embedded in technological and economic decisions, even people of obvious good will are inclined to leave things to the "experts." These experts may encourage others in this; only they know the full facts of the matter, they say. The issues are too complex to allow free and open discussion, they say. We know best, they say. It is undeniable that experts do know a great deal about "current" technology. The problem is that many of these experts tend automatically to favor the status quo, and may foreclose options just because "it isn't done that way." Involved and informed outsiders can force people to re-examine their preconceptions. People are so much creatures of habit that this process can be extremely uncomfortable.

It is simply not true that laymen cannot grasp the issues or the technology involved in matters of social interest. In this book I attempt to inform people in a fashion that will allow them to use their good sense when they approach such questions as the application of technology to agriculture. We must have citizens who are not willing to "leave things to the experts." Responsible societal decisions cannot be made without broad involvement.

If our children were cold, we would wish to keep them warm. If we ever thought about it, we would wish the same for our grandchildren and greatgrandchildren. The responsibility for their well-being and for the equitable distribution of resources between present and future—for example, for their access to the dowry of the last several hundred million years of stored solar energy (coal and oil underground)—rests clearly with all of us at the present time.

Since I have been at the Ohio State University, I have been involved in another physics course, Physics 100.02, which was generated in the early 1970s in response to the same pressures that led to Physics 114 at the University of Oregon.

Through this course I became more involved in the experimental and phenomenological details of matters pertaining to energy. Many of these insights were incorporated in subsequent drafts of the book.

The purpose of this book is to give the reader some understanding of the important decisions that must be made soon; of the tradeoffs between risks and benefits as well as we are able to define them; of the consequent social and political choices; of the role the reader, as an informed citizen, is able to play. I have attempted to indicate in the text where my own value judgments have intruded. I have attempted to tell the story in a balanced way, and to present a lot of data in graphs and tables. I hope there is enough information given to allow the reader to decide which issues are important for the future. Hard decisions are looming and they will be made either in full knowledge of the issues involved or by default. My prayer is that we take the former course.

For the teacher: These issues must involve your value system; they cannot be addressed in a moral vacuum. Level with your students. Let them know where you stand.

For the student: Question, question, question! Question my values as presented here, your teacher's as given in class, and most of all, your own.

Acknowledgments

It is impossible to thank everyone who had an effect on this book—that would involve my entire life. On technical matters, I'd like to thank Seymour Alpert for arguing with me about the prudence, or lack of it, involved in burning petroleum in the late 1960s (he made me get my facts straighter); Amit Goswami for his advice and encouragement as I took on Physics 114 at Oregon; E. Leonard Jossem and the late James Harris for helping a klutzy theorist do class experiments which sometimes work; John Harte for interesting conversations on the Lotka-Volterra equation; Richard Bower for information on Canadian energy policy; and Bunny Clark, for making me revise many examination questions I'd written until they were reasonably comprehensible.

Thank you, former students! You know who you are. It is your interest that helped me be enthusiastic class after class, your questions that set me off in different directions and made me think, your patience with my jokes that helped us groan together as we grew together. I trust you are informed, responsible citizens.

During part of the time I worked on this book I benefited from the support of the Aspen Center for Physics and the Alexander von Humboldt Foundation. I am deeply indebted to AvH for allowing me the opportunity to study energy conservation in West Germany firsthand.

I wish to thank the following for reviewing all or portions of the manuscript: Marvin W. Baker, Jr., University of Oklahoma; Philip E. Barnhart, Otterbein College; Lon D. Drake, University of Iowa; Lynn W. Glass; Iowa State University; John A. Jones, Miami-Dade Community College, South Campus; George W. Rayfield, University of Oregon; William C. Rense, Shippensburg University; Donald L. Rice, Chesapeake Biological Laboratory, University of Maryland; K. R. Gina Rothe, California State University, Chico; Gerald R. Taylor, Jr., James Madison University; and Antoni Wodzicki, Western Washington University.

I thank my good friend Bob Friedman for support and encouragement about the book, which heartened me to the gargantuan task of revising and updating what you now read. Judy Aubrecht read several drafts of the manuscript and proferred helpful advice. Thank you. I want to thank my daughter, Laurie Wagner, for a thorough and critical reading of the manuscript. My wife, Michelle, did a splendid job of typing the manuscript and dealing with a sometimes unreasonable author. Last, but certainly most important, Michelle and our daughter Katarina put up with my foibles during the writing of the book and helped me in so many ways that it is impossible to express my abiding gratitude and love for their support.

Gordon Aubrecht

CONTENTS

25
SAFETY AND
NUCLEAR ENERGY   468

26
TOCSIN   509

INTRODUCTION

Once upon a time, people thought that the supply of energy available to do useful work was inexhaustible. Once upon a time, fossil fuels—oil and coal—were so cheap that no one was concerned if they were wasted. Once upon a time, people thought they could throw away anything, and it would never contaminate the air, the water, or the land, and there would always be room. More recently, we have come to realize that these are only fairy tale ideas. Fossil fuel resources on Earth are limited. The environment cannot be abused with impunity when the population in an area grows too large.

What should we, as citizens, do about energy? What are the long-term prospects for energy and resource supplies? Should we just enjoy the benefits of low gas prices? Gas was cheap in the 1950s and 1960s, but it was very expensive in the 1970s and early 1980s. Is it responsible to buy gas-guzzling cars now that gas prices are lower again? Should we concern ourselves about the health risks of carbon monoxide emissions from cars? Should government act on acid rain? Is climate changing, and if so, is the change due to human actions?

These questions are the concern of all citizens in a democracy. Many of them are discussed by our elected representatives in the political forum. We must consider what advice to give these representatives. Citizens and policymakers need to consider the long-term results of our actions and to construct long-term solutions. Unfortunately, the desire of officials to be reelected often inhibits the government's ability to deal well with these situations. Twenty years seems infinitely long to someone elected for two, or four, or even six years. There must be some way for the system to respond to pressing questions that have factual answers, or to set priorities after determining the best available answer when the facts are in dispute. Issues of energy and environment are of concern to everyone. The world is, after all, finite.

Some of the questions concerning the future of our planet and our lives may be answered from the store of scientific knowledge available now. Politicians should solicit scientific advice and pay attention to it. For other questions, we need further research to determine the right answers. In some cases, we may never know enough to be able to answer the questions in detail. We should realize that the scientific community disagrees about the results of research and may give conflicting advice.

In this volume, all of these energy issues are addressed. Like Caesar's Gaul, the book is divided into three parts: general background (Chapters 1–10); resource availability (Chapters 11–19); and consequences of current practice (Chapters 20–

1

25). In the first section, we examine energy as the ability to do work, power as the rate of doing work, related physics concepts, constraints on heat engines from the laws of thermodynamics, manner of generation of electric current, distribution of electricity, exponential growth, the making of projections for demand, the logistic curve, the pitfalls of projecting, and the role of population demand. In the second section, we look at the availability of water, mineral resources, fuel supplies, and solar energy technology and biomass resources. Finally, the last part explores consequences for health, society, and the environment stemming from our present attitudes and practices.

What is known is that no method of generating energy is without risk. In obtaining and using energy, people are doing things that could result in events such as catastrophic climatic change; local pollution of air, water, and ground water, which could be hazardous to the health of large numbers of people; and long-distance pollution that causes harmful effects on agriculture, silviculture, and aquaculture.

Of course, we cannot give up the use of energy. To do so would return us to the sort of life characterized by the philosopher Hobbes as "poor, nasty, brutish, and short." People do not want to give up their daily energy "fix." To have the energy to run businesses, hospitals, and schools, and to make life more satisfying, we have to accept the problems associated with generating energy.

In order to address these daunting problems, and problems of the Earth's finite resources, I believe that we, ordinary citizens of the world, must first acknowledge these problems if we are to succeed in overcoming them. There are many reasons for *concern,* but concern need not be synonymous with despair. It is by facing problems, not by denial, that we find solutions. I am optimistic that solutions will be found if the gravity of the problems of pollution, overpopulation, and irresponsible use of energy are acknowledged by the public and our politicians. Then we

can work on solutions and find better ways to use our resources and preserve our environment.

As you read this book, I hope you will be sharpening your critical faculties. I hope that in the future, as you read articles in newspapers and popular magazines, you will be able to distinguish the overstated and the untrue, the persiflage and the balderdash, from accurate information. Your responsibility is to make the future world one in which you wish to live, and one in which you would wish your descendants to live.

OVERVIEW: CHAPTERS 1–10

Energy cannot be discussed without consideration of what it is physically. Sources of energy are many, and a knowledge of physics is necessary to understand how most sources can be tapped. Chapters 1 and 2 discuss some constraints on energy policy and use. Chapter 3 defines energy physically and discusses its various forms. Chapter 4 introduces electricity; the electric utility system in the United States is also discussed in Chapters 8 to 10: predictions of energy demand (Chapter 8), the history of energy demand and utility development (Chapter 9), and an introduction to the consequences of electrical energy (Chapter 10). Chapter 5 describes chemical energy, and Chapter 6, nuclear energy. Chapter 7 considers the heat engine and thermodynamics, the study of thermal energy in transit.

OVERVIEW: CHAPTERS 11–20

Supplies of energy and related minerals form the basis of our economy. In these chapters, the various kinds of energy resources and resources in support of the energy economy are considered. Solar energy is the focus of this group of chapters. Forms of solar energy are considered in Chapter 11 (hydroelectricity), Chapter 13 (stored solar energy), Chapter 15 (wind, tides, and photovoltaics), Chapters 16 and 17 (biomass energy), and Chapter 18 (satellite capture of solar energy). Resources in support of energy are considered in Chapter 11 (water resources) and Chapter 12 (mineral resources). Energy resources themselves are discussed in Chapter 13 (coal, oil, gas), Chapter 14 (nuclear), Chapter 18 (energy alternatives), Chapter 19 (conservation), and Chapter 20 (recycling).

OVERVIEW: CHAPTERS 21–26

The final six chapters are concerned with the consequences of current activities in the transportation sector of the economy (Chapter 21), acid rain (Chapter 22), climate (Chapter 23), human intervention in the Earth's weather system (Chapter 24), and nuclear energy (Chapter 25). Chapter 26 attempts to tie together all the issues involving energy in context.

1
GENERAL CONSIDERATIONS

The study of energy involves scientific, economic, political, and social elements. Everything affects everything else. The story of humanity's interaction with energy begins with examples of some unintended consequences of political and social decisions (both made and unmade).

KEY TERMS   *TANSTAAFL*
catalyst
ozone
chlorofluorocarbon
"tragedy of the commons"

We live in a world in a state of flux: Things are always changing. Jokes change, politics change, the definitions of words we use to express ourselves change. One of the unhappy results of this process in our milieu is that words are debased.

Which word is this year's fashion? Are we "escalating" the decline in our "infrastructure"? Whatever happened to the "energy crisis"? Code words such as these are used to try to make complex issues simpler, but they often distort rather than simplify. Take *energy crisis*: Cheap gasoline is available now, right? The experts, the newscasters, those who told us in the early 1970s that the days of cheap energy were over—now it looks like they were all wrong. We think we needn't worry. We are led to this conclusion because we were earlier seduced into equating *energy crisis* with a temporary gasoline shortage caused by political decisions.

One of the purposes of this book is to dampen the rhetorical extravagance of popular "energy language" in order to examine the underlying issues. We will first have to come to some agreement on definitions. That is, we have to agree to speak the same language.

The topic of energy and pollution is an inextricable mix of *physical* and *social* (or societal) questions. Energy is a physical entity; pollution, on the other hand, requires some value judgment to define, some recognition of the societal cost of an energy strategy. One person's pollutant may be another's raw material. Much of the book will explore the physical aspects of energy generation and the ensuing pollution. Inevitably, value judgments will be made. I will try to warn you when I make them. We will also explore *how* to use information without being overwhelmed by it. We must consider how to interpret data, and how to judge what other people have done with these data.

To make these ideas more concrete, I want to introduce two constraints to be remembered in dealing with our problems. One constraint is both physical and social; the other is a purely social phenomenon.

TANSTAAFL

John W. Campbell, editor of *Astounding Science Fiction Magazine,* developed a group of writers in the 1940s who wrote "hard science" science fiction. Some authors, such as Robert Heinlein and Eric Frank Russell, applied engineering principles to society. Many of these authors wrote stories of worlds that ran on the principle of TANSTAAFL (*tan-staff-el*): "There Ain't No Such Thing As A Free Lunch." The stories were antiutopian in that nothing in future societies was free, and everyone realized this and lived accordingly. In other words, if people want something, they have to pay for it.

We are certainly conversant with this principle in everyday life. We say: "You get what you pay for." The point is that this fact holds for physical phenomena as well. Life on Earth would not exist if the planet's energy deficit were not paid by the sun. (We will explore this dependence on solar energy later.)

Everything done by people has an effect on everything else. For example, communications satellites (orbited with the laudable intention of allowing poor countries such as India to broadcast to the entire subcontinent with only one transmitter) interfere with radio astronomy. *SMS-1,* a synchronous meteorological satellite, also causes problems for astronomers (1).

The Ozone Layer and TANSTAAFL

Even large-scale meteorological features of our planet, such as the protective ozone layer, can be altered by people. As is now well known, the Earth possesses this layer of ozone in the upper stratosphere (upper atmosphere). Normal oxygen (O_2) has two atoms in its molecule. Ozone is a form of oxygen in which *three* oxygen atoms (O_3) are bound together. Earth's ozone layer is unique, at least in our own solar system. This layer is primarily due to the presence of life here. The ozone layer, formed because life on Earth emitted oxygen into the atmosphere, has in turn molded life on Earth.

The ozone layer absorbs energy from the sun and heats the air. The sun emits enormous amounts of energy, which is spread over a very wide band (spectrum) of energies. Most of this energy reaches the Earth in the form of light. Ozone absorbs certain bundles of energy from the sun's ultraviolet (UV) radiation. It then re-emits this energy in smaller bundles. The smaller bundles of light that get into the atmosphere below the ozone layer can no longer escape into space. The absorption of radiant energy by the ozone layer heats up the atmosphere.

The ozone layer protects living things. If the ozone layer is damaged, our eyes and skin would get more UV radiation than they are adapted to, causing increases in blindness and skin cancer. Skin cancer affects fair-skinned Caucasians most. Even now, there are between 200,000 and 600,000 cases of skin cancer in the United States each year. A 5 percent ozone depletion could cause about a 10 percent increase in radiation and perhaps an additional 8000 cases of cancer yearly among U. S. whites (2). If the ozone concentration in the ozone layer decreased greatly, even people with the blackest of skins would suffer.

There are two atmospheric cycles that can reduce ozone. These cycles involve nitrogen oxide (NO) and chlorine (Cl). These chemicals interact with ozone to produce the normal molecular oxygen, destroying the ozone.

The ozone layer formed in part because high-energy radiation from the sun can break up molecular oxygen into single oxygen atoms. Some remain as oxygen atoms (O); most combine with oxygen molecules (O_2) to form ozone (O_3). When an NO molecule interacts with an ozone molecule, it forms nitrogen dioxide (NO_2) and a normal oxygen molecule (O_2). If NO_2 encounters a single oxygen atom (O), it forms NO and O_2. The whole cycle can then begin again. NO acts as a catalyst—although it makes the reaction occur, the NO remains unchanged and thus can repeat its performance.

Nitrogen oxide gets into the atmosphere naturally when lightning breaks up molecular oxygen (O_2) and molecular nitrogen (N_2). Sometimes a nitrogen atom and an oxygen atom get together. In fact, this process is responsible for the major part of all the nitrogen fertilizer applied each year to the world's soil. Very little of the lightning-formed NO gets into the ozone layer and so is little threat.

Supersonic transport (SST) airplanes, such as the *Concorde,* burn kerosene in the upper atmosphere to produce carbon dioxide (CO_2), nitrogen oxides, and water. The nitrogen oxides are created because some atmospheric nitrogen is broken up in the combustion chamber, where it combines with oxygen. In addition, when the plane flies faster than the speed of sound (supersonic speeds), the shock wave causes most normal nitrogen molecules (N_2) and normal oxygen molecules (O_2) to break up; many do not recombine but form nitrogen oxides instead. Since SST exhaust is emitted in or near the ozone layer, SSTs pose an immediate threat to the integrity of the ozone layer itself. The ozone is eaten up by nitrogen oxides from SSTs.

There are very few civilian supersonic planes now flying. In 1974, an estimate was made that a fleet of five hundred continuously operating supersonic transports would result in a permanent reduction of ozone by 16 percent in the Northern Hemisphere and 8 percent in the Southern Hemisphere (3). This estimate convinced the U. S. Congress that it was unwise to subsidize American civilian SSTs and thereby prevented major SST production. Another disincentive is that the *Concorde,* an SST subsidized by the British and French governments, has not been a commercial success.

Nuclear weapons tests have provided information on this ozone-depleting effect because explosion plumes from atmospheric tests raise huge amounts of nitrogen oxides into the stratosphere. The heavy testing of U. S. and Soviet bombs between 1948 and 1961 resulted in a 4 percent ozone depletion that took 2½ years to regenerate (3). The maximum effect of a bomb test on the ozone layer occurs several months after its detonation. (It is possible that high-level nuclear explosions could cause nitrogen oxide to be fed into the stratosphere over a period as long as ten years.)

Chlorofluorocarbons and TANSTAAFL

Another illustration of TANSTAAFL concerns the effects of chlorofluorocarbons (mainly Freon, CF_2Cl_2, which is a common refrigerant and is used widely in aerosol sprays). Just as with nitrogen oxide, if a free chlorine atom is loose in the ozone layer, it can act as a catalyst. It can destroy two ozone molecules and reconstitute itself afterward. The chlorofluorocarbons are problems because as they rise into or through the ozone layer, they break up, liberating atomic chlorine. Below the ozone layer, they are inert (they were chosen as propellants in spray cans because they did not react with the contents). But above the ozone layer, these gases encounter UV radiation, which results in their breakup. The chlorine is then free to destroy ozone.

By 1974, over a million metric tons of chlorofluorocarbons per year were being released. It was at that time that Molina and Rowland (4) realized the possible destructive effects of chlorofluorocarbons on the ozone layer. There were fears at that time of an eventual 20 percent depletion in ozone, but little was known then about the chemistry of the atmosphere. It appears that chlorofluorocarbons remain in the air somewhere between 40 and 150 years. Because it takes about 15 years for the material now entering the atmosphere to get into the ozone layer, it will stay long enough to get its chance to wreak havoc with the ozone.

Chlorine chemistry in the atmosphere is now much better understood (5), and the best current prediction is that there will be an eventual 3 to 5 percent depletion based on the 1977 worldwide production rate. Ozone is predicted to increase near the ground (where it is a pollutant that causes plants to sicken and die), but to decrease by as much as 20 percent in the stratosphere (6). A decrease of 3 percent in the stratosphere has already been observed; the projected 20 percent decrease would cause the skin cancer increase described earlier. This is TANSTAAFL with a vengeance. Scientists and technologists involved in the development of spray cans never conceived that a danger to the entire human race could follow from their contributions to convenience and progress.

Conditions may be improving. An international agreement under the sponsorship of the United Nations Environment Program was signed by major producers in 1987; the agreement may result in a cut in production levels up to 50 percent. The projected cuts are still no cause for joy, for production is currently rising again after hav-

FIGURE 1.1

This measurement of ozone depletion during the Antarctic spring (October 1987) was obtained by the Total Ozone Mapping Spectrometer aboard NASA's *Nimbus* 7 satellite. (NASA)

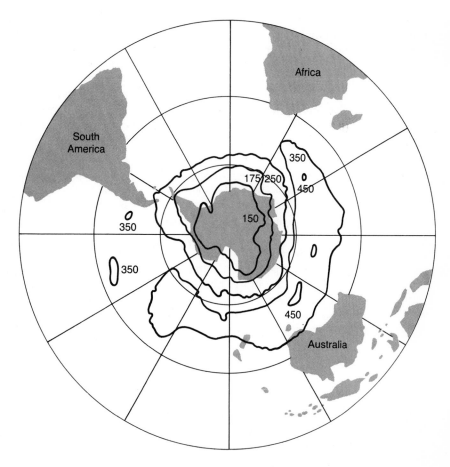

ing dropped in the years from 1975 to 1980. Europeans still use Freon as a propellant in spray cans; in North America, its use for this purpose has essentially stopped. It is difficult to halt production entirely because chlorofluorocarbons have been the predominant refrigerants used in air conditioners and because they are used to produce hundreds of millions of foam hamburger cartons each year.

Vivid evidence of ozone layer depletion was seen in measurements made by the *Nimbus 7* satellite (7,8), as shown in Figure 1.1. The South Pole has a visible "hole" over it. This Antarctic ozone deficiency was first brought to public attention by Farman, Gardiner, and Shanklin (9) and has been shown to be occurring at heights between 10 and 20 km (10). Thinning of the ozone has occurred each year since 1979 from September to November, during the Antarctic late winter (11). The severity of the effect has been increasing since the early 1970s, when it began to occur (8). The cause is still a mystery, but most speculation centers on the interaction of bromine and chlorine (12), on sodium chemistry in the polar atmosphere, on upward air transport and dilution of ozone, and on the existence of conditions near the poles that allow very slow reactions to occur (13).

More experimental evidence in support of a chemical cause for the Antarctic hole continues to be found (14); there is also theoretical support for this hypothesis, involving reactions of normally inactive compounds on the surfaces of atmospheric aerosol particles (15).

More worrisome are reports of a rapid decrease in atmospheric ozone in the mid-1980s. After discovery of the Antarctic ozone hole, scientists checked and reanalyzed the data from the solar backscatter ultraviolet detector aboard *Nimbus 7* (7,16). If the data are verified, as the respected Ozone Trends Panel believes, the problem will have to be addressed immediately by the manufacturing nations. It was the reporting of these data that put sufficient pressure on manufacturers to allow producing countries to reach their modest agreement to cut production.

THE TRAGEDY OF THE COMMONS

Twentieth century socioeconomist Garrett Hardin has brought the concept of the "tragedy of the commons" to public attention through his investigations of the world population problem. In contrast to the views of the *laissez-faire* economist Adam Smith, who argued that an individual intent on personal gain would be led to benefit the public interest by an "invisible hand" (17), Hardin (18) holds that individuals behave in a selfish fashion that ultimately leads to societal destruction.

To illustrate his ideas, Hardin uses a village *commons*—a field open to all inhabitants of a village for use as pasture, park, or whatever they please. Each herdsman of the village is entitled to use the commons. Of course, each herdsman tries to keep as many cattle (or other domestic animals) on the commons as possible. This method might work well for a time because war, famine, and disease hold both human and animal populations in check. However, the day would eventually come when carrying capacity—the maximum number of cattle that could be supported on the commons indefinitely—would be reached.

Let us suppose that ten villagers each had 20 cattle on the commons. In this case, there would be 200 cattle, which we take to be the carrying capacity. If another cow were added, each of the 201 cattle would get 200/201 of its requirements. Thus, the herdsman who owns the extra cow would see new assets: an additional cow; but he also would see new debits: each of his 21 cattle gets only 200/201 of its requirements, and consequently is a little more bony, produces slightly less milk, etc.

We might call his assets +1, and his debits −21/201 = −1/10, giving net assets of +9/10. Figure 1.2 shows how the situation appears to the herdsman. The rational herdsman would then add another animal, and another, and another. When carrying capacity is exceeded, the entire herd is more poorly fed. Eventually the limit of permanent damage is attained, and *all* cattle (not

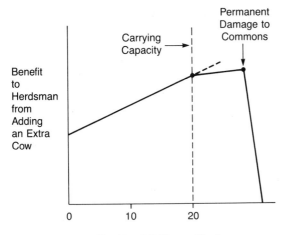

FIGURE 1.2
Benefit is perceived by the rational herdsman as he
increases the number of cattle he runs on a commons. Note that overuse will cause the system to
collapse eventually.

just the added one) starve, become very sickly,
or die. The entire herd (and its economic benefit)
is lost. "Each man is locked into a system that
compels him to increase his herd without limit—
in a world that is limited . . . ." "Freedom in a
commons brings ruin to all" (18). Because of its
remorselessness, Hardin calls this process the
"tragedy of the commons."

Air is known in economics as a "free good."
It is breathed by everyone; everyone uses it. If
an industry used air and polluted it without cost,
all people would suffer because the air would
be polluted, but the industry's owner would reap
economic gains. Such a case is an example of the
tragedy of the commons.

Hardin believes that tragedy of the commons
will operate in every case in which personal gain
is attained by losses for all, even if the losses may
seem negligible to the losers. The wealth (or
other advantage) of the gainers will lead them
to destroy what all possess in common. If all
possess some thing (the air, for example), the
thing is perceived as having no individual economic value.

The Earth is finite and thus can support only
a finite population. Certain groups see advantages
in increasing their numbers, and so encourage
people to have many children. The Earth in this
case is similar to an English commons, where
people have use of common goods: air, water,
land, resources.

If population grows without restraint, the Earth
may not be able to support that increased population. Hardin would coerce countries with exploding populations to limit the number of
children born in each family, with the reasoning
that "injustice is preferable to total ruin" (18).

We are led to the conclusion that, if we exploit
resources in a *laissez-faire* manner, we favor people or institutions who focus narrowly on profits
and distribute losses as widely as possible. We
are practically forced to foul our own nest. The
cost of dumping waste, as far as a profit-oriented
company is concerned, is less than the cost of
cleaning up. When the waste is discovered in the
environment, the companies are no longer iden-
tifiable, and the public must pay for any cleanup
to be done.

Let us study a modern example that illustrates
the "tragedy of the commons" principle. Power
companies are constantly besieged by customers
complaining of the high cost of energy. Obviously, such a company wants to buy its fuel (for
example, coal) as cheaply as possible. It shops
around.

Coal is mined in two different ways: from deep
underground and in strip mines. Deep mines are
reached through a shaft into the ground, through
which miners gain access to the coal seam and
out of which comes the coal. Strip mines are
used when the coal seam is near the surface. The
topsoil and other overburden materials are removed by giant power shovels and piled out of
the way. The shovels then dig out the coal. After
the coal has been mined, there may be reclamation. This involves refilling the hole, spreading
topsoil, and then planting new vegetation on it.

We now imagine three coal companies:

X Coal Company mines coal in a deep mine
at a price of $35.69/ton at the mouth of the mine

(that is, not including the cost of transportation from the mine).

Y Coal Company mines coal in a strip mine at $10.54/ton, but reclaims the stripped land at a cost of $3.00/ton. The total cost for Y Coal Company is then $13.54/ton.

Z Coal Company mines coal in a strip mine and sells it at $11.17/ton, with no reclamation of land.

Clearly, the strip mining company that tears off the overburden, heaping it in unsightly spoil banks, and then leaves everything there without treatment (causing erosion and contaminating ground water with acid leached from the spoil) has an advantage in selling at a lower price. Z Coal Company may sell its coal at $11.17/ton (including a handsome profit), while to make the same profit, Y Coal Company has to charge $13.54/ton. Of course the power company will buy from the Z Coal Company. (The transportation costs, averaging $3.01/ton in 1968 and $27.77/ton in 1980, may be different enough to favor the Y Coal Company or even the X Coal Company. We have *assumed* here that all other costs are equal [(19,20)]).

We may well imagine the effect at the next stockholders' meeting of the Y Coal Company. Most likely, their reclamation practices would cease. By use of reasoning such as this, environmental groups were successful in convincing the U. S. Congress that outside pressure on strip miners was necessary to achieve reclamation. Congress passed the Strip Mining Act of 1977, which mandates reclamation. Such an approach equalizes costs for all producers, as well as protecting society as a whole from greed. While the Act has not been totally successful, the situation has improved since its passage.

An indirect effect of a strip miner's decision not to reclaim land is the underpricing of coal in terms of its true societal cost, thereby encouraging unnecessary energy use. Cheap energy would be gained at a cost to society as a whole; there is no incentive to curb consumption if energy is too cheap. When energy is underpriced,

it becomes advantageous for a builder to construct and sell houses with minimal insulation, because they are cheaper for consumers to buy and it is so inexpensive to heat them (this actually occurred in the 1950s and 1960s).

And, of course, the power companies buying from strip miners who do not reclaim would not see themselves as responsible for the unreclaimed strip-mined land. It was those irresponsible strip miners, after all!

One aim of this book is to try to explain the physical and social *cost* associated with each strategy for generating energy. No choice connected with energy use is free of cost. Most choices involve benefits to some people. We must know the ultimate gains and costs, as well as who wins or loses, in order to make informed decisions as to future energy strategy. Our analysis will, of necessity, be incomplete. Society must make the best of the situation and make decisions from the best presently available knowledge. Decisions will have to be made. Even doing nothing involves making a decision!

When speaking of costs and benefits, we must *quantify*. In Appendix 1, we discuss some of the background information needed in this book: powers of ten and scientific notation. We will use this material to practice the art of estimation in Chapter 2.

SUMMARY

Everything we do extracts a price, and everything is connected to everything else. There is no such thing as a free lunch (TANSTAAFL). When people exploit resources held in common, "free goods" in economic parlance, they are usually aware only of their own immediate gain or loss. They often are not aware of (or ignore) the costs to everyone. The inexorability of the consequences of such selfish actions constitutes the tragic character of the tragedy of the commons.

Because of the costs associated with every action, and because the tragedy of the commons illustrates a way in which organizations or individuals avoid paying for the social costs inflicted by their actions, leaving others to bear that burden, society at large has a legitimate interest in regulation of these actions. Such regulation is best pursued when the citizenry is aware of the costs and benefits associated with the action, and a general framework equitable for all is set up.

REFERENCES

1. *Physics Today* 27, no. 9 (1974): 77.
2. W. Sullivan, *New York Times,* 7 Sept. 1975.
3. F. S. Rowland, *New Scientist* 64 (1974): 717; A. L. Hammond and T. H. Maugh, *Science* 186 (1974): 335.
4. M. J. Molina and F. S. Rowland, *Nature* 249 (1974): 810.
5. P. M. Solomon, R. de Zafra, A. Parrish, and J. W. Barrett, *Science* 244 (1984): 1210.
6. T. H. Maugh, *Science* 223 (1984): 1051.
7. D. F. Heath, Senior Scientist at NASA, Greenbelt, Md., private communication.
8. R. A. Kerr, *Science* 232 (1986): 1602; W. Sullivan, *New York Times,* 7 November 1985; J. Gleick, *New York Times,* 29 July 1985.
9. J. C. Farman, B. G. Gardiner, and J. D. Shanklin, *Nature* 315 (1985): 207.
10. S. Solomon, R. R. Garcia, F. S. Rowland, and D. J. Wuebbles, *Nature* 321 (1986): 755.
11. M. R. Schoeberl and A. J. Krueger, *Geophys. Res. Lett.* 13 (1986): 119; the entire issue, no. 12 of vol. 13, was devoted to papers about the ozone hole.
12. M. B. McElroy, R. J. Salawitch, S. C. Wofsy, and J. A. Logan, *Nature* 322 (1986): 759.
13. R. J. Cicerone, *Science* 237 (1987): 35.
14. P. Solomon, A. Parrish, and R. deZafra, quoted in R. A. Kerr, *Science* 236 (1987): 1182.
15. M. Ko, J. Rodriguez, and N. D. Sze, quoted in R. A. Kerr, *Science* 236 (1987): 1182.
16. R. A. Kerr, *Science* 239 (1988): 1489. See also J. Glieck, *New York Times,* 20 March 1988.
17. A. Smith, *Wealth of Nations* (New York: Oxford Univ. Press, 1976), 423.
18. G. Hardin, *Science* 162 (1968): 1243. Hardin attributes the idea to W. F. Lloyd, whose *Two Lectures on the Checks to Population* (New York: Oxford Univ. Press, 1833), is reprinted in Hardin, ed., *Population, Evolution, and Birth Control* (San Francisco: Freeman, 1964).
19. *Mineral Facts and Problems,* 1980 ed. Bu. Mines Bull. 650, U. S. Dept. of the Interior, Washington, D.C.: GPO, 1980.
20. H. Perry, *Science* 222 (1983): 377. Eastern coal cost $35.69/ton in 1980, while Western coal cost $11.17/ton on average. Wyoming coal cost $10.54/ton. Strip mine reclamation costs were estimated to run between $1 and $5 per ton of coal strip mined. The delivered cost of coal in Illinois was $38.94/ton for Wyoming coal.

PROBLEMS AND QUESTIONS

True or False

1. Chlorofluoromethanes are dangerous because of the very thing that made them useful in the first place: they do not react in the lower atmosphere. Therefore, they persist long enough to get through the ozone layer, where they then attack the ozone.
2. Nothing attacks the ozone layer when the Earth is in its natural state. Ozone molecules are neither created nor destroyed.
3. The installation of smoke scrubbers in coal-fired power plants required legislation because the utilities could not justify the expense so long as other utilities were free not to install them.
4. One molecule of nitrogen oxide or chlorine can incapacitate only a few ozone molecules before being removed from the ozone layer.
5. Chlorofluoromethanes such as Freon can be broken up only above the ozone layer. This occurs because the layer does not let light, which would have enough energy to break up Freon, through to the lower atmosphere.
6. It is possible to construct energy-generating facilities without any cost being borne by society at large.
7. The ozone layer created the conditions under which life exists on Earth at the present time.
8. Blacks would suffer more than whites if the ozone layer were destroyed.
9. Economic forces of competition assure that "free goods" such as air and water will be in danger of degradation.
10. Coal from deep mines would never be competitive with coal from strip mines because of its greater cost of extraction.

Multiple Choice

11. Mercury had long been used to cure felt for hats. People who are exposed to mercury may become "mad as a hatter." In the form of methyl mercury, the element may be concentrated in the food chain. A chemical plant exhausts mercury to the environment. This is an example of:
    a. tragedy of the commons.   c. both a and b.
    b. TANSTAAFL.   d. neither a nor b.
12. The "tragedy of the commons" is a reflection of
    a. physical phenomena.   d. neither physical nor social
    b. social phenomena.   phenomena.
    c. both physical and social phe-   e. political stupidity.
       nomena.
13. The principle of TANSTAAFL is a reflection of
    a. physical phenomena.   d. neither physical nor social
    b. social phenomena.   phenomena.
    c. both physical and social phe-   e. political stupidity.
       nomena.
14. SSTs are a danger to the ozone layer because
    a. shock waves from the plane   d. they cause production of ni-
       break up ozone directly.   trogen oxides, which act to
    b. they burn ozone in the air to   break up ozone.
       attain their high speeds.   e. SSTs punch holes in the
    c. they spray chlorofluorocar-   ozone layer as they rise
       bons into the atmosphere.   through it.

15. A paper factory fitted with no expensive pollution control devices can cause massive damage to streams or rivers. According to the "tragedy of the commons," such damage will occur because it
   a. is inexorable.
   b. is inevitable.
   c. results from the paper factory's loss due to polluted water being less than its gain in profits.
   d. can only be reversed by governmental regulation.
   e. does not cost anyone anything.

16. The hole in the ozone layer over the Antarctic is
   a. not dangerous, because it occurs only for a few months of the year.
   b. worrisome, because it seems to be increasing in size.
   c. inconsistent with models of the stratosphere.
   d. not interesting, because it has happened for some years already.
   e. proof that spray cans must be banned worldwide.

Discussion Questions

17. Is it a good idea to write "scare" books with titles such as *Plague 1989*? Argue both sides of the question.

18. What are some examples of the operation of TANSTAAFL or "tragedy of the commons" different from those given previously?

19. What are some other cases of seemingly innocuous developments that have had tragic consequences (that is, inexorable consequences) for at least some people?

2

A DIGRESSION ON THE NECESSITY OF A FINITE WORLD POPULATION

As we shall see, the number of people who must be supported by energy or any resource determines to a large extent the amount that must be produced or extracted. In this chapter, we use some ideas from Appendix 1 to develop estimates of the human carrying capacity of the Earth. The examples used here illustrate the consequences of unimpeded geometric, or exponential, growth.

Since Earth is finite, there are limits to growth. We begin with an absurd assumption about external conditions limiting growth, and end with a fairly realistic set of conditions limiting growth in the number of humans.

KEY TERMS   *mass*
            doubling time
            linear thinking
            trophic pyramid
            terraformation

"The whole world is watching," chanted antiwar demonstrators at the 1968 Democratic National Convention in Chicago. Even twenty years ago, they knew that what happens half a world away is as immediate as what happens on the other side of the city. They were the vanguard in a shift of awareness—for, from time immemorial, humankind has thought in terms of an infinite world. Is it only accidental that concern about overpopulation, lack of resources, and the environment in general has followed hard upon the photographs of our blue-white planet (Figure 2.1) suspended atop the lifeless lunar plain?

MASS IS LIMITED ON EARTH

How might we begin? Each of us has intrinsic attributes; one such measurable one is our mass. Mass is the agent responsible for the force we feel pulling us toward the ground (see Chapter 3). A person having twice the mass of another person would be correspondingly pulled down toward Earth's surface with twice the force. In the metric system, the mass unit is the kilogram (1 kilogram of mass experiences a force of about 2 pounds on the Earth). Suppose that a typical person has a mass of about 80 kg (a weight of about 175 lb). To begin, imagine some kind of upper limit to the population's mass.

FIGURE 2.1
The Earth looms over the lunar surface. Color photos of scenes such as this may have sparked people to become more aware of the fragility of their environment. (NASA)

It is clear that the Earth's entire mass could never be turned into people: the Earth's mass is about 6×10^{24} kg. The ultimate number of typical people (of mass 80 kg) allowed if they had as much mass as the Earth would be given by

$$N = \text{mass of the Earth}$$

or, defining the symbol N to be the maximum number of typical people, this is solved by

$$N = \frac{\text{mass of the Earth}}{\text{mass of a typical person}}$$
$$= \frac{6 \times 10^{24} \text{ kg}}{80 \text{ kg}} = 7.5 \times 10^{22}.$$

While this is a finite number, it is a mind-boggler. If we asked that many people to go jump in a lake, and 100 of them did just that each second, it would take over 20,000 billion years for them all to do it. Of course, if a million people were to jump into the lake each second, it would take only a bit over 2 billion years.

What would the Earth look like if such a population could exist on it? How long would it take us to get to that level?

How Fast Will the Limits of Population Growth Be Reached?

The present doubling time of the Earth's population is about thirty-five years (1). If we know the amount of time it takes the number of people to double, we are able to estimate how long it will take the population to grow to any size. For the purposes of our estimate, it is necessary to assume that the doubling time is not changing; that is, that it will take thirty-five years to double in 2500 A. D. also. This is *not* a good assumption if we look at the history of population increase over the last four hundred years. The doubling time has been decreasing; that is, each succeeding generation has been doubling itself faster than the previous one, despite famine, pestilence, and war. If we do use the doubling time as it is now, though, we will find an *upper limit* on the time it takes to reach a certain value.

There has been a crucial assumption made here that has not yet been made overt. This estimation procedure as described has *assumed* that what we presently see will continue to be the same in the future as it has been in the past. This assumption may or may not be true, but we must recognize that we are making the assumption. People exhibit a bias in their belief that the future will be more of the same despite all of history to the contrary. Given our lack of information about the future, we have to estimate. However, we must not lull ourselves into thinking that we have described what *must* occur. This approach to the estimation problem is called "linear thinking," and we shall see much more of it and its variants elsewhere in this book, where the same cautions will apply.

To recapitulate, we asked how long it would take the present population of Earth, 5×10^9 people, to reach our maximum estimate of population (N) at the present rate of increase. If we denote the present population by the symbol N_0, then in 35 years, there will be twice that many people ($2N_0$); in 70 years, there will be $2(2N_0) = 2^2N_0$; in 105 years, there will be $2(2^2N_0) = 2^3N_0$; and so on. After m doubling times, the number of people is 2^mN_0. Solving with our numbers for the number of doublings, m, we have

$$2^m = \frac{N}{N_0} = \frac{7.5 \times 10^{22}}{5 \times 10^9} = 1.5 \times 10^{13}.$$

With this number and the approximation $2^{10} = 1024 \cong 1000 = 10^3$, we obtain an m of about 44. Since each doubling time is only 35 years, 44 doubling times is only 1540 years. That is, if humanity insists on doubling itself every 35 years, the mass of people alive in 3500 A. D. would have used up the entire mass of the Earth. (Of course, Earthlings would have to bring mass from elsewhere if even a small fraction of such a number of people were living.) In fact, with this sort of proliferation, the mass of humanity would equal the mass of the known universe in about 4500 years!

To put this number, 7.5×10^{22} people, into context, note that the total world surface area (including oceans) is about 5.2×10^{14} m². The gross estimate of population density is

$$7.5 \times \frac{10^{22} \text{ people}}{5.2} \times 10^{14}\text{m}^2 = \frac{150 \text{ million people}}{\text{m}^2}!$$

This would be something like squeezing the entire present population of Earth into a house trailer!

ENERGY LIMITS ON WORLD POPULATION

To make a more realistic estimate, it must be recognized that (at least crudely speaking) the ultimate factor limiting world population is the energy budget enforced by the sun. Solar energy allows plants to grow; it is this energy that is stored up as deposits of natural gas, coal, and oil. This energy is also responsible for the Earth's relatively high temperature. This discussion about energy anticipates our definition of energy and energy units in the next chapter. For our purposes here, it is enough to state that an important energy unit is the kilojoule (kJ); the average person on Earth uses about 8400 kJ of energy per day to function normally (that's about 2000 kcal—called "calories"—per day to those who read diet books). The typical American diet is about 14,000 kJ/day, while someone from Bengal in India or from Bangladesh might use less than 6000 kJ/day.

Each minute, the surface of the Earth receives about 10^{16} kJ (2) from the sun. Since the sun shines an average of twelve hours a day, the Earth on the average gets about 7.2×10^{18} kJ/day. If we imagine that people could make total use of this solar energy directly, the Earth could support

$$\frac{7.2 \times 10^{18} \text{ kJ}}{8400 \text{ kJ/person}} = 8.5 \times 10^{14} \text{ people.}$$

This is about eight orders of magnitude (eight factors of 10, or 100 million) fewer than in our first estimate. Following the same reasoning as in the previous example, the number of doubling times is about eighteen; in about six hundred years at present population growth rates, the energy-sustainable limits of population would be reached. By this estimate, the ultimate population density of the Earth is about 1.6 people/m²—somewhat more plausible than before. This population density would be like crowding about 180 people into a typical middle-class American house.

Of course, this assumes that we could use the sun's energy directly, over water and land surfaces, and that we would kill off all competing flora or fauna. Without the algae in the oceans, and the plants on land, no oxygen would be produced! In fact, only 70 percent of the sun's energy even reaches the surface of the Earth. The rest is reflected from the clouds, the atmosphere itself, and the surface. Moreover, we should not count the energy falling on the ocean (fish are only minor contributors to human nutrition) or on Antarctica. Let us ask what the human population could be if we killed off all land animals and plants and got our requirements directly from the sun.

In that case, using a ratio of land area to total area of 26 percent (3,4) and taking into account the factor of $\cong \frac{1}{2}$ due to reflection of sunlight by the Earth, the maximum number of people possible would be

$$(1/2)(0.26)(8.5 \times 10^{14} \text{ people}) = 1.1 \times 10^{14} \text{ people}.$$

Attainment of this population would take about fifteen doubling times (about 2500 A.D.). The population would have a *land-area* density of 0.8 people/m² in this case, which would be like fitting about ninety people into the aforementioned typical house.

The Trophic Pyramid   The next problem is that we cannot directly use the sun's energy for food. Plants use energy from the sun to grow roots, pump water, and so on. Thus, plants can pass only about 10 percent of the energy they receive as food value to their consumers. Plants are called *producers*. Animals preying on the pro-

ducers get only about 10 percent of the plant energy. They are called *primary consumers*. Animals use food energy to keep themselves warm and perform normal bodily functions; the food that is not used is excreted as waste. If we ate a cow, getting only about 10 percent of its energy, we would be called *secondary consumers*. If we were eating salmon that had fed on small fish, which in turn had fed on plants, we would be *tertiary consumers*. As you see, food consumption follows a series of steps resembling a pyramid of lowered available energy with each upward step (see Figure 2.2). This pyramid is called the *trophic pyramid,* or occasionally, the *trophic chain*. (The word *trophic* refers to nutrition.)

In America, people are about a third primary and two-thirds secondary consumers. Suppose, for the sake of our population estimate, that people are primary consumers. Let us also take into account that photosynthesis (the process by which plants make sugars and oxygen out of sunlight and carbon dioxide) is about 6 percent = 0.06 = 6×10^{-2} efficient (5). Therefore,

$$\begin{aligned} N_{max} &= (6 \times 10^{-2})(10^{-1})(1.1 \times 10^{14} \text{ people}) \\ &= 6.7 \times 10^{11} \text{ people} \\ &\cong 134 \times \text{(current population)} \end{aligned}$$

so that the corresponding m is about 7; in about 345 years, we will have attained maximum population.

Finally, not all land is arable (capable of producing crops). If it were, our preceding calculation would be correct (assuming no other animals or insects competed for food, there were no weeds, and so forth). In fact, only about 25

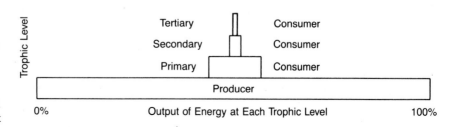

FIGURE 2.2
A trophic pyramid. The original level represents captured solar energy. Only about 10% of the energy available at any level in the pyramid can be turned into energy at the next higher level.

percent of Earth's land is arable (5–7). (The cost of increasing arable land is about $500–$1000 per hectare. The hectare is a metric unit of area, 10,000 m², or a bit under 2½ acres). Thus only about a quarter of the 670 billion can be accommodated unless we are able to increase arable land drastically, which seems unlikely—at least until large-scale weather control is possible. We probably do not want to kill off every other land animal and insect. We should take into account another factor of, say, ½ to share food with these other animals

$$N_{max} = (1/4)(1/2)(6.7 \times 10^{11} \text{ people})$$
$$= 8.4 \times 10^{10} \text{ people.}$$

This number is not very far from the maximum population estimate made by the Club of Rome study—15 to 20 billion (7). Thus,

$$2^m = \frac{N_{max}}{N_0} = \frac{8.4 \times 10^{10}}{5 \times 10^9}$$
$$= 1.68 \times 10^1$$
$$= 16.8.$$

Since 2^4 is 16, we see that m is about 4 (it is actually 4.1 by calculation). The human race will reach that limit (on Earth) in only 140 years at the present rate of increase.

The density of population is, in this case

$$\frac{8.4 \times 10^{10} \text{ people}}{1.33 \times 10^8 \text{ km}^2} = \frac{630 \text{ people}}{\text{km}^2} = \frac{1600 \text{ people}}{\text{mi}^2},$$

which is that of a suburban town. In comparison, the present world average population density is

$$\frac{5 \times 10^9 \text{ people}}{1.33 \times 10^8 \text{ km}^2} = \frac{40 \text{ people}}{\text{km}^2} = \frac{100 \text{ people}}{\text{mi}^2},$$

which is roughly the population density of Indiana.

The foregoing discussion has focused on outlining the barest essentials of survival for the maximum population. It has ignored the grim political, social, and psychological price all people would pay in order for Earth to support such a population (8).

Since the energy available to each person would shrink (by some near-future time, the stored energy in our planet's crust will have been

totally squandered), the amenities available to each person would be fewer. People would be crowded together in nonproductive areas such as deserts or mountainsides in order to maximize productive land. Presently, good agricultural land is being paved over or built over to support increasing populations and to facilitate intercommunication.

The basic question remains: Even if Earth could support 84 billion people, should it? What sort of price—physical, psychic, and spiritual—would be acceptable? Thoughtful science fiction authors have addressed this problem, generally showing us chilling visions of the future (9).

Agricultural Limits to World Population As we shall see in Chapter 17, hand labor is the most energy-efficient means of grain and vegetable production. Yields of about 4 million kJ/hectare/year are typical. We could reformulate our population question to ask how many people the Earth could support at an adequate diet (say, 10,000 kJ/person/day, or about 4 million kJ/person/year). Using an estimate of 1.8×10^9 hectares for the current area under cultivation, the yield of 6.3×10^6 kJ/hectare will feed about 5.4×10^9 people (10). This number is just about the Earth's current population. Of course, more energy input would allow a greater number to be fed, but our question here concerns the ultimate situation, after all fossil fuels have been used. We could increase the area under cultivation, as previously noted, but the land in use now is the better half of all the world's arable land.

HUMANITY IN SPACE

There is serious scientific speculation about a plan that could provide an outlet for our increasing population for some time, while holding out the promise of a high standard of living.

G. K. O'Neill (11) has proposed building cylinders several kilometers long and wide to provide homes for millions. The cylinders would be located at a point called Lagrange point 5 (L5), where gravitational tugs of nearby bodies are minimal. Built of moon material, orbiting the Earth–Moon system, and powered by the sun's energy, the cylinders would be large enough to contain mountains and weather systems (Figure 2.3). If each cylinder built others with part of its production, population growth could be accommodated for a time. This sort of thing could continue until the entire Moon was sacrificed; thereafter, asteroids and other planets could be used for construction materials. The energy cost of transporting people to the O'Neill habitats would be substantial, and only a few thousand people per year would be able to go there. Of course, such solutions are at most short-term palliatives.

Another feasible (if short-term) measure would be to make Venus inhabitable by people. This could be done at little cost over a period as short as a few centuries. We need only introduce simple plant life into Venus' upper atmosphere, where temperatures are not extremely high (12). Abundant sunlight and carbon dioxide are available there, so conditions are ideal for these organisms, and the population should burgeon. The presence of life would alter conditions on Venus, since the life forms would incorporate carbon and release free oxygen. The decrease in the amount of carbon dioxide would tend to cool Venus' atmosphere; thus life could penetrate further down in an ever-accelerating process. The only possible difficulty is a shortage of hydrogen and/or water; however, this could be rapidly remedied since the outer planets in our solar system are hydrogen-rich. Solid hydrogen could simply be dumped down Venus' gravitational well. If there were any free oxygen, the hydrogen would burn and form water, which could then be used by the plants.

The process of making a planet habitable for human beings is called *terraformation*. All the elements described are within the reach of present technology (12), and Venus is close enough for it to be conceivable that large numbers of people (perhaps tens of thousands per year over a long time period) might be transported there. Again, the energy cost of transportation is substantial, and the diminution of population growth, short-term.

FIGURE 2.3
An artist's conception of a space colony at Lagrange point 5. (NASA)

FIGURE 2.4
A Mariner photo of
Mars' surface. (NASA)

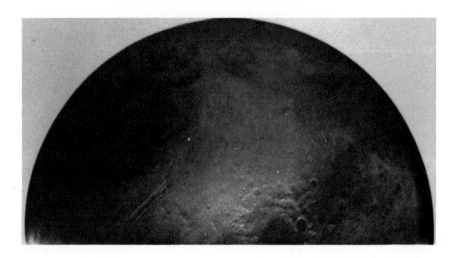

Lest you imagine that these scenarios are pure wish-fulfillment in a science fiction dream of mine, a scientific study of the possibility of colonizing Mars was carried out over a decade ago (13). If we merely seeded (Figure 2.4) Mars with the "proper" microorganisms, it would generate sufficient oxygen by photosynthesis within about 100,000 years to make the planet livable. Vaporization of the polar caps could shorten the interval to 10,000 years. Genetically engineered "super microbes" could be designed to make the change even sooner (13).

SUMMARY

World population cannot grow indefinitely. The Earth does not absorb all solar energy, and what energy is absorbed is absorbed by plants. People eat plants (are primary consumers), or eat animals that eat plants (are secondary consumers), or eat animals that eat animals that eat plants (are tertiary consumers). We estimated that Earth *could* support ten times as many people as now live on it, but most of these people will probably not find life very pleasant by our standards, and no other animals of any size could exist. A more realistic limiting estimate is about eight times the current population, a number that would be reached in a mere 105 years at the present doubling time of 35 years.

The growing human population could people surrounding planets in time. It seems possible to make Mars and Venus habitable, and economically and physically feasible to construct giant space cylinders at Lagrange point 5. None of these possibilities, however, could absorb a large population of immigrants from Earth.

REFERENCES

1. R. Freedman and B. Berelson, *Sci. Am.* 251, no. 3 (1974): 31.
2. F. Daniels, *Direct Use of the Sun's Energy* (New York: Ballantine, 1974); *Handbook of Physical Constants,* 51st ed. (Cleveland: C. R. C. Publ. Co.).
3. *Encyclopedia Americana.* The ratio is 28.3 percent, including Antarctica.
4. J. P. Holdren and P. R. Ehrlich, *Am. Sci.* 62 (1974): 282.
5. I. Asimov, *Life and Energy* (New York: Avon, 1972), 362.
6. R. Revelle, *Sci. Am.* 231, no. 3 (1974): 161.
7. D. H. Meadows, D. L. Meadows, J. Randers, and W. W. Behrens III, *The Limits to Growth* (New York: Signet, 1972), 60.

8. R. Dubos, *The God Within* (New York: Charles Scribner's Sons, 1972).

9. J. Brunner, *Stand on Zanzibar* (New York: Ballantine, 1976); H. Harrison, *Make Room, Make Room* (New York: Ace, 1984); and C. M. Kornbluth, *The Marching Morons* (New York: Ballantine, 1959).

10. F. S. Roberts, *Modules in Applied Mathematics*, vol. 4, pages 250–285 of *Life Science Models,* H. Marcus-Roberts and M. Thompson, eds. (New York: Springer-Verlag, 1984).

11. G. K. O'Neill, *Phys. Today* 27, no. 9 (1974): 32. *See also* G. K. O'Neill, *New York Times Magazine,* 18 Jan. 1976, 10.

12. J. Pournelle, *Galaxy,* 36, no. 8 (1975): 65. *See also* S. L. Gillett, *Analog* 104, no. 12 (1984): 64.

13. A. L. Robinson, *Science* 195 (1977): 662.

PROBLEMS AND QUESTIONS

True or False

1. Less than a quarter of the Earth's land area (excluding Antarctica) is arable or potentially arable.

2. The energy source ultimately limiting life on Venus is the sun.

3. Since some city areas in the United States have population densities as large as 10,000 people/mi² (and over small areas such as lower Manhattan, as large as 50,000 people/mi²), we could easily adjust to life at population densities of 6300 people/km² (16,000 people/mi²).

4. If a population is originally 2000, after one doubling time the population would be 4000, and after two doubling times the population would be 6000.

5. It will be possible to transport most of Earth's current population to other planets.

6. The ultimate population that could exist on Earth at sufficient nutritional levels is probably about 20 billion.

7. The land area of Earth provides the ultimate limit on population growth.

8. People occasionally are quaternary consumers (feeding on tertiary consumers).

9. The amount of arable land on earth is roughly half the total planetary land area.

10. Venus and Mars could not possibly ever support a human population.

Multiple Choice

11. Whale eat krill (small crustaceans). When people eat whales, they are
    a. producers.
    b. primary consumers.
    c. secondary consumers.
    d. tertiary consumers.
    e. none of the above.

12. Cannibals would be
    a. producers.
    b. primary consumers.
    c. secondary consumers.
    d. tertiary consumers.
    e. none of the above.

13. The energy flux from the sun is about 1400 watts/m² = 1.4 kW/m². If the Earth's radius is 6.4 Mm, so that its apparent area is 1.3×10^{14} m², the amount of power received is about
    a. 9.2×10^{10} W.
    b. 2.9×10^{15} W.
    c. 1.8×10^{17} W.
    d. 2.5×10^{20} W.
    e. 2.4×10^{31} W.

14. In a country in which the population doubles every twenty years, a 1980 population of 10 million will reach 80 million in
    a. 1990.
    b. 2000.
    c. 2020.
    d. 2040.
    e. 2060.

15. The ultimate source of most energy used by people is energy from
    a. the sun.
    b. old fossils under the ground.
    c. the Earth's internal heat.
    d. starlight.
    e. the oil and gas made when the Earth was formed.

16. Cornell astronomer Thomas Gold has proposed a theory that natural gas deposits were formed with the Earth (see Chapter 13). How would this affect the chapter's argument on energy limits to human population if he were correct?
    a. It would remove all limits on population growth.
    b. It is irrelevant to the basic argument, that mankind's numbers are limited because the amount of solar energy is limited.
    c. The Earth would be able to support more than 84 billion people.
    d. The Earth would be able to support even fewer people than now because natural gas poisons the atmosphere.
    e. The quality of life of all people alive will increase.

17. The reason that each successive step in the trophic pyramid stores only about 10 percent of the energy of the preceding step is that
    a. much food energy is used to support the metabolism.
    b. most food energy is excreted.
    c. most food energy goes into building muscles.
    d. consumers are very active and burn off the food energy to support continuing high levels of activity.
    e. social factors cause wastage of most food.

Discussion Questions

18. How would you order your priorities if it came to a choice among animals, plants, and people? You might wish to read the interesting discussion in *The God Within* (see Reference 8).

19. How long would it take for the orbiting cylinders to have some effect in reducing the population on Earth? For instance, you might want to consider how long it will take to build the first one. How big would it be? How many people do you think could live on a cylinder such as the one described? Try to estimate all of the above.

20. Think of other "technological fix" solutions to the burgeoning population. Evaluate the realism of any proposal you make.

21. Give other examples of population densities from those given in the text.

22. Is it possible to reconcile the idea in this chapter that there are limits to human population growth with the view adopted by some religious and minority population groups that large families are desirable? What issues of agreement or disagreement are there? What possible solutions are there?

3

WORK, ENERGY AND POWER

So far, we have discussed matters relating to a vague concept of what is meant by energy. *The definition of energy is central to a book on the subject. Energy has to do with work, and work, with the force and movement of the object upon which a force is exerted. These topics, as well as the definition of various types of energy and the definitions of efficiency and power, form the basis of this chapter.*

KEY TERMS　　*work*
velocity
acceleration
force
displacement
vector
gravity
weight
gravitational field
inertia
mass
thermal energy
kinetic energy
potential energy
friction
conservation of energy
efficiency
basal metabolic rate

Leete's eyes flashed. "Set the divariable veeblefurtzer on that Seltz coil," he shouted to Igor, "or we shall all be blown to atoms!" As the snap of electric discharges continued, something dark began to swirl at the bottom of the vat . . . .

This sounds like a 1930s pulp novel or horror flick. Yet even today this sort of image of the scientist and his language colors our attitudes. No horror movie is complete without a mad scientist mouthing pseudoscientific jargon. While true-to-life scientists often do introduce technical words, their purpose in doing so is to make their communications more precise. Often a scientific word strays into general usage (usually with its meaning somewhat altered) and sometimes an ordinary word is given a specific scientific meaning. A recent example of the former phenomenon is "radar," and of the latter, the use of "color" and "charm" as whimsical descriptions of the esoteric properties of the elementary particle building blocks of nature.

To enable us to communicate in this book (and to open the wider sphere of the scientific literature), we must agree on the use of certain words. These words are universally used in scientific prose, so no hint of misunderstanding can be allowed. Many of the problems encountered in communications between scientists and the public can perhaps be attributed to their different definitions of the same words.

The word *work* is generally understood to mean toil, labor, or employment (it also means a military fortification). If you and I were speaking in a casual way, we would use one of these meanings, and each would know what the other meant. If, on the other hand, I spoke to you *as a physicist* about work, we could have some serious misunderstandings. Suppose we were watching an old movie in which a bride is carried over the threshold by the groom. Of course, many things happen when one person carries another: The person is picked up, then carried, then put down. Work is done in the picking up and putting down and in the starting and stopping of the walk. However, if we consider only the part during which the groom walks at a steady speed on a level floor carrying his bride, no work is done, in a technical sense. Perhaps this distinction seems nonsensical to you. The problem arises because the word *work* has different meanings as used in casual conversation and in its technical sense.

BASICS OF MOTION: SPEED, VELOCITY, AND ACCELERATION

To define the scientific use of *work,* we must first consider the concepts of distance (and displacement), speed (and velocity), acceleration, and force.

Distance is a familiar quantity. We measure distances with tape measures, meter sticks, or other measuring devices. Distance measures how far away things are. While *displacement* seems less familiar, displacement merely adds information on direction to that of how far away something is. For example, a newcomer to a neighborhood may be told the distance to the local school. Such information is incomplete unless the school is in sight. It would be far more useful to know in addition to the distance the direction to the school. Distance is just a number; when both the distance and the direction from one point to another are specified, we have the information contained in the vector known as *displacement*. To illustrate how distance differs from displacement, consider an object that travels

in a circular path from some starting point and returns to that point. The object will have traveled a nonzero distance, the circumference of the circle, but it will not have been displaced from its starting point, and so its displacement will be zero.

This additional information about direction, essential for everyday life, distinguishes quantities known as *vectors* from quantities specified only by numbers. *Vectors* are quantities that have a direction specified in addition to their numerical values. Displacement, velocity, acceleration, and force are all vectors. These physical quantities have a basis in our direct experience, and these common perceptions provide a basis for our mutual understanding.

The speedometer in a car tells us how much territory we can cover in a given time. A reading of 50 miles per hour on the speedometer tells us that, if the car were to be kept moving at that speed, it would have gone 50 miles after an hour had passed, 100 miles after two hours, and so on. If we were to double the speed, we would double the distance covered in a given time (go 100 miles an hour), or cover the same distance in a shorter time (go 50 miles in half an hour).

People often use the terms *speed* and *velocity* interchangeably in common speech. Speed is different from velocity, however, because speed is a scalar—a number—and velocity is a vector. Specifying the velocity of an object imparts not only the information on speed but also the information about the direction in which the object moves.

Speed is not a vector, but a scalar. Velocity, a vector, gives more information than just speed: it also specifies a direction. We might say, "I drove east at 55 miles per hour from Cheyenne to Omaha," or "I drove north at 80 miles per hour from San Francisco to Portland on I-5." In the former case, the speed is 55 mi/h, and the velocity is 55 mi/h, east. In the latter case, the speed is 80 mi/h, and the velocity is 80 mi/h, north. Velocity is a vector specifying both speed and the direction in which the object of interest moves. Of course, we would not make it to Portland without having paid several speeding tickets!

CALCULATING AVERAGE SPEED FROM DISTANCE TRAVELED AND TIME

We define the average speed, v_{av}, as the total distance traveled/time required to travel that distance. The unit of speed is the meter per second (m/s), the kilometer per hour (km/h), the foot per second (ft/s), or the mile per hour (mi/h). If an object travels 50 meters in 10 seconds, it has an average speed of 50 meters/10 seconds, or 5 meters/second. If an object travels a distance of 100 kilometers in the course of an hour, it has an average speed of 100 kilometers/hour. If an object travels 45 miles in the course of an hour, it has an average speed of 45 miles/hour.

Suppose a pig arises from his mud puddle and waddles 3 meters eastward in a straight line, 4 meters northward in a straight line, then 5 meters in a straight line back to the original spot from which he started. Suppose also that the 3 meter walk took 12 seconds, the 4 meter walk 20 seconds, and the 5 meter walk 28 seconds. During the first part of the walk, the average speed is 3 meters/12 seconds, or ¼ meter/second (0.25 m/s). During the second part of the walk, the average speed is 4 meters/20 seconds, or ⅕ meter/second (0.20 m/s). On the pig's return to the mud puddle, the average speed is 5 meters/28 seconds, or ⁵⁄₂₈ meter/second (0.18 m/s). Overall, the average speed for the trip was

$$\frac{(3 \text{ meters} + 4 \text{ meters} + 5 \text{ meters})}{(12 + 20 + 28) \text{ seconds}} = \frac{12 \text{ meters}}{60 \text{ seconds}} = ⅕ \text{ meters/second} = \frac{0.20 \text{ m}}{\text{s}}$$

Note that the average speed for the entire trip is *not* the average of the average speeds during each of the three parts, $1/3(0.25 + 0.20 + 0.18)$ meters/second (which is 0.21 m/s), because each of the time intervals is different.

Average speed is useful in describing the motion of a body, but it does not tell everything about the motion. For instance, a car traveling 100 kilometers in an hour could conceivably have traveled at 150 km/h for forty minutes, then halted. Its average speed really does not describe such an hour's journey very well. The way to deal with such a situation is to take the average over a shorter interval, and do this several times. In the example given above, if the speed were sampled every fifteen minutes, we would see average speeds of 150 km/h, 150 km/h, 0 km/h, and 0 km/h. If the speed were sampled every five minutes, we would have average speeds of 150 km/h, 150 km/h, 150 km/h, 150 km/h, 150 km/h, 150 km/h, 150 km/h, 150 km/h, 0 km/h, 0 km/h, 0 km/h, and 0 km/h. While such a list may be tedious to read, it certainly represents the car's journey more accurately than an hour's or fifteen minutes' average.

For an object's motions that are more variable than in the idealizations so far presented, we are forced to take samples over smaller and smaller time intervals, the more varied the motion is, to get an accurate record of that motion. Of course, no matter how small the time interval, we can find the speed if we can determine the distance traveled. The *instantaneous speed* is the speed of any object at an instant, which we visualize as being the distance traveled in a very short time divided by that time interval. As we make the time interval shorter and shorter, we approach nearer and nearer to the true speed in an interval, the instantaneous speed.

CALCULATING AVERAGE VELOCITY FROM NET DISPLACEMENT AND TIME

The average velocity is a vector quantity, which means that it has a direction as well as a numerical value. We define average velocity, $\vec{v}_{av}$, as the net displacement/total time required to achieve the displacement. For example, a car traveling 100 kilometers northward in an hour will have an average velocity of 100 km/h, north. A car traveling 55 miles northwest in an hour would have an average velocity of 55 mi/h, northwest.

Consider again the example of the pig waddling 3 meters eastward in 12 seconds, 4 meters northward in 20 seconds, and 5 meters at 53.1 degrees south of west back to the original spot in the mud puddle in 28 seconds. Since the motion in each part of the pig's walk was in a straight line, the velocities for each part have lengths given by the respective average speeds. Thus, the velocities are: 0.25 m/s, east; 0.20 m/s, north; and 0.24 m/s, 53.1° south of west. The average velocity for the trip is, however, given by the net displacement divided by 60 seconds, since the motion is not just in a straight line. The net displacement of the ending point with respect to the starting point is zero, even though the pig travels a path that is a total of 12 meters long. As a result, the average velocity is zero.

Anything that returns to its starting position after some specified time will have an average velocity of zero over that time interval, because its net displacement is zero, no matter how great a distance it traveled in the meantime. The Earth travels in an almost circular orbit around the sun and takes exactly one year to return to its starting position relative to the sun. Thus the Earth's average velocity with respect to the sun, measured over one year, is zero, despite the fact that the Earth has traveled a distance of about 940 million kilometers during that time.

We may define the instantaneous velocity of an object as the velocity at a particular instant of time.

Acceleration

Acceleration is simply a measure of how rapidly the velocity changes. To say that an MG sports car has better acceleration than an antique Studebaker means that the MG can change its speed in a shorter time interval than can the Studebaker. An engine that can make the speed change from 0 to 60 miles per hour faster than another is said to produce better acceleration. An "accelerator" is a device that allows you to change the speed of your car. The harder it is pushed, the greater the rate at which your speed changes (given the limitations of your car). The car's brake is also a sort of accelerator, since applying the brake changes the car's speed. Since acceleration happens when velocity changes, a car can accelerate even when the speed does not change. For example, if a car is driven along a circular track at a fixed speed, the direction of the velocity vector changes, and since there is a change in velocity, there is acceleration. In this sense, the steering wheel is also an accelerator of sorts.

FORCE AND MOTION

We intuitively recognize that a *force* is a push or a pull. If you try to force someone to do something, it implies that you change a state of rest into a state of motion, or a state of motion into a state of rest. This intuitive understanding nearly

CALCULATING AVERAGE ACCELERATION
FROM VELOCITY CHANGE AND TIME

Since we define average acceleration, $\vec{a}_{av}$, as the change in velocity/time required to change velocity, the unit of acceleration is (meters per second) per second, (m/s)/s, or m/s². In the English system, the unit of acceleration is the (foot per second) per second, or ft/s².

If we suppose that an object changes its velocity from 50 m/s to 0 m/s along some direction in ten seconds, the average acceleration has a magnitude of 5 m/s²; alternatively, we may say that it has a net average acceleration along the direction specified of -5 m/s². The minus sign indicates that the velocity has decreased along the direction given, in the specified interval.

Suppose we look upward and see an object that has zero speed at some instant of time and a speed of 10 m/s downward one second later. The average acceleration can be expressed as 10 m/s², downward, or, equivalently, as -10 m/s², upward. For this reason, physicists do not usually speak of deceleration; acceleration may take on both positive and negative values.

By considering an object traveling along an arc of a circle at a constant speed, it is possible to show that the instantaneous acceleration has a magnitude of speed times speed divided by the radius of the arc of the circle at any point along its path ($a = v^2/R$), and that the direction of the acceleration is toward the circle's center. This is the radial acceleration associated with circular motion.

---

describes the scientific use of the word as well. An existing system continues to exist (particles moving apart from one another, or approaching one another, or whatever) unless an outside agent affects the bodies in the system. Such an outside agent is known as a *force*. There is always a direction associated with a force, so force is a vector quantity.

If I were to jump off a table, I would not expect to fly upward (especially since I am not a mild-mannered reporter, or even a mild-mannered physicist!), nor would I expect to wing to the right. I know I would plunge downward, my speed increasing as I fell. I admit to having tested this time and time again during my earlier years.

Because speed changes as an object falls, it must be subject to an acceleration; this acceleration due to gravity is approximately 10 m/s² in magnitude. Objects not supported by other objects near the Earth's surface feel a downward force; the direction *down* (toward the center of the Earth) characterizes the gravitational force.

Forces may be exerted by objects such as walls and floors, or by rods or blocks, or by wires or ropes, or by "force fields" such as are exerted by the sun on the planets (gravitational) or the nucleus of an atom on the atomic electrons (electromagnetic).

Consider a table sitting on the floor. How is it influenced by the forces acting on it? We know that the table would fall downward if the floor's support were removed, because of the gravitational force on the object, its weight. It does not fall because of the countervailing effect of the floor. The floor appears immovable to us, but the weight of the table acting on the floor causes the floor to bend, or deform. The floor underneath a heavy desk can often be seen to bear the permanent traces of such prolonged deformation. The deformation causes forces inside the floor that oppose further deformation. These forces ultimately act on the desk, exerting a force equal in magnitude to the weight of the desk. Such a force, perpendicular to the supporting surface, is called the normal force. Because the floor exerts an upward force on the desk and gravity a downward force, and both forces are the same size, the net effect of the two forces is zero. Here we see an important consequence of the vector nature of the force: adding two forces equal in magnitude may produce a net force anywhere from twice as great as either one to zero.

The unit of force is named after Sir Isaac Newton, who did so much to explain motion: it is called the *newton*. A typical apple weighs about a newton. In the English system, the unit for force is the pound. A pint of water weighs about a pound.

Bodies move when net nonzero forces act upon them. So this idea of force as something that changes a body from a state of rest to a state of motion, or vice versa, must be connected to the idea of the acceleration experienced by that body. Imagine standing on roller skates. Suppose that, during some short time interval, you were pushed by a certain force. After the force ceased, you would be moving at some speed (Figure 3.1a). Now imagine starting all over again. This time, suppose you are pushed by twice the force that acted the first time, during a time interval of the same length as that before. The speed at which you would be skating after this second experiment is greater than the speed you would have attained after the first (Figure 3.1b). The greater the force, the greater the acceleration; the greater the acceleration, the greater the ultimate speed. In the same way, the greater the acceleration of a car, the greater the force the passengers would experience "pushing them into their seats." (Actually, the seat, as part of the car, is exerting a force on the passengers to accelerate them. They are being forced to change their states of motion along with the car, and they experience the force the seat exerts on them to change this motion as that of some agent pushing them into their seats.)

There is another factor involved here. Return to the thought experiment with the roller skates.

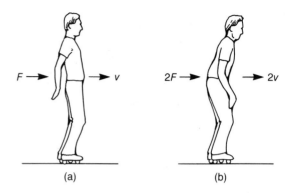

(a)   (b)

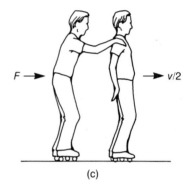

(c)

FIGURE 3.1

(a) A skater is pushed by a force F for a certain time; after the force is removed, the skater is moving at speed v. (b) If the force is doubled (to $2F$) and applied for the same length of time, the final speed is doubled (to $2v$ relative to the situation in part a). (c) If the skater and the skater's clone are pushed with force F applied for the same length of time, the final speed is halved (to $v/2$ relative to the situation in part a).

In a third experiment, suppose you were teamed up with an identical cloned you. Both of you together are pushed with the same force as was used the first time, for the identical time interval. The two of you would be skating after the force ceased at a slower speed than before (Figure 3.1c). The greater the number of clones holding together, the slower the ultimate speed after application of identical forces over identical time intervals upon the greater amount of stuff there.

This reflects the idea that the more stuff—matter—there is, for application of a given force, the smaller the resultant acceleration. This quality of matter—that it resists changes in being put into motion—is called *inertia*. The quantitative measure of the inertia of matter is called *mass*: Mass has to do with the amount of matter present. Two identical objects would have twice the mass of a single one. Five such objects would have five times the mass of one object. The mass of an object can be defined in some absolute way only by arbitrarily choosing a standard mass. This arbitrary standard mass is then duplicated and subdivided, and these subdivisions and duplicates can be used to define other masses by comparison. The standard mass in the metric system is the kilogram.

Very often people confuse the idea of *mass* with that of *weight*. The weight of a mass is the force exerted by gravity on that mass. A nonliving body's mass is fixed once and for all the time it exists as that body. Its mass is "stamped upon it." Its weight is the gravitational force exerted on it; this force depends upon the position of the mass in the universe. We call the ratio of gravitational force to the mass at any point in space the gravitational field at that point. (We can think of the Earth's gravitational field as the reason that bodies released near the Earth's surface fall.) Your weight on the moon is not the same as your weight on Earth, although your mass would be the same on the moon as it is in your living room. Your weight on Mars is different both from your weight on the moon and on Earth, but you would still measure the same mass there.

Both mass and the strength of a force influence the consequent acceleration. Acceleration increases if the force is increased for a given mass. Acceleration decreases, for application of a given force, if the mass is increased. The simplest hypothesis relating force, mass, and acceleration that reflects these realities is

force = mass × acceleration.

This equation is a simplified way of stating Newton's second law of motion. It is valid over an immense range of conditions. For example, if you are in an accelerating car, the force you experience pushing on you as the speed changes is simply your mass multiplied by the car's acceleration.

WORK

We can now define *work*: the *work* done by any force is the product of the force and the distance moved *in the direction of the force*. If the force is exactly along the direction of motion, then the work done is just

work = force × distance moved.

If the force is exactly perpendicular to the direction of motion, there is no motion in the force's direction, and consequently, no work at all is done by the force.

To return to the example of the groom carrying his bride over the threshold, suppose he picks her straight up and then walks steadily into the house (Figure 3.2). In picking her straight up, he is moving her in the direction opposite to the gravitational force on her, which is her weight. He therefore does an amount of work on her equal to her weight (W) times the change in height (h):

work by groom on bride = weight
                                   × change in height
                                = $W \times h$.

If we could ask the groom, we would both agree that he worked to pick up the bride. He might also insist that he did a lot of work carrying her at constant speed across the threshold. However, in a technical sense, the only work he did resulted from the force he applied (parallel to the horizon) to accelerate his bride up to the constant speed, and the force he applied to stop. (These latter forces are of the same order of magnitude as the bride's weight, but applied over a very short distance.)

Assume that the bride's weight is directed downward, and the groom carries her horizontally across the threshold walking at a steady speed on a level. Since the groom is not accelerating at any time during the carry except at the very start and the very end of the carry (let's agree to discuss only the carry; the work done in starting

FIGURE 3.2

A groom carries his bride over the threshold. (The author and his wife re-enact this here.)

and stopping is probably at most a few percent of that done in lifting the bride). For most of the carrying time, there is no force being exerted on the bride in the horizontal direction. Our conclusion is that no work is done during the part of the carry that took place at constant speed, because the distance moved *in the force's direction* is zero. The groom's perception seems to contradict the statement that no work is considered to have been done if the force is perpendicular to the direction moved.

We would also agree with a student pulling an all-nighter to study for a physics exam that the student was working. If someone pointed out that he or she sat in a chair and was not moving, and

claimed that therefore no work was being done, the student would probably become indignant.

In fact, work is being done by the principals in both these examples. In the case of the student, the body is constantly forcing blood to move through capillaries and veins, forcing the rib cage alternately to expand and shrink, moving lymphatic fluid, forcing blood through a filter, and so on. All these things constitute work, in both senses of the word. This is only a small part of the work that the body continuously does. For example, the muscles in the body are always moving slightly, preparing to be used, even when we think those muscles are perfectly still. Try holding your arm straight out from your body, still. After

DERIVING KINETIC ENERGY FROM WORK

Kinetic energy is the energy of motion of a mass. Remember that the force applied equals (mass) × (acceleration). We further said that acceleration was the increase in velocity in a given time.

Speed is the change in distance experienced by an object in a given time. If the *change* in distance is called Δd (delta *d*) and the change in the time Δt (delta *t*), then the speed (*v*) is given by

$$v = \frac{\Delta d}{\Delta t}.$$

(The Greek symbol Δ means "change.") Now, acceleration is the change in speed in a given time (taking the same change in time). Calling the acceleration *a*,

$$a = \frac{\Delta v}{\Delta t}.$$

Now, force (*F*) is just (mass) × (acceleration), or

$$F = ma = m\left(\frac{\Delta v}{\Delta t}\right).$$

In this time, Δt, the object has gone a distance Δd. Let us assume that the movement is along the force's direction. This is a "free fall" problem, which means that nothing stops the mass from moving in the force's direction. So, the work done is

$$\text{work} = F\Delta d = m\left(\frac{\Delta v}{\Delta t}\right)\Delta d.$$

Let us write the speed before as v_b and the speed after as v_a. Then

$$\text{work} = m\left(\frac{v_a - v_b}{\Delta t}\right)\Delta d.$$

But Δd is just the (average speed) × (change in time). In the same way that the average of 4 and 6 is (4 + 6)/2, or 5, the average speed during the time interval Δt is ½($v_a + v_b$). Therefore,

$$\Delta d = \frac{1}{2}(v_a + v_b)\,\Delta t$$

and

$$\text{work} = m\,(v_a - v_b) \times \frac{1}{2}(v_a + v_b)$$

$$= \frac{1}{2}m\,(v_a - v_b)\,(v_a + v_b)$$

$$= \frac{1}{2}m\,(v_a^2 - v_b^2).$$

Since this has to do with *motion* it is *kinetic energy*. If v_b is zero, then

$$\text{kinetic energy} = \frac{1}{2}mv_a^2.$$

a short while, you will notice your arm moving perceptibly.

The groom in our example is doing work because the bride is being moved slightly up and down as the groom's muscles twitch and he tries to compensate for it. He is doing work on her by raising her, just as the gravitational force is doing work on her as she is lowered. Assuming success in keeping her at about the same height, the *total amount of work done on her* in carrying her is zero. Because of the inevitable twitching of his muscles, the groom has done work, and we might understand his feeling tired.

ENERGY AND ENERGY TRANSFORMATIONS

In all the discussion thus far, only work has been mentioned. Work is just one form of *energy*. This word, *energy*, was first coined by Thomas Young in 1807 (1). It is from the Greek and means roughly "work within." As we shall use it, energy is the capacity to do work. Work is central to the idea of energy. Now that we have some idea of what work is, we can consider without confusion the many forms of energy. As long as something can do work, it is able to use energy.

Recall our groom. He did work against the gravitational force to lift the bride. Because of this input of work, the bride has energy associated with her position above the surface of the ground. Any object in a gravitational field possesses energy by virtue of its position relative to the ground. If the object is allowed to fall, this energy has the potential to change into other forms of energy and to do work. For this reason, energy of position in the gravitational field is known as *gravitational potential energy*, potentially work. What if the groom had stumbled on the threshold? As he stumbles, the gravitational potential energy is converted into *energy of motion*—the bride falls, her speed increasing. As she hits the floor, her energy of motion is transformed into *energy of deformation*, her own and the floor's. This energy of deformation of these two bodies is transformed (finally) into two forms

of energy: *sound energy*, which propagates as a sound wave through the air (eventually heating up the air), and *thermal* or *internal energy* spread throughout the two bodies (see Figure 3.3).

In our example, then, *work* becomes *gravitational potential energy* becomes *energy of motion* becomes *deformation energy* becomes *sound energy* and *thermal energy*. The final result is the heating up of the environment (and perhaps the relationship as well!).

Kinetic energy is synonymous with *energy of motion*. The word *kinetic* comes from the Greek word *kine*, which means "motion." One of our words for movies, cinema, comes from the same root. The distinction between forms of energy is sometimes obvious and sometimes a matter of convenience. For example, consider some solid material. It is made up of an array of atoms (or molecules) bound together by electric forces (see Chapter 4). Each atom remains in its place relative to the others. The atoms are vibrating in place. If we heat this solid material, and then could magnify the structure enormously, we would see that the distance moved by the atoms would on average be greater. So an increase in temperature is reflected in an increase in the kinetic energy of the atoms in the solid, spread out among all of them. Temperature is simply a measure of the average energy of vibration of the material's atoms. Thermal energy, what is colloquially called heat energy, is something like socialized (randomly distributed) kinetic energy (2).

Examples of several sorts of energy are listed in the accompanying box. As we shall see, any sort of energy can be transformed to any other sort.

Suppose a brick is carried up a mountain. Work is done in carrying it up the mountain. The increase in potential energy in the gravitational field is the amount of work done in lifting the brick to the top of the mountain by some path. One such path might be as shown in Figure 3.4. We can imagine breaking up *any* change in height into segments along the direction of *W*, the weight (that is, parallel or antiparallel to it), and segments having no length in *W*'s direction (that is,

FIGURE 3.3
What might have been
had the groom stum-
bled.

FIGURE 3.4
How an arbitrary path is
broken into segments
(for the sake of clarity,
we have used steps that
are actually much too
large to reproduce the
path). The path seg-
ments parallel to the
weight contribute to the
work. Path segments
perpendicular to the
weight require no work
to traverse.

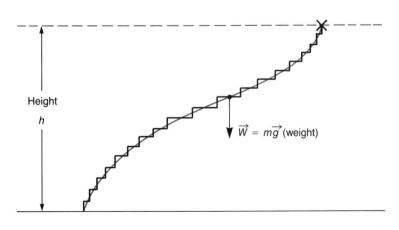

DIFFERENT FORMS OF ENERGY

Mechanical energy is the kind of energy used by machines. The category of mechanical energy lumps together kinetic energy and potential energy (that is, energy with levers, gears, and wheels that use kinetic or potential energy to convert into work).

Examples of mechanical energy are the pendulum, a levered-up rock, and similar devices. The pendulum has already been discussed. At the instant the pendulum is released, it has gravitational potential energy and no kinetic energy. At the bottom of the swing, it has kinetic energy but no potential energy. It just keeps going, trading energy back and forth between these two forms. A levered-up rock can fall, thus doing work.

Other forms of mechanical energy include landslides, avalanches, and rain. To make rain, the sun evaporates water, which condenses into clouds. This water then falls as rain. It does work if it can be captured behind a dam, after which its gravitational potential energy is converted to electrical energy as water falls through turbines in the dam.

Earthquakes, which release stored rock strain energy along fractures, exhibit large-scale land motion. Tsunamis (so-called tidal waves) are water waves caused by underwater earthquakes and are also forms of mechanical energy.

Nuclear energy becomes available when unstable nuclei spontaneously change by throwing off particles. The decay of the neutron into a proton, an electron, and an antineutrino is an illustration of the conversion of nuclear mass energy into kinetic energy. This form of energy is discussed in detail in Chapters 6 and 14.

Thermal energy is energy internal to the body, due to motion of the atoms making up the body. Hot springs such as those found in Yellowstone Park, in the Geysers, California, and in the city of Klamath Falls, Oregon, are of high temperature because of contact with the hot rock making up the basement of the continent. Volcanoes, such as Mauna Loa, are manifestations of thermal energy. The steam that drives the piston in a steam engine has thermal energy.

Electrical energy is energy stored in the back-and-forth motion of electrons in electric utility lines. Lightning involves the transformation of electrical energy into heat energy and light energy.

Chemical energy is energy that has been stored in chemical form, such as in fuels or sugars or as energy stored in car batteries. Gasoline is a chemical that combines with oxygen and a little heat to release the great amount of heat stored in the chemical structure of the gasoline. Other such chemicals include sucrose, methane, ethanol, and methanol.

Radiant energy, or electromagnetic energy, is simply the energy carried by light. It is this energy that makes the chemical storage of photosynthesis possible. Chemical energy can be converted into light energy by phosphorescent organisms, some deep water fish, and fireflies (3).

Most of these forms of energy are treated in subsequent chapters.

perpendicular to it). Thus we have done work only along those segments parallel to W, because of our definition of work. Those segments along W finally add up to h, the height above the ground. So

$$\text{(gravitational potential energy)} = \text{work done} = W \times h,$$

which is the same result we found for the bride.

Kinetic energy, or energy of motion, would be zero for a body at rest. The kinetic energy of a moving body must depend on the speed. We define a body of mass m moving with speed v to have kinetic energy $\frac{1}{2} mv^2$.

A pendulum can be made by hanging a mass from a ceiling by a string. When the mass is pushed aside and released, it undergoes a regular to-and-fro motion. Because the mass has to be pushed, so that a force is exerted on it, and it moves as a result, work is done, giving the pendulum potential energy. As the mass is released, it falls, losing some of this potential energy and increasing its kinetic energy. At the bottom of its swing, it has no potential energy (relative to a hanging mass), and all its energy is kinetic energy. Potential energy has been converted to kinetic energy. This kinetic energy converts itself into potential energy, the potential energy converts itself into kinetic energy, *ad nauseam*. Actually, in the operation of a pendulum, some energy is transferred from the pendulum to the air by collision with air molecules, and some energy heats up the pivot due to friction there. Thus, if we considered *only* kinetic and potential energy, we would not have the whole story.

FRICTION AND HEAT

One of the difficulties we have in understanding motion is that objects moving near the Earth's surface eventually stop. This led the ancient Greeks to the false idea that motion is only possible when a force is continually applied: We can observe motion at constant speed only when we apply a force to an object. Surfaces in contact grip one another rather like Velcro as they move rel-ative to each other. This phenomenon is called *friction*.

There are several kinds of friction: friction between bodies moving on one another; friction between motionless bodies in contact, preventing their motion, called *static friction;* and friction as a solid object moves through a fluid, called *drag*. Air resistance, a form of drag, is discussed in more detail in Chapter 21.

When objects in contact move, work is done by the frictional force. This work is lost, as work, to the system and heats up the components, appearing as thermal energy. For this reason, mechanical systems produce less output work than the energy they take in.

What this really means is that some forms of energy, which can easily be transformed into useful mechanical work, are worth more (or are of higher quality) than forms of energy it is difficult to transform, such as internal energy. Equivalent amounts of energy do not necessarily have the same capacity to do work. This topic is discussed in more detail in Chapter 7.

CONSERVATION OF ENERGY

In real situations, then, energy appears to be lost. This is only chimerical, however, because what actually happens is that the energy is changed into a different form. Energy is neither lost nor gained in any process. We see examples of this in our everyday lives. In a car, the chemical energy of gasoline becomes thermal energy and work (there is so much heat given off in a gasoline engine that it must forcibly be cooled). Water is caught in a dam to be transformed into mechanical kinetic energy in a turbine. And, since work can produce thermal energy, thermal energy can be made to do work.

The guiding principle in our approach to energy is that it can never be gained or lost. This tenet is called the *principle of conservation of energy*. The great physicist Hermann Helmholtz is credited with first formulating the principle in a useful way in the 1880s.

The amount of energy of the universe does not change. Whatever ways energy appears in the universe, the total amount is the same as it was ten minutes ago, or 10^8 years ago, or 10^8 years from now. This was essentially the principle enunciated when we defined TANSTAAFL. We may have energy in any of its forms, and the amount of any sort changes from instant to instant perhaps, but the total energy is the same.

As an illustration of the principle of conservation of energy, let us look at the pendulum again. Recall that gravitational potential energy (PE) is given by Wh ($=$ mgh), and that kinetic energy (KE) is $\frac{1}{2} mv^2$. We have used the fact that W, the gravitational force (weight), is given by (mass) times (gravitational acceleration), which we write: mg. The symbol g stands for the acceleration due to gravity, 9.8 m/s².

The maximum potential energy of an object that falls is transformed into kinetic and thermal energy as the object falls. We will consider an idealized pendulum (one in which we ignore heat produced at the pivot, and friction losses in collisions with air molecules); since energy is conserved, in this case

KE + PE = constant.

Therefore, there is a relation between the maximum height, h, and the maximum speed, v, that is

$$mgh = \frac{1}{2} mv^2, \text{ or } v^2 = 2gh$$

no matter what the mass is! Suppose the maximum height, h, is ½ m. Then, making the approximation that g, the acceleration of gravity at Earth's surface, is 10 m/s² (or 32 ft/s²)

$$v^2 = 2(10 \text{ m/s}^2) (1/2 \text{ m}) = 10 \text{ m}^2/\text{s}^2.$$

To determine v, we must take the square root, giving

$$v = 3.1 \text{ m/s} (= 19.5 \text{ ft/s}).$$

We may thus use the principle of conservation of energy to find the speed of the pendulum mass at the bottom of its swing.

The notion that work and thermal energy ("heat") are interchangeable took a long time to become accepted. Most scientists had thought that thermal energy was a fluid; this fluid was called the *caloric*. However, Benjamin Thompson, an American Tory who became the German Count Rumford, noticed in 1798 that a tremendous amount of heat was produced in boring out cannon. Water was poured on the cannon to soak up the heat, and after a time, the water boiled. The more water put in, the more boiled off. Rumford was forced to conclude that the amount of "caloric" was infinite. Rumford recognized that the motion of the boring instrument became the motion of small pieces of borer, the cannon itself, and the water. In this conclusion, Rumford was far ahead of his time and was ignored.

It was not until the 1840s that the so-called mechanical equivalent of heat was measured by James Prescott Joule. He found that a fixed amount of mechanical energy was converted to a fixed amount of thermal energy. This set the stage for the much later acceptance of the principle of conservation of energy.

Simple Machines

As another example of the conservation of energy, consider the jack and the lever. Note that, with our definition of work, we could do the same amount of work by applying a large force through a small distance as by applying a small force through a large distance. This is the principle of the jack and the lever. For the lever of Figure 3.5, an amount of work done is

$$\text{work} = F_1 d_1 = F_2 d_2.$$

Thus, if d_2 is large, F_2 can be small.

Pulleys are made of wheels, with ropes strung over them, as shown in Figure 3.6. Fixed pulleys, which are attached to walls or ceilings, act to change a force's direction. Movable pulleys are used to change the force's direction and to decrease the amount of force necessary to move an object. In Figure 3.6, which shows two pulleys, the force F needed to raise weight W at a constant

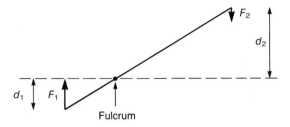

FIGURE 3.5

The lever is a simple machine. F^2 is the input force, which is applied through a distance d^2; F^1 is the consequent output force, acting through a distance d^1. The lever must be supported at one point (called the fulcrum) in order to work. Other simple machines include the wedge and the pulley.

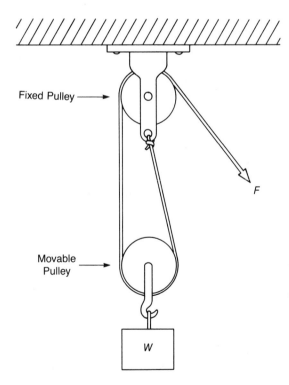

FIGURE 3.6

In a system containing one fixed and one movable pulley, a weight W may be lifted by application of a force F equal to $W/2$.

speed is less than W (it is ½ W). For an ideal pulley, one with no friction in the pulleys and no inertia, the amount of work done to raise the weight a distance d (Wd) must equal the amount of work done by the force F. Conservation of energy guarantees that the amount of work coming out of a machine equals the amount of work put in. In the case of the pulley of Figure 3.6, the distance through which the force F must act is $2d$, since F is ½W and

$$(F) \text{ (distance through which } F \text{ acts)} = Wd.$$

In this case, the ratio of the output force to the input force (known as the mechanical advantage) is

$$\frac{W}{F} = \frac{W}{(W/2)} = 2.$$

Each additional set of pulleys added to such a system reduces the force required to move an object, but at the price of increasing the distance through which the force is exerted. We cannot get something for nothing, but we may reduce the force necessary to do a job (or achieve a mechanical advantage) by increasing the amount of rope that needs to be pulled through the hands. The key point to remember is that, while force can be multiplied by use of a machine, the amount of work that has to be done is the same whether or not the machine is used (if an ideal frictionless machine is assumed).

Efficiency

Real machines do have friction in them as well as the possibility of deformations. Ropes stretch, levers bend. In this real world, some input energy must either be used to battle friction or be stored in a stretched rope or bent rod. Consequently, the transformation of energy by a real machine involves a loss in available work to thermal energy (friction) or deformation energy (stretched string or bent rod). The ratio of work output to work input is exactly one for an ideal machine, and less than one for a real machine, because of the losses mentioned above. This ratio is called the *efficiency*:

$$\text{efficiency} = \frac{\text{work output}}{\text{work input}}.$$

Consider a pulley system, as in Figure 3.6, that has negligible friction in the pulleys, but in which the rope stretches by 10 percent when the weight is 500 N. Because the force F is determined solely by the fact that there are now two ropes holding up the weight W, F is still ½ W. Now, however, to raise the weight a distance d, we have to pull in a 10 percent longer rope, because of rope stretching, than we would have pulled in for a stretchless rope: $2.2d$. The work output is still Wd; the work input is now $(F) \times (2.2d) = (½ W) \times (2.2d) = 1.1\,Wd$. The efficiency is therefore

$$\frac{\text{work output}}{\text{work input}} = \frac{Wd}{1.1\,Wd} = 0.91,$$

or 91 percent. Similar calculations could be done for the other examples.

ENERGY IN HISTORY

For our early ancestors, who lived by hunting, ate their kill raw, and were much smaller than we are, there was an energy budget of about 6000 to 8000 joules per day. (Or, for those more familiar with food intake terms, this amounts to 1500 to 2000 kcal per day.)

When people learned how to use fire, the amount of energy used per day was probably about double or triple the former amount: about 12,500 to 17,000 joules per day. (I estimate about 20 MJ/kg heat value for wood, that is, a use of about 3 to 4 kg of wood per day.)

When human beings settled down to a more sedentary life of agriculture using oxen or horses, energy usage probably tripled per person; thus, a person used about 40,000 kJ per day. I obtain this by assuming that an animal uses about 100 terajoules (TJ) per year, based on an analysis (4) that estimates that 1 kg of feed must be supplied

for each 50 kg of body mass and taking an average mass of 750 kg. At 16.75 MJ/kg feed, this is 250,000 kJ/day. Assuming one horse or ox to about ten people, this is an additional 25,000 kJ per day per person.

In Europe during the early Renaissance, windmills and coal began to be used, and wood was used in large amounts for heating, cooking, metalworking, and other such tasks. Water power was also used to grind grain. Thus, energy use per capita probably doubled again, to 80,000 to 100,000 kJ.

The United States in the mid-nineteenth century used about 400 MJ/capita/day, and today as an industrial power, we use about 1000 MJ/capita/day (see Table 3.1). With only about a twentieth of the world population, we use about a third of the world's total manmade energy (5). Some important energy equivalents and units are listed in the following box.

In modern times, we have substituted other forms of energy for human labor. The chemical energy of gasoline is transformed into kinetic and heat energy. The chemical energy of wood (which is actually stored solar energy) powers boilers that generate electric energy. This electric energy can become mechanical energy (in machines), light energy (from light bulbs), thermal energy (in ovens or water heaters), and so forth.

TABLE 3.1
Daily energy use per person.

	(MJ/Capita/Day)	(kcal/Capita/Day)
Hunters	8	2,000
Use of fire	17	4,000
Domestication of animals	40	10,000
Renaissance	100	25,000
1850[a], U.S.	420	100,000
1900[a], U.S.	460	110,000
1950[a], U.S.	690	166,000
1973, U.S.	1030	245,000
1986, U.S.	900	214,000

a) Reference 4.

VARIOUS ENERGY UNITS

The British thermal unit (abbreviated Btu), the amount of heat required to raise the temperature of one pound of water 1° F, is equivalent to 778 foot-pounds. Remember that, since work is force times distance, the English system, with distances measured in feet and force in pounds, states work in the unit of foot-pounds.

In the *MKS* (meter-kilogram-second) system, the length unit is the meter. The unit of area is the square meter or the hectare, 10^4 m². The unit of volume is the cubic meter, or, for a liquid, the liter, 10^{-3} m³. The liter is also 10^3 cm³, sometimes denoted 10^3 cc (a cm³ is a cubic centimeter, or cc). The force experienced by an object, its mass times its acceleration, has therefore these appropriate units: Speed has units m/s, so acceleration has units (m/s)/s = m/s², and mass has units of kg, so that force has units kg m/s², which unit of force has the name the newton (N). The unit of work or energy is the unit of force times the unit for distance: the N m, otherwise known as the joule (J). The joule is named for James Prescott Joule, who showed that an MKS unit of heat, the kilocalorie, is the amount of heat required to raise the temperature of 1 kilogram of water by 1° C. The kilocalorie is best known to dieters as the "calories" they eat. A calorie is actually the amount of heat necessary to raise 1 gram of water 1° C.

We have many energy units—ft-lb, N m, J, cal or kcal, Btu. Relations among them are listed here for comparison.

1 Btu	= 0.252 kcal	= 252 cal
	= 778 ft-lb	= 1060 J
1 J	= 0.734 ft-lb	= 0.239 cal = 2.39×10^{-4} kcal
1 ft-lb	= 1.362 J	
1 kcal	= 4186 J, 1 cal = 4.186 J	

POWER

So far we have discussed work and energy. In our personal lives it is relevant, not only *how much* work was done, but also *how fast* it was done.

Suppose it snows, and I decide to shovel the walk. Once the snow has stopped falling, there is a certain amount of work that must be done. If I were able to shovel all of the snow from my walk in one hour, while my neighbors took two, obviously I can do work faster. This is true even though the *same* amount of work would have

been done by me and my neighbor. The rate at which work is done is called *power*. A hair dryer set to operate at 1200 W dries your hair twice as fast as one set to 600 W.

We define *power* as work done/time needed to do the work. This differs somewhat from the colloquial meaning of the word. In the international system, the unit of work is the joule, and the unit of time is the second. In that system, the unit of power is thus the joule/second. This unit is given the name watt (W), in honor of James Watt, the inventor of the modern steam engine. Since one watt is one joule per second, the joule could also be written in terms of watt-seconds. Since an hour is 3600 seconds, a watt-hour (Wh) is 3.6×10^3 joules.

The kilowatt-hour (kWh), a widely used unit of energy, is equal to 3.6×10^6 J. When you pay your electric bill, you pay for energy by the kWh. In this book, the kWh will be used as the energy unit as much as is possible.

Many electrical devices are labeled by the amount of power that they draw. A 100 W light bulb can use electric energy only at that rate—100 J/s. A 1200 W electric heater uses energy at 1200 J/s, and cannot use it any faster than that (or if it does, it will burn up). Most electric devices list the number of watts the device should use.

Metabolic Power

I can do work at a rate of around 75 J/s. I get this estimate by imagining that I am digging. I can lift a 5 to 10 kilogram (10 to 20 pound) clod of dirt about a meter to throw it on a pile, and do it steadily. Thus, the potential energy changes by ≈5 to 10 kg × 9.8 m/s² × 1 m, or 50 to 100 J. Since the process of lifting takes about one second, I would estimate that I could do work at the rate of 50 to 100 J/s. My estimate—75 J/s—is in the middle of this range.

I also can walk up a flight of stairs about 3 meters (≈10 feet) high easily in about twenty seconds. In that case, I do an amount of work equal to the product of my weight and the change in the height: 75 kg × 9.8 m/s² × 3 m = 2250 J. From the definition, the power developed is then

$$\text{power} = \frac{2250\,\text{J}}{20\,\text{s}} = \frac{110\,\text{J}}{\text{s}} = 110\,\text{W}.$$

In the English system, this is about 170 lb × 10 ft/20 s = 85 ft-lb/s.

A horse can do work at a rate of about 550 ft-lb/s. We call 550 ft-lb/s one horsepower; a person works at about 1/10 hp. In fact, at maximum a person can do work at the rate of about 1 hp for very short times. If I were to run up the stairs full speed, I could probably get to the top in about three seconds. Thus

$$\text{maximum power} = \frac{1700\,\text{ft-lb}}{3\,\text{s}} \approx \frac{570\,\text{ft-lb}}{\text{s}} \approx 1\,\text{hp}.$$

Since one calorie is 4.186 J, 1 calorie = 4.186 J/(3.6×10^6 J/kWh) = 1.16×10^{-6} kWh, and so one kilocalorie is 1.16×10^{-3} kWh. A human being must expend energy at a rate of about 2×10^3 kcal/day, or about 2.32 kWh/day. In the United States, the daily energy use is somewhat higher, about 3.7 kWh/day. This amount of energy would burn a 100 W bulb for 37 hours, or a 150 W bulb for a day. A typical monthly food bill for one person in the United States is in the neighborhood of $150. The cost of personpower at 3.7 kWh per day for 30 days is

$150/111 kWh = $1.35/kWh.

My electricity costs about 8 cents/kWh, still an order of magnitude cheaper than personpower. You are a lot more expensive to supply with energy than a 150 W bulb! That is because most of this energy goes into supporting body functions—heart, liver, and so on. As we found earlier, people are capable (over extended times) of an ordinary power expenditure of 50 ft-lb/s = 68 W.

Measures of metabolic power requirements of people at rest can indicate how much of this power is required for the body. This is a measure of the *basal metabolic rate* (BMR). Roughly, BMR is 45 to 70 kcal/h, depending on body mass. (It turns out that the relationship between the BMR and mass is about 0.6 kcal/h/kg body mass; if the body mass = 50 kg, the BMR is 30 kcal/hour) (6). A person with a 70 kg mass (a weight of slightly more than 150 lb) has a BMR of 43 kcal/h = 5×10^{-2} kWh/h. Thus, the body of a 70 kg person uses energy at the rate of 50 W. With an average input of 150 W, we must either expend energy at an average rate of 100 W or get fat!

Of course, we are not totally efficient in using this excess power. Moreover, it is used in only about two-thirds of the day. During waking hours, about 150 W must be burned. Assuming energy usage in the body to be about 33 percent efficient, this means that a person can do ordinary work at a power output of 50 W, which agrees well with the 70 W estimate previously obtained.

SUMMARY

In this chapter, the definitions crucial for understanding energy basics are given. Speed measures the rate at which distance is covered. Velocity measures the rate of change of displacement and is a vector quantity, by which we mean that we specify not only how fast, but also in what direction. The change of velocity with time is called *acceleration,* which is also a vector.

In order to make an object's acceleration increase, we have to push harder on it. We find that this force equals the product of the acceleration and the object's mass. The relationship $F = ma$ is known as Newton's second law. A body's mass is characteristic of the body. A body's weight is the gravitational force on the body, and so the weight of an apple on the moon is different from the weight of that same apple on Earth or on Mars, although the mass of that apple remains the same regardless of its interplanetary travels!

For work to be done, there must be a force on an object, and the object must move: work = (force) × (distance moved in the direction of the force). The product of force and distance moved perpendicular to the force is not work. It is not even relevant. Objects capable of doing work (or exerting a force) are said to have *energy*. Familiar sorts of energy include energy of motion (kinetic energy), energy of position (gravitational potential energy), energy of internal rearrangement of atoms in a material (deformation energy), energy stored in atomic compounds (chemical energy), energy stored in atomic nuclei (nuclear energy), energy carried by light (radiant energy), or energy stored in a car battery (electric energy). Energy is measured using many different units. We prefer to use the joule or the kilowatt-hour wherever possible. Other commonly used units are the kilocalorie, the British thermal unit, and the foot-pound.

The total amount of energy in the universe in all its various forms does not change. We speak of this as the principle of conservation of energy. Energy is not created or destroyed; it is transformed, or changed from one form to another. Real machines are less than 100 percent efficient. Power measures the rate at which work is done. Common units for power measurement are the watt and the horsepower.

REFERENCES

1. I. Asimov, *Words of Science* (New York: Mentor, 1969), 110. *The American Heritage Dictionary of the English Language* points out that Aristotle used the word *energia* (*en*-at + *ergon*, work).
2. E. M. Rogers, *Physics for the Enquiring Mind* (Princeton, New Jersey: Princeton Univ. Press, 1960).
3. M. J. Cormier, *Natural History* 83, no. 3 (1974): 26.
4. J. C. Fisher, *Energy Crises in Perspective* (New York: John Wiley & Sons, 1974).
5. C. Starr, *Energy and Power* (San Francisco: Freeman, 1971), 3.
6. I. Asimov, *Life and Energy* (New York: Avon, 1972), 182.

PROBLEMS AND QUESTIONS

True or False

1. Conservation of energy is a proven fact of the universe.
2. Sometimes a tornado-like cone forms over water and produces a windspout. This is an example of mechanical energy.
3. Only fast-moving objects have kinetic energy (slow-moving objects do not).

4. It takes "The Flyer" one-third the time to climb a particular hill as a freight train. This gives enough information to calculate the power developed by the two diesel engines.

5. When a 10 N weight falls 10 m, its gravitational potential energy changes by 10 J.

6. It is an extremely complicated problem to determine the rate of energy use of a 150 W light bulb.

7. Machines can amplify input forces.

8. The efficiency of any real machine is less than 100 percent.

9. The greater the number of fixed pulleys in a pulley system, the smaller the input force necessary to move an object.

10. In the lever, the input force usually moves through a greater distance than the output force.

11. The use of a rope attached to a winch to lift a boat out of the water is an example of a case in which some of the input energy is lost in stretching the rope.

Multiple Choice

12. In going from the top of a hill to the bottom, the gravitational potential energy of Jack and Jill
   a. became the same as their kinetic energy.
   b. increased.
   c. decreased.
   d. stayed the same.

13. Let m be mass; v, velocity; a, acceleration; F, force; h, height above ground; g, the gravitational acceleration of the Earth. To determine the *total energy* of an object that is falling without spinning, I must know
   a. F only.
   b. m, g, and h only.
   c. m, g, h, and v only.
   d. all quantities listed above.
   e. none of the above.

14. The production of 1 MW of power for 1 s is equivalent to the production of energy of about
   a. 1,000,000 J.
   b. 947 Btu.
   c. 239 kcal.
   d. all of the above.
   e. none of the above.

15. A 500 N object is pulled up a 20 m ramp that rises from ground level to a height of 3 m. By how much does the gravitational potential energy change in this process?
   a. 25 J
   b. 167 J
   c. 600 J
   d. 1500 J
   e. 10,000 J

16. Mr. P would like to be able to lift a 300 kg refrigerator, but he is capable of lifting only 75 kg. Which of the following methods could Mr. P use to lift the refrigerator?
   a. A lever
   b. A pulley system with all pulleys fixed
   c. Ask Mr. T for help, since Mr. T is capable of lifting 125 kg.
   d. Disassemble the refrigerator with a hacksaw.
   e. Wet the compressor so that the working fluid would evaporate, making the refrigerator lighter.

17. Two bricklayers use two different winches to lift identical loads of bricks to the roof of the same building in the same time. Winch *A* turns 50 times to lift its load, while winch *B* turns 100 times to lift its load. If you neglect friction, which of the following is *true*?

a. The person using winch *A* supplies the same energy and the same power as the person using winch *B*.

b. The person using winch *A* supplies twice the energy but the same power as the person using winch *B*.

c. The person using winch *A* supplies twice the energy and twice the power as the person using winch *B*.

d. The person using winch *A* supplies the same energy and twice the power as the person using winch *B*.

e. The person using winch *A* supplies the same energy and half the power as the person using winch *B*.

18. A ball rolls 5 m along a table at constant speed. The ball weighs 20 N. The work done on the ball by gravity is

a. 0 J.
b. 4 J.
c. 5 J.
d. 20 J.
e. 100 J.

19. The potential energy of a 5 N weight is 20 J with respect to a table top. How far above the table top is the mass?

a. 1 m
b. 4 m
c. 20 m
d. 100 m
e. It is impossible to determine from the information given.

20. A weight is lifted by a real lever in which the distance from the point the force is applied to the support point (fulcrum) is five times as great as the distance from the support point to the weight. A 100 N weight is lifted, and the weight moves upward 8 mm when a 20 N force applied to the lever's other end moves the lever's end down 50 mm. What is the efficiency of this real lever?

a. 100%
b. 80%
c. 50%
d. 20%
e. 0%

21. The rate of energy utilization by technological people is about how many orders of magnitude larger than that of early people?

a. 1
b. 2
c. 3
d. 4

Discussion Questions

22. Consider raising weights. Start by moving 1 pound a distance of 1 foot. Try to see what happens if the weight is raised several feet; or see what happens if 2 pounds are raised 1 foot. From this, move to considering 5 pounds, 10 feet, or other weights, other heights! How does this generalize?

23. Your friend says that the principle of conservation of energy must mean that no energy can be lost. His conclusion is that batteries cannot possibly work, there is no such thing as friction, and that our technological world is an hallucination. What does the principle of conservation of energy actually imply about such processes?

24. Another friend devises a clever scheme to get an unending source of energy. This friend will carry a ball up a hill on a gently winding path, and then roll the ball down a steep path downhill. The crux of your friend's argument is that this extra energy comes because it takes less energy to get the ball up the hill than down the hill. Explain whether your friend is right or wrong, and why.

25. Your boy or girl friend bests you at Indian wrestling, and taunts you: "Nah-nah-nah-nah-*nah*-nah—I'm more *power*ful than you are, you 90 pound weakling!" Would he or she be correct if the *physics* definition of power is used?

26. Why is it impossible to construct real machines to be 100 percent efficient?

4

ELECTRICITY

When the word energy *is mentioned, most people think of electricity. While this association overstates the case, electricity* is *an important component of energy use and generation worldwide. In this chapter, electricity and magnetism are described from the physicist's point of view. The electric motor, generator, and electric transformer are explained; their respective roles in the electrical energy distribution are also described.*

KEY TERMS
   electric charge
   atom
   proton
   electron
   quantum
   Coulomb's law
   electrical potential energy
   potential
   voltage
   current
   circuit
   resistance
   Ohm's law
   Joule heating
   domain
   magnetic field
   electromagnetic induction

generator
induction motor
transformer

Our society uses many different fuel sources. Electricity comes mainly from burning coal or oil, or from nuclear energy. However, for some applications, electric energy itself is a preferred "fuel" source for various reasons. Its main advantages are its cleanliness, convenience, safety, and ease of control. Its main disadvantage is its greater cost in comparison to coal, oil, or gas. Before we can understand the advantages and disadvantages of electric energy use, we must discuss electricity and how we get energy from it.

The electric plug and electric batteries in various forms dominate our experience with electricity nowadays, but people have coexisted with electricity and electric energy from that prehistoric time when humans first used lightning-kindled wood to keep watchfires and campfires burning. Lightning has fascinated people for centuries. Benjamin Franklin's well-known interest in lightning led to his invention of the lightning rod.

ELECTRICAL CHARGE

Franklin supposed that there was some electrical "fluid" that could be separated from (and distinguished from) some other electrical "fluid." Today we call the "fluid" electric *charge,* and follow Franklin's definitions of positive electric charge and negative electric charge. When two units of equal size but opposite charge combine, they form an electrically neutral material. When the charge is separated—accumulating in clouds, for example—it builds up until there is a breakdown in air as the charge tries to reach "the ground" to neutralize itself. Franklin, in his famous kite experiment, was able to drain off some of the cloud electricity into a Leyden jar (storage cell for charge) and show that it was the same as the so-called static charge produced by rubbing cat's fur on quartz, silk on rubber, or a shoe on carpet. Most of us have been jolted at one time or another when an accumulation of charge produced a spark from us to a door handle.

We are not accustomed to thinking about electric charge because virtually everything in our experience is electrically neutral. Each positive charge has a corresponding negative charge bound to it. Atoms ultimately make up all material objects. An atom contains a *nucleus* of tiny particles ($\approx 10^{-15}$ m diameter) called protons and neutrons. The protons and neutrons have about the same mass and behave in similar ways, except that the proton carries a small unit of positive charge, while the neutron is electrically neutral. Outside the nucleus of the atom is a "sheath" of electrons at a radius of about 10^{-10}m.

We often refer to the electrons as a cloud, because we cannot actually specify the position of any electron (Chapter 5). The electrons have only 1/1830 the proton mass and carry small units of negative charge (which happen to be of exactly the same size as the proton's positive units of charge). The electron has a size so tiny that it can be considered a point. The total positive nuclear charge is offset by the equal amount of negative charge carried by the atom's electrons. The result is that most atoms of our acquaintance

have no net charge. However, since the electrons are really all that is seen of the atom itself (the nucleus is so small), electrons may rather easily be removed from the atom. These electrons may even be removed by rubbing, as we experienced in the case of a spark between us and a door handle. Since electrons may be removed relatively easily from atoms, and they have a smaller mass by far than protons, electrons move relatively more freely within and between materials than protons.

It appears from our discussion that electric charge comes in bundles of specific size, since the size of the proton's charge is the same as the size of the electron's charge. The size of this bundle, or *quantum* (basic unit, from the Latin, meaning "how much"), of charge was first measured by the American physicist Robert Millikan in the early 1900s. All subsequent experimentation has confirmed that all observable charge comes in integer multiples of this charge quantum. It is conventional to reserve the symbol e for the quantum of charge. Then a proton has a charge of e, a neutron has a charge of $0\,e$, and an electron has a charge of $-e$. Charges can come in groups of $0\,e$, $-5\,e$, or $27,500,430\,e$, but never $\frac{1}{2}\,e$ (although particles called *quarks,* which exist only as constituents of the more familiar particles, have a quantum of charge of 1/3 e).

We measure charge in coulombs (C), honoring Charles Augustus Coulomb, the first physicist to state the law of electric force. The coulomb is very large compared to the quantum of charge, which is $e = 1.6 \times 10^{-19}$ C. This means that an object carrying a charge of -1 C has a net surplus of about 6×10^{18} electrons!

Electric Force

We have considered examples of charge separation leading to sound and motion (lightning), or doing work (starting a fire). Since energy is changed into other forms, it is clear that these phenomena result from forces exerted by electric charges. When the charge moves under the influence of the electric force, work is done (see Chapter 3).

We already know what happens if positive and negative charges are separated: The charges are drawn toward or attracted to one another. This is illustrated in Figure 4.1a. By experience, we find that two positive charges or two negative charges repel one another, as is illustrated in Figure 4.1b.

An amusing experiment can be constructed to illustrate both forces. A charge that we may suppose to be positive can be built up on a rod, call it rod A, which is placed on an insulating stand. An insulator such as rubber prevents charges from moving. Ball B, made of a conducting material, is hung as shown in Figure 4.2 on page 48. A conductor is a material that allows the carriers of electric charge, electrons, to move relatively freely. Gold, silver, and copper are good conductors. When the rod is placed near the ball, ball B, being free to move, is pulled toward rod A until the two touch. After a short time in contact, ball B is repelled from rod A.

We may explain our observation by noting that, since charge carriers are free to move in ball B, negative charges in the ball will be attracted toward the positive charge on rod A. This means that electrons will build up on the side of the ball that faces rod A. Ball B was originally neutral, so positive charge is left behind as the electrons move away. Therefore, a positive charge

builds up on the side of the ball facing away from rod A. The negative half of ball B is attracted to rod A, and the positive half is repelled. The negative half, because it is slightly closer to rod A than the positive half, wins, and the balls move together. While the two balls are in contact, electrons flow from ball B to rod A in an attempt to neutralize rod A's positive charge. Soon, however, enough charge is drained so that ball B, which now has a net positive charge, can push away from A. Then rod A and ball B repel one another.

The reason that B's negative half wins the tug-of-war with B's positive half was discovered by Coulomb in his famous force experiments. Charges are usually assigned the symbol q. Coulomb found that the electric force, F_E, between two charges q_1 and q_2 was given by

$$F_E = (\text{constant})\, q_1 q_2 / (\text{distance of separation})^2.$$

This force law is now called Coulomb's law in honor of his work. In the international system of units

$$F_E = (9 \times 10^9 \, \text{N m}^2/\text{C}^2) \, \frac{q_1 q_2}{R^2},$$

where R is the distance between the centers of the two charges.

We now apply Coulomb's law to the tug-of-war between B's positive and negative halves. Since B was originally neutral, the negative and positive charges on B are equal in magnitude. The numerator of the electric force equation is thus the same for both charges. The negative charge is slightly closer to A than the positive charge is, so the value of R corresponding to the negative charge is smaller than that corresponding to the positive charge. From the Coulomb's law equation, the force of attraction of the negative charge is thus larger than the force of repulsion of the positive charge on A. B consequently accelerates toward A.

Of course, Coulomb's law is of much broader scope than this particular example. It explains the electric force between any static charged bodies, or bodies with distributions of charge.

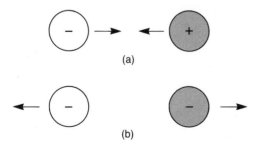

(a)

(b)

FIGURE 4.1
The effect of electric charge. (a) Unlike sign charges attract each other. (b) Like sign charges repel each other.

FIGURE 4.2
A conducting ball, *B*, is
attracted to a charged
ball, *A*, because some
charges from *B* are at-
tracted toward *A*. In
moving closer to *A*, ball
B's charges become
more strongly separated,
increasing the effect.

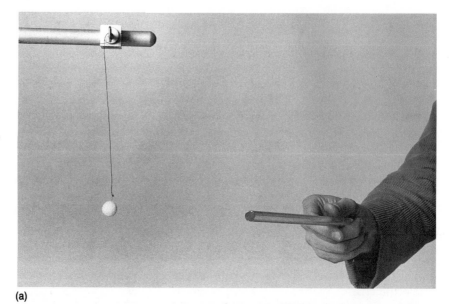

(a)

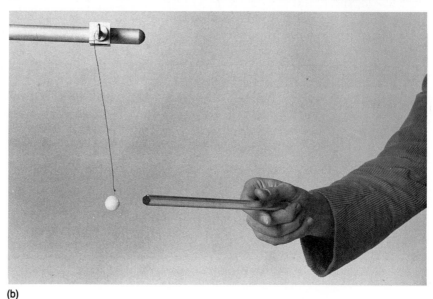

(b)

Electric Energy

Now that we know what the electric force is, we
can simply apply the formula for work we found
in Chapter 3

work = (force) × (distance moved along the force).

Notice that, by Coulomb's law, the expression for
the force changes as the distance between two
charges changes. To determine the electrical work
actually done by moving charges, we should calculate the tiny amount of work done at each point
in moving a tiny distance, and add up all the tiny
amounts of work. (This calculation is not difficult
but requires a technique learned in higher mathematics courses; we will not treat that subject
here.)

By doing work on charges by moving them in a region of electric force, we can build up *electrical potential energy,* which can be recovered as kinetic energy or as some other form of energy. A point charge has a potential energy relative to another one, since we know that, if we let one of the charges go, it will move; that is, it will gain kinetic energy. Imagine a very massive collection of negative charges. If a positive charge some distance away is released, it will move toward the negative charge. In this case, its electrical potential energy is converted into kinetic energy. Were a negative charge some distance away, it would flee the vicinity upon release. Its potential energy, too, is converted into kinetic energy. If we imagine a pair of charges to have an electrical potential energy of 100 joules with both charges fixed in place, we may get 100 J of kinetic energy from the charge we release. If the charge that moves does work as it moves from a position with potential energy of 100 J to one with potential energy of 50 J, then 50 J of work may be obtained. In determining the net final kinetic energy or the net work done, only the difference between the original value of the potential energy and the final value of the potential energy is important.

Since only differences in potential energy are of any interest, we may set the "scale" of this potential energy any way we want. This means we can conspire together to set the electrical potential energy to zero (or any other desired value) anywhere we please. For collections of point charges, it is usual to take an infinite separation of charges to correspond to zero electrical potential energy. This is a reasonable choice since the charges should not know of one another's existence if they are infinitely separated.

Work is done when objects having forces exerted on them move. If we have a charge at some position, it takes work to move another charge of the same sign (positive with positive, or negative with negative) in from infinity. To move a charge of opposite sign (negative with positive) in from infinity would liberate work. The work required or liberated can, in principle, be calculated from the basic definition of work.

In many applications, it is convenient to use the electrical potential energy difference per unit charge between two points in space. This is sometimes simply called the *potential,* symbolized V

V = potential
   = (electrical potential energy difference)/charge.

The unit of electrical potential is the *volt* (V), named for Alessandro Volta in honor of his pioneering electrical research in the eighteenth and nineteenth centuries. A volt is the potential difference in which a charge of 1 C changes its energy by 1 J in moving between two points. Because the units of potential difference are volts, it is common to refer to the potential as *voltage.*

CURRENT AND CIRCUITS

The existence of differences in electrical potential at different points in space may result in the motion of charges. A net motion of a particular sign of charge in some direction is called an electric *current.* Current is measured by counting the number of charges going by in a given unit of time. You might imagine an elf entering in a great ledger the number of charge quanta going past him to right or left. If he counts more charge going one way than the other, a current is flowing. Since the quantum of charge is so small, the current in practice measures the number of coulombs passing each second. A current of one coulomb per second is called an *ampere,* in recognition of Ampere's contribution to the understanding of the interconnection of electricity and magnetism

1 ampere = 1 coulomb/second (1 A = 1 C/s).

In most applications of electricity, it is the electron that moves. Thus, the electron is the energy carrier that supports the transfer of electrical energy in the medium. There are several reasons that the energy carriers are electrons. One reason is the small mass of the electron in comparison to the proton, or in comparison to an atomic nucleus that may contain many protons and neutrons. We saw the reason for this in Chapter 3:

Since force is given as mass times acceleration, for a given force, the larger the mass, the smaller the acceleration. The other reason is that most conductors of electricity are metals, and metals are used in electric wiring. In the metal, the atoms of the metal bind themselves together by letting one or two of their electrons free to wander randomly throughout the interior of the metal. Each of the myriads of atoms frees an electron, and the atoms are left bound to one another into the metallic structure. If the atoms were to move, they would essentially have to move as the whole substance. In response to an electric force, the electrons do accelerate. In many practical applications, this motion of electrons is achieved by making the metal into wire, and then creating a potential difference between the ends of the wire.

These electrons responding to an applied electric force can neither be destroyed nor created. If I stuff electrons into one end of a wire, an equal number must exit from the other end of the wire, say into the ground. But the electrons I originally stuffed into the wire had to come from somewhere; that somewhere is the ground, too. *Ground,* in electricity jargon, represents the reservoir of electrons that are sent jaunting around doing the tasks we require of them in our transmission and use of electricity. It happens that the Earth is a reasonable conductor, so that we really can literally get electrons from the ground and dump electrons into the ground. It is from this fact that the terminology springs. Because no electrons are gained or lost, because electrons are found in the ground and returned there, there is a wholeness to the process, a circulation. In a sense, the electrons make a closed loop.

In an electric distribution system, as well as in devices run from cells or batteries, there are also closed loops. Because it is possible to circle from the starting point through electrical elements back to the starting point, such a closed loop is called a *circuit*. In simple electric circuits, as well as in the electric distribution system, the closed loop is made completely of wire. In a circuit having a potential difference supplied by a battery (Figure 4.3), the lowest potential in the circuit is sometimes referred to as the ground.

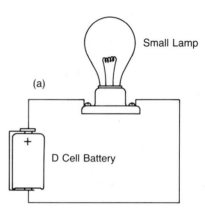

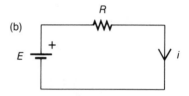

FIGURE 4.3
A simple electric circuit, with a source of potential difference E, and a lamp that provides resistance R to the flow of electric current I. (a) A diagram of an actual apparatus (a small lamp, for example). (b) A more schematic representation.

In the electric distribution system, the two prongs in the outlet are for the "hot wire" (smaller opening) and the reference wire (larger opening), which functions as a ground for the circuit. It is possible for a potential difference to exist between this reference potential and true Earth ground; for this reason, to provide absolute safety to people operating them, three-prong devices have a connector between the metal parts and the Earth built in.

In the electric generator or battery, charge is taken from the virtually infinite reservoir, and work is done on the charge. Because work is done on the charge, it gains energy. Because it has energy, when it travels in an electric circuit, it can give that energy up again as work wherever we want the work done.

We can picture the generating plant as a factory. Let the electrons be represented by little balls. These little balls roll out of the generator high in the air, in a gutter-like channel out into the distribution system. The gravitational potential energy of the balls in the gutters is analogous to the electric potential energy of the electrons leaving a power plant. The gutters we imagine are sent out to houses and factories. The balls roll there, and at these places, they fall through downspouts, turning their gravitational potential energy into kinetic energy that can do work. This is analogous to the electrons falling in potential inside an electric device. Finally, the balls roll in low-lying gutters back to the generating plant. There, our elf (the one counting charges and entering them in a ledger) takes the balls from the low-lying gutters, climbs a ladder, and places the balls in the gutters high in the air, where the cycle can begin again. This is analogous to the return of the electrons to ground at low potential, and to the raising of electrical potential in the generator.

The system is that work is done on the electrons in the plant, where a lot can be done at once, and then the electrons are sent out as messengers bearing their gift of work to be used wherever it is convenient to the user. This analogy best represents a direct-current generator, rather than the generators we use to supply electricity to our homes and schools. Even then, it is flawed as a description of what really occurs. Nevertheless, this analogy supplies a useful characterization of the process of energy transfer in the electric distribution system.

Resistance

Various materials resist the flow of these electrons by different amounts. Materials such as silver, copper, or aluminum offer little resistance to the movement of electrons; consequently, they are said to exhibit low electrical *resistance*. Other materials, such as rubber or carbon, offer a great resistance. Actually, resistance depends on the size and shape of the material, as well as on which material it is. *Resistivity* characterizes the resistive properties of materials in a way depending only on the material. All copper has the same resis-

tivity, but a 2 m long copper wire has twice the resistance of a 1 m long copper wire of the same shape.

Almost all materials exhibit some resistance to current flow. A very few materials, called *superconductors,* show no measurable resistance whatsoever at very low temperatures ($-250°$ C). These are discussed in more detail in Chapter 18.

We can easily understand why most materials do have resistance to the flow of current. Electrons in a wire across which there is a potential difference are accelerated. Before they can move too far, however, they slam into one of the atoms making up the wire, transferring some kinetic energy to it, making it vibrate faster. Then the electron, which still feels the potential, starts accelerating again. Again it collides with a wire atom. This will happen countless times as an electron wends its way through the wire. Thus the wire's internal energy increases; it heats up. Some materials stop the electrons very efficiently, and so offer high resistance. The coil of an ordinary toaster is made of such material: It stops the electrons so well that the wire glows red hot and cooks the toast. Incidentally, this analysis implies that an electron's velocity of progression (*drift velocity*) through a wire is quite slow. Typical speeds of drift are only tenths of millimeters per second!

Ohm's Law

Georg Ohm discovered that it was usually possible to relate voltage (potential difference), current, and (a constant) resistance in electric circuits. If we use the symbol I for current, and the symbol R for resistance, the relation (called *Ohm's law*) reads

$V = IR.$

This resistance to flow of current can be quantified by the definition: the resistance offered a 1 A current between two points that differ in potential by 1 V is defined as 1 ohm (symbolized by Ω, capital Greek omega). For a circuit with a voltage of 5 V and a current of 1 A, the resistance must be 5 Ω, for example. A current of 10 A moving through a resistance of 100 Ω drops 1000 V in potential.

Recall from Chapter 3 that power is the work done per unit time. Since the potential is the work done per unit charge, and the current is the charge moving per unit time, the product of potential (work/charge) and current (charge/time) (*VI*) gives the amount of power (work/time) dissipated in the circuit. That is, in a time *t,* an amount of energy *VIt* goes into the circuit. We write

$$P = VI.$$

The unit of power is the watt (1 W = 1 J/s), so volts times amperes must equal watts. If Ohm's law holds, *V = IR,* and therefore

$$P = I^2R.$$

For this case, in which the entire power supplied to the circuit is dissipated as heat, we speak of *Joule heating* (again, named for James Prescott Joule). Fuses (or circuit breakers) use this Joule heating to prevent circuits from drawing too much current for the wiring. If too great a current travels through the device, a wire melts from the heat inside the fuse, breaking the circuit.

MAGNETISM

As children, we all played with little magnets. We found we could exert forces and pick up some sorts of metals. We also saw that the end labeled *N* (north) repelled other *N*s and attracted the end labeled *S* (south), and likewise that *S*s repel one another.

While magnets have had use for centuries as compasses, the modern uses of the magnet and magnetic forces date to Christian Oersted's discovery in 1819 that a wire carrying an electric current can act as a magnet. The greater the current used, the stronger was the magnet created. This was the beginning of the recognition that electricity and magnetism are aspects of the same phenomenon. Since J. Clerk Maxwell (a nine-

teenth century English physicist), we have known that we deal with something called electromagnetism, and that electric and magnetic force are just different aspects of the same phenomenon. Magnets exert forces on moving charges. Moving charges (electric currents) produce magnetic fields.

At the microscopic level, circulating electrons in atoms can cause magnetic fields, as can circulating protons within the nucleus. In materials known as ferromagnets—iron or chromium, for example—these magnetic fields line up in the same direction. Many materials, such as aluminum, are not ferromagnetic.

We all know that ordinary iron is not a magnet. This is because the lining up involves small volumes of the material (known as *domains*). A piece of iron of a size that we can see has many domains; in each domain, the magnetic fields line up. We can think of a domain as a little magnet. An ordinary piece of iron is made up of myriads of little magnets oriented at random, so that the fields cancel, and the iron is not a magnet itself. In the presence of an external magnetic field, the little magnets are all forced to point in the same direction. The piece of iron becomes a big magnet.

An electromagnet is constructed in this way: We wrap a wire around a piece of iron. The wires are connected to a battery, so that a current flows in the wires. The current produces a magnetic field. The magnetic field causes the domains to line up, and presto! We have a magnet. When the current is turned off, the piece of iron will return to its random domain orientation. The amplification of magnetic field possible in electromagnets is very large. Electromagnets are used to pick up entire cars in junkyards.

In permanent magnets, the domains are lined up. This can be done in a number of ways. The rocks of Magnesia in Asia Minor (from which the term "magnet" is derived) became magnets by being heated in the Earth and then cooled. The Earth's magnetic field caused the domains in the iron-rich rock to line up. Another way to make a magnet is to subject a piece of iron to a strong magnetic field. When the field is turned off again, there is a remnant ordering of the domains, and the iron has become a permanent magnet. If I could very carefully saw such an iron bar in half perpendicular to the direction in which the magnet points, I would get two magnets, head to tail, each pointing in the same direction. If they re- main in this orientation, they will attract one an- other and reconstitute the original magnet. If I saw the iron bar in half parallel to the direction of the magnet, I get two magnets, side by side, each pointing in the same direction. These mag- nets will not attract one another. Will they repel one another?

(a)

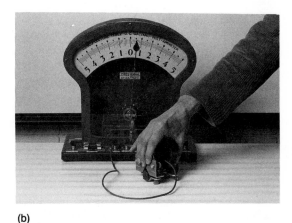

(b)

(c)

(d)

FIGURE 4.4
Magnetic induction. (a) A power supply causes a galvanometer to deflect. (b) A magnet moving through a loop of wire causes a galvanometer to deflect. (c) A moving loop of wire in the vicinity of a magnet causes a galvanometer to deflect. (d) Moving a piece of iron through a current loop causes a galvanometer to de- flect.

Since an electric current is able to create something that acts like a magnet, it was inevitable that someone would try to see if a magnet can create a current. In 1831, Michael Faraday, the English physicist, found that a magnet moving through a wire coil will start a current flowing in the wire (see Figure 4.4b). Faraday called this *electromagnetic induction*. When the magnet is

FIGURE 4.5
Iron fillings line up on a glass plate because of the magnet lying under the plate.

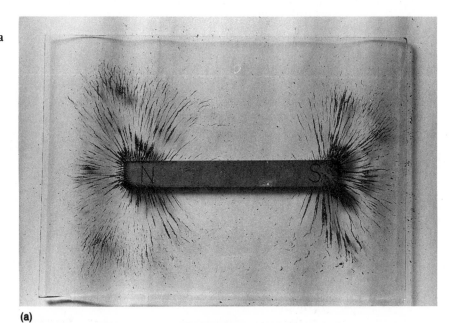

(a)

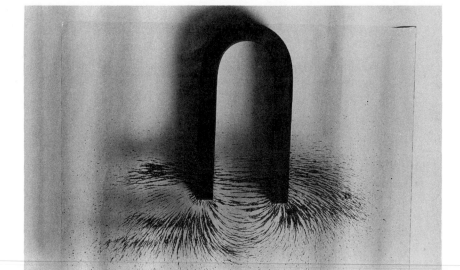

(b)

moved in one way, a current flows as shown in a galvanometer similar to the one in Figure 4.4. The galvanometer is a device having a moving pointer; the pointer moves in response to current flow in such a way that a greater current causes a greater deflection (Figure 4.4a). If the magnet stops moving inside the coil, the current stops. When the magnet is withdrawn, a current flows in the opposite direction from that originally exhibited. Thus, one key to generating a current by using a magnet is to have the magnet move.

We find we can also generate a current by moving the coil while we keep the magnet stationary (Figure 4.4c). It is necessary only that the magnet and coil move with respect to each other to start the flow of current.

In the region about a magnet, there is some effect of the magnet. If we move a permanent magnet near a nail, the nail will be attracted to the magnet even if it does not touch it. If iron filings are placed on a piece of paper and a magnet is brought near it, the filings rearrange themselves in patterns such as that shown in Figure 4.5. It is as if there were lines in space connecting one end of a magnet to the other: north pole to south pole. Those lines would be there even without the filings to show us that they are there. They are results of the magnet itself. The abstract lines we imagine to run through space we call *lines of magnetic field*. Outside a single magnet, they run from north pole to south pole. If we were to put two magnets south pole to north pole, we would expect magnetic field lines to go straight from the north pole to the south pole (with perhaps a few straying off if the magnets are not the same strength). For magnets of equal strength, there is a region of uniform magnetic field between the poles.

Now consider a coil put into such a region of uniform field. If it is inserted as in Figure 4.6a, with the plane of the coil parallel to the faces of the magnets, magnetic field lines penetrate the coil. If the coil is as arranged in Figure 4.6b, with the plane of the coil perpendicular to the magnet faces, there is no magnetic field line to penetrate

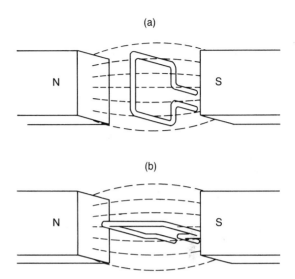

(a)

(b)

FIGURE 4.6
The generator consists of a coil which can be made to rotate and magnets which are fixed in place. (a) A coil in a magnetic field with the coil plane parallel to the magnet faces. (b) A coil on a magnetic field with the coil plane perpendicular to the magnet faces.

the coil. If we could turn the coil from parallel to perpendicular, we could change the amount of magnetic field cut by the plane of the coil. But we know that when the magnetic field inside the coil changes, a current is produced. We can obtain a current simply by turning the coil.

This is a simplified description of an electrical generator. In an electrical generating facility, the coil between the magnets is turned by water falling through a turbine, or by steam from a boiler moving through a turbine. The device is then called a generator. As the coil rotates, the amount of magnetic field through the coil changes very little at first, as the coil moves from parallel configuration. The change grows faster as the coil approaches the perpendicular configuration. As the coil moves on past the perpendicular, the change grows slower again until the coil is in the parallel configuration again, when the change gradually stops. As it moves through the parallel configuration, the coil is moving in an opposite

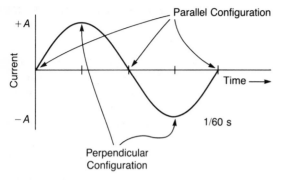

FIGURE 4.7

Alternating electric current resulting from coil rotation in a fixed magnetic field is plotted against time. One complete cycle takes ⅟₆₀ s, for North American generators; it takes ⅟₅₀ s for one complete cycle in Europe and much of the rest of the world.

sense to that originally moved, so that the change continues through zero into negative values. The current generated in a complete cycle is shown in Figure 4.7. As the coil rotates, the charge moves back and forth, so this kind of current is called *alternating current* (AC); the kind of steady-flow current we considered previously is called *direct current* (DC).

Electric Motors

The same principle, worked in turnabout fashion, can produce rotating rods from the flow of electric current. Consider a wire in a magnetic field. As we have seen, as long as the wire is stationary, the charges in it are stationary and the magnetic field does not change, so there is no magnetic force. Magnetic forces occur only when there is motion of charges in regions of magnetic field. Now suppose that the wire is in a region of changing magnetic field. As a result of the change in magnetic field, a current is induced in the wire. Because there is a current in the wire—that is, since there are moving charges in a region of magnetic field—there is a magnetic force on the wire causing it to move. If the wire is bent into a squared-off U shape and the ends of the U are held on supports, the wire pivots around the supports.

Suppose another U is connected rigidly 180 degrees away from the first. If the magnetic field is uniform, there will be the same induced current and the same force on each. They would then try to swivel in opposite directions, and no net motion would be possible. If the magnetic field is stronger near the bottom wire, for example, the magnetic force will be stronger on the wire on the bottom, causing it to begin to rotate (Figure 4.8a). Now, if similar U-shaped pairs are placed around the periphery, as in Figure 4.8b, the wires will start to rotate and will continue to do so. This is the principle of operation of the squirrel-cage motor shown in Figure 4.8c. The squirrel-cage motor is an example of a shaded pole induction motor. The shading of the pole supplies the nonuniformity of the magnetic field. Even if the squirrel-cage were a solid cylinder, it would rotate in the same way.

The most common AC motor achieves rotary motion by setting up two alternating magnetic fields at right angles to one another. This changing field causes the cylinder to rotate. A picture of such a motor is shown in Figure 4.9. The section of the cylinder inside the region of magnetic field is often made of laminated segments to reduce Joule heating (I^2R) losses from the induced currents.

An induction motor we all know is the watt-hour meter on each house that is read by the electric company every month. The motor is fed a small current proportional to the line voltage, and the magnetic field is set up by coils fed from the line current. As a result, the rotation speed is proportional to the product of current and voltage—that is, to the power. The total amount of rotation is speed times time, or total distance traveled by the disk. We therefore *pay* for power times time, or energy.

There are also direct-current electric motors. The principle of DC operation is that current-carrying wires in a region of magnetic field will feel a force. In the DC motor, there is a pair of magnets and a series of wires wound about the

FIGURE 4.8

The principle of electric motors. (a) A wire loop in a nonuniform magnetic field. (b) Several wire loops in a nonuniform magnetic field. (c) Squirrel-cage configuration, for a squirrel-cage motor.

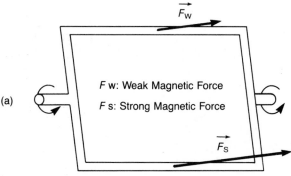

Region of Weak Magnetic Field

$\overrightarrow{F}_W$

(a)

F w: Weak Magnetic Force

F s: Strong Magnetic Force

$\overrightarrow{F}_S$

Region of Strong Magnetic Field

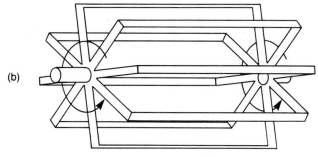

Region of Weak Magnetic Field

(b)

Region of Strong Magnetic Field

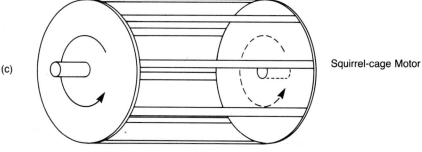

Region of Weak Magnetic Field

(c)

Squirrel-cage Motor

Region of Strong Magnetic Field

central cylinder connected to the voltage source in such a way that, as the cylinder rotates, the current can change direction (through a device called a commutator), as illustrated in Figure 4.10. Note that it is always necessary to have two or more magnetic fields to have an electric motor.

Transmission of Electric Energy

The output of generating plants is run into transmission lines for distribution to schools, homes, and businesses. In all these lines, energy is lost to resistive, or Joule, heating. Long-distance transmission lines are of high voltage because of Joule heating. A generating plant produces a certain amount of power, given by *IV,* where *I* is the current generated in the plant and *V* is the potential difference between the ends of the wire. (It is actually somewhat more complicated for AC than for DC, but our conclusions will be the same, so we will use the DC calculation.) The power output of the plant must all be transmitted. The relation $P = IV$ tells us that we may change the current and voltage of the transmitted electric energy together as long as the product of current and voltage, the power generated at this plant, is a constant:

FIGURE 4.9
A squirrel-cage motor in use.

FIGURE 4.10
Diagram of a direct current electric motor showing the two sides of the commutator, and the connections to the battery.

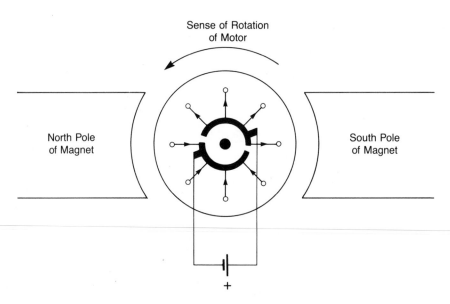

$$P_{\text{power plant}}$$

$$= I_{\text{power plant}} V_{\text{power plant}}$$

$$= I_{\text{transmission line}} V_{\text{transmission line}}.$$

Since the transmission line has a certain fixed resistance, R, we may minimize Joule heating losses (I^2R) in the transmission wire by minimizing the current that the wire carries. This implies that the voltage of the transmission line should be very high, since the power transmitted is constant. Then, Joule heat dissipated in transmission

$$= (I_{\text{transmission line}})^2 R$$

$$= (P_{\text{power plant}}/V_{\text{transmission line}})^2 R.$$

So, the higher the voltage, the lower the current and the smaller the Joule heating losses. About 10 percent of electric energy generated is lost in the transmission process to Joule heat. Only about 2 percent is lost in long-distance high-voltage transmission.

Transformers

The voltage is raised from that of the power plant by a device called a step-up transformer (defined below). In your home, you use low-voltage power because it is safer. The electric company transmits (locally) a voltage that is made even smaller as you use it. This is accomplished by a step-down transformer, which does the opposite of the step-up transformer.

Suppose we do an experiment in which we wrap a wire many times around a length of iron rod. If the ends of the wire are connected to the wall socket, an alternating current flows in the wire. Since the current changes with time, it causes a changing magnetic field in the rod. A wire wound once around the rod will enclose a changing magnetic field, inducing a current to flow in the wire and therefore inducing a voltage across the wire's ends. Suppose that voltage has a value of 0.1 V. If the same wire is wound twice around the rod, each loop will have a voltage 0.1 V induced; since there are two loops, the net voltage across the wire's ends is 0.2 V. Likewise, wrapping the same wire eight times around the rod will produce 0.1 V in each loop, or 0.8 V overall.

The transformer consists of two separate unconnected wires wound around a common core, usually of iron. One of the wires is wound around the core many times; the other has fewer turns of wire. An AC current in one of the wires causes a magnetic field (which changes as the current changes) in the iron. The changing magnetic field in the iron in turn induces an AC current in the other coil of loops.

The voltage in any loop wrapped around an iron core will be the same, whether it is a loop on the input, or primary, coil or a loop on the output, or secondary, coil. This means that the voltage on the primary coil is divided among the loops (if there are 100 loops on the primary coil, each loop will have a voltage 1/100 of the input voltage). If the number of loops on the primary and secondary coils is the same, then the device makes no change in the voltage across the device. If the number of loops on the primary is greater than the number on the secondary, the voltage on the secondary will be lower (that is, the voltage will be stepped down). If the number of loops on the primary is smaller than the number on the secondary, the voltage on the secondary will be higher than that on the primary (it will be stepped up). This phenomenon can occur only with AC. For DC, a magnetic field will be created, but the magnetic field does not change, and consequently, no current is induced in the secondary coil.

The transformer is a clever device (and is probably the reason that Westinghouse's advocacy of AC bested Edison's of DC). It provides another example of conservation of energy. In the ideal case, the amount of energy going into a transformer in a given amount of time is equal to the amount coming from the transformer in the same time. Of course, in real life, some energy is lost to Joule heating, but we can say that, as a good approximation, $(VI)_{\text{output}} = (VI)_{\text{input}}$.

Power transmission is often accomplished by constructing large pylons to carry the wires in the air (Figure 4.11). The pylons are built in this

FIGURE 4.11
Transmission lines in
rural Ohio.

manner because air acts as an insulator. Thus, the bare wires are protected from breakdown by the air itself because air is a good insulator (poor conductor). If the wires were to be buried underground, we would need to wrap them in large amounts of glass and plastic to protect them from breakdown. This alternative is obviously much more expensive because of the cost of excavation and insulation.

When the transmission lines enter your house, they go through a watt-hour meter, the induction motor described earlier. The meter is geared (much like the odometer of your car); the turns of the disk are recorded on dials, one for each power of ten recorded. Since the stored information is the number of turns of the disk, what has been recorded is *energy* use. The disk speed is proportional to power consumption, but you do not pay for the rate of energy use. You pay for the *total amount of energy used*.

In your home, the transmitted energy is finally transformed into light energy, thermal energy, mechanical energy, and so on. Motors run from electricity by running induction in reverse. In Chapters 7–9, we discuss where this energy goes.

SUMMARY

Charged objects can exert forces on other charged objects. Charge comes in two forms: positive and negative. Two positive charges repel each other; two negative charges repel each other; positive charges attract and are attracted by negative charges.

Since charges exert forces, and objects move, work can be done by them. The amount of electrical work done per unit charge is a measure of the electrical potential energy per unit charge, otherwise known as the potential, or the voltage. A voltage can cause charges to flow—that is, cause a current. Ohm's law states that current is proportional to voltage. The proportionality constant connecting them is called the resistance: $V = IR$.

The electric power produced by a source having potential V and delivering current I is $P = IV$. This is also the power delivered to something drawing current I at voltage V. In the latter case, if the object is a simple resistance (not a motor), then $V = IR$, so that $P = I(IR) = I^2R$. This produces heat in the resistor, which is called Joule heat.

Magnets also exert forces on other magnets or on materials such as iron. The magnets arise at the nuclear and atomic levels from the circulation of currents. A current produces a magnetic field. Contrariwise, a current can be made by moving a magnet near a coil of wire, or by moving a coil of wire around a stationary magnet. That is, the changing magnetic fields cause charges to move.

These principles allow construction of electric motors that turn as a result of input of electric current; electromagnets, in which a current loop surrounds a piece of iron and the domains in the iron line up to produce a strong magnet; and electric generators, in which a coil rotating in a magnetic field produces a current by induction.

Transmission lines operate at high voltage in order to minimize Joule heating. Voltages may be increased by using a device called a step-up transformer and decreased by using a step-down transformer. The transformer works only with AC and is another example of the utility of magnetic induction.

PROBLEMS AND QUESTIONS

True or False

1. Since protons are so much more massive than electrons, there have to be more protons than electrons in atoms to make the total electric force zero.
2. The ampere is an MKS unit of potential difference.
3. A wire loop moving in a region containing magnetic field carries an induced current.
4. A heating coil on an electric stove is a place where some electrons are lost in supplying the heat.
5. It is not necessary to have a complete circuit with AC because the electrons just move back and forth.
6. A transformer changes the power in an electric circuit.
7. A substantial portion of the power leaving the generating plant is lost to Joule heat during transmission.
8. Magnetic fields from magnets differ in principle from the magnetic fields produced by currents.
9. A light bulb rated 120 W will carry a current of 1 A when plugged into a 120 V outlet.
10. All materials have magnetic domains such as the ones discussed for iron; as a result, all metallic materials are ferromagnetic.

Multiple Choice

11. A 100 W light bulb, when plugged into a 120 V outlet, produces what current?
    a. 0.875 A   d. 144 A
    b. 1.000 A   e. not enough information given
    c. 1.200 A   to answer the question
12. A long magnet is cut into three pieces as shown: What do the pieces look like afterwards?
    a. [S ___ S] [S ___ N] [N ___ N]
    b. [S ___ N] [N ___ S] [S ___ N]
    c. [S ___ N] [S ___ N] [S ___ N]
    d. [S ___ S] [S ___ S] [S ___ N]
    e. [S ___ N] [N ___ N] [N ___ N]

13. A flashlight bulb has a (rated) voltage of 2 V and a (rated) current of 0.06 A. These numbers indicate an electrical power requirement of
    a. 33.3 W.　　　　　　　　　　　c. 0.12 W.
    b. 2.06 W.　　　　　　　　　　　d. 0.03 W.

14. A 100 MW power plant sends its power out on a high-voltage line at 5×10^5 V. The Joule heat generated per ohm is
    a. 5×10^8 W/Ω.　　　　　　d. 4×10^4 W/Ω.
    b. 100 MW/Ω.　　　　　　　　e. none of the above.
    c. 2×10^2 W/Ω.

15. Consider two copper wires that are identical except that one is twice the length of the other. The resistance of the longer one, as compared to the shorter, is
    a. greater.　　　　　　　　　　　d. impossible to determine be-
    b. smaller.　　　　　　　　　　　　cause of lack of information.
    c. the same.

16. Electric transformers can work because
    a. alternating currents jump　　　d. parallel wires induce currents
    　　around.　　　　　　　　　　　　in each other.
    b. direct currents produce mag-　　e. Westinghouse was a better in-
    　　netic fields.　　　　　　　　　　ventor than Edison.
    c. changing currents cause
    　　changing magnetic fields.

17. A 120 V house circuit is protected by a 15 A circuit breaker. You are shivering in the cold spring weather, and so you turn on a 1200 W heater in addition to your 400 W television and your 150 W light bulb (all on the same circuit). What happens if you turn on a 100 W light, which is on the same circuit?
    a. Nothing.　　　　　　　　　　　d. The heater will get less hot
    b. The 100 W bulb burns out　　　　to compensate.
    　　from the stress.　　　　　　　　e. The circuit breaker will inter-
    c. A fire starts in the overloaded　　rupt the circuit.
    　　circuit.

18. If you have an American-made electric razor with two settings, 110 V and 220 V, then it is likely that inside the razor is a
    a. step-up transformer.　　　　　　d. commutator.
    b. step-down transformer.　　　　　e. electromagnet.
    c. permanent magnet.

19. In a certain region of space, point A is at a potential 200 V higher than point B. If we move a charge of 2 C from A to B, then we
    a. must do 400 J of work.　　　　　d. obtain 100 J of work.
    b. must do 100 J of work.　　　　　e. obtain 400 J of work.
    c. do no work.

20. A device draws 10 A from 120 V circuit. How much energy does the device use in twelve seconds?
    a. 10 J　　　　　　　　　　　　　d. 1440 J
    b. 12 J　　　　　　　　　　　　　e. 14,400 J
    c. 1200 J

Discussion Questions

21. A physics experiment in the scientific literature reports to have seen evidence for a charge 2/3 *e*. How would you, as a physicist, respond to such a report?

22. Why should electric motors need at least two magnetic fields?

23. We mentioned the phenomenon of superconductivity (resistanceless current transmission) in this chapter. This would clearly be a good way to transmit power. What are the difficulties with this approach at the present time?

24. How can any energy be transmitted by AC when the electrons just slosh to and fro sixty times a second?

25. The actual average speed of electrons in a circuit may typically be 10 mm/s. Yet your light bulb lights the instant you throw the switch. Is this possible? (Hint: The effect of the electric force travels at the speed of light, 3×10^8 m/s.)

5

ATOMS AND CHEMICAL ENERGY

In the preceding chapters, we have looked at the definitions of work, energy, and power, and described some of the various forms of energy and how they may be converted into other forms of energy. The electric energy we use comes in large part from chemical processes (this is discussed in more detail in Chapters 9,10, and 13). Burning—a chemical reaction—supplies energy and allows production of materials. Solar energy often causes chemical changes. Thus, a basic understanding of the process of chemical change is essential to any understanding of energy and energy production.

KEY TERMS states
Pauli exclusion principle
energy levels
periodic table
exothermic
activation energy
endothermic
catalyst
photosynthesis
enzyme
battery
electrode
electrolyte
cell
fuel cell

In this chapter, we shall study the process of chemical change. Many of the chemical compounds used to produce energy are burned and involve several chemical recombinations. Gasoline powers automobiles and trucks; kerosene serves as a jet fuel. Natural gas may heat our homes directly, or generate electricity. Coal and fuel oil are burned to generate electricity as well. The root source of the energy used for heat, transportation, and industry in most of the world is chemical energy. Table 5.1 presents the energy available in combustion from several important sources. Of course, that chemical energy originally came from the sun, which operates on nuclear energy (we discuss photosynthesis in this chapter as well as in Chapter 16).

If we were able to look inside a material and shrink our field of vision again and again, we would eventually be able to identify an impenetrable shell. We would be looking at the outer surface of the atom or molecule. This surface is provided by the atom's electrons. A typical atomic "size" (diameter) is 0.1 nanometer (10^{-10} m).

Chemistry is the basis of life. Cells are made up of molecules, and molecules are made of atoms. Molecules and atoms and their properties are the study of chemists, physical chemists, chemical physicists, and biochemists. Atoms consist of the central atomic nucleus with electrons around it. The nucleus itself is made of protons

TABLE 5.1

Heats of combustion of various substances.

Substance	In MJ/kg	In kWh/kg
Fuels		
Crude oil	45.0	12.5
Fuel oil (mid-continent)	45.1	12.5
Gasoline	46.9	13.0
Kerosene	46.7	13.0
Average coal (bituminous and lignite)	26.0	7.2
Ohio coal (bituminous)	29.5	8.2
North Dakota coal (lignite)	13.9	3.9
Fuel alcohol	27.5	7.5
Hydrogen, to liquid water (gas)	141.9 (122.0)	39.4 (33.9)
Foods		
Butter	38.5	10.7
Mean animal fats	37.8	11.0
Egg (white)	23.9	6.6
(yolk)	33.9	9.4
Linseed oil	39.5	11.0
Olive oil	39.3	10.9
Other		
bagasse (sugar cane refuse, 12% water)	16.9	4.7
Oak wood (13% water)	16.7	4.6
Pine wood (12% water)	18.5	5.1
Dynamite (75%)	5.4	1.5
Iron	6.6	1.8

SOURCE: Reprinted with permission from *Handbook of Chemistry and Physics*, 32nd ed. Copyright 1959 by CRC Press, Inc., Boca Raton, FL.

and neutrons, which are the subject of the next chapter.

PROPERTIES OF ATOMS

The chemical properties of atoms are determined by the configuration of the electrons in the atoms. Atoms are held together by electric forces. The atom, as we have seen in Chapter 4, has no net charge. It has positive charge in its tiny nucleus, and the electrons carry a counterbalancing negative charge in the space of the atom.

While the electrons in the atoms are wavelike and are not localized in space, they have a high probability of being near positions they would have occupied if they were only points of mass. The electrons repel each other electrically, just as they are attracted electrically to the nucleus. Furthermore, they cannot all get close to the nucleus. We find experimentally that no two electrons may exist in the same *state* (defined by their mean distances, energies, etc.) in an atom. This phenomenon is referred to as the *Pauli exclusion principle*. Because of this fact of nature, electrons from one atom do not interpenetrate those of another atom. They provide what we consider to be the outer surface of the atom, the "size."

The electrons thus take certain "positions" around the atomic nucleus. The first electrons occupy the lowest, most stable, of the possible "positions." The closest-in electrons have given up the most energy to become part of the atom. Subsequent electrons joining the atom give up less energy, and are then less stable and more easily removed from the atom.

Figure 5.1 shows what we might see if we looked at a glow discharge from a tube containing a pure vapor that is excited by a high voltage. To observe more than a general colored glow, we use a device called a diffraction grating, which is a piece of glass or plastic on which tiny parallel ridges separated by a distance of ≈ 1 μm cause different colors of light to bend by different amounts. Only discrete lines are seen. Electrons joining an atom can give up only certain discrete amounts ("quanta") of energy.

Each filled "position" corresponds to a certain amount of energy that the electron gave up to join the atom; we speak of the allowed energy values as *energy levels*. Figure 5.2 shows the first few allowed electron energy states for hydrogen, labeled by their quantum numbers (numbers which together define the state of the electron). Hydrogen has only one proton in its nucleus and one electron. An electron may transfer from one energy level to another by emission or absorption of light that has energy corresponding to the specific energy difference between the two lev-

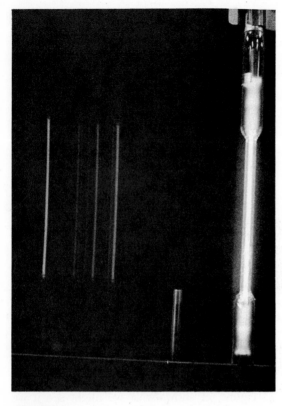

FIGURE 5.1
A look at a glow discharge tube through a diffraction grating.

sublevel because electrons can take on two separate guises: "up" electrons, or "down" electrons. No sublevel can have two "up" or two "down" electrons, but if one electron is "up" and one is "down," the two electrons are in different states for the same energy level, which *is* allowed by nature.

As an analogy, consider a eucalyptus tree with pairs of thin leafy branches going up the tree

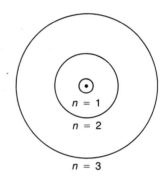

Relative Radii of First Three Hydrogen Orbits

(a)

els. We speak of quantum numbers, numbers that characterize the discrete states of the electrons. The energy is mostly determined by the principal quantum number, denoted n. In ordinary hydrogen, the electron is in the state $n = 1$. Higher values of n denote excited states, which are unstable, and emit light energy until the state $n = 1$ is reached.

In helium, there are two protons in the nucleus of the atom. There are thus two electrons. Both electrons can fit in the $n = 1$ level. (The $n = 1$ level has sublevels defined by quantum numbers known as the orbital angular momentum—l—and its projection in a fixed direction—m.) In more complex atoms, states of higher n are filled. In each, two electrons can fit in each sublevel. There can be two electrons in each

$E = 0$ $E > 0$ Unbound Electrons

$E = -1.5$ eV _____ $n = 3$ Bound Electrons

$E = -3.4$ eV _____ $n = 2$

$E = -13.6$ eV _____ $n = 1$

Energy Levels in the Hydrogen Atom

(b)

FIGURE 5.2
(a) A sketch of the first few allowed electron orbits as allowed in the Bohr model (1912) for hydrogen. R, the Bohr radius, is 52.9 pm. Allowed orbits have radii R, $4R$, $9R$, and so on. (b) Diagram of energy levels for hydrogen. The Bohr model ultimately led to the development of quantum mechanics, which in principle can describe all atoms.

trunk, and koalas trying to eat the leaves. Since the koalas have some weight, we imagine that they would break the thin branches if more than one koala were to sit on each branch. Since each branch can hold only one koala, a new koala climbing the tree would have to climb to the next available free branch to have a safe perch on which to eat eucalyptus leaves. The electrons in atoms are forced by the Pauli exclusion principle to take "positions" higher in energy than those already there. In this way, each atom has its own configuration of "branches" on the "tree" (available electron energy levels), and if one sees this configuration, it is possible to identify the type of atom. Groups of levels sharing a certain characteristic (called *orbital angular momentum*) are known as shells.

In fact, most of the chemical properties of atoms are dependent on the configurations of the *outer electrons*. If the outer electron configurations of two atoms are the same, they will combine with still other atoms (which basically "see" only the outsides of these atoms) in approximately the same way. Therefore, they will form similar chemical compounds, have similar physical characteristics (boiling or melting points), and so on. It will take similar amounts of energy to remove an outer electron from, or *ionize*, atoms with similar outer shell configurations. This leads to recurring similarities among atoms. The atoms appearing in the columns of the periodic table of the elements exhibit similar melting points and have similar chemical behavior (Figure 5.3, page 68).

Atoms possessing certain configurations of electrons are quite resistant to indulging in chemical interactions. The electrons in these atoms are not easily removed, nor can an extra electron easily be added to these atoms. Such atoms in nature are gases at room temperature and are called *inert* (or Noble) gases. Such atoms have a full outer shell. They make up the last column in Figure 5.3. If an atom, in order to balance the positive charge of the protons in the nucleus, has such a configuration *plus* one additional electron, it is easy to get it to give up that extra electron and become a positively charged, or positively

ionized, atom. These atoms are in group I A in Figure 5.3. If an atom is lacking one electron of such a full outer shell, it is easy to get it to accept an extra electron and become a negatively charged, or negatively ionized, atom. These atoms are in group VII A of Figure 5.3.

Many chemical combinations occur because an atom of the first type (such as sodium) gives up one of its electrons to an atom of the second type (such as chlorine). Since the ionized atoms have opposite charges, they attract one another (forming compounds such as sodium chloride—salt). In other sorts of compounds, electrons are shared among the constituents in such a way as to make the environment of each atom more like the "stable" configuration.

The other atoms shown in Figure 5.3 have two "extra" electrons (group II A) or three "extra" electrons (group III A); or they lack two electrons of a full shell (group VI A), or lack three electrons of a full shell (group V A). The atoms of group IV A can be thought of as either possessing four extra electrons or lacking four electrons. The first element in the group IV A is carbon. Because of the many ways carbon can combine with other elements, molecules containing carbon can exist in countless forms. There is a branch of chemistry that deals only with carbon compounds; it is called *organic chemistry*. This name recognizes the importance of carbon for life. The most common other form of life imagined by science fiction writers is one modeled after the second element in column IV A, silicon. Silicon binds in ways similar to carbon. However, a silicon life form is difficult to imagine, and the proposed chemistry seems a little farfetched. Silicon and the rest of the group IV A elements (except lead) have found use in semiconductor technology (see Chapter 15).

CHEMICAL COMBINATION OF ATOMS

Formation of compounds from atoms occurs, generally, when the compound is more stable than the constituents were when they were alone. The constituents become more stable by giving up some of their energy to become a compound.

Atomic number
Symbol of element
Atomic weight

1
H
1.0080

Radioactive state at 1 atmos. and 20°C

Inert gas
Gas
Liquid
Solid—all others

Light Metals

	I A	II A
1	1 **H** 1.0080	
2	3 **Li** 6.939	4 **Be** 9.012
3	11 **Na** 22.990	12 **Mg** 24.31
4	19 **K** 39.102	20 **Ca** 40.08
5	37 **Rb** 85.47	38 **Sr** 87.62
6	55 **Cs** 132.91	56 **Ba** 137.34
7	87 **Fr** (223)	88 **Ra** 226.05

Transitional Elements

Heavy Metals

III B	IV B	V B	VI B	VII B	VIII B			I B	II B
21 **Sc** 44.96	22 **Ti** 47.90	23 **V** 50.94	24 **Cr** 52.00	25 **Mn** 54.94	26 **Fe** 55.85	27 **Co** 58.93	28 **Ni** 58.71	29 **Cu** 63.54	30 **Zn** 65.37
39 **Y** 88.91	40 **Zr** 91.22	41 **Nb** 92.91	42 **Mo** 95.94	43 **Tc** (99)	44 **Ru** 101.1	45 **Rh** 102.90	46 **Pd** 106.4	47 **Ag** 107.870	48 **Cd** 112.40
57 TO 71	72 **Hf** 178.49	73 **Ta** 180.95	74 **W** 183.85	75 **Re** 186.2	76 **Os** 190.2	77 **Ir** 192.2	78 **Pt** 195.09	79 **Au** 197.0	80 **Hg** 200.59
89 TO 103									

Nonmetals

III A	IV A	V A	VI A	VII A	VIII A
					2 **He** 4.003
5 **B** 10.81	6 **C** 12.011	7 **N** 14.007	8 **O** 15.9994	9 **F** 18.998	10 **Ne** 20.183
13 **Al** 26.98	14 **Si** 28.09	15 **P** 30.974	16 **S** 32.064	17 **Cl** 35.453	18 **Ar** 39.948
31 **Ga** 69.72	32 **Ge** 72.59	33 **As** 74.92	34 **Se** 78.96	35 **Br** 79.909	36 **Kr** 83.80
49 **In** 114.82	50 **Sn** 118.69	51 **Sb** 121.75	52 **Te** 127.60	53 **I** 126.90	54 **Xe** 131.30
81 **Tl** 204.37	82 **Pb** 207.19	83 **Bi** 208.98	84 **Po** (210)	85 **At** (210)	86 **Rn** (222)

Rare Earth Elements

Lanthanide series

57 **La** 138.91	58 **Ce** 140.12	59 **Pr** 140.91	60 **Nd** 144.24	61 **Pm** (147)	62 **Sm** 150.35	63 **Eu** 151.96	64 **Gd** 157.25	65 **Tb** 158.92	66 **Dy** 162.50	67 **Ho** 164.93	68 **Er** 167.26	69 **Tm** 168.93	70 **Yb** 173.04	71 **Lu** 174.97

Actinide series

89 **Ac** (227)	90 **Th** 232.04	91 **Pa** (231)	92 **U** 238.03	93 **Np** (237)	94 **Pu** (242)	95 **Am** (243)	96 **Cm** (247)	97 **Bk** (249)	98 **Cf** (251)	99 **Es** (254)	100 **Fm** (253)	101 **Md** (256)	102 **No** (254)	103 **Lw** (257)

FIGURE 5.3

The periodic table of the elements. Elements falling in the columns have similar physical and chemical properties.

USING THE PERIODIC TABLE

The periodic table can be used to predict the way compounds form from constituent atoms. Here we concentrate on the groups identified by an A in Figure 5.3. The column headings I through IV can be taken to represent the number of extra electrons; column headings IV through VII can be subtracted from eight to determine the number of electrons an element is lacking. The resultant number is known as the chemical valence of the atom.

Thus, atoms from group I A combine one to one with atoms from group VII A (for example, LiF); an atom from group II A will combine with two atoms from group VII A ($MgCl_2$, where the subscript indicates that two chlorine atoms combine with one magnesium), or with one atom from group VI A (CaO). All atoms from such groups can be combined in this way.

Such chemical reactions, which give off energy as they proceed, are called *exothermic*. The word *exothermic* really means "heat out" (*ex* in Latin means "out" or "from," and *therm* is from the Greek for "heat"). As an analogy for such a reaction, consider billiard balls on a pool table. It is easy for the balls to roll into holes and fall below the surface, where they can mix with the other balls already down there. To unmix the billiard balls, we have to give them back the gravitational potential energy they gave up in falling down into the hole by putting them back up on the table. Similarly, atoms give up some of their energy when they share or exchange electrons to form a compound; the constituents may be reformed only if energy is added to the system. Since the electrons are in something analogous to a hole, we characterize the atom as an "energy well."

In most cases, some energy is necessary to get the atoms to give up or share their electrons. Of course, they then give up a great deal more energy as they form the compound. The small amount of energy necessary to start the reaction of formation of compounds is called the *activation energy*. Something as simple as one atom's bumping into another can supply the activation energy necessary to start a reaction. Activation energy is supplied for macroscopic (large)

amounts of chemicals by methods as varied as rubbing them together with a pestle in a mortar, by heating them with a match, by exposing them to light (as in photographic film), or by heating them in a crucible.

Once the activation energy starts a reaction that is exothermic, or energy-producing, more energy is produced in the chemical process than was put in as activation energy. This energy surplus can be a source of activation energy to spread the reaction. If the energy liberated is much greater than the activation energy, the reaction may take place so rapidly that an explosion occurs.

For example, if gaseous hydrogen and gaseous oxygen are mixed, water is formed very slowly at normal room temperatures. The reaction may occur because the hydrogen and oxygen molecules are colliding with one another. Occasionally, there will be a random collision between molecules of such high kinetic energy that they are able to supply the activation energy. Energy is released (as constituent kinetic energy) but is spread out through the entire gas mixture by the random collision process.

If a flame is inserted into the mixture of hydrogen and oxygen, an entire region of molecules gets enough energy to become water. At that point, the constituents' kinetic energy represents a concentrated supply of gas molecule activation energy so that molecules adjacent to the first molecules also combine, producing more energy, causing more water to be produced, causing more energy to be released, causing . . . . This reaction is exothermic, and the reaction rate is high because the energy produced so far outruns the activation energy.

Much of the time, chemicals combining in exothermic reactions give off their energy in just this way: by increasing their kinetic energy (and thereby their speed). When molecules speed up, the temperature of the compound increases (as we shall see in Chapter 7). Thus, energy is transferred *to* the substance itself. Its temperature increases. The energy evolved from a reaction need not *necessarily* cause a temperature increase, so we speak of exothermic reactions in general as energy-producing chemical reactions. The energy may be drawn off as electrical energy instead of heat, as in the battery (in which chemical energy is transformed to electrical energy). Or it may be produced in the form of light energy. The light sticks and light necklaces you may have seen for sale in amusement parks produce light—but no heat—from the chemical reaction taking place between the two liquids inside the tube.

Since the combination of two chemicals can often evolve energy, we might imagine that we could subsequently break the chemical bonds if we paid a high enough energy price. Such a reaction sucks in energy from its surroundings, perhaps even lowering the temperature, in order to break the chemical bonds. The word *endothermic* (*endo* is from the Greek for "within") characterizes such reactions. When ammonium chloride or ammonium nitrate are dissolved in water, the temperature of the mixture decreases. The mixing of ammonium chloride with water absorbs so much energy from the surroundings that water touching the vessel may actually freeze. Water can be broken up into its constituents, hydrogen and oxygen, when an external energy source is available. If electrical energy is supplied, the process of water breakup is called electrolysis.

So far, I've discussed simple compound formation or simple compound breakup. However, many chemical reactions involve a multitude of steps, some of which may be exothermic and some of which may be endothermic. The overall reaction is classified as exothermic if the *net* result of all steps in the reaction is an energy release and as endothermic if the *net* result of all steps in the reaction is energy absorption. If a reaction takes place in which 30 joules of energy are absorbed in one step and 50 joules of energy are produced in another step, the net result is that $50 J - 30 J = 20 J$ of energy is produced; this is an exothermic reaction. The reverse reaction (if it is possible) would be an endothermic reaction, and in the two parts, 30 J would be produced while 50 J would be absorbed because of conservation of energy; thus a net of $30 J - 50 J = -20 J$ would be produced.

Catalysis

It is possible to change the rate at which reactions occur. Electron exchanges can occur when electron shells are in the vicinity of one another. Suppose we add to a reaction materials that allow the appropriate surfaces in the constituents to come into contact with one another. The materials serve to enhance the speed with which the chemical reaction occurs, but the added materials are not used up or even affected in the process. Materials that have this property are known as *catalysts;* the process is called *catalysis*. Examples of catalysts include platinum powder, used to increase the rate of combination of hydrogen and oxygen to form water; finely ground ashes, used to allow a cube of sugar to burn (without the ash, the sugar simply melts and does not burn); and, more important from our standpoint, enzymes used with the chlorophyll in plants to enhance the rate at which carbon dioxide and water are converted into sugars and oxygen.

All of the reactions listed as examples would occur without catalysts. They would simply take a longer time to complete. The catalysts are unchanged because, after the reaction occurs, the constituents leave the surface as a compound and allow their places to be taken by new sets of constituents.

I have implied that catalysts are essential to life. In living things, catalysts called enzymes make life possible. Enzymes are proteins in living things that act to speed up reactions. Many chemical reactions occur quite slowly. Iron rusts (combines with oxygen) very slowly at normal temperatures. In the presence of water, especially salt water, the reaction occurs more rapidly. People who live in areas in which salt is spread on roadways for ice control can observe their cars corroding after only a few years' exposure. Nevertheless, rusting and corroding do not happen as rapidly as a wood fire reduces a log to ashes. Paper is made from wood; it is burning just by existing. The burning is slow. In several hundred years, the paper in this book you are reading will have disintegrated. You have probably seen old books with their brittle, yellowed, half-burnt paper. We can hasten the process by supplying the activation energy for a chain reaction, and burn the book in minutes.

Many reactions in the human body are of this "burning" kind (called *oxidation*) and usually proceed slowly. The enzymes may catalyze specific reactions, allowing them to occur rapidly. Of course, as with any catalysis, reactions cannot be made to happen that could not have otherwise occurred; the catalyst merely hastens them.

Photosynthesis

The body of an animal mines the lode of energy deposited in plant cells by the process of *photosynthesis*. This process is not yet fully understood. In photosynthesis, energy-carrying sunlight is absorbed by chlorophyll in plants and allows an endothermic reaction of water and carbon dioxide to take place to produce oxygen and sugars, which store much of the energy. A green cell can produce up to thirty times its volume of oxygen each hour (1).

Photosynthesis proceeds by using enzymes to make oxygen, then transferring hydrogen atoms using the light energy to effect the transfer, and then finally combining carbon dioxide with the hydrogen to form the sugars. In this process, energy of about 6.2 kJ/kg (1474 kcal/kg) of carbon dioxide is stored as chemical energy—the sugars—in plants (1).

Many millions of years ago, plant life in swamps used photosynthesis to produce stored energy. Some of the plant life fell into the muck, making the swamp into a peat bog. Geologic processes may have led to the laying of sediment over the peaty material. Then biological action, the heat of the Earth, and the pressure of the overlying sediment (turned to rock) resulted in the formation of a coal deposit. Therefore, coal is merely stored solar energy. A similar process, involving other kinds of organic matter, leads to formation of oil and gas deposits; they too are stored solar energy.

Coal and oil resources for 50 million to 100 million years from now are currently being formed here on Earth. These *are* renewable resources, but on a time scale somewhat long for humanity. Wood and plants and ocean algae form a much shorter term reservoir of stored solar energy, and it is this reservoir that animals tap to utilize the energy necessary to sustain their daily existence.

Carbon Compounds

As mentioned previously, carbon compounds play a vital role in the existence of life on Earth. Plants "inhale" carbon dioxide and "exhale" oxygen. Animals reverse this process, burning oxygen with the carbon compounds in their food to produce the energy for life, and producing carbon dioxide as a waste product. Foods consumed by animals are often carbohydrates (compounds containing carbon, oxygen, and hydrogen). Life also plays a role in the formation of the fossil fuels (see Chapter 13), which are deposits of hydrocarbon compounds (compounds containing carbon and hydrogen). These hydrocarbon compounds can produce energy as they burn in the presence of oxygen.

Carbon can bond easily with itself and other compounds, and its symmetrical configuration of electrons allows it to bond in many different ways. Many carbon compounds form closed rings and other interesting three-dimensional shapes. Natural gas is mostly methane, CH_4. The other fossil fuels have a more complicated structure than methane, with carbons bonding to other carbons. Petroleum's chemical composition is approximately CH_2 because of this carbon–carbon bonding, and so produces relatively more carbon dioxide and less water than methane. Coal has a chemical composition with even more carbon–carbon bonds. Its composition is approximately $CH_{0.8}$; it produces relatively more carbon dioxide and less water than petroleum.

When methane burns completely in oxygen, only carbon dioxide (CO_2) and water may be produced. Carbon monoxide is partially burned carbon; it is dangerous to animals because it binds with red blood cells (as does oxygen) but cannot be removed (in contrast to oxygen, which is easily removed). The red blood cells transfer oxygen to the rest of the body from the lungs; hence, carbon monoxide can prevent transfer of oxygen from the lungs (see Chapter 22).

When methane burns in air, a mixture of nitrogen and oxygen, it can produce nitrogen compounds as well as carbon compounds. In a real combustion situation, methane and air can produce carbon monoxide (CO), carbon dioxide (CO_2), and various compounds of nitrogen and oxygen: N_2O, NO, NO_2, N_2O_2, and so on (these compounds are known collectively as NO_x). The

effects of these and other combustion products are discussed in Chapter 22.

LEMON POWER, OR THE BATTERY

When I was a child, I found out about lemon power. We would take pieces of two different metals (say, copper and zinc) and stick them into opposite ends of a lemon. Then we would stick our tongues onto both metal pieces at the same time. Presto! We experienced a distinct tingling in the tongue due to a flow of current—lemon power.

We did not really know it, but we were discovering the principle of the battery. Batteries need two *electrodes* made of different materials and a fluid, the *electrolyte,* to conduct electricity between the electrodes. The lemon could work as a battery because lemon juice is an acid and conducts current relatively easily. Any fruit acid will do as an electrolyte—acetic acid (cider vinegar), wine vinegar, and so on.

The battery is so named because any one *cell* (composed of the two electrodes and the electrolyte) does not deliver much voltage; a battery, or series, of cells connected together as in Figure 5.4 could deliver an appreciable voltage between the ends. Such a battery is not very useful because the acid will spill if one attempts to move it. Several methods have been used to overcome these problems. For example, a container could be built so that a series of bimetal strips is held closely in the container, acid is put in between the strips, and the top is closed. This describes

FIGURE 5.4
A battery of simple cells.

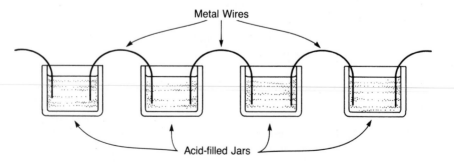

Metal Wires

Acid-filled Jars

the essentials of an auto battery. A dry cell flash-light "battery" (Figure 5.5) is made so that the outer container acts as one electrode, the other is replaced by a carbon rod in contact with the metal top, and a paste electrolyte fills up the rest of the space. A dry cell is not really dry, but moist; however, the acid contained in it will not spill.

There are two general types of batteries: primary and secondary cells. A primary cell is a throwaway (nonrechargeable) battery; a secondary cell is designed to be recharged.

FUEL CELLS

Another source of electrical energy from chemical energy is found in fuel cells. Fuel cells were first developed in 1839 by Sir William Grove. In the hydrogen–oxygen fuel cell, hydrogen is fed to one porous electrode and oxygen to the other. The electrodes are separated by an electrolytic material. The hydrogen in the electrode is converted catalytically to hydrogen ions, releasing

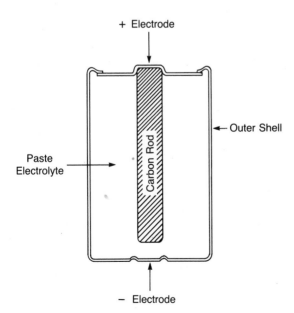

FIGURE 5.5
A dry cell. The outer shell is the negative electrode. The carbon rod is the positive electrode.

electrons to run through an external electrical circuit. The electrons then combine with oxygen at the other electrode, producing oxygen ions. As the ions travel through the electrolyte, they meet and react to produce water. The only outputs of fuel cells are water and electrical energy, and the output continues as long as hydrogen (or other fuel) and oxygen are supplied to the cell. This process is electrolysis run backwards.

A practical fuel cell first became possible in the 1950s with the development of suitable ceramics. Rapid development followed in the 1960s, as fuel cells were designed for use in the space program. Most current models use natural gas from which hydrogen is stripped. Large-scale fuel cell power plants for electricity are being tested at present in New York City (4.8 MW) and Tokyo (11 MW), although both plants are experiencing some problems (2). Smaller scale fuel cells are undergoing tests in various places around the country. For example, a 40 kW prototype was installed by Southern California Edison on an old landfill in City of Industry, California, which is now the site of a hotel complex and golf course. Anaerobic digestion of the garbage in the landfill produces methane. The gas from the landfill supplies hot water for the complex and fuel for the fuel cell. Fuel cells are attractive because they have proved in prototype to be up to 80 percent efficient (see Chapter 7), and their installation price is about the same as that for coal or gas power plants.

SUMMARY

Chemical energy arises from the exchange or sharing of electrons among atoms. When chemical bonds are made in a reaction, energy is given off, and we say the process is *exothermic*. Such reactions occur because the compounds are more stable afterward. When chemical bonds are broken in a reaction, energy must be absorbed from the surroundings, and we say the process is *endothermic*. Most reactions involve both the making and the breaking of chemical bonds. If the overall effect is the making of bonds, the reaction is exothermic; if the overall effect is the breaking of bonds, the reaction is endothermic.

Activation energy is the energy necessary to begin the process of the formation of compounds. *Catalysts* are materials that can change the rate of reactions without themselves being consumed in the reaction. *Enzymes* are biological catalysts.

Some exothermic chemical reactions can be arranged to produce electric current. To make an electric cell, we need two dissimilar metals and an *electrolyte*. A battery consists of a series of cells. A fuel cell produces electricity from the combination of hydrogen and oxygen to form water.

REFERENCES

1. Govindjzee, "Photosynthesis," in Vol. 10 of *McGraw-Hill Encyclopedia of Science and Technology,* (New York: McGraw-Hill Book Co., 1974).
2. E. Marshall, *Science* 224 (1984): 268.

PROBLEMS AND QUESTIONS

True or False

1. The process of photosynthesis would be possible even if no enzymes were involved.
2. In one part of a reaction, 72 kJ of energy is released, while in a second part of the reaction, 80 kJ of energy is absorbed. The two-part reaction is endothermic overall.
3. In amusement parks, it is possible to buy a plastic stick or necklace that glows for some hours. The stick never gets hot but it does give off light. This is an example of an exothermic chemical reaction producing light but no heat.
4. The inner electrons are the most important in determining how atoms react chemically.
5. A reaction in which several subprocesses are taking place, some of which are endothermic and some of which are exothermic, cannot be classified as either endothermic or exothermic itself.
6. There is no essential difference between carbohydrates (plant protein) and hydrocarbons.
7. Electrons in atoms circle the nucleus at distances hundreds of times larger than the nuclear diameter ($\approx 10^{-15}$ m).
8. The heat of combustion of various foods is roughly the same as the heat of combustion of many fossil fuels.
9. Enzymes are organic catalysts.
10. Photosynthesis is a reaction through which solar energy in the form of light is stored in plants.

Multiple Choice

11. In the laboratory, fine iron wires of mass 50 g are burned in pure oxygen, producing energy as heat and light. When 50 g of fine iron wires is put outside and allowed to rust, we would expect

    a. no net energy to be given off.
    b. a smaller amount of energy to be given off.
    c. a greater amount of energy to be given off.
    d. exactly the same amount of energy to be given off.
    e. rust to occur explosively outside just as it does in the laboratory.

12. In a reaction of a certain amount of chemical *A* and a certain amount of chemical *C*, there is a phase in which 50 kJ of energy is given off, another phase in which 20 kJ of energy is absorbed, and a third phase in which 10 kJ of energy is given off. Overall, the reaction is

   a. exothermic, with 60 kJ of energy given off.
   b. endothermic, with 60 kJ of energy absorbed.
   c. exothermic, with 40 kJ of energy given off.
   d. endothermic, with 40 kJ of energy absorbed.
   e. exothermic, with 80 kJ of energy given off.

13. A catalyst is a chemical that

   a. causes cats to tip over.
   b. can be added to reactants to "spice up" the reaction.
   c. can be added to reactants to make the reaction faster, and is left unchanged after the reaction is over.
   d. will react chemically with other chemicals, causing more bonds to be made or broken, and ends up in the final compound.
   e. allows reactions that never could occur otherwise to take place in its presence.

14. Electrons in atoms may move among the energy levels. When a transition occurs,

   a. any amount of energy greater than some minimum will allow the move to occur.
   b. any amount of energy up to some maximum will allow the move to occur.
   c. only energy from external sources can make it happen.
   d. only internal energy will be involved.
   e. only discrete amounts of energy can be either emitted or absorbed.

15. The periodic table shows that only certain chemical reactions may occur. Which of the following reactions should *not* occur?

   a. $H + H \rightarrow H_2$
   b. $Na + F \rightarrow NaF$
   c. $Na + Cl \rightarrow NaCl$
   d. $4 Ca + Cl_2 \rightarrow 2 Ca_2Cl$
   e. $Sr + 2S \rightarrow SrS_2$

16. The periodic table shows that only certain chemical reactions may occur. Which of the following reactions should *not* occur?

   a. $C + 2 H_2 \rightarrow CH_4$
   b. $C + 2 Cl_2 \rightarrow CCl_4$
   c. $Na + C + F + N \rightarrow CNaFN$
   d. $Mg + Cl_2 \rightarrow MgCl_2$
   e. $2 K + F \rightarrow K_2F$

Discussion Questions

17. How is it possible for *endothermic* reactions to run in nature?

18. How much energy is released by an amount of iron oxidizing slowly compared to the situation in which the same amount of iron is oxidizing rapidly (burning)? Explain.

19. Explain how an exothermic reaction would need heat or outside energy to start the reaction when an exothermic reaction must produce energy.

20. Will chemical properties of two atoms with identical outer electron shells be very similar, even if the nuclei of the atoms differ substantially in their makeup?

21. Explain why coal burning produces more carbon dioxide than burning of natural gas (methane).

6
NUCLEAR ENERGY

In the previous chapter, we examined chemical processes that can be used to generate energy. Another process that is important for generating electric energy is the process of nuclear fission (nuclear breakup). Nuclear reactors supply about 15 percent of energy needs in the United States and over 50 percent of France's energy needs. This chapter provides the introduction necessary to understand how reactors work. We return to this subject in Chapter 14.

KEY TERMS   *nucleus*
             half-life
             nucleon
             mass number
             atomic number
             binding energy
             electron-volt
             fissile nucleus
             nuclear activation energy
             chain reaction
             moderator
             alpha particles
             beta particles
             gamma radiation
             big bang
             nucleosynthesis

THE NUCLEUS

If we examine a material on a much smaller scale than the atom, we find the atom composed mostly of empty space. At the atom's core lies the atomic nucleus, which has a diameter measured in femtometers (10^{-15} m).

If we imagine a nucleus to have a diameter of about 10 mm, the atom itself would have a diameter of about 1 km—100,000 times greater. An atomic nucleus is built from protons and neutrons. These particles can also exist outside a nucleus. A proton is the nucleus of a hydrogen atom; since the universe is mostly hydrogen, many protons exist in the universe. If a neutron remains outside the nucleus of an atom, it will eventually decay (see box). Inside a nucleus, it generally does not decay. Despite their propensity to break apart, there are many neutrons around us coming from the sun and from the breakup of radioactive materials.

Protons and neutrons are almost indistinguishable. The fact that the proton has one unit of electric charge and the neutron has none is responsible for the difference between them. Because of this similarity, we often speak of *nucleons* in cases where the charge is not an important difference. A nucleon is a proton or a neutron; nucleons are either or both.

THE NEUTRON

The neutron is unstable; that is, it decays. This process is caused by the so-called weak interaction, which is one of the four interactions known to exist. If one defines dimensionless numbers to illustrate the respective sizes of these interactions, they are in the ratio

strong interaction : electromagnetic interaction
: weak interaction : gravitational interaction : : $1 : 10^{-2} : 10^{-5} : 10^{-39}$.

At the present time, work is continuing on the task of unifying the four interactions, and great progress has been made in tying the first three together into a theory known as Quantum Chromodynamics. Sheldon Glashow, Steven Weinberg, and Abdus Salam shared the 1979 Nobel Prize for the electroweak theory, which ties the weak and electromagnetic interactions together into one theory. A further Nobel was awarded in 1984 to Carlo Rubbia and Simon van der Meer for their experimental discovery of two particles (W and Z) predicted by this theory.

The neutron mass is 1.6768×10^{-27} kg (939.6 MeV), while that of the proton is 1.6727×10^{-27} kg (938.3 MeV), and that of the electron is 9.11×10^{-31} kg (0.5 MeV). Since m_n (mass of the neutron) is greater than $m_p + m_e$ (mass of the proton plus mass of the electron), it is possible for the neutron to decay into a proton and an electron, while conserving mass-energy. Another particle, the elusive massless neutrino, is also produced in this decay. The excess energy, 0.8 MeV, becomes kinetic energy of the proton, electron, and neutrino.

Given that the neutron will decay if it is outside the nucleus of an atom, there is no knowledge that some particular neutron will decay after a given time. However, we may say that half an assemblage of many neutrons will indulge in decay in a time interval characteristic of neutrons. This time is called the *half-life* for the decay, and is related to the decay time (lifetime) of the neutron. The neutron half-life is about ten minutes. Half lives, in the statistical sense described here, may be defined for all unstable elementary particles and nuclei.

Recall that in exponential growth, the growth rate depends on the number of things or particles at the beginning. Doubling times may be defined for all exponentially growing quantities. The rate of decay of particles depends on the number at the beginning—the greater the number of particles, the greater the number of decays. The number of particles is exponentially decreasing. Exponential decay is the inverse of exponential growth, and half-lives are the flip side of doubling times. In exponential growth, it takes the same time—the doubling time—to increase from 2 million to 4 million as from 1 million to 2 million. In exponential decay, it takes the same time—the half-life—to decay from 4 million to 2 million as from 2 million to 1 million.

Because protons and neutrons have practically the same mass, the mass of a nucleus containing a number A of nucleons has a mass roughly A times the average proton-neutron mass. The number of nucleons is called the *mass number* and is given the symbol A. Of course, the number of protons is important for an atom because it determines the number of electrons necessary

B efore we continue, we should define an energy unit that is convenient for discussing nuclear processes: the MeV, previously mentioned in the discussion of free neutron decay. The size of the charge on the electron and proton, denoted e, was found to be 1.6×10^{-19} coulombs. A charge e falling through an electrical potential difference (energy per unit charge) of 1 volt (1 joule/coulomb) gains 1.6×10^{-19} J in the process. This amount of energy is defined to be 1 *electron-volt* (1 eV). Most atomic processes take place on the eV scale. The ionization energy of the electron in a hydrogen atom is 13.6 eV (2.18 attojoules).

An MeV is a million electron-volts (160 femtojoules). Most nuclear processes take place on the MeV scale. In these energy units, the total energy release in an average fission is about 200 MeV or 30 picojoules. The energy equivalent of a proton is about 940 MeV, almost a GeV. The newest particle accelerators produce particle energies a thousand times greater, over a TeV (about 1/6 microjoule).

to make the atomic charge zero. These electrons are all that we "see" of ordinary atoms, and so the number of electrons determines how a particular atom acts chemically. That is, the number of protons in an atom determines what kind of an atom it will be. Thus atoms of the same element (same proton number) can have different nucleon numbers: helium-3 and helium-4; carbon-12 and carbon-14; and iron-56 and iron-58. Such nuclei are called *isotopes*. The number of protons is called the atomic number and is given the symbol Z. The number of neutrons is given the symbol N; clearly $A = Z + N$.

While Z and A totally specify an atom (and its nucleus), it is common to give the elements names and symbols. Hydrogen has $Z = 1, A = 1$, and has the symbol H; helium has $Z = 2, A = 4$, and has the symbol He; and so on. It is customary to specify an element by using all of this information: ^A_Z (element name). Thus, ordinary hydrogen is also known as $^1_1$H; helium as $^4_2$He; and so on.

One interesting fact, illustrated in Figure 6.1, is that as Z (the atomic number) gets larger, N (the neutron number) gets larger faster. That is, the higher Z is, the greater the number of neutrons for each proton. The reason for this is that the protons repel one another electrically. The

nucleons are bound together by what we call the strong force, which is greater than the electrical force of repulsion, when the protons are close together, at distances less than 10^{-15} m. Despite this, the electrical repulsion forces can get quite large. The neutrons act to screen the protons somewhat from one another. In order that there be enough neutrons to be effective as a screen, there should be more neutrons than protons in a nucleus.

Binding Energy

When two nucleons interact, they generally "bounce" off one another because of electrical repulsion or because there is a hard core of "protection" surrounding the nucleon. Only when two nucleons really hit one another hard—at high energy—can the two cores interpenetrate. When this happens, the two nucleons can manage to be attracted to one another. By emitting some energy as light (gamma rays), they can remain together. They act as if they have fallen into a hole or a "well." That is, by giving up energy, they can become bound together. The amount of energy they give up to remain together is called the *binding energy*.

FIGURE 6.1
The neutron number, N, is plotted in terms of atomic number, Z. Note that the number of neutrons in a nucleus increases faster than the number of protons.

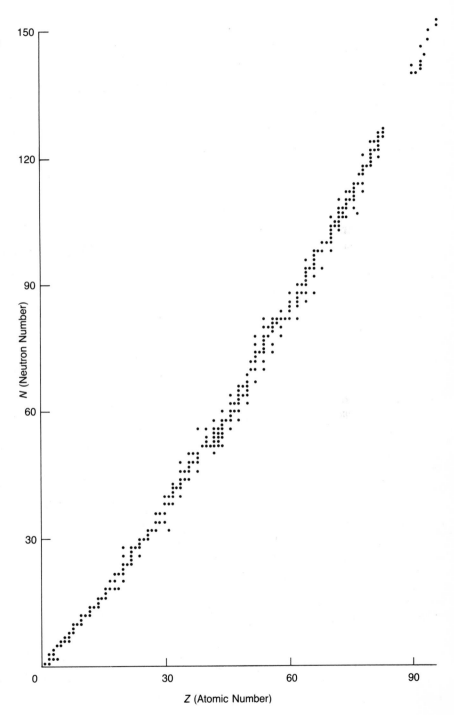

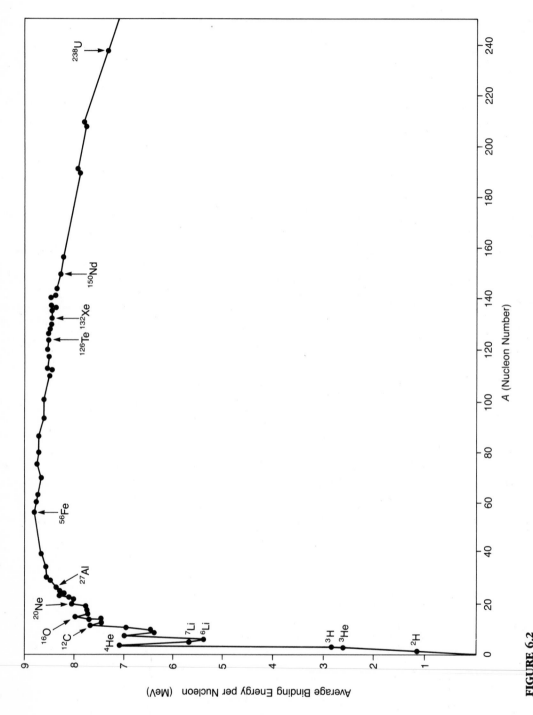

FIGURE 6.2

The average binding energy per nucleon is plotted in terms of the mass number, or nucleon number, $A = N + Z$.

As the number of nucleons increases, there is a rapid increase in the average binding energy per nucleon. This comes about because, as more and more neutrons are taken in, they get between the protons to shield them from one another. The protons then do not need as much energy as before to repel other protons electrically, so they can give up more energy as they gather together to form a nucleus. At high mass number (high A), the neutron's shielding effect cannot save any more energy; after this we would expect the binding energy per nucleon to be approximately constant. The curve of the actual average binding energy per nucleon as a function of A, the nucleon number, is shown in Figure 6.2. Note that the general features are *approximately* as we discussed. However, for A greater than that of iron-56 (^{56}Fe), the average binding energy per nucleon slightly decreases. It is precisely *this* characteristic that permits fission in nature to occur.

FISSION AND THE LIQUID DROP MODEL

Fissile nuclei—that is, nuclei that can undergo fission, or break into two pieces—are generally stable until something happens to give them a little energy (for example, when they absorb another neutron). This is called *nuclear activation energy,* and is analogous to the chemical activation energy discussed in Chapter 5. At this stage, the nucleus seems to act something like a drop of liquid that oscillates between a spherical shape and a dumbbell shape (or like an accordion being opened and closed rhythmically). If the two ends of the dumbbell get far enough apart as they oscillate, the nucleus can separate into two parts, which become nuclei of lower-A elements. The ratio of neutrons to protons decreases as Z decreases (see Figure 6.1), so that, when fission into two pieces occurs, there are extra neutrons in the product nuclei. These are rapidly emitted to allow these nuclei to become stable. Most fissions end up producing two or three extra neu-

trons. These neutrons can then interact with other fissile nuclei, causing additional fissions. This is the essence of the idea of the *chain reaction* (note that each neutron going in produces two or three new neutrons in causing a fission).

There is no way to predict exactly which nuclei will be produced in a fission reaction; this depends on the details of how the "drop of liquid" oscillated. There is a distribution of fission product masses. That distribution of product mass numbers for the fission of a uranium-235 ($^{235}_{92}$U) nucleus that absorbs a neutron is shown in Figure 6.3 (1). The mass of each of the fission products is roughly half the original mass. From the diagram it is obvious that what typically occurs is the production of one nucleus with $A \approx 90$ and one nucleus with $A \approx 140$.

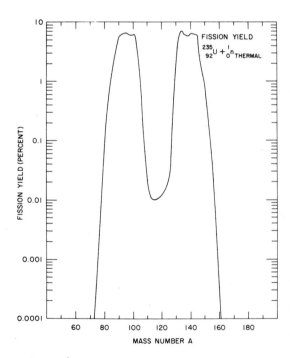

FIGURE 6.3
The yield of fission products is plotted in terms of the mass number for fission products resulting from the fission of an atom of ^{235}U, which has absorbed a thermal neutron. (Reprinted by permission from *The Physics Teacher*.)

Typical fissile nuclei such as ^{235}U, ^{239}Pu (plutonium), and ^{232}Th (thorium) react with slow incident neutrons,

$$n + {}^{235}_{92}U \rightarrow {}^{236}_{92}U$$
$$n + {}^{239}_{94}Pu \rightarrow {}^{240}_{94}Pu$$
$$n + {}^{232}_{90}Th \rightarrow {}^{233}_{90}Th,$$

which then decay into two fission fragments, neutrons, and radiant energy.

If there is a heavy nucleus, Figure 6.2 shows that the nucleons in it could give up energy—that is, be more tightly bound into the nucleus containing them—if the heavy nucleus were to break up into two lighter fragments. This additional energy can then appear as radiant energy (gamma rays) from decay of highly excited states, as well as kinetic energy of the fission fragments. The energy distribution from nuclear fission is shown in Table 6.1. All processes in which mass is transformed into energy and energy into mass obey Einstein's mass–energy relation $E = mc^2$, where c is the speed of light, 3×10^8 m/s.

It is the extra neutrons produced in fission that allow uncontrolled chain reactions (the "atomic" bomb) and controlled chain reactions (fission reactors in nuclear power plants). It takes a neutron to initiate the sequence, and, on the average, for each neutron going in, 2.43 come out. If these cause other fissions and enough fissile uranium is present, a chain reaction can ensue, producing a lot of energy at once as a bomb. If the uranium is instead put into water, and carbon rods are intermingled with the uranium, the neutrons produced in fission are slowed

TABLE 6.1
Distribution of energy release in typical fission.

Source of Energy	Typical Energy
Kinetic energy of fission fragments	167 MeV
Kinetic energy of fission neutrons	5 MeV
Instantaneous gamma rays	7 MeV
Emissions from fission fragments	
Gamma rays	7 MeV
Beta particles	7 MeV
Neutrinos	11 MeV

SOURCE: Reference 2.

down by collisions with the atoms in the water and the carbon rods. The water molecules can absorb neutrons; the carbon cannot, but they act to slow down the neutrons. These materials that slow neutrons down are called *moderators*. Some neutrons escape the reactor and some are captured by the nonfissile uranium atoms such as $^{238}_{92}$U. It is thus possible to make certain that there is always the same number of neutrons crossing a unit area in a given time. This system can then persist for lengthy periods of time in a state of dynamic equilibrium.

Energy From Fission

We shall investigate two representative decays of $^{235}_{92}$U $+ n$, $^{236}_{92}$U, using the binding energy curve, Figure 6.2

i.) $^{236}_{92}U \rightarrow {}^{90}_{36}Kr + {}^{143}_{56}Ba + 3n$

(This is the decay into krypton-90 and barium-143.) From Figure 6.2, we read that the average binding energy per nucleon is about 0 MeV for n, 7.6 MeV for ^{236}U, 8.8 MeV for ^{90}Kr, and 8.4 MeV for ^{143}Ba. The total binding energies for the respective nuclei are therefore

$$236(7.6 \text{ MeV}) = 1794 \text{ MeV, for } {}^{236}U,$$
$$90(8.8 \text{ MeV}) = 792 \text{ MeV, for } {}^{90}Kr, \text{ and}$$
$$143(8.4 \text{ MeV}) = 1201 \text{ MeV, for } {}^{143}Ba.$$

The net energy emitted in the fission is the difference between the binding energies of the fission products and the binding energy of the original nucleus:

$$BE({}^{90}Kr) + BE({}^{143}Ba) - BE({}^{236}U)$$
$$= 792 \text{ MeV} + 1201 \text{ MeV} - 1794 \text{ MeV} = 199 \text{ MeV}.$$

ii.) $^{236}_{92}U \rightarrow {}^{97}_{38}Sr + {}^{137}_{54}Xe + 2n$

(This is the decay into strontium-97 and xenon-137.) Again reading from Figure 6.2, the binding energy per nucleon is about 0 MeV for n, 7.6 MeV for ^{236}U, 8.6 MeV for ^{97}Sr, and 8.4 MeV for ^{137}Xe. The total respective binding energies are

$$236(7.6 \text{ MeV}) = 1794 \text{ MeV, for } {}^{236}U,$$
$$97(8.6 \text{ MeV}) = 834 \text{ MeV, for } {}^{97}Sr, \text{ and}$$
$$137(8.4 \text{ MeV}) = 1151 \text{ MeV, for } {}^{137}Xe.$$

Geiger-Müller tubes consist of a central wire anode at high voltage and a cylindrical metal cathode sealed in a glass tube containing a gas. A charged particle will ionize the gas in the tube, and the high voltage pulls the particles to the wire. The electrical signal can be amplified and made to cause a speaker to turn on momentarily.

In cloud chambers, a saturated vapor is in a container. As ionizing particles traverse the container, they leave ions behind, which causes little clouds to form along the track. The heavier the tracks, the greater the ionization. Hence, one would expect alphas to leave heavier tracks than betas in cloud chambers.

Bubble chambers work in a similar way, but they are filled with a liquid that is made supersaturated when the volume of the container expands. As in the cloud chamber, ions left behind cause drops to condense. The tracks can be photographed to make permanent records for later analysis.

There are many other observation devices: sodium iodide crystals that flash as particles travel through; acoustic hodoscopes and multiwire proportional chambers, which read out electrical signals from electrodes; spark chambers that make visible a series of sparks to mark a particle's passage; nuclear emulsion film, and so on. Most modern detectors used in high-energy physics research nowadays record directly onto computer storage rather than making photographic records.

This time the net energy emitted in the fission is

$$BE(^{97}\text{Sr}) + BE(^{137}\text{Xe}) - BE(^{236}\text{U}) =$$
$$834\,\text{MeV} + 1151\,\text{MeV} - 1794\,\text{MeV} = 191\,\text{MeV}.$$

These numbers are in general agreement with the sum of all the energies listed in Table 6.1—204 MeV—for the "typical" fission.

Radioactive Decay

Fission—breakup into two roughly equivalent pieces—is not the only way to change one atom into another sort of atom. Many elements exhibit a process called radioactive decay. There are three usual ways for particles to decay, called *alpha-emission, beta-emission,* and *gamma-emission.* Alpha particles are helium nuclei (^4_2He) and so consist of two protons and two neutrons, and are rather massive. Alpha decay is the process by which a nucleus emits an alpha particle spontaneously and becomes a nucleus of an atom lower in atomic number by 2 and nucleon number by

4 than the "parent" nucleus. The protons can ionize air fairly effectively. Alpha particles cannot penetrate very much matter because they interact so effectively, and can be stopped even by a thin sheet of cardboard.

Beta particles are electrons (or positrons, anti-electrons) and thus are much less massive than alpha particles. Beta decay occurs when a neutron inside a nucleus spontaneously decays, emitting an electron; the atomic number increases by 1 in beta decay. Since electrons and positrons are singly charged, they are not as effective at ionizing air as the alphas. Betas can penetrate matter more easily than the alphas, and need aluminum or even lead to make them stop. Alpha and beta particles, as well as more exotic particles, can be detected by use of geiger counters, or by observation in cloud chambers, bubble chambers, or more sophisticated particle detection devices.

Gamma decay occurs when a nucleus in an excited state falls to the ground state and emits electromagnetic radiation equal to the energy difference in the process. In order for gamma radiation to be detected, it must be converted into electron–anti-electron pairs. Thicknesses of lead cause pair production; the electrons are then easily detected.

Gamma rays are very energetic forms of electromagnetic radiation or light. They do not ionize air very well at all, but do cause electron–positron pairs to form out of pure energy (another example of the Einstein mass–energy relation, $E = mc^2$). Gamma radiation is the most penetrating of all, requiring great thicknesses of lead in order to be stopped.

When nuclei emit alpha particles, they are transmuted into other sorts of nuclei because $Z = 2$ and $A = 4$ for the alpha particle (a helium nucleus). Thus $^{232}_{90}$Th decays by alpha emission to $^{228}_{88}$Ra, or

$$^{232}_{90}\text{Th} \rightarrow {}^{4}_{2}\text{He} + {}^{228}_{88}\text{Ra}.$$

A nucleus that emits a beta particle changes its Z by 1, but not its A, as in the decay

$$^{3}_{1}\text{H} \rightarrow e^- + {}^{3}_{2}\text{He}.$$

In this case, conservation of charge and several more exotic conservation rules require that one of the neutrons in the tritium nucleus decay into a proton, an electron, and an antineutrino. Gamma transitions occur when excited states of nuclei decay into the ground state. Table 6.2 presents several interesting half-lives.

Still another process can change atomic identity. This is electron capture. As explained in Chapter 5, electrons fill energy levels outside the nucleus of an atom. While it is often convenient to visualize the electron as a "planet" orbiting a nuclear "sun," electrons or atoms cannot be located in one spot the way we usually think of locating objects. The electrons are spread out in space somehow—they are even in the space "occupied" by the nucleus. Under certain circumstances, a nucleus will absorb one of its electrons, changing a proton into a neutron and emitting a neutrino. The new nucleus will have a Z of 1 less than the parent, but the A will be the same.

NUCLEAR FUSION

In order to understand how nuclear fusion differs from nuclear fission, it is useful to return to Figure 6.2, the curve of the average binding energy per nucleon versus A. Note the steep rise in average binding energy per nucleon for small A. This implies that if we knock together low-A nuclei to form higher-A nuclei, we can obtain energy release also. However, in order to get that energy, we must put a lot of energy in. We must put in energy to overcome the internuclear repulsion that exists when the nuclei are not close together. This introduces a considerable complication in implementation of any scheme for fusion power. Fusion reactions include

$$^{1}_{1}\text{H} + {}^{2}_{1}\text{H} \rightarrow p + {}^{3}_{1}\text{H} + 3.25 \text{ MeV (22 GWh/kg)}$$

$$^{2}_{1}\text{H} + {}^{2}_{1}\text{H} \rightarrow n + {}^{3}_{2}\text{He} + 4.0 \text{ MeV (27 GWh/kg)}$$

$$^{2}_{1}\text{H} + {}^{3}_{1}\text{H} \rightarrow n + {}^{4}_{2}\text{He} + 17.6 \text{ MeV (94 GWh/kg)}$$

$$^{2}_{1}\text{H} + {}^{3}_{2}\text{He} \rightarrow p + {}^{4}_{2}\text{He} + 18.3 \text{ MeV (98 GWh/kg)}$$

$$^{6}_{3}\text{Li} + {}^{2}_{1}\text{H} \rightarrow 2{}^{4}_{2}\text{He} + n + 22.4 \text{ MeV (75 GWh/kg)}$$

Here, $^{2}_{1}$H is hydrogen with both a proton and a neutron in its nucleus, the sort known as deuterium; $^{3}_{1}$H is hydrogen with a proton and two neutrons, known as tritium. Deuterium is stable and thus can be found at some small concentration in ordinary hydrogen; tritium decays rapidly (Table 6.2) and is not found in nature. Deuterium ($^{2}_{1}$H) is found instead of hydrogen ($^{1}_{1}$H) in one out of every 6500 hydrogen atoms (3).

This curve of the binding energy is useful for explaining how the elements we know got here. If we accept the "big bang theory" as correct, then the universe formed in an explosion, and there was only radiant energy in it. As the universe expanded, it cooled. When it cooled far enough, matter could exist.

The matter in the early universe was mostly hydrogen with a small admixture of helium. As matter collected, stars grew. Inside stars, the temperature can climb to tens of millions of kelvin (our sun's core is at a temperature of about 10 MK). At these temperatures, nuclei can get to within 10^{-15} m of one another, and fusion re-

TABLE 6.2

Typical half-lives for alpha emission.

Element	Half-life
$^{3}_{1}H$	12.3 years
$^{15}_{8}O$	126 seconds
$^{14}_{6}C$	5730 years
$^{29}_{13}Al$	6.6 minutes
$^{31}_{14}Si$	2.62 hours
$^{32}_{14}Si$	650 years
$^{37}_{19}K$	1.3 seconds
$^{52}_{26}Fe$	7.8 hours
$^{90}_{38}Sr$	28.1 years
$^{197}_{83}Bi$	2 minutes
$^{226}_{88}Ra$	1620 years
$^{232}_{90}Th$	13.9 gigayears
$^{232}_{92}U$	73.6 years
$^{238}_{92}U$	4.5 gigayears

SOURCE: Reprinted with permission from *Handbook of Chemistry and Physics,* 32nd ed. Copyright 1959, 1974 by CRC Press, Inc., Boca Raton, FL.

actions can occur. Since the curve shows that particles fused to other particles can be more deeply bound until the product mass reaches 56, fusion in a massive star can proceed until a layer of iron is in the core, surrounded by layers of lighter elements.

If this were all, our universe would be uninteresting and humanless. However, as the fusion fires are banked (that is, more iron is formed in the stellar core), the heat output goes down. If the temperature goes down, the outer layers of the star can fall inward (because the pressure is related to the temperature, as we see in the next chapter). In certain circumstances, this can cause a supernova, which is a catastrophic explosion in which the collapsing outer layers of a star pump incredible energy into the central core.

With so much energy available, endothermic fusion reactions can occur. Such reactions are responsible for the synthesis of all elements having an A greater than 56. The supernova explosions spread the elements around the universe. When new stars form from interstellar gas, some heavier elements are incorporated. Our own planet is made mostly of such recycled stardust.

This is possible because our solar system formed some 10 billion years after the big bang, leaving plenty of time for many supernova events.

Much progress has been made in our understanding of the formation of nuclei—*nucleosynthesis*—in stars during the last several decades. William Fowler was awarded the 1983 Nobel Prize for his work on the problem of formation of heavy elements inside stars. We think we understand the general way elements are formed, but many exciting problems remain to be solved.

Once the elements are formed, many decay into lighter elements, or undergo fission into lighter elements. Some of the original energy available in the supernova can be recovered by allowing a large-A nucleus to fission. We already know how to recover this energy. Nuclear energy generation is the subject of Chapter 14.

SUMMARY

Nuclear energy arises from the binding of protons and neutrons into a nucleus. In going into the nuclear energy well more deeply, each nucleon gives up some energy. As a result, the larger nuclei have masses smaller than the sum of the masses of their constituent nucleons. The difference is the binding energy of the nucleons in the nucleus.

Since larger nuclei have more nucleons, they have a greater binding energy simply by virtue of the greater number of nucleons. We remove this spurious dependence by looking at the average binding energy per nucleon. When this quantity is plotted against the total nucleon number, the curve reveals regions where fusion allows nucleons to become more stable (A<56), and regions in which fission allows nucleons to become more stable (very large A).

When a large-A nucleus fissions, it breaks into two smaller pieces. Since the number of neutrons per proton increases with increasing A, a large-A nucleus that breaks into two smaller-A nuclei produces a few extra neutrons. This allows further fissions of fissile nuclei to occur, producing more neutrons, producing more fissions, producing more fissions ... all in an eyeblink. This is an uncontrolled fission reaction, a nuclear explosion. In reactors, moderators and control rods absorb extra neutrons, allowing a controlled nuclear reaction. Controlled fusion in the stars provides light and energy to the universe (and to incidental planets). Uncontrolled fusion in supernovae leads to nucleosynthesis.

REFERENCES

1. F. J. Shore, *Phys. Teacher* 12 (1974): 327.
2. S. Glasstone, *Sourcebook on Atomic Energy* 3rd ed. (New York: Van Nostrand Reinhold Co., 1967).
3. R. F. Post and P. L. Ribe, *Science* 186 (1974): 397.

PROBLEMS AND QUESTIONS

True or False

1. It is likely that both nuclei from a fission will have about the same mass.
2. There are never more protons than neutrons in a nucleus.
3. Elements such as xenon, cadmium, and plutonium are produced in the centers of stars in the normal nuclear processes taking place there.
4. Since the strong force is so strong, it must be stronger than all other forces affecting nucleons, even when the nucleons are meters apart.
5. The binding energy of each succeeding nucleon added to a nucleus is the same (as given in Figure 6.2).
6. The electron-volt (eV) is a useful energy unit to use when discussing ionization of atomic electrons.
7. The MeV is the appropriate energy unit to use in discussing nuclear fission processes.
8. Most energy from a nuclear fusion reaction appears as kinetic energy of the nuclear particles.
9. The total binding energy of heavier elements is much greater than that of lighter elements.
10. The half-life of an element that decays is the time required for half of the element to have decayed.

Multiple Choice

11. Fissile uranium that is now on Earth originally came from
    a. normal chemical reactions in stars.
    b. normal nuclear reactions in stars.
    c. events known as novas.
    d. events known as supernovas.
    e. the vicinity of the galactic center.
12. About how much energy is released in a fission?
    a. 10 MeV
    b. 20 MeV
    c. 20 J
    d. 200 MeV
    e. 200 J
13. Which of the following best estimates how much energy is typically released in fusion of hydrogen nuclei with other hydrogen nuclei or other close-lying nuclei?
    a. 0.1 MeV
    b. 10 MeV
    c. 20 J
    d. 200 MeV
    e. 200 J

14. Which of the following is *not* used to detect radioactivity?
    a. Unexposed film   d. Van de Graaff detectors
    b. Geiger-Müller tubes   e. All of the above
    c. Cloud chambers

15. Which of the following types of radiation is the most penetrating?
    a. Microwave radiation   d. Gamma radiation
    b. Alpha radition   e. Dead skunk radiation
    c. Beta radiation

16. Some nuclei are changed by a reaction known as electron capture, in which
    the nucleus absorbs one of the atomic electrons. Which of the following cor-
    rectly characterizes the atomic number of the new nucleus with respect to that
    of the present nucleus?
    a. Increases by 2   d. Decreases by 1
    b. Increases by 1   e. Decreases by 2
    c. Stays the same

17. Some nuclei decay by alpha particle emission. Which of the following correctly
    characterizes the atomic number of the new nucleus with respect to that of the
    present nucleus?
    a. Increases by 2   d. Decreases by 1
    b. Increases by 1   e. Decreases by 2
    c. Stays the same

18. Which of the following would probably *not* be very dangerous to ingest in
    very small amounts?
    a. tritium, ^3_1H; half-life, 12.3 years   d. $^{15}_8O$; half-life, 124 seconds
    b. $^{90}_{38}Sr$; half-life, 28.9 years   e. $^{235}_{92}U$; half-life, 4.5 billion years
    c. ^9_4Be; half-life, 53.4 days

19. Which region of the curve of Figure 6.2 is most likely to be involved in a fu-
    sion reaction?
    a. $A > 20$ d. $A < 200$
    b. $A < 50$ e. $A > 200$
    c. $100 < A < 150$

20. Designate the region for $A < 56$ as X, for $56 < A < 140$ as Y, and the region
    for $A > 140$ as Z in Figure 6.2. Which of the statements given below is true?
    a. In fission, elements from X d. Elements in region X can
       combine to form elements in   give up energy by combining
       Y.   to form other elements also
    b. In fission an element from Y in X.
       breaks up into two elements,   e. In fusion, elements in Z
       one from X and one from Z.   break up into two elements,
    c. Elements in regions Y and Z one from X and one from Y.
       are produced by fusion reac-
       tions in the sun.

Discussion Questions

21. Explain how it is advantageous energetically for a nuclear fusion process to
    occur, and indicate the region of the binding energy curve involved.
22. Explain why it is so difficult to develop fusion reactors.
23. What effect would you expect the perfection of an energy-cheap process for
    enriching uranium to have on the economy of a nuclear fission energy genera-
    tion facility?

7

THE EFFICIENCY OF ENERGY GENERATION
AND THERMODYNAMICS

In our world, machines are not 100 percent efficient. Most machines evolve thermal energy through friction. The study of thermodynamics allows an understanding of the limitations of efficiency for machines. Thermodynamics deals with the study of thermal energy and the transfer of energy to and from particular systems. An understanding of energy conservation and entropy is essential in the choice of suitable strategies for producing energy.

KEY TERMS   *thermodynamics*
temperature
Celsius scale of temperature
thermal energy
internal energy
Kelvin scale of temperature
first law of thermodynamics
conduction
convection
radiation
specific heat
latent heat of fusion
latent heat of vaporization
Brownian motion
statistical phenomena
evaporation
second law of thermodynamics

order
disorder
entropy
thermodynamic efficiency
Carnot cycle
second-law efficiency
waste heat

Chemical change by burning is involved in most of the methods we use to generate energy. We burn oil and gas to heat our homes, we burn gasoline in our cars, and we burn coal, oil, and gas in our power plants. Most of our energy-generating methods are inherently inefficient. Some ways are less inefficient than others. To understand the limitations on efficiency in such processes, we must study the effects of temperature changes—*thermodynamics.*

TEMPERATURE

We know that the temperature of a body is related to the "hotness" of the body. Temperature is a useful concept because it can be defined macroscopically (in the large): if two systems are at the same temperature as a third system (say, a thermometer), then they are at the same temperature as each other and are in thermal equilibrium with each other. We therefore have a way to *measure* temperature. (Measurement is near and dear to the scientist's heart!) We can bring a thermometer to a system, allow it to reach a state of equilibrium with the system, and then read the temperature.

Temperature is normally measured in Celsius degrees, which are of such a size that the interval between the temperatures at which water freezes and boils at sea level is exactly 100 degrees. The freezing temperature is set to be 0°C in this scale. In the United States, temperature is also measured in Fahrenheit degrees; water freezes at 32°F and boils at 212°F at sea level on this scale.

Since temperature is related to the amount of "hotness" of a body, and since heat transfer can do work (for example, a metal at high temperature can boil water), we must consider energy in connection with heat. By *heat* we refer to energy in transit from one object to another. According to this definition, an object cannot have a "heat content." When people use this expression, they are probably referring to how hot—what temperature—a body is. What does the temperature measure, then, if not "heat content"?

An object at the South Pole is most definitely a cold object. But, by comparison to an object on the planet Pluto, it is hot. The difference between a wooden block at the South Pole and that same wooden block on Pluto would not be obvious to an observer who only looked at it. Its color and shape would be the same. The difference lies inside the blocks. In wood, as in other solids, the atoms are fixed relative to other atoms by electric forces between them. However, if there is energy available inside the wood block, it will appear as kinetic energy of the atoms; they can move back and forth as if connected to springs. Each atom in the wood is moving. Some atoms have more kinetic energy than others. The total energy inside the block is distributed among all its atoms. We call this randomized kinetic energy *internal energy* or *thermal energy*. Temperature is a measure of this internal energy. A wood block at the South Pole has more internal energy, and

thus a higher temperature, than a similar block on Pluto.

An object at 0°C has thermal energy, because we know that colder temperatures exist. We shall use a scale having degrees of the same size as Celsius degrees, but beginning from the place that, in theory, should have zero internal energy. (Actually, this zero temperature is not attainable because substances change their form—the scientist would say change their *phase*—when sufficiently cold, and because there are subtle distinctions between forms of particles, which results in the existence of what is called *zero-point energy*.) This temperature scale we have chosen to use is called the *Kelvin*, or absolute, temperature scale: its zero is at −273°C. On the Kelvin scale, ice freezes at 273 K (0°C), room temperature is near 300 K (27°C), and water boils at 373 K (100°C).

First Law of Thermodynamics

We saw that heat is energy in transit from one body to another at a different temperature. If we push on a piston to compress a gas, we do work on it. A gas that expands can push on a piston, and in so doing, works on its surroundings.

If we transfer heat to the system but do no work on it, we increase the amount of energy in the system—its internal energy. If we do work on the system, while adding no heat, we also increase the internal energy. If we let the system do work while adding no heat, or if we take out heat and do no work, or if we take out heat and do work, the internal energy of that system must decrease. We summarize this statement of energy conservation as

(heat absorbed by the system) −
(work done by the system)
= change in internal energy of the system.

The signs shown are important. They are chosen assuming that heat is absorbed and that work is done *by* the system. If heat is given off *by* the system, the first term in parentheses is negative; if work is done *on* the system, the second term in parentheses is negative. Denoting heat absorbed by ΔQ, work by W, and internal energy by ΔU, the relation just expressed in words may be written

$$\Delta Q - W = \Delta U.$$

This particular statement of energy conservation is known as the *first law of thermodynamics*.

HEAT TRANSFER

We know that if two glasses are stuck, one inside the other, we can get them apart by running hot water on the one on the outside. Why? When we increase the temperature of the glass, the atoms vibrate faster. They also travel out further from their position of equilibrium. Since each atom takes up a little more space, the glass expands when heated. Thus, the atoms in the glass are forced to move through a certain distance, and work is done. *Because* there is a difference in temperature, heat can be converted into work, or internal energy, or both. This is a general characteristic of heat transfer: There must be a temperature difference.

The transfer of heat by *conduction* occurs in situations in which there is a temperature difference and there is a solid material between the two temperatures. A window during winter is roughly at room temperature on the inside and outdoor temperature on the outside. (There is a "boundary layer" of cooler air just inside the window and warmer air just outside the window. The boundary layer acts as an insulator and cuts down the rate of heat transfer to a certain extent. The actual conduction depends on the size of the boundary layer, which is determined by the amount of wind outdoors and air circulation indoors.) Even if there are no air leaks, heat would be conducted through the window from the inside to the outside of the room because of the temperature difference. Heat may also be conducted through walls, through the bottom of a pan from a heating element or flame, through a coffee cup, and so on.

Heat may also be transferred by hot material as it moves; in this case, we say it is transferred by *convection*. In forced air furnaces, gas or oil is burned to heat air, and the hot air is pushed by a fan out into the heat registers. When this hot air touches other air or material objects, heat is transferred from hot air to colder air or colder objects. If a person wears a strong perfume and walks through a room, the person moving through the air moves the air as well. In this case, the moving air carries the smell with it. The smell is mixed by convection.

The third (and final) way to transfer heat between bodies at different temperatures is by *radiation*. In contrast to the other two methods, transfer of heat by radiation takes place even in the absence of material. The sun transfers heat to the Earth by radiation. We shall discuss this method of heat transfer in more detail in the first chapter on solar energy, Chapter 15.

Specific Heat Capacity

The heat capacity of an object is the ratio of the amount of heat transferred to an object to the rise in the temperature of the object. For example, a pot of water on a stove, a bathtub full of water, and Lake Erie are all water. Naturally, Lake Erie has a greater heat capacity than the bathtub, which in turn has more than the pot of water, simply because it contains more water. Said another way, an amount of heat that could boil the pot of water would hardly raise the bathtub water's temperature and would disappear into Lake Erie with barely a trace. Each body has its own heat capacity. But as scientists, we believe that fresh water should be the same the world over (ignoring minor variations in dissolved mineral content), just as iron should be the same the world over. Differences arise from differences in the amount of matter, which as we saw in Chapter 3 is measured by the mass. The greater the mass, the greater the heat capacity.

If we divide the heat capacity of any object by its mass, we should have something that is characteristic of the composition of the object, specific to it. The heat capacity of Lake Erie divided by the mass of water in Lake Erie should be the same as the heat capacity of the water in the pot divided by its mass. We call this quantity the specific heat capacity (or *specific heat* for short). The specific heat of water is 1 kilocalorie per kilogram per Celsius degree (or 4186 joules/kg/C°, or 1.16 × 10⁻³ kWh/kg C°). Most substances have specific heat capacities much smaller than that of water. In Table 7.1, we present several specific heats of substances. (These are specific heats taken for fixed pressure.)

Latent Heat

Matter usually appears to us in one of three states or phases: solid, liquid, or gas. Atoms in a solid are fixed in place relative to other atoms. Atoms in a liquid, while they are pulled toward adjacent atoms, are not locked into place. Atoms in gases are essentially free from the influence of one another.

When we have a glass of ice water, the water and the ice (solid water) are at the same temperature, 0°C, because they are in contact. The two phases differ but can coexist at the same temperature. We know that the water will remain very cold until all the ice is gone. The reason is that it takes energy to "unlock" the solid, to pull the molecules far enough apart so that they are somewhat free to move. We must pay an energy price to free the molecules: This price, measured in energy per kilogram required to change the phase from solid to liquid, is known as the *latent heat of fusion*. The latent heat of fusion of water is 333 kJ/kg.

In a similar fashion, steam (water vapor) and liquid water may coexist at 100°C at sea level. The energy price paid per kilogram to unbind the liquid molecules and make them free gas molecules is called the *latent heat of vaporization*. Water's latent heat of vaporization is 2.25 MJ/kg.

Latent heats and specific heats are very important to know in dealing with the measurement or calculation of heat transferred when objects of different temperatures are brought together and finally reach some equilibrium temperature.

TABLE 7.1

Mean specific heats of water and various elements (at constant pressure).

Element	Specific Heat, kJ/kg°C
Water	4.186
Aluminum	0.900
Carbon (graphite)	0.712
Carbon (diamond)	0.519
Copper	0.268
Helium	5.191
Hydrogen	14.274
Iron	0.444
Lead	0.159
Mercury	0.139
Oxygen	0.917
Platinum	0.133
Silver	0.237
Tin	0.226
Zinc	0.388

SOURCE: Reprinted with permission from *Handbook of Chemistry and Physics*, 32nd ed. Copyright 1959 by CRC Press, Inc., Boca Raton, FL.

THE STATISTICS OF HEAT TRANSFER

Energy conservation still does not tell us all we must know about thermal systems. To understand the crucial concept of time variation in the system, consider the following example: You have prepared your supper and just put it on the table. The phone rings, and a long-lost friend launches a reunion with you over the phone. Half an hour later you sit down at the table. What does your knowledge of thermodynamics tell you?

1. Objects have a measurable temperature.
2. Energy is conserved.

As far as these precepts are concerned, you might expect to find your potatoes, coffee, and roast beef hot, and your ice cream cold. Energy is certainly conserved this way.

The dismal truth is that the potatoes, coffee, and roast beef are now rather cold, and the ice cream is a sort of cool puddle in the dish. If we left it long enough, everything on the table would be at the same temperature.

To see what is happening, consider the roast beef. It sits on a plate in the atmosphere. The roast beef and the plate are made of atoms. These atoms are chemically bound or strung together in such a way that we know the plate is a plate and the roast beef is roast beef. But some of the atoms of the plate and the roast beef touch each other. Since the roast beef is hot, the roast beef atoms will be vibrating at a greater speed than the plate atoms. When the atoms of the roast beef and the plate touch, the roast beef atoms, which are moving faster *in general,* hit atoms in the plate, which are moving more slowly *in general.* When they collide, the roast beef atom *generally* rebounds slower and the plate atom faster than each was moving originally. The plate atom, in turn, hits other slower plate atoms, *in general* ending up going more slowly; the newly faster plate atom transfers some of its vibrational energy to other slower plate atoms, and the cycle continues to spread energy throughout the material. A similar exchange occurs with the atoms of the atmosphere.

Meanwhile, the slower roast beef atoms near the plate edge get hit by faster roast beef atoms, *in general* picking up vibrational energy again. Subsequently, the now-faster roast beef atom hits the now-slower plate atom. Again, *in general,* the roast beef atom comes off slower, the plate atom faster. And so it goes.

You must have noticed by now that I keep saying "in general." This is for several reasons. Not all atoms in a hotter material are faster than all atoms in the cooler material. As we have seen, the atoms keep colliding with one another, trading their energy. Thus, in our example, there are a few plate atoms faster than most roast beef atoms. Further, we cannot be absolutely certain that energy is transferred from the faster atom to the slower atom. Since all the atoms are vibrating, or moving back and forth, they must stop occasionally. An atom of lesser energy at its maximum speed could collide with an atom of greater energy when it is at rest, although this is unlikely

to happen. Thus, in most collisions, energy is transferred from the more energetic to the less energetic material (from hot to cold). In many materials (such as metals), electrons are "free" to move about within the material and can hasten such a transfer of energy.

The point to be made here is that the cooling of the roast beef by transfer of vibrational energy to the plate is not inevitable. But with the large number of collisions, it is extremely unlikely that the roast beef atoms would end up vibrating faster (that is, become hotter) than they were at the start, and that the plate atoms would become slower on average than they were at the start (that is, colder). The transfer of heat by contact, or conduction, is seen to be a *statistical* phenomenon.

Any *particular* outcome of any *particular* collision will not result in an inevitable transfer from faster to slower. However, because it is so much more probable that such a transfer will happen, we can say that in virtually all cases it *will* happen, since a material is composed of a *large number* of atoms. For each atom of this large number, there is a high probability that energy will be transferred from faster to slower at each collision. With all atoms taken into account and all collisions occurring each instant, there is an overwhelming tendency to transfer energy to the colder material.

In the same sort of probabilistic way, we can turn to insurance tables to find that there is a certain risk of heart attack in men of age 55. This does not mean that we can say that any *particular* man will have a heart attack at age 55. The collection of all men of that age will have approximately the predicted number of heart attacks, though, because there are so many men in the age 55 category. Because of the *large numbers* involved, we can make firm macroscopic predictions.

In the transfer of heat, the original energy gets randomly spread throughout a substance. Each atom gets its bit of additional vibrational energy (on average) or loses its bit of extra vibrational energy (on average). There is no way to distin-

THE MAXWELL-BOLTZMANN DISTRIBUTION

Gases are composed of atoms or molecules. These atoms or molecules do not really interact with each other except through collisions. In many cases, we may think of a gas as a collection of tiny billiard balls flying through space, hitting one another again and again. Even if we were to think that all atoms or molecules had the same speed to begin with (although we do not), the constant collisions would result in a spread of many speeds. Some atoms could have very high speeds, others low ones. In real gases there is a distribution of speeds (Figure 7.1). This distribution is called the Maxwell-Boltzmann distribution, and it depends on temperature, as shown. The high temperature curve has proportionally many more fast molecules or atoms than the low temperature curves. As the temperature rises, the highest point on the curve is pushed out to higher v, and the maximum is pushed down toward the axis. All curves shown in the figure have a similar shape.

It is also possible to show that the average kinetic energy of an atom or molecule, $<\frac{1}{2}mv^2>$, is directly proportional to the absolute temperature T: $<\frac{1}{2}mv^2> = 3/2\ kT$, where m is the atomic or molecular mass and k is a constant known as Boltzmann's constant. The average of the square of the speed increases with temperature, just as is seen in the curves.

We can actually experience the effects of the Maxwell-Boltzmann distribution by performing a few evaporation experiments.

What is evaporation, and why does it require energy? Recall that the atoms or molecules of all materials are in constant motion. Materials of course differ. We are generally familiar with three states, or phases, of material: solid, liquid, and gas. In order to understand evaporation, we must look at the differences among the three phases.

In solids, the molecules are cemented into their relative positions despite their incessant motion. There is a structure that is maintained. We are familiar with many solids that have crystalline structure. The symmetry of the crystal reflects the symmetry of the bonds between the atoms or molecules composing the crystal. Most

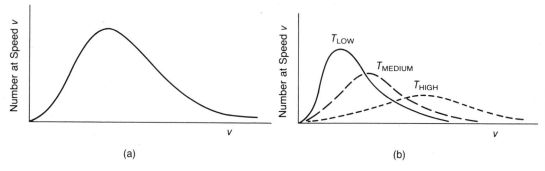

(a) (b)

FIGURE 7.1

(a) The distribution of proportions of the total number of particles in terms of the speed. (Maxwell-Boltzmann distribution). (b) The distribution of proportions of the total number of particles in terms of the speed for different temperatures. The *shape* of the distributions is identical, but distorted because the curves must cover identical areas.

solids do not exhibit such a clearly ordered structure, but they are structured nevertheless.

In liquids, the interatomic or intermolecular forces attracting the atoms or molecules to each other are weaker than those in solids. This allows the atoms or molecules to move somewhat in relation to one another, while still keeping the liquid together. This weakened interatomic force is the reason that liquids can take on the shapes of their containers. The attractive force between liquid molecules is evidenced by the phenomenon of surface tension. The beading of water, for example, is due to surface tension. Inside a liquid, and away from the boundary, molecules are surrounded by other molecules of the same type, and are thus attracted in all directions equally. This is another way of saying that there is no net force at all on molecules *inside* the liquid; those molecules are in equilibrium. For the molecules on or very near the liquid surface, however, attractive forces come only from inside the liquid to act upon them. Thus, they are attracted back to the liquid surface should they try to depart.

In gases, such as water vapor, the molecules are essentially free to move wherever they are going. That is, the attractive forces are essentially negligible. There is no surface tension.

In all these phases, molecules are dancing about in some way. In the solid, the molecules vibrate in various ways about their fixed positions in the solid lattice. In liquids, this dance of the molecules also occurs. Even in transparent liquids, we are not able to see the motion of the liquid molecules. In some liquids that contain very large molecules, the large molecules can be kicked about randomly under the impact of the smaller, invisible liquid molecules. If the molecules are large enough to be seen, we can then see the effect of the random motion on the larger molecules. This visible motion is termed *Brownian motion*.

The properties of materials at temperatures above absolute zero—that is, those materials exhibiting the molecular motion—are described by distributions giving the relative numbers of molecules at the different speeds of motion possible for the material. The Maxwell-Boltzmann distribution of Figure 7.1a illustrates the properties of the distribution of speeds in typical gases or liquids: There is a range of speeds of molecules, and there is a most probable speed at a given temperature. Thus, in any gas or liquid, there are many molecules going fast and many going slow. In fact, the distribution shows that we can measure the temperature by determining the average molecular speed (see Figure 7.1b).

Molecules approaching the boundary of a liquid generally are trapped back into the liquid by the surface tension. However, the very fast molecules may be able to penetrate the surface barrier and escape the liquid altogether. The more surface available, of course, the faster the rate of escape. The rate also increases as the temperature increases because of the boosting of molecules to higher speeds.

We may distinguish two sorts of evaporation. In liquids with a relatively small surface area, the liquid absorbs heat from the surroundings to keep the *temperature* of the liquid fixed as the faster molecules escape. The absorbed heat speeds up some molecules, restoring the original Maxwell-Boltzmann distribution. The total number of molecules in the liquid decreases, but the distribution—which depends only on temperature—remains the same. In the second sort of evaporation, the surface area is relatively large. Thus, the faster molecules leave the surface rapidly enough that the liquid cannot absorb enough heat to restore the original distribution. In this case of non-equilibrium evaporation, the distribution quickly becomes truncated. (It looks as shown in Figure 7.2). Since the absolute temperature measures the average kinetic energy of a gas, a departure of high-kinetic-energy molecules, shown in Figure 7.2, will cause a thermometer to record a lower temperature.

This effect is enhanced when the material in question has a smaller surface tension. For example, the effect is much greater for alcohol than for water. Conversely, the temperature drop is much smaller for a liquid with a large surface tension. Lubricating oil exhibits a tiny temperature drop compared to water.

Gasoline spilled on our hands makes the skin feel cool because the gasoline molecules evaporate rapidly. This effect can easily be seen with tap water, a paper towel, and a thermometer able to register temperatures around 10°C to 20°C. Wet the towel and wrap it around the bulb of the thermometer. You should be able to observe a decrease in temperature of a few degrees.

The Maxwell-Boltzmann distribution can also be used to explain how the hydrogen and helium originally in Earth's atmosphere disappeared, while the nitrogen and oxygen show little sign of imminent departure. Recall that $<\frac{1}{2} mv^2> = 3/2 \, kT$. The atmosphere's temperature changes little on average from year to year; for a fixed T, the smaller m is, the larger $<v^2>$ will be. The square root of $<v^2>$, called the root-mean-square speed (v_{rms}) is a measure of how fast the atoms or molecules are whizzing around. The root-mean-square speed for hydrogen is $\sqrt{14}$ times as great as that for nitrogen because the mass of the nitrogen molecule is fourteen times that of hydrogen. Similarly, the root-mean-square speed of helium is $\sqrt{7}$ times as great as that for nitrogen because the mass of a nitrogen molecule is seven times that of a helium atom. In fact, the root-mean-square speeds in Earth's atmosphere of hydrogen, helium, nitrogen, and oxygen are, respectively, 1.93 km/s, 1.37 km/s, 0.52 km/s, and 0.48 km/s. Escape speed for the Earth is only 11.2 km/s. *Escape speed* is the speed an object has to go to escape Earth's gravity. A rocket ship to Mars must go at a speed greater than 11.2 km/s to escape from Earth. If a gas molecule has a speed in excess of 11.2 km/s, it will escape totally from Earth. For this reason, there is almost no hydrogen or helium in Earth's present atmosphere.

Because of the rapid (exponential) falloff in number of particles with increasing speed, the distribution tells us that there are proportionally about a million times more hydrogen molecules with speeds exceeding 11.2 km/s than nitrogen molecules. Because of the large volume of the atmosphere, the loss of hydrogen molecules proceeds in a quasi-equilibrium fashion, with slower molecules gaining energy and the distribution preserving its shape. After a geologically short time, almost all hydrogen will have escaped, while practically no nitrogen will have been lost.

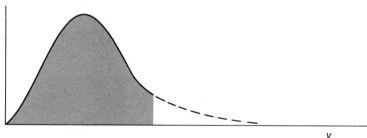

Relative
Number
of Liquid
Molecules
in a
Speed
Interval

v

FIGURE 7.2
This illustrates what happens to a distribution when almost all the faster particles are suddenly removed. The effect is a cooling, since temperature is a measure of average kinetic energy of the particles.

guish or idiosyncratically label a constituent atom or molecule. The beef cools, the ice cream melts, and the plate and the air become slightly warmer.

Without this large number of atoms, or the large number of collisions, we could not be certain of the outcome. That is why our discussion focuses on *statistics*. In fact, it is this statistical probability phenomenon that allows us to distinguish future from past—to label "time's arrow."

Several phenomena prove beyond any reasonable doubt that collisions occur between atoms (at least in gases and liquids). In 1821, Robert Brown, a botanist, observed under a microscope that pollen particles in water jiggled. The explanation of this *Brownian motion* was provided by Albert Einstein (see box). The jiggling of the pollen particles, which are visible, is due to their collisions with water molecules, which are invisible. Smoke particles in still air also exhibit Brownian motion, reflecting smoke particle collision with air molecules. The phenomenon of *diffusion,* in which small amounts of impurities mix thoroughly in gases or liquids, takes place because the molecules are in constant motion. An example of this is provided by the release of a smelly substance (for example, cheap perfume) in a still jar. Even if there is no gross air motion—caused, say, by someone moving through the room dragging the smell along—the smell will slowly but surely penetrate every corner of the room. Another example of diffusion is the mixing of a drop of food coloring put into still water. The color will spread out from the drop and, after some time, produce water of uniform color. In both of these examples, collision with the gas or liquid molecules is responsible for the mixing. The drop will not spontaneously unmix, nor will the smelly substance become concentrated in one part of a room. Again, we can see "time's arrow."

THE SECOND LAW OF THERMODYNAMICS

Heat transfer from a warmer to a colder body resembles transfer of water from a reservoir to a lower level. Work can be obtained from water dammed in a reservoir by allowing it to fall the height of the dam and run turbines in powerhouses, thereby producing electricity. On top of the dam (in the reservoir) all the water is at one level. There is *no way* to have water in the reservoir do work while remaining at its single level. There must be a difference created in height to allow gravitational potential energy to be converted into work.

We can imagine that a large body at a given temperature is a temperature reservoir. Work is produced by heat flowing "down" from a reservoir at a higher temperature to a reservoir at a lower temperature (just as water flows from a higher to a lower elevation). It is not possible to use the water in the top reservoir to get work, even though it contains gravitational potential energy in immense quantities; likewise, it is not possible to get work from only one temperature reservoir. This temperature reservoir contains gigantic quantities of thermal energy, but it is not available for use.

Air molecules form an immense reservoir of energy. However, it is not possible to run cars on this energy. We must transfer energy from a higher to a lower temperature in order to do work. We therefore burn gasoline to produce a high enough temperature. We have explained that temperature is a measure of internal energy or of randomized vibrational energy. Recall that energy is conserved: If work is done, then heat has been transferred, or internal energy has been changed or both (first law of thermodynamics). Our discussion of the irreversible changes of the roast beef dinner has prepared us for one way of stating the *second law of thermodynamics*: Systems isolated from the rest of the universe will move toward equilibrium with their surroundings.

When there is equilibrium, there can be no temperature difference. If there is no temperature difference, there is no transfer of heat. There is, then, another way to state the second law: It is impossible for a machine to take heat from a reservoir at a certain temperature, to produce work, and to exhaust heat to a reservoir at the *same* temperature.

ORDER AND DISORDER

Still another way to phrase the second law is to say that processes go in such a way that disorder increases. Before this statement makes sense, we must decide what we mean by *order*. Order is set by rules. The fifty-two playing cards in a deck, for example, look very similar to one another to an untrained eye. As we learn more, we are able to distinguish two sets of equal size: red cards and black cards. You could order by color: black, red, black, red; or black, black, red, red. Closer inspection reveals that there are two sorts of red cards—diamonds and hearts—and two sorts of black cards—clubs and spades. There could be an arbitrary decision that, say, spades come before hearts, hearts before diamonds, diamonds before clubs. This is an ordering of the suits you would use in playing the game of bridge. You could now order by the "value" of the suits: clubs, diamonds, hearts, spades, clubs, diamonds, hearts, spades, etc.

Someone could also notice that most cards have numbers. The cards may be ordered according to number. As we know, all these orderings are important if we are to be able to play cards well. The orderings are completely arbitrary, and the rules are based on the arbitrary orderings; once we accept them, we can play bridge, poker, or whatever.

A newly opened deck of cards has the cards in a standard order. By moving just one card, we have disordered the deck. If we then move another card, chosen at random, the deck will become even more disordered. Only if we chose the card first moved and happened to restore it to its original position would the deck become more ordered (this would happen by chance only about once every 2500 tries). As more cards are moved, the probability for restoration of the original order gets even smaller. This is another example of the statistical basis of the second law of thermodynamics.

In talking about the second law of thermodynamics and order or disorder, we must define our (arbitrary) order. Only when this is done can we discuss changes in order. There is an overwhelming probability that order will be lost, whether we are talking about the moving of cards or the collisions of molecules. In every real-world case—in which the number of ways to increase disorder is so much greater than the number of ways to increase order—disorder will increase.

Remember the roast beef dinner? In the beginning, there was an order, a hierarchy, of objects by temperature. The roast beef was hot, the plate and air were at room temperature, and the ice cream was cold. At the end, all of the vibrational energy was randomly spread out, with the ice cream being about the same temperature as the roast beef and the air and the plate. There was no longer a temperature order or hierarchy. That original *order* was now lost; in its place was *disorder*. And, so it is with other, potentially more complex, systems.

Entropy

By defining order, we have also defined disorder. It is possible to identify "disorder" quantitatively with something known as *entropy*. We can then say that *the entropy in the universe always increases* in any process. (The only exception occurs when a system is already at equilibrium. In this case, any process taking place within the system will leave it in equilibrium and thus leave its entropy unchanged.) Just because entropy increases does not mean that there can never be a localized entropy decrease. Such local decreases are possible as long as the entropy of the universe as a whole increases in the process. The inexorability of the increase in entropy in the universe is still another way to express the second law of thermodynamics.

Examples of a large entropy increase occur in Humpty Dumpty's fall, or in the breaking of an egg yolk, or in the mixing of cookie batter. Note that all these processes are irreversible. Some deposits of natural gas contain substantial amounts of helium (as high as 8 percent in one case). Helium occurs naturally in the atmosphere in a concentration of only five helium atoms per million molecules of air. When this high-helium natural gas is burned and the helium is dissipated in the atmosphere, it goes from a readily-available resource to one that is extremely hard to obtain. This increase in entropy sparked the U.S. Interior Department to experiment with a helium conservation program (1).

It is possible to get helium from air. One must use enough energy to liquefy the oxygen and nitrogen in air to retrieve the helium. This means that air must be cooled to $-193°$ C. It requires much more energy to obtain helium from air than from the high-concentration natural gas (5300 kWh/m^3 as against 3.5 kWh/m^3) (2). The dissipation is virtually irreversible.

The system goes from concentrated (high order, low disorder) to dilute (low order, high disorder). That is, its entropy increases. The helium conservation program stored 793 Mm^3 before it was terminated in 1971; about 560 Mm^3 has been wasted since. Sales of helium have been growing because of the special uses in superconductor technology and pure research. In 1982, about 37 Mm^3 of helium was sold commercially (2). You can see how costly in energy it is to undo the entropy increase caused by mixing the helium into the air when the helium-rich natural gas is burned.

How can we have cold ice cream if all processes involve an increase in entropy? A freezer surely has a lower entropy than its surroundings. But the freezer has to be cooled by a motor that exhausts heat into the air. It takes energy input to decrease entropy locally. Generation of this energy causes an entropy increase in the universe, which more than compensates for the entropy decrease in the freezer, and increases the *total entropy* of the entire universe.

Entropy may be decreased in a system by input of energy from outside the system: A system in equilibrium may be put into a state of disequilibrium by doing work on it. Likewise, as systems return from a state of disequilibrium to one of equilibrium with their surroundings, useful work may be done. For example, when we compress air into a tank, we create a pressure disequilibrium by doing work on the air. We now have a pressure difference where there was none before, at an energy cost. This pressure disequilibrium could now provide useful work by running a power tool, raising a hydraulic lift, or whatever.

Another familiar example of entropy increase as a transition from order to disorder is that of a sponge. A sponge is an object riddled with tiny holes. When it is slightly wet, water can travel inside the sponge by a phenomenon known as capillary action. In capillary action, water in thin tubes rises due to water surface tension, and water therefore can be drawn up in the sponge. Suppose there is a big spill and a slightly damp sponge. While this might seem to be a disordered system, in fact it is *ordered* with respect to the degree of wetness. As water flows into the sponge, the *wet–dry order* vanishes and everything becomes uniformly wet. At this point, no more water (or milk or whatever) is absorbed by the sponge. Nothing more happens until energy is used to restore the wet–dry order (by manually wringing out the sponge, for example). Then the sponge can be used again in the same manner.

MAXIMUM THERMODYNAMIC EFFICIENCY OF A SYSTEM

It is now possible to speak about the theoretical efficiency of a process. At a given temperature, there is a total amount of energy involved in the vibrations of atoms in a material. In general, the ratio of the work done in some process to the total energy available is the *efficiency* of that process:

$$\text{efficiency} = \frac{\text{work output}}{\text{total available energy}}$$

Here the total available work is the total amount of internal (vibrational) energy in the material above what is there at absolute zero (the so-called zero-point energy).

There are exact methods for proving the relation of efficiency to temperature; these were delineated by the founder of the science of thermodynamics, the French engineer Sadi Carnot. Recall that the internal energy of a substance is proportional to temperature. The constant of proportionality varies from substance to substance (it is called the heat capacity at constant volume, as previously discussed). We usually assume that the heat capacity for a certain substance is constant for all temperatures. The internal energy is proportional to kelvin temperature:

$$\text{internal energy} = (\text{heat capacity}) \times (\text{absolute temperature}).$$

It is then plausible that efficiency may be expressed as a ratio of temperatures.

Using the absolute temperature scale, we can consider a special process (called a Carnot cycle) in which work is done in such a way that all parts of the process can be reversed and all intervening steps retraced. A prototypical heat engine is shown in Figure 7.3. The Carnot cycle is an idealization and is never encountered in the "real world." In particular, no real engine is reversible. This is still another way to express the second law of thermodynamics.

The cycle (see box for a complete description) consists of absorbing heat from a high temperature reservoir, B, at absolute temperature T_B, doing work, and disposing of the remaining heat at the low temperature reservoir, A, at absolute temperature T_A. This very special cycle has an efficiency of

$$(T_B - T_A)/T_B.$$

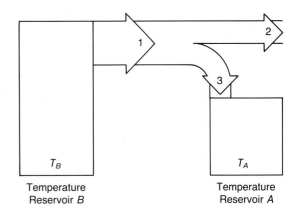

1 Energy Absorbed from Reservoir at High Temperature

2 Output of Useful Work

3 Energy Rejected to Reservoir at Low Temperature

FIGURE 7.3
Schematic representation of a process by which thermal energy is absorbed from the high temperature reservoir, some thermal energy is rejected to the low temperature reservoir, and work is done. The device illustrated works as a heat engine.

Even though it is an *ideal* cycle, the efficiency is not 100 percent, because some energy must be rejected at the lower temperature. Only if all the original internal energy that is available is used can the Carnot cycle be 100 percent efficient; in this case, the low temperature must be 0 K.

In terms of machines, the second law of thermodynamics says that it is impossible to devise a machine that takes heat from a reservoir and produce only useful work with it. In a similar way, a boat could not run on the gravitational potential energy stored behind a dam in a reservoir.

The Carnot engine is a special case of a heat engine. In all heat engines, work is done as the material expands and then is compressed again. In the Otto engine (used in a car), the gasoline burned in one particular cylinder causes the volume of the cylinder to expand and contributes part of its resultant energy to compressing the air in other cylinders. A little later in the cycle,

THE CARNOT CYCLE

The Carnot cycle consists of four successive steps:
1. From its initial state, the working fluid expands at fixed temperature T_B, absorbing heat from its surroundings.
2. The working fluid is removed from temperature reservoir B, and insulated from all outside influence. No heat can be absorbed or given off; such a process is termed *adiabatic*. The working fluid expands adiabatically, and the temperature decreases to T_A.
3. The working fluid is placed on temperature reservoir A. It is compressed at fixed temperature T_A, exhausting heat to its surroundings.
4. The working fluid is removed from temperature reservoir A. It expands adiabatically until it returns to the initial state.

The Carnot cycle is a closed cycle. In the process, energy is absorbed at the high temperature reservoir (B), and a smaller amount of energy is given off at the low temperature reservoir (A). The difference between the energy absorbed and the energy rejected is the amount of work produced by the Carnot engine. If we run the Carnot engine in reverse, the result is a Carnot refrigerator. Real engines are less efficient than Carnot engines. Real refrigerators or air conditioners require more energy input for a given amount of cooling than a Carnot refrigerator.

another cylinder will fire and cause the air in the first cylinder to be compressed in preparation for the gasoline to be burned again. In the Otto engine, the low temperature reservoir is the outside air, and the high temperature reservoir is the combusted hot air and gasoline mixture just after the spark ignites it.

The Carnot efficiency is the maximum *theoretical* efficiency of a heat engine operating between temperature reservoirs at temperatures T_B and T_A. No engine may exceed this efficiency; however, there is no guarantee that it can be attained. The degree of attainment depends on the particular engineering of the heat engine. The automobile engine is about half as efficient as it theoretically could be. The second law of thermodynamics guarantees that the efficiency $(T_B - T_A)/T_B$ is the *absolute maximum* efficiency attainable by any possible contrivable heat engine. This constraint has only to do with the second law of thermodynamics and could be said to preclude the existence of perpetual motion machines otherwise allowed by energy conser-

vation. We shall encounter many examples of Carnot efficiency in later chapters.

If we reverse the cycle for the heat engine just discussed, we have a process that takes in heat from a low temperature reservoir, does work on it, and exhausts the original heat plus the additional work to the high temperature reservoir. Such a process can be thought of as cooling the low temperature reservoir or as heating the high temperature reservoir. A refrigerator is a device for cooling a low temperature reservoir; a prototypical refrigerator (or air conditioner) is shown in Figure 7.4. When such a device is used to heat the high temperature reservoir, we call it a heat pump.

An electric heat pump is more efficient than resistive electric heating. In resistive heating, the thermal energy transferred is Joule heat. In the heat pump, thermal energy is taken from outside air or water (remember, since they are far above absolute zero in temperature, they have a lot of thermal energy). Electric energy is put in to run the pump and compressor, and the sum total of

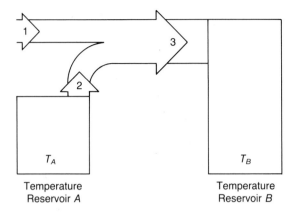

1 Work Input

2 Energy Extracted from Reservoir at Low Temperature

3 Energy Rejected to Reservoir at High Temperature

FIGURE 7.4
A schematic representation of a refrigerator, in which thermal energy is absorbed at the low temperature reservoir, work is done on the system, and the total (thermal energy plus work) is rejected at the high temperature reservoir.

the outside thermal energy plus the work done is delivered to heat the room, the high temperature reservoir. Hence, the electricity provides only a small part of this thermal energy, and we pay only for that small part. Thermal energy from the great outdoors does not directly cost money.

It is clear from these explanations that the higher the temperature of the heated material, the more work that can be done by a given amount of thermal energy (we are here comparing high and low temperatures—a greater amount of material at a low temperature can contain as much thermal energy as a smaller amount at a higher temperature—see box on the Carnot cycle).

SECOND-LAW EFFICIENCIES

Most definitions of efficiency, in contrast to the one we have just developed, do not take the dissipative effects of entropy into account. Such

definitions can be very misleading; they can severely misstate the efficiency of a given process. This problem led a group of physicists involved in a project under the auspices of the American Physical Society to propose a different definition of efficiency, which they call *second-law efficiency* (3). This definition is equivalent to the efficiency as calculated for thermal systems (such as power plants), and it takes the "quality" of the energy into account in other applications. This idea of "quality" of energy is meant to clarify how a given amount of energy differs in its ability to do useful work. Electric energy is of high quality because it may be used easily to do work of many kinds. Room temperature thermal energy is of low quality; it may be used to heat a room, but it is difficult to use it to raise a weight or run a stereo.

The authors of the American Physical Society study define a quantity they call *available work*: the "maximum work that can be provided by a system (or by fuel) as it proceeds (by any path) to a specified final state in thermodynamic equilibrium with the atmosphere" (3). Available work is consumed during a process. The second-law efficiency is then defined as the available work divided by the actual work used in the device for some specified purpose. With this definition of available work, we can make quantitative the distinction between high-quality energy (great deal of available work), and low-quality energy (much smaller amount of available work). Work is of highest quality, as is electric energy. Thermal energy is of low quality.

There are great differences between conventional measures of efficiency (which we may call first-law efficiency), and the second-law efficiencies. The definition of efficiency as the ratio of the work output to the total available energy is a definition of a first-law efficiency. Table 7.2 presents a comparison of first- and second-law efficiencies from Reference 3.

The calculation of second-law efficiencies is an important component of any national policy of energy conservation because it shows how much room is left for improvement (4). A max-

TABLE 7.2
Comparison of efficiencies of household heating systems at typical ambient and heating or cooling temperatures.

Task	First-Law Efficiency	Second-Law Efficiency
Water heat		
Electric	0.75 (0.25)	0.015
Gas	0.50	0.029
Space heat		
At room	0.60	0.028
At register	0.60	0.074
At furnace	0.75	0.145
Air conditioning	2.00 (0.67)	0.045

NOTE: Numbers in parentheses refer to efficiencies, taking into account the loss of a factor of 3 in converting from thermal to electric energy production (see text for explanation of the factor of 3 in efficiency).

imization of the second-law efficiency provides a standard against which proposed improvements can be measured. This standard was used in an action by the Federal Power Commission (5), in which only part of a proposal by an oil company consortium to use offshore gas to run refineries was granted (on the basis of a second-law argument).

For a furnace in a power plant, the available work is the heat of combustion, so that our thermodynamic efficiency and the second-law efficiency are identical. For a solar hot water heater, in contrast, the first-law efficiency is given as the ratio of the heat Q_2 added to the warm reservoir at temperature T_2 to the heat Q_1 taken from the hot reservoir at T_1: Q_2/Q_1. The second-law efficiency is

$$\frac{Q_2}{Q_1} \times \frac{(1/T - 1/T_2)}{(1/T - 1/T_1)},$$

where T is the atmospheric absolute temperature.

Let us calculate the second-law efficiency of a power plant operating at an upper temperature of 500°C. A typical river temperature is about 20°C, and river water will probably be used as a coolant in our imaginary generating facility, so we will take 20°C as the temperature of the lower temperature reservoir. The thermodynamic efficiency is calculated with the absolute temperature, so first we must change Celsius to kelvin temperatures: 500°C is $(500 + 273)$ K = 773 K; 20°C is $(20 + 273)$ K = 293 K. We may then write that the thermodynamic efficiency

$$= \frac{(T_B - T_A)}{T_B} = \frac{(773\text{ K} - 293\text{ K})}{773\text{K}}$$

$$= \frac{480\text{ K}}{773\text{ K}} = 0.65.$$

The idealized, or Carnot, engine operates with a maximum efficiency of 65 percent between the typical generating facility temperatures. All actual engines are far less efficient than the Carnot engine. The actual efficiency of all U.S. generating facilities rose from 30.8 percent in 1960 to 32.3 percent in 1968 (6). As a good approximation, we may take 33 percent (i.e., one third) as the actual power generation efficiency of American fossil-fuel plants; this is only half the theoretical maximum efficiency. The second-law efficiency of a nuclear generating facility is also about 33 percent.

The first-law efficiency is defined differently for different processes, which is one of the reasons that the definition of second-law efficiency is superior. For example, for an electric motor, the first-law efficiency is defined as work out/work in.

A power plant burning fuel energy at a rate of 3000 MW to produce electricity for sale transmits only (⅓) × (3000 MW) = 1000 MW to its customers. The other 2000 MW is exhausted to the environment and turns up as *waste heat*. This is heat of low thermodynamic quality—that is, it will not do work. We will discuss this waste in some detail in later chapters.

We have given energies in kilowatt-hours (kWh) in many places in the text. However, as we have just seen, there is a distinction between electric energy and fuel (or thermal) energy of a factor of 3. We will avoid any confusion in the rest of the book by distinguishing between electric kilowatt-hours (kWh_e) and thermal kilowatt-hours (kWh_t) where that distinction matters:

$$1 \, kWh_t = \tfrac{1}{3} \, kWh_e \,; 1 kWh_e = 3 \, kWh_t.$$

SUMMARY

Heat is energy in transit from higher to lower temperature. Heat may be transferred by conduction, convection, and radiation. In conduction, heat moves through a material that does not move itself (a metal knife held in boiling water will eventually cause burning pain in the fingertips of the knife-holder). In convection, hot material carries heat along with it; for example, hot air from a jet engine can make people in the backwash hot. In radiation, no material medium is required to transfer heat.

Heat is work and work is heat: We call this restatement of the principle of energy conservation *the first law of thermodynamics*. Heat is transferred only if there is a temperature difference. While we cannot measure the "heat content" of a body, we can measure the internal energy, which is the sum of the kinetic energy of all atoms and the potential (binding) energy of the atoms in a material. Absolute temperature is a measure of the internal energy, or "degree of hotness." Atoms in materials may have any energy. The average energy is measured by the absolute temperature, but the individual atoms have energies that fall on a characteristic distribution, the Maxwell-Boltzmann distribution, which describes the number of atoms at a particular energy as a function of its energy.

Entropy is a measure of *disorder*. To define disorder, we must first agree on what we will call *order*. There are many ways possible to define order. Once the order is defined, there is only one way for something to be ordered, and many possible ways for it to be disordered; any random change is likely to decrease order. This tendency to disorder or entropy is called the second law of thermodynamics. The heart of the second law is its statistical character. The second law may be stated in many different ways:

a system tends toward equilibrium with its surroundings;

the entropy of the universe increases or remains the same in any process;

it is possible to decrease entropy locally only if work is brought in from outside the system;

it is impossible to build a machine that takes heat from a temperature reservoir and produces only useful work;

it is impossible for any machine to absorb heat from a temperature reservoir, do work, and exhaust heat to a reservoir at the same temperature; and

all real processes are irreversible.

(This list by no means exhausts the possible alternative statements of the second law.)

Even the Carnot engine, theoretically the most efficient heat engine, must exhaust waste heat to the low temperature reservoir. The efficiency, which is the ratio of the work done (equal to the difference between heat absorbed at the high temperature reservoir and heat rejected at the low temperature reservoir) to the heat absorbed at the high temperature reservoir, is therefore always less than one. In fact it is possible to show that the efficiency of the Carnot cycle is given by $(T_H - T_L)/T_H$, where T_H is the absolute temperature of the high temperature reservoir, and T_L is the absolute temperature of the low temperature reservoir. The second law of thermodynamics guarantees that any real engine has an efficiency less than that of a Carnot engine operating between the same high and low temperatures.

Electric energy can do many things; it is an energy source of great versatility, and we therefore say its quality is high. Thermal energy is much more difficult to use; due to its lack of versatility, we say it is of low quality.

REFERENCES

1. C. A. Price, in *Patient Earth,* ed. J. A. Harte and R. H. Socolow (New York: Holt, Rinehart and Winston, 1971), 70; and J. A. Harte and R. H. Socolow, *Patient Earth,* p. 84.
2. E. F. Hammel, M. C. Krupka, and K. D. Williamson, *Science* 223 (1984): 789.
3. R. H. Socolow et al., eds., *Efficient Use of Energy, The APS Studies on the Technical Aspects of the More Efficient Use of Energy* (New York: American Institute of Physics, 1975).
4. R. U. Ayres and I. Nair, *Phys. Today* 39, no. 11 (1984): 62.
5. W. D. Metz, *Science* 188 (1975): 820.
6. Stanford Research Institute, *Patterns of Energy Consumption in the United States* (Washington, D.C.: GPO, 1972).

PROBLEMS AND QUESTIONS

True or False

1. The inside of an oven at 300°C shows a great deal of thermodynamic *order* because everything in it is the same temperature.
2. Since the spreading of heat is *statistical* in nature, there are no general principles we can enunciate concerning it.
3. Temperature reservoirs must be taller than surrounding objects in order to be useful, since heat flows down.
4. Since the second law of thermodynamics says that all physical processes take place in such a way that disorder increases, life should be physically impossible.
5. Most substances should expand as the temperature rises.
6. Lake Michigan water has a smaller specific heat than Lake Superior water because Lake Michigan holds a smaller amount of water.
7. It requires energy added to water at 0°C to make ice at 0°C.
8. The definition of *heat* is that it is the energy content of a body.
9. All atoms in a particular gas are moving at the same speed, one that is characterized by the absolute temperature.
10. It is never possible to transfer thermal energy from a colder body to a hotter body.
11. A pot handle heats up because of heat transfer by radiation after the pot is put on the stove.

Multiple Choice

12. Suppose an electric company's boilers run at a temperature of 600°C. The facility uses river water to cool its spent steam. Assume that the mean river temperature is 20°C. Suppose that a twin boiler is air-cooled on Baffin Island (mean temperature −20°C). The power station on Baffin Island in northern Canada is theoretically about
    a. 100% more efficient than the electric company's.
    b. 100% less efficient than the electric company's.
    c. 5% more efficient than the electric company's.
    d. 5% less efficient than the electric company's.
    e. none of the above.

13. Substance *A* has twice the heat capacity of substance *B*. If equal masses of both substances are raised 10°C in temperature, and substance *A* gains 30 kJ of internal energy, substance *B*
    a. gains 60 kJ of internal energy.
    b. gains 30 kJ of internal energy.
    c. gains 15 kJ of internal energy.
    d. does none of the above.

14. Copper at 90°C is put into an equal mass of water at 10°C. The final temperature is
    a. 50°C.
    b. 80°C.
    (Hint: refer to Table 7.1.)
    c. 100°C.
    d. impossible to calculate with the information given.

15. If food coloring is put into a container of water, the water will be uniformly colored after some time. Which physical concept best describes the phenomenon?
    a. Diffusion
    b. Conduction
    c. Convection
    d. Radiation
    e. Brownian motion

16. Which involves the greatest increase in entropy?
    a. A person cleaning house
    b. Physics students taking a final exam
    c. Cooking a fast-food hamburger
    d. A rock concert by the Rolling Stones
    e. Driving a car five blocks

17. Systems tend to become more disordered with time. Why is this?
    a. Once order is defined, the number of ordered states of the system is very small compared to the number of disordered states of the system.
    b. The systems lose entropy, which is responsible for keeping them ordered.
    c. Disorderly conduct is a state of nature.
    d. We take energy out of systems, so they cannot keep up their order.
    e. To keep a system in a steady state always requires energy.

18. Which of the following systems will not be able to produce work?
    a. A charged capacitor connected to a motor
    b. A cylinder full of compressed air exiting through a turbine
    c. Water behind a reservoir falling to a lower level
    d. The mixing of fresh water and salt water
    e. A gas at 300 K

19. Which of the following state(s) the second law of thermodynamics?
    a. Any system left to itself tends to equilibrium with its surroundings.
    b. The entropy of a system in equilibrium with its surroundings remains constant.
    c. It takes energy to change a system from its equilibrium configuration.
    d. All physical processes are irreversible.
    e. All the other statements are statements of the second law of thermodynamics.

20. Which of the following is probably *most* efficient?
    a. An engine operating between
       300 K and 600 K
    b. A 100 W light bulb
    c. A plant operating in the
       ocean between the tempera-
       ture reservoirs at 4°C and
       20°C to produce electricity
    d. An engine operating between
       800 K and 1000 K
    e. An engine operating between
       1000 K and 1500 K

21. In Germany at Christmas time, one often sees "Christmas pyramids," which
    contain creche scenes on wooden platforms. The platforms turn because they
    are surrounded by burning candles and are connected to wood vanes above
    the candles. Why does the platform turn?
    a. Conduction of heat through
       air makes the wood vanes
       lighter.
    b. Convection currents cause
       the air to rise past the vanes,
       causing them to rotate.
    c. There is a hidden electric
       motor in the pyramid.
    d. The heavy candle wax from
       the burning candles strikes
       the underside of the vanes,
       causing the rotation.
    e. Colder air is forced to rise
       because of the burning can-
       dles, which make the vanes
       rotate.

22. All materials
    a. exhibit temperature changes
       during changes of phase.
    b. expand when they are
       heated.
    c. are good conductors of heat.
    d. increase their temperatures
       when heat is added.
    e. are made of atoms that are in
       constant motion.

23. Pans on stoves often have wooden handles. Why are wood handles used?
    a. Wood does not burn.
    b. Wood is a poorer heat con-
       ductor than metal.
    c. Wood handles are not so slip-
       pery as metal handles.
    d. Wood is lighter than metal.
    e. Wood is more decorative
       than metal.

24. It takes the same amount of energy to raise the temperatures of bodies A and
    B by the same degree. We *must* conclude that
    a. Bodies A and B are identical.
    b. Bodies A and B have the
       same heat capacity.
    c. Bodies A and B have the
       same latent heat.
    d. If body A is more massive
       than body B, its heat capacity
       is greater than that of body B.
    e. If body A is more massive
       than body B, its heat capacity
       is less than that of body B.

25. A skunk that has just sprayed walks past your open door. The odor is trans-
    ported to your nose mainly by
    a. convection.
    b. conduction.
    c. diffusion.
    d. radiation.
    e. evaporation.

26. Friction
   a. causes the efficiency of real machines to be less than 100%.
   b. operates in all real machines.
   c. involves conversion of mechanical energy to random motion of the molecules in the substances.
   d. would not operate were it not for electric forces between adjoining surfaces.
   e. is involved in all of the phenomena listed above.

Discussion Questions

27. Why isn't it possible to run a car on the energy in the air? Or does this mean that the air does not contain energy (since it cannot run the car)?
28. Suppose there were an audience of 200 for a physics lecture. How many ways are dead silence or the simultaneous clamor of 200 voices possible? How many possible ways are there for one person to talk? On the basis of the second law of thermodynamics, in the most general way possible, describe the actual state of the audience.
29. Enumerate the possible states of an audience of five persons. What is the *most likely* state here? (A state consists of a list of all five people and the information as to whether they are speaking or not.)
30. Do most processes occur with their theoretical maximum efficiencies?
31. Discuss the idea that the assertion "entropy tends to a maximum" must be wrong because there are many occurrences here on Earth that lead to increases in order (petroleum formation, plant growth, and so on).

Problems

32. Suppose a power plant producing 1000 megawatts is 40 percent efficient. Water (1000 kg/m³) passes through the condensers at a rate of 30 m³/s. What temperature increase (in °C) is experienced by this water?
33. Suppose the flow rate in Problem 32 were doubled. What is the temperature increase in this case? Note: in deciding flow rates, one must consider the tradeoffs among the amount of water transferred, the effect on organisms, and the effect on power plant efficiency (see Chapters 8, 9, and 11).

8

CONSUMPTION OF ELECTRICAL ENERGY: PROJECTIONS AND EXPONENTIAL GROWTH

In the last four chapters, we have discussed some means for generating energy for human use. Electricity is one of the most common forms of energy we encounter in our everyday lives. In this and the next several chapters, we consider specific details of electric generation and distribution systems.

In the current chapter, we consider some ways people have developed for predicting the future: future population, future energy use, future electricity demand. The most important modes of thinking about the future are linear extrapolation and exponential extrapolation. Various extrapolation methods and the consequences of using each method are discussed. Each extrapolation method has its own drawbacks.

KEY TERMS   *projection techniques*
            growth rate
            doubling time
            geometric progression
            exponential growth
            logarithm
            logistic curve
            price elasticity

Immense expansion in the use of energy, especially electricity, and huge increases in the size of generating facilities characterize our century. The utilities loom as the fastest growing energy users (1,2). Electric power consumption in the United States grew at a rate of 7 percent annually from 1961 to 1965, at 8.6 percent from 1965 to 1969, and at 9.25 percent in 1970 (1), while the population grew by only about 1.6 percent each year. This trend did not continue through the 1970s or 1980s; two successive energy crises caused a falloff in consumption, first after 1973 and again in 1979 (Figure 8.1) (3).

MAKING PROJECTIONS

A question we may fairly ask when presented with the current world energy picture is: What will happen to energy demand in the next few decades? This is a legitimate question, and there are many answers. Every few months someone else is telling us what will happen (or perhaps more modestly, what they predict will happen) at the turn of the century (Table 8.1).

With such a rash of predictions, most with different outcomes, it is useful to know just how people go about making predictions about energy. Before we evaluate the plausibility of others' predictions, let's take a look at how to go about making some of our own.

FIGURE 8.1
(a) Electricity generation from 1950 to 1985. (b) Component growth in electricity generation, 1960 to 1982.

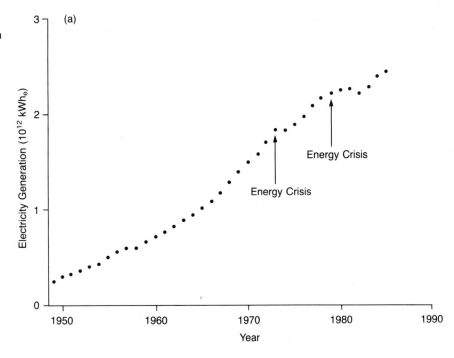

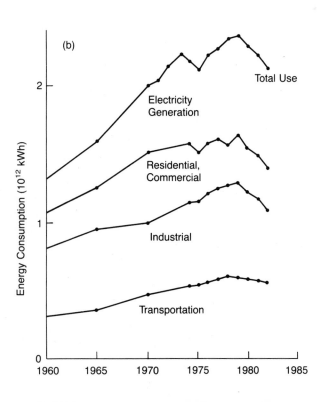

TABLE 8.1a
U.S. energy use in
TkWh$_e$ (1 TkWh$_e$ =
3.6×10^{21} J).

Year	Source	Projection
1960	Actual use	13.0
1970	Actual use	19.7
1980	Actual use	22.4

TABLE 8.1b
Projections of U.S.
energy use in 2000 (in
TkWh$_e$).

Year	Source	Projection	Remark
1972	D. C. Chapman et al.	24–123	Scaled from electricity generation
1972	A. Lovins	36.8	
1972	Sierra Club	41.2	
1973	J. C. Fisher	39.0	
1974	Off. Tech. Assmt.	39–130	Scaled from residential
1974	S. Freeman	50–100	
1974	W. Häfele	47	
1976	A. Lovins	22.1	
1976	F. von Hippel	26.2–28.0	
1978	Nat. Acad. Sci.	19.7–28.3	CONAES study
1978	Inst. En. Anal.	28.3–29.7	
1979	Dept. Energy	35.9	
1979	Exxon Corp.	28.0	
1981	Edison Elec. Inst.	34.5	
1981	Dept. Energy	30.0	
1981	Mellon Inst.	25.9	
1981	Audubon Soc.	23.6	
1981	M. Ross, R. Williams	18.8	
1981	Solar Energy Res. Inst.	18.3–19.4	

Linear Projections

If a given quantity is fixed in value, that means it will not change. Most quantities of interest to us—money in the bank, for example, or the amount of beer sold—do change. Suppose we observe a change in a quantity during some given time interval. For instance, suppose that in 1987, 500 million pizzas were sold nationwide, and that in 1988, 506 million pizzas were sold nationwide. We might assume that in 1989 (and even onward), the increase in the number of pizzas sold would

be 6 million. In 2000, we would predict, 578 million pizzas would be sold. This is an example of *linear projection*. Figure 8.2a shows such a projection. This linear projection is conceptually the simplest way to predict the future. We say that people making such extrapolations are using "linear thinking."

Projections, however, cannot be made in a vacuum. It is often helpful to bring outside information to bear in evaluating a projection. In Table 8.2, for example, we see the Federal Power Commission's 1972 predictions of utilities' en-

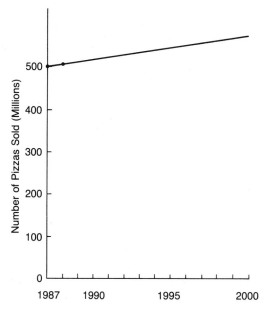

FIGURE 8.2
The number of pizzas sold illustrates linear growth in this hypothetical example.

TABLE 8.2
Utilities' energy requirements (in TkWh$_e$).

	1970	1975	1980	1985	1990
Fossil					
Coal	0.75	0.99	1.16	1.38	1.61
Gas	0.40	0.42	0.39	0.40	0.43
Oil	0.21	0.35	0.40	0.45	0.50
Subtotal	1.36	1.76	1.95	2.23	2.54
Hydro	0.28	0.29	0.32	0.35	0.37
Nuclear	0.02	0.24	0.96	1.73	3.10
Total	1.66	2.29	3.23	4.31	6.01

ergy requirements (4). Table 8.2 looks absurd in hindsight. It was probably not realistic to expect half the energy generated in 1990 to come from nuclear installations, even from the perspective of 1972. Already by that time, there had been a consistent trickle of antinuclear referenda on state ballots starting in the late 1960s, long before the incident at Three Mile Island in 1979 galvanized opposition to nuclear energy, or the disaster at

Chernobyl in 1986 convinced even more people of the hazards of nuclear energy. In this chapter, we shall try to give you an idea how to address some of these concerns.

Exponential Projections

Linear projections often fail. One reason they fail is that many quantities grow or decay in a way that depends on their initial size. For example, when people catch a cold, they somehow have been exposed to a virus, and the virus has found conditions in the body favoring its rapid growth. Viruses increase in number more rapidly as there are more of them in the bloodstream. The number of viruses "explodes" and affects the sufferer with typical cold symptoms—sneezes, a runny nose, and so on. If it takes one day for the number of viruses in our body to double, then after a week, *one* invading virus becomes 128; after two weeks, 16,384; after three weeks, 2.1 million; after four weeks, 256 million; and so on. This sort of growth is known as *exponential* or *geometric* growth.

Many quantities exhibit exponential growth or decay in nature. In Chapter 2, we discussed the exponential growth of population—we use the population example again shortly—and in Chapter 6, we discussed radioactive decay. These are just two of many examples of the ubiquity of exponential effects, that the number of things at any time determines the rate of growth or decay in that thing.

Projection of the demand for electricity (our main concern in this chapter) certainly depends on the growth of population, since the more people who can use electricity, the more is used. We would expect this projection to involve exponential growth.

Growth rate and doubling time are intimately related. Table 8.3 presents the doubling times for given growth rates. Note that an investment of money at 3 percent per year interest would double its original value in only 23.4 years. This table allows us to discover how long something will take to double if we know its current rate of

TABLE 8.3
Growth rates and doubling times (compounded annually).

Growth Rate (%)	Doubling Time
0	Infinity
1	69.7
2	35.0
3	23.4
4	17.7
5	14.2
6	11.9
7	10.2
8	9.0
9	8.0
10	7.3
11	6.6
12	6.1
14	5.3
16	4.7
18	4.2
20	3.8
30	2.6
40	2.1
50	1.7
75	1.2
100	1.0
200	0.6
500	0.4
1000	0.3

increase. In this way, we can discover how long, at a certain compound interest rate, a bank will take to double the money we have deposited in it, or how long it will take a mosquito-infested lake to become doubly unbearable. As an example, if the mosquito population grows at a rate of 7 percent every week, the number of mosquitoes will double in a little over ten weeks.

This is a very important aspect of growth, but how precise can we be? We do not know the best way to judge what will occur in the future. As we will see, the doubling time (or growth rate) in real-life situations can change. How do we decide what growth will occur?

Exponential change can be surprising to people because the final stages of growth seem to

occur so rapidly. A jar that is a quarter full of fruit flies (doubling time, say, three days) will be totally full six days from now. The first changes, though at the same rate, are perceived as slower (and thus nonthreatening to the environment). The final filling of the jar seems to happen entirely too rapidly.

Consider the example of growth in human population. For something that progresses geometrically, the new amount depends multiplicatively on the old amount. For a population, the number of children born (and hence the added population) depends on the number of women of childbearing age. The number of women of childbearing age is some proportion of the total population. The number of people dying also depends on the population. We can therefore say that the change in population (call this ΔN) is proportional to the population itself (call this N). Any proportionality can be written as an equation. The equation resulting from the proportionality for population is

$$\Delta N = a\,N\,\Delta t.$$

Here a is the net rate of addition to the population. This is the birth rate (number of children born per capita per unit time) minus the death rate (number of people dying per capita per unit time). The time interval involved in going from a population N to a population $N + \Delta N$ is Δt.

Let us assume temporarily that the birth and death rates are constant. For this case, the equation just given for ΔN has a solution for the population N as a function of time $[(N(t)]$. If the population at some starting time ($t = 0$) is N_o, then

$$N(t) = N_o\,e^{at}$$

or

$$N(t) = N_o\,10^{ct}$$

where c is related to a (see Appendix 2). Here we consider a situation in which the power of ten is not an integer but is a real number, ct, that changes with time. The properties of powers we discuss in Appendix 2 continue to hold. It is just

GEOMETRIC GROWTH

L et us take a, the net rate of addition, to be unchanging. We shall take the time interval to be one year, and suppose that a is 10 percent per year. We will then calculate the new population after passage of a year's time. This, symbolically, is

$$N + \Delta N = N + aN \Delta t = N(1 + a \Delta t).$$

$N + \Delta N$ is the new beginning population for the next year, replacing N. The starting population, which we have denoted by N_o, has some specific value. Let us use 1 million as the starting population and tabulate the numbers year by year:

Year		Population
0	$N_0 = 1,000,000$	
1	$N_1 = N_o(1 + a \Delta t)$	$= 1,100,000$
2	$N_2 = N_1(1 + a \Delta t) = N_o(1 + a \Delta t)^2$	$= 1,210,000$
3	$N_3 = N_2(1 + a \Delta t) = \ldots = N_o(1 + a \Delta t)^3$	$= 1,331,000$
4	$N_4 = N_3(1 + a \Delta t) = \ldots = N_o(1 + a \Delta t)^4$	$= 1,610,710$
5	$N_5 = N_4(1 + a \Delta t) = \ldots = N_o(1 + a \Delta t)^5$	$= 1,610,710$
6	$N_6 = N_5(1 + a \Delta t) = \ldots = N_o(1 + a \Delta t)^6$	$= 1,771,781$
7	$N_7 = N_6(1 + a \Delta t) = \ldots = N_o(1 + a \Delta t)^7$	$= 1,948,959$
8	$N_8 = N_7(1 + a \Delta t) = \ldots = N_o(1 + a \Delta t)^8$	$= 2,143,755$
9	$N_9 = N_8(1 + a \Delta t) = \ldots = N_o(1 + a \Delta t)^9$	$= 2,358,130$
10	$N_{10} = N_9(1 + a \Delta t) = \ldots = N_o(1 + a \Delta t)^{10}$	$= 2,593,944$
11	$N_{11} = N_{10}(1 + a \Delta t) = \ldots = N_o(1 + a \Delta t)^{11}$	$= 2,853,294$
12	$N_{12} = N_{11}(1 + a \Delta t) = \ldots = N_o(1 + a \Delta t)^{12}$	$= 3,138,623$

[handwritten margin notes: 1.4641, 1.61051, "✗ signif. Fig?"]

....

....

....

....

| 20 | $N_{20} = N_{19}(1 + a \Delta t) = \ldots = N_o(1 + a \Delta t)^{20}$ | $= 5,678,836$ |

Note how quickly the total mounts. This is the geometric, or compounding, feature of this sort of growth.

that the power (of e or ten, respectively) is continuously changing.

If we were to try to graph the function $N(t) = N_o 10^{ct}$ ($N_o e^{at}$) versus t on graph paper with one logarithmic scale and one linear scale (semilogarithmic graph paper), it would appear as a *straight line*. This follows because the logarithm of 10^{ct}, which is ct, is a straight line when plotted against t. It is easier to extend a straight line with a ruler laid to a piece of paper than it is to extend an exponential curve.

People are relatively comfortable making projections once geometric growth is presented as a straight line on semilogarithmic graph paper because it is possible to visualize the increase as a straight line. This feels comfortable because it is linear (which is why linear thinking is so attractive). Several examples of geometric, or exponential, growth are given in the accompanying boxes.

Let's recall the example of the growth of the U.S. population and its projections. This will reinforce some important concepts about exponential growth. The population of the United States is measured by the Census Bureau every decade (5). Table 8.4 and Figure 8.3 show this historical growth. In the discussion of population doubling in Chapter 2, it was shown that the first measured doubling took about twenty years. This doubling time characterized population growth up until the time of the Civil War (1860s). By the turn of the century, the doubling time had reached thirty years. The last doubling has occurred from about 1925 to 1980—that is, fifty-five years. For the United States, as opposed to the world as a whole, the population growth rate has been dropping (with one very notable "bump"—the baby boom—between about 1945 and 1962).

If we wish to project the U.S. population into the future, the simplest way is to continue the

TABLE 8.4
History of U.S. population growth.

Census Year	Population (millions)	Annual Growth Rate (%) Between Censuses
1790	3.929	
1800	5.308	3.05
1810	7.240	3.15
1820	9.638	2.90
1830	12.866	2.93
1840	17.069	2.87
1850	23.192	3.11
1860	31.443	3.09
1870	39.818	2.39
1880	50.156	2.34
1890	62.948	2.30
1900	75.995	1.90
1910	91.972	1.93
1920	105.711	1.40
1930	122.775	1.51
1940	131.669	0.70
1950	151.326	1.40
1960	179.323	1.71
1970	203.302	1.26
1980	226.505	1.09

FIGURE 8.3
The U.S. population as recorded by the Census Bureau.

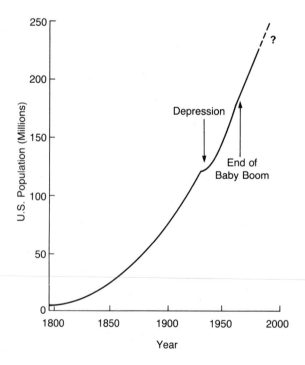

There is an old story of a boy who played chess with the king and beat him. The king offered him his choice of payment; the boy chose to be paid on the chessboard in the following way. On the first day, the king would put one grain of wheat on the first square of the chessboard's sixty-four squares. On the second day, he would double the number of grains on the first square and place them on the second square. On the third day, the king would double the number of grains on the second square and place them on the third square. And so it was to go.

The king was pleased at the boy's generosity, for he imagined that he was getting away easily. As we saw in Chapter 2, however, if we double something ten times we multiply it by about a factor of 1000. This means that on the sixty-fourth day, the boy would be entitled to approximately 1.6×10^{19} grains of wheat. This is more than has been harvested in all human history.

line in the graph of Figure 8.3. The trouble is that we really cannot judge very well which way the line should be extended. This is because of the characteristic steep rise that is occurring, what economist Thomas Malthus referred to as "geometric progression." We need a better way to make judgments about growth.

Using Semilog Graph Paper to Make Projections

In Figure 8.4, we replot the curve of Figure 8.3 on semilogarithmic ally. Note first that it is *not* a straight line (if we extend the original growth of the 1790–1850 era by a straight line, the pop-

FIGURE 8.4
The U.S. population as recorded by the Census Bureau and plotted semi-logarithmically.

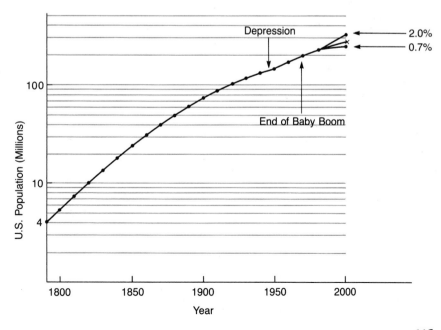

ulation in 1990 would be predicted at about one billion). This means that the net rate of population increase is declining for the United States. For any small segment of the curve, though, a straight line gives an adequate description. Note also the projections of population increase indicated. The projections follow from hypothesizing various straight-line increases from the graph. They may then be read in tabular form, as indicated in Table 8.5 (6).

Projections are usually made, therefore, by graphing the past behavior of some quantity (say, demand for electricity) on semilog paper, and then drawing the best possible straight line through the past behavior on into future behavior. This is a plausible way to project short-term growth. The Edison Electric Institute in 1960 made a ten-year projection of electricity demand, predicting 1.31 TkWh$_e$ demand (7); the actual 1970 demand was 1.39 TkWh$_e$. This agreement of projection with reality is quite good; here, the straight-line method worked well.

Suppose that on the basis of the 1959–1964 figures for electricity demand (7.35 percent sales growth per year), you predicted the 1969 and 1970 sales by extending a straight line on semilog graph paper (7). The straight line represents an annual increment of 7.35 percent. Six eventful years pass from 1964, with military buildup in Vietnam, a U.S. invasion of the Dominican Republic, a six day war in the Middle East, riots in the centers of many American cities, and so on.

TABLE 8.5
Population projections.

	1970	1980	1990	2000	Children per Family
A (1.8%)	208.9	249.4	298.1	356.1	3.35
B (1.6%)	207.1	241.9	284.4	331.6	3.10
C (1.3%)	205.5	233.6	268.8	304.4	2.775
D (1%)	204.1	226.0	254.0	280.1	2.45
UN	205	251		304	
Actual	203.3	226.5			

SOURCE: U.S. Census Bureau projections, 1960s.

In 1970 you recall your predictions:

1969	1.28 TkWh$_e$
1970	1.37 TkWh$_e$

Actual energy sales were 1.31 TkWh$_e$ in 1969 and 1.39 TkWh$_e$ in 1970. In 1972, you look at your prediction for 1971: 1.47 TkWh$_e$. This was the actual sales figure for 1971. It is this kind of experience, in all sorts of applications at all levels, that has reinforced the predominance of this method of prediction.

Lacking other input, this is a perfectly acceptable way to make projections. However, we must be aware of the pitfalls involved in making projections when we are in the process of interpreting them. The greatest pitfall is the belief that the future will bring the same as the past, only more so. This is a reasonable enough belief for projections of a few years, but the premise can fail miserably when extended indefinitely (remember our already-mentioned projection of a U.S. population of a billion for 1990). Circumstances that are assumed to remain steady change all too soon, vitiating even the best-informed predictions. This is true in longer term projections, whether our predictions are linear or exponential.

If the logarithmic scale is used, the resulting graphs may be misleading. The "agreement" of the data with the prediction is emphasized in a plot on full logarithmic graph paper as compared to a linear scale. Individual points would appear closer to the line than they would with a linear plot because of the compression resulting from logarithmic rescaling. (A straight line on a graph on full logarithmic graph paper is a straight line on linear paper only in special circumstances.)

MAKING THE ACTUAL ELECTRIC CONSUMPTION PROJECTIONS

There is much uncertainty in population growth projections, even for the United States, as we have just seen. The conservation ethic has become more pervasive as a result of the energy crises of the 1970s. Sales of appliances that use large amounts of energy are not rising much, as we shall see in Chapter 9. For such reasons, projections that consist in drawing straight lines on graphs can lead to estimates that may be far too high. An illustration of how expert opinion can differ is given in Figure 8.5 (8), which shows a wide range of energy demand projections for 1980 made during the 1970s.

The amount of energy sold by utilities and the rate of growth are historically quite large. One reason for the tremendous growth is that electricity, from World War II onward, has been a better and better buy, as shown in Figure 8.6 (9). The trend of reduced real cost, which is shown on the graph to extend until about 1970, could not continue indefinitely; in fact, the real price has risen steadily since 1971. The costs of pollution control are mainly responsible for the price rise between 1970 and 1973 (U.S. utilities spent only 3.8 percent of their total capital budget for pollution control, as compared to the 10.3 percent budgeted by the iron and steel industry).

Price Elasticity

The main cause of utility price increases is the swift rise in utilities' fuel prices after decades of little change. The historical picture of real fuel costs, shown in Figure 8.7 (9), verifies the extreme stability of fuel prices until 1973. The meteoric rise of adjusted fuel prices is astounding.

Real price is the most important determinant of demand (7); there is substantial *elasticity* (that is, price rises cause reduced demand, and price decreases cause increased demand) in electricity use. A Canadian study (10) shows energy price elasticity in Canada in 1984 to be -0.3 to -0.6 (and about -0.7 for gasoline and other oil products). These numbers imply that a price increase causes a consumption decrease. Elasticities in the United States should be very similar to those in Canada. Thus, raising the price of electricity can be said to promote conservation of energy.

Various studies of the effect of cost on use have led to different predictions for the amount of electricity to be used in the future. A "think

tank" analysis by RAND Corporation (11) of energy demand in California concluded that the use

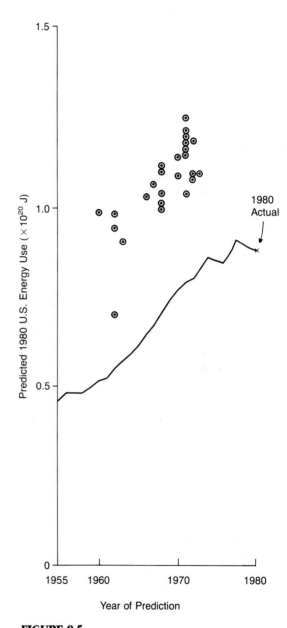

FIGURE 8.5
Predictions of energy use in 1980 as compared to actual production of energy are plotted in terms of the year the projection was made. (Adapted from Ref. 8 by permission, Duxbury Press.)

growth rate will be between 3.4 and 6.3 percent for all reasonable assumptions (a doubling of energy cost would lower the growth rate from 6.3 percent to 4.7 percent). A study of national energy requirements by another group attempted to provide perspective by varying the possible population growth rates and prices assumed (7). Possibilities for electricity use in the year 2000 range from a low of 2 TkWh$_e$, if the price of electricity doubles, to a high of 9.9 TkWh$_e$, if costs decrease approximately as they did prior to 1970. An Environmental Defense Fund study in Wisconsin indicated that industrial demand would slacken if industrial users paid a true share of the cost (as we shall see in the next chapter, the utilities signed up industrial users by giving them the lion's share of the savings realized by

increasing the base load). This reflects one of the major recommendations of the Ford Foundation Energy Policy Project (12): that promotional rates and subsidies be eliminated.

As we have previously discussed, the U.S. population growth rate is dropping spectacularly. The U.S. market should approach the saturation point soon in major appliances. Fisher (13), in 1973, analyzed U.S. energy use in the year 2000, taking into account revised population growth estimates, saturation, projected increases in industrial efficiency, and different growth rates of different sectors of the economy. His conclusions were shown earlier in this chapter, in Table 8.1b. Each successive lowering of the estimate followed upon the use of more care in the choice of assumptions, recognition of changes in energy-

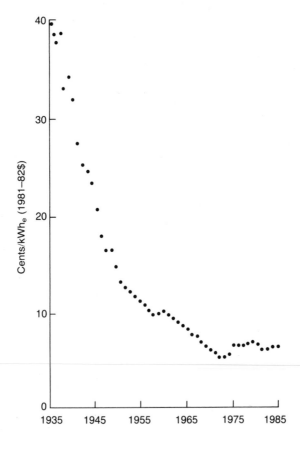

FIGURE 8.6
The real (deflated) cost of energy in central Ohio, 1935–1985. (Data courtesy of Columbus Southern Power Company.)

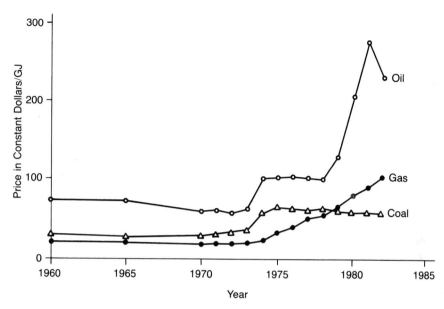

FIGURE 8.7
Real (deflated) fuel costs, 1960–82. (U.S. Department of Commerce.)

use habits of individuals and industry (14), and more experience in seeing the effects of energy conservation measures in everyday life.

Predicting into the Far Future

Thoughtful scenarists of future energy demand are worried. In 1975, the average per capita energy use was 2 kWh$_e$ per person. However, the world's use was not evenly distributed: Americans use about 11 kWh$_e$ per person; Europeans use about 5 kWh$_e$ per person; and Third World countries use less than 1 kWh$_e$ per person (15). Shall we assume a future world of wastrel Americans, or a modest rise in per capita power? Even if the per capita energy use does not rise, the number of people using energy will rise exponentially. Merely running the Red Queen's race (16) will involve global energy production of almost incomprehensible proportions and the investment of hundreds of billions of dollars to supply the necessary energy (15,17).

Sassin and Häfele see the need for drastic changes in the world's economic structure. As

we shall see in Chapter 13, fossil fuels are inexorably running out. As this happens, there must be a shift to some other resource, and such changes in the United States have historically taken about sixty years (in the rest of the world, about a hundred years) to achieve a 50 percent penetration for the new fuel (17). We do not even know now what the new fuel will be. If it is solar energy (discussed in Chapters 15, 16, and 17), the only realistic scenario at present involves a tremendous civil engineering undertaking, using the bulk of cement now available in world production (17). If it is nuclear energy (discussed in Chapter 14), the experience of the last decade indicates that a "transition from fossil-fuel energy to nuclear energy will probably be an uphill fight in terms of cost and effort" (15).

Any transition to a sustainable energy future will become more difficult the longer it is put off. People would like a world with a stable price structure, full employment, and increasing wealth for the people in it. These goals are often mutually exclusive, except for short timespans such as the three decades following World War II in the developed world (18). The decline in the ratio of American energy use to gross national product, which many see as a hopeful sign for the future, is almost entirely attributable to three factors (18)

> the proportion of petroleum used,
>
> the proportion of hydroelectric and nuclear energy generation, and
>
> the proportion of direct fuel use in the ratio of household to manufacturing energy use.

Much of the decline is due to a shift to fuels having a greater energy return on investment (EROI) (18). Not all fuels are equal; some take much more energy than others to find, exploit, and refine to a handy form. Per joule, oil is worth 1.3 to 2.45 times as much as coal (18). Any transition will involve a shift to fuels with a smaller EROI. People are content with the status quo, and want to stave off installing a new energy infrastructure. Problematically, the longer we put off the change, the less rewarding the change will appear to be, and so the speed of the change will be even slower (15,18).

The economic recovery from the recession of 1980 to 1983 is in part due to declining OPEC oil prices, which dropped due to decreased oil demand caused by the recession. As the experience of very low oil costs in 1986 shows, very cheap oil may be disadvantageous as well because it leads to increased dependence on imported oil and encourages wastefulness on the part of consumers. In any case, cheaper oil will again become expensive oil. Soon, oil will be gone forever (see Chapter 13). The cost of extraction of energy resources, which hovered around 3 percent for most of this century, rose by 1982 to 10 percent (18). The switch to the sustainable energy future should not be put off until we have squandered our one-time access to inexpensive, useful fuel.

Fisher looked even further into the future than the turn of the century: He assumed U.S. population growth to drop steadily to zero in the 2200s, with a stable population of 1 billion, and a stable

consumption to be reached ultimately of about 11 $TkWh_e$. These predictions should serve to remind us again that, no matter how thoughtful the predicter, any but short-term predictions are fraught with difficulties. We must look on *any* two-century prediction with extreme skepticism.

The Logistic Curve

The exponential method of prediction fails to recognize that all systems of our acquaintance are finite. Consider the example of a new species exploiting a hitherto untapped ecological niche. At first, with superabundant resources, there are no checks on the growth of the new species. Individuals of the species are too few to have attracted predators, for example. As a result, the growth in numbers of this new species is exponential. As the numbers increase, the species becomes more interesting to predators and exploits more and more of the finite resources available to it. At some stage, the growth in numbers must cease. The population thereafter remains essentially stable unless the environment is somehow altered. The phenomenon of untrammeled growth in the beginning, approaching a steady-state situation at the end, occurs in many facets of our experience. (See the saturation of appliances curve in Chapter 9 and the yield versus energy-input curve in Chapter 17, for example.) The curve describing this behavior is known as *logistic* (or sigmoidal); a comparison of the logistic curve and its corresponding exponential (for small *t*) is shown in Figure 8.8.

In a finite world, the logistic curve describes many situations in which there is, for a time, what appears to be exponential growth. Further out on the logistic curve, we could mock up exponential growth by a slightly different exponential curve. The logistic curve, as opposed to the exponential curve, begins after a certain point to turn over and gradually decrease until it reaches a constant value. While the logistic curve is below its turnover, one cannot be certain whether it is exponential or logistic. It is only after the turnover has occurred that the logistic curve clearly reveals its logistic character.

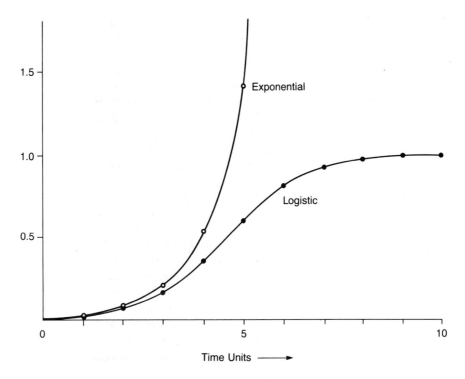

FIGURE 8.8
The logistic curve and exponential curve compared.

The exploitation of mineral resources often follows a logistic function. In many cases, however, economic activity suppresses production when it would have risen without economic interference. This makes the time past the turnover (fall time) longer than the time it takes to rise to turnover (rise time). The amount produced per unit time in such a case may be described by what is called a *Verhulst function*. Mineral resources the production of which is described by the Verhulst function seem to have ratios of fall time to rise time clustered at 1 (logistic curve), 5, and 15 (19).

The reason that care is needed in interpreting projections is that the straight-line semilogarithmic projections are often approximations to the logistic curve; the turnover is ignored. If the data projected are far below turnover, the projection may be valid over broad swaths of time.

If the data to be projected are near turnover, the "straight-line" prediction is grossly wrong.

SUMMARY

Exponential growth occurs if the growth in a quantity during a particular period depends on how much of that quantity there was at the beginning of the period. The rate of change in such a quantity is the growth rate times the quantity. Exponential or geometrical growth occurs much faster than linear growth, in which a fixed amount is added to a quantity in every unit of time.

Exponential growth is characterized by its doubling time; exponential decay, by its halving time (half-life). Specification of either of these times is equivalent to specification of the growth or decay rate, respectively. Exponential growth catches most people unaware because, by the time the problems associated with this type of growth become obvious, the time available to do anything about it is very short. If it takes a week for water hyacinth to double the area of a lake it covers, it may take until the lake is a quarter (or more) covered before the observer realizes that there is a problem with the plants; only two weeks remain in which to prevent the complete choking of the lake!

Special graph paper (semilogarithmic graph paper) can be used to show exponential growth as a straight line. Most projections of future energy use (or consumption of any sort) consist in extending such a straight line to future times. Such projections work well if trends are extrapolated for short intervals (one or two doubling times) into the future. No such projection can be trustworthy for predictions many doubling times hence.

REFERENCES

1. E. Cook, in *Energy and Power* (San Francisco: Freeman, 1971), 83.
2. C. M. Summers, in *Energy and Power,* 95.
3. Department of Commerce *Statistical Abstract of the United States, 1984* (Washington, D.C.: GPO, 1984).
4. Office of Emergency Preparedness, *The Potential for Energy Conservation: A Staff Study* (Washington, D.C.: GPO, 1972).
5. An interesting discussion of the movement of the center of population in the U. S. over the years is found in A. A. Bartlett and R. L. Conklin, *Am. J. Phys.* 53 (1985): 242.
6. N. Keyfitz, in *Resources and Man* (San Francisco: Freeman, 1969).
7. D. Chapman, T. Tyrrell, and T. Mount, *Science* 178 (1972): 703.
8. C. Steinhart and J. Steinhart, *The Fires of Culture* (N. Scituate, Mass.: Duxbury, 1974).
9. R. Jones, Columbus and Southern Power Company, private communication. The prices have been adjusted for the rise in the CPI.
10. Economic Council of Canada, *Connections* (Ottawa: Canadian Government Publishing Centre, 1985).
11. W. E. Mooz and C. C. Mow, as quoted in A. Hammond, *Science* 178 (1971): 1186.
12. *A Time to Choose: America's Energy Future* (Cambridge, Mass.: Ballinger, 1974).
13. J. Fisher, *Energy Crises in Perspective* (New York: John Wiley & Sons, 1974).
14. E. Marshall, *Science* 208 (1980): 1353.
15. W. Sassin, *Sci. Am.* 243, no. 3 (1980): 118 describes the results of the IIASA study of world energy growth.
16. L. Carroll, *Through the Looking Glass* (New York: Clarkson N. Potter, 1960).
17. W. Häfele, *Science* 209 (1980): 174.
18. C. J. Cleveland et al., *Science* 225 (1984): 890.
19. L. D. Roper, *Am. J. Phys.* 47 (1979): 467.
20. G. Aubrecht, *The Physics Teacher* 20 (1980): 444.

PROBLEMS AND QUESTIONS

True or False

1. The growth rate of electric power consumption is about the same as the growth in population in the 1970s.
2. One must take the effects of inflation into account when looking at energy costs.
3. Experts generally agree on the shape of future demand.
4. A population growing at a rate of 2 percent per year will add *exactly twice* as many people as a population growing at a rate of 1 percent per year.
5. The amount of time it requires to double a population depends on how many people there were originally and on the rate of growth.
6. The further in the future one predicts, the more likely it is that the prediction will be fulfilled.
7. If you invested your money in the bank at 8 percent interest, compounded annually, it would take nine years for your investment to double.
8. One use of logarithmic graph paper is to make apparent future growth in a quantity that grows geometrically.
9. The cost of electricity decreased between 1940 and 1960.

Multiple Choice

10. In a country in which the population doubles every twenty years, a 1970 population of 10 million will reach 80 million in
    a. 1990.
    b. 2010.
    c. 2030.
    d. 2050.
    e. 2070.
11. A certain piece of land can support up to one hundred cattle. If the number of cattle doubles every three years, how many years will it take for an original cattle population of twenty-five to reach the capacity of the land?
    a. 3
    b. 6
    c. 9
    d. 12
    e. 15
12. On the basis of the experience of the last twenty years, one wishes to make projections for some future period. About how far into the future should one have confidence in the projections?
    a. 1 week
    b. 1 month
    c. 1 year
    d. 1 decade
    e. 20 years
13. The real price of electricity
    a. will probably fall substantially in the future.
    b. will probably remain about the same in the future.
    c. will probably rise substantially in the future.
    d. is unlikely to drop much further.
    e. is unlikely to rise much further.
14. Water hyacinths can double their number in one week. If a lake is about a quarter covered by water hyacinths, how long will it be until the entire lake chokes on water hyacinth (barring outside intervention)?
    a. 1 week
    b. 2 weeks
    c. 3 weeks
    d. 4 weeks
    e. 6 weeks

15. For fixed amounts of material being used at constant rate, the amount of material decreases exponentially with time. This leads to the idea of the inverse of doubling time: halving time. Suppose the halving time of a stockpile of steel is one year. What proportion of the steel will remain in the stockpile after three years?

a. 1/2
b. 1/3
c. 1/4
d. 1/8
e. It will all be gone.

16. An amount of money is deposited in a bank for ten years at fixed annual interest rate. At the end of this time, four times the original amount deposited is withdrawn. What is the growth rate? (Use Table 8.4 to estimate the rate).

a. 7%
b. 15%
c. 17.7%
d. 19%
e. Not enough information is given.

17. Suppose the amount of fish caught worldwide, assumed to be a small proportion of the total fish population, grows at an annual rate of about 3 percent. About how long will it take for the fish catch to double?

a. 3 years
b. 13 years
c. 23 years
d. 33 years
e. Not enough information is given.

18. Which of the functions listed below is most likely to describe the exploitation of a resource through the entire history of its exploitation?

a. Linear growth function
b. Exponential growth function
c. Logistic function
d. Logarithmic function
e. None of the above can describe the exploitation of a finite resource.

19. Elephants are being hunted illegally in Africa for ivory. If current poaching experience is any guide, the number of elephants will decrease by half in about 25 years. If the elephant population of a certain African country is currently (1990) 16,000, by what date would we expect the number to have decreased to 1000?

a. 2015
b. 2040
c. 2065
d. 2090
e. 2115

20. Which of the following statements about linear growth is true?

a. Linear growth is characterized by a fixed doubling time.
b. During the early stages of growth, linear growth could produce greater gains than exponential growth.
c. Linear growth describes most systems in nature.
d. The growth rate increases as time passes.
e. The growth rate is the same for all different species.

Discussion Questions

21. Why is extrapolation by the method of drawing straight lines on semilogarithmic paper often wrong in predicting the future?

22. Apply some of the ideas in your answer to question 21 to the radio and television industry. Consider the inventions of vacuum tubes (1920s), transistors (1950s), integrated circuits (1960s), field effect transistors (1960s), and light-emitting diodes (1970s). How would a 1920s straight-line extrapolator be wrong? How would such a company fare in the marketplace?

9

TO YOUR SCATTERED ELECTRIC OUTLETS GO

Electricity is central to national energy policy. In this chapter, we study the history of the utilities, and show how economy of scale encouraged monopoly. In Chapter 8, we saw how utilities have gone about projecting future demand. This provides the basis for a discussion of base load, peak load, and the choice of generating facility. The electric grid and transmission of electricity also play an important role. Cogeneration and conservation (see Chapter 20) are also discussed briefly.

KEY TERMS   base load
            intermediate load
            cycling load
            reserve capacity
            economy of scale
            screening curve
            hard and soft energy paths
            cogeneration
            district heating
            total energy system
            electric grid

ENERGY END USE AND ENERGY SECTORS

People use energy for many different things: to heat themselves, to cool themselves, to cook their foods, to move from one place to another. Natural gas, fuel oil, or electricity may be used to heat a home; natural gas or electricity to heat foods; and gasoline, kerosene, and diesel fuels to power our transportation systems. The major sources of energy for the U.S. economy are shown for 1981 in Figure 9.1 (1).

As energy is used, it may be employed in a variety of ways. It may be used to heat a home or a place of business, for example. Although the end use is for the purpose of heating in both cases, the energy is used to different economic purpose. Heat in business is used as part of the cost of production of goods or services.

It is conventional to allocate energy use to four sectors of the economy: industrial, transportation, residential, and commercial. The sources shown in Figure 9.1 can be misleading. In the transportation sector, of course, all energy is fuel energy. In the residential sector, coal, oil, and natural gas are used directly to generate space heating, and they are also used indirectly. That is, they are used to generate electricity that is used in various ways in the home. At present,

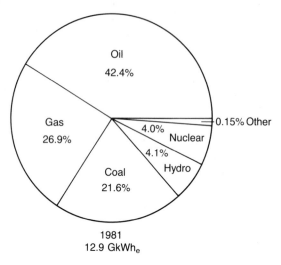

FIGURE 9.1
U.S. energy sources, 1981.

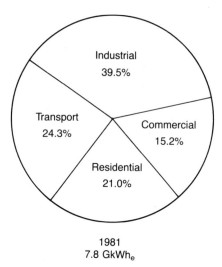

FIGURE 9.2
U.S. energy uses, 1981.

over a quarter of residential energy use is electrical energy. Plants built for generating this electricity are among the most concentrated sources of air and water pollution in the world. In this chapter, we study the electric generation and distribution system. This will lead us in succeeding chapters into discussion of the importance of water and of other forms of resources.

Figure 9.2 (1) presents the 1981 energy picture from the end-use perspective. You may notice that the total energy numbers given in Figures 9.1 and 9.2 do not match. The discrepancy arises from the distinction between kWh_e (kilowatt-hours, electric) and kWh_t (kilowatt-hours, thermal) we discussed in Chapter 7. We must choose one of the two ways to report the energy, which can cause problems because most reports do not distinguish between them. In Figure 9.1, energy sources are given exactly as reported in Table 979 of Reference 1. In Figure 9.2, electricity allocation in each sector has been used to "correct" the figures (since it costs 3 kWh of thermal energy to produce 1 kWh of electric energy). Another problem with the reported figures is that, in some cases, the use of fossil fuel as a raw material (in so-called feedstocks as, for example, in plastics)

is counted as energy use. This energy does appear as product, but it is not direct energy use and so should not be tallied.

It is not appropriate to discount, or correct, fossil-fuel energy used for space heat in Figure 9.2 because the energy is used directly to heat and not indirectly to make electricity. It is just as wrong to count hydroelectric energy as equal, kilowatt-hour for kilowatt-hour, to coal burned in a generating plant. Recall that a 1000 MW hydroelectric installation would put 1000 MW into the nation's energy grid; it takes 3000 MW of thermal power to put 1000 MW into the grid. Hydroelectricity or photovoltaic electricity (see Chapter 15) should be accounted for in an appropriate fashion. This means that we should either multiply hydroelectricity by 3 to express as equivalent thermal kilowatt-hours, or we should divide all thermal energies by 3. We should also remove feedstocks from the list. Once this internal accounting has been done, we are free to express this energy in any way we wish: kilowatt-hours (thermal); kilowatt-hours (electric); joules; or even as quads (quadrillion Btu).

Space heat looms as the largest share of both the residential and commercial sectors, as we see from the figures. In Chapter 19, we shall discuss some useful economy measures in the space heating of residences, as well as other conservation measures applicable to residences, commercial establishments, and industrial use.

The residential sector is probably the most familiar to most of us. Aside from space heating, water heating demands the most energy, followed by cooking, refrigeration, and air conditioning. Other uses, such as for washing machines and TV, are minor constituents of the total demand (2).

EVOLUTION OF THE UTILITIES

The electric industry has grown mightily in this country since Thomas Edison began formation of generating companies in 1881. Edison, as we have mentioned, had embraced direct-current (DC) generation of electricity. Because the voltage was only 110 V (about that used today), and because transformers could not be used with DC, Joule heating losses in the system were high. Electricity could be transmitted only a few miles; DC generating facilities had to be widely scattered. Edison began his DC system before George Westinghouse initiated his alternating-current (AC) system, so that in the beginning many sources of electricity were scattered around the countryside. It was the usual practice for industry to generate its own energy as needed rather than buying expensive energy from a generating company.

As AC came to be accepted, it became feasible to have larger facilities to generate energy. This led the many scattered electric companies to the path they have followed ever since: their growth as utilities rather than as competing businesses. Samuel Insull, of Chicago Edison, led the utilities onto this path (3). By obtaining monopoly rights as public utilities, they were able to trade competition for regulation by state public utilities commissions (PUCs). Since utilities are usually allowed profits as a fixed percentage of costs, this brought about disincentives for innovation. The utility will make its profit no matter what.

An additional advantage for the utility in its monopoly position is that it has a steady supply of captive customers to provide the capital necessary for expansion. Expansion increases capital costs; since PUCs give profits based on costs, expansion means increased profit. The extra energy must be sold to customers who are encouraged to use it in ever more frivolous ways. This provides more expansion capital, and so on.

The amalgamation of the small utilities into giant energy conglomerates proceeded apace since the last decade of the nineteenth century for essentially two reasons: the difficulty of storing electricity, and economy of scale.

Because storage of electric energy is difficult, the generating facility must have on hand enough capacity to supply all its customers when they want electricity. When an electric switch is flicked, the generator puts out a little more current, which is used as it is generated. If the electric company has only residential users, it has to supply almost all of them in the early morning and then again in the evening when most are at home. The capacity is virtually idle the rest of the time. It is abhorrent to a business to tie up capital in capacity that is sporadically needed. Of course, even with long spans of low use, less capacity is needed than would be the case if each residence had its own generator. A family's peak use may be 10 kW, and the family next door may also have a peak use of 10 kW; however, they probably operate on different schedules. One family may be sitting down to dinner just as the other is getting up from the table. Thus, the peak use from the two families together is considerably less than 20 kW. This was illustrated by Samuel Insull in his speeches on behalf of utility monopolies (3): An apartment house that would need 68.5 kW capacity if each user were to be able to draw at will had an actual maximum demand of 20 kW. The more people there are using electricity, the more the utility can smooth out its demand. The larger the utility, the more customers it can serve and the smaller the minute-to-minute fluctuations in demand.

To keep their expensive generating equipment in use, the electric companies began an aggressive selling campaign to woo industrial users away from generating their own energy. An industrial user will continue to generate energy unless a utility can sell its energy very cheaply. Since industrial users tend to have a steadier demand, utilities can further smooth out demand by having many industrial users. For these reasons, the utilities set industrial rates lower than residential rates.

A typical large-city utility daily energy use profile is shown in Figure 9.3. Only about half the capacity is in constant use; this constant demand is called the *base load*. Another 40 percent or so of the demand cycles on and off slowly; this is known as *intermediate,* or *cycling, load.* About

10 percent of capacity is used only a few hours a day; this is called *peak load*. The peak load comes around dinner time.

Base load actually changes from month to month. In the southern states, for example, energy demand peaks in the summer months because of air conditioner use. Air conditioning may be the most important contributor to the explosive growth of population in the Sunbelt. Who would have considered living in Houston or Phoenix without it? Further north, in Canada, a typical annual load curve will peak in the winter months (as shown in Figure 9.4) (4). Such variation is also of concern to the utilities, and another important reason for soliciting industrial business.

Base load is most economical for the utility to supply. Base load is generated all day, is always running, and maintenance can proceed contin-

FIGURE 9.3
Winter and summer
load curves for a city in
the northeastern U.S. are
schematically indicated.

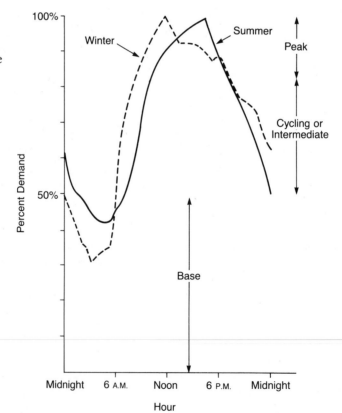

FIGURE 9.4
An annual load curve is shown. This curve is for a Canadian utility, and shows peak annual demand in winter. A curve from Texas would show peak annual demand in midsummer. (Reproduced with permission of the Ministry of Supply and Services Canada.)

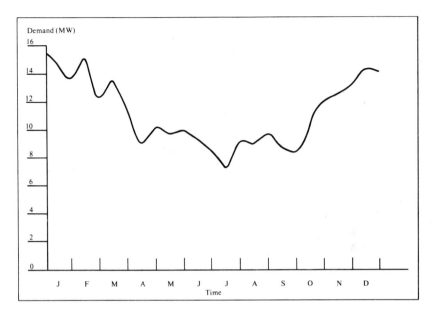

uously. The utilities' largest generating facilities are dedicated to base load production. Since it is difficult to start up and shut down nuclear plants, they are used to provide base load energy. The utility usually has special additional generators that are turned on for only a few hours a day for the peak load; this is the most expensive energy generated by the utility. Figure 9.5 (4) presents a different view of the load curve—an annual perspective. The figure is indicative of the situation in Canada, where hydroelectricity is more important than in the United States. For this reason, hydroelectricity in the United States would be baseload rather than baseload and peak, as in Canada. Nuclear electricity generation is a somewhat smaller proportion of total electricity generation in Canada than in the United States.

Table 9.1 (1,4) shows the mix of sources of electric energy in the two countries; note the heavy Canadian dependence on oil and hydroelectricity and the heavy U.S. emphasis on coal as the sources of electricity. While it is important to be able to meet peak demand, it is also important not to have too much reserve capacity. If the reserve capacity is too large, equipment sits idle, not earning a return on the capital in-

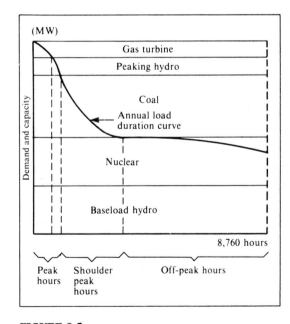

FIGURE 9.5
A load duration curve for calculating minimum cost of generation for a utility (Reproduced with permission of the Ministry of Supply and Services, Canada.)

TABLE 9.1
Electricity production by source, 1982, percent.

	U.S.	Canada
Hydroelectricity	13.8	25.8
Coal	53.1	11.0
Oil	6.5	36.9
Gas turbine	13.6	18.6
Nuclear	12.6	3.8
Internal combustion, wood, waste, geothermal	0.2	4.0

SOURCE: See Table 7-1 of Reference 4 and Table 998 of Reference 1.

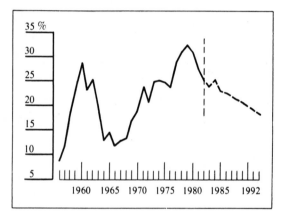

FIGURE 9.6
Canadian reserve capacity as proportion of peak demand for the years 1956 through 1982, as well as projections to 1992. (Reproduced with permission of the Ministry of Supply and Services, Canada.)

vested in it. Figure 9.6 (4) shows that Canadian utilities have a peak demand over capacity averaging 20 percent over the last three decades. This is about where the rule of thumb in the industry sets reserve margins. The high reserve capacity of the late 1970s in Canada is probably due to predictions in the early years of the decade of an impending shortage of electricity. Such faulty prediction has also been a problem in the United States. Of course, without the energy crises of

the 1970s, the predictions might have been correct.

Economies of Scale

There are efficiency problems with the use of electricity as compared to other "fuels." Inappropriate use of electricity has been responsible for some diseconomy in certain industrial applications, where direct heat use would have been more appropriate (5). To an extent, economy of size (or scale) may counterbalance some of these inefficiencies. By economy of scale for a generating facility, we mean that the larger the capacity of a facility, the lower the cost per kilowatt of capacity. For example, doubling a boiler's size does not double the amount of steel needed or the amount of fabrication necessary. Tripling the volume of a building does not triple the architect's fees or the cost of erecting it. Erection cost is roughly proportional to the outside surface area of a building. Suppose the building were a cube. If I triple the length of each side, the surface area would increase by a factor of 9 and the volume would increase by a factor of 27. This holds true for other components in a generating facility as well. Table 9.2 shows how power plant size is affected by economy of scale. (Another example, that of the economies of scale in the transport of crude oil by tanker, is presented in

TABLE 9.2
How economies of scale affect power plant size in West Germany.

Capacity (MW)	Number	Total Production (MW)
<1	409	119
1 to 10	263	982
10 to 50	106	2663
50 to 200	89	9926
200 to 500	52	16,883
500 to 1000	29	21,002
>1000	13	23,113

SOURCE: *die Zeit*, 21 September 1984.

FIGURE 9.7

Economies of scale in production of ammonia. Source: *Industrialization and Productivity*, U.N. Bulletin 10, 1966; Fig. VII, p. 16. Reproduced by permission.

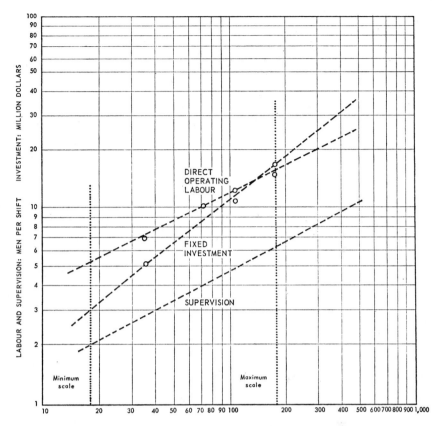

PRODUCTION SCALE: THOUSAND TONS PER YEAR

the box on page 133.) Figure 9.7 illustrates economy of scale in industrial production of ammonia.

THE MIX OF GENERATING FACILITIES

Most utilities use a mixture of generating facilities from fossil fuel and from nuclear fission; that is, some energy is supplied by using turbines fired on natural gas, some by oil or coal, and some by nuclear reactor. If the utility is burning fossil fuel, it might use an advanced design boiler (discussed in Chapter 13) instead of the standard grate furnace. Why would a utility want to have such a diversified mix of generating facilities?

As pointed out earlier in this chapter, the utilities are faced with a fluctuating demand that

depends on the hour of the day. Only about 50 percent of the total demand is constant, all-day demand. Figure 9.8 is a schematic showing the daily winter and summer demand for energy in southern California (7). Note that, in either case, peak demand lasts only several hours a day. It would be wasteful for the utility to be capable of generating peak demand at all times during the day, but the capacity to handle peak demand must of course be there, in order to allow the generators to work harder when necessary as the demand arises.

It is reasonable for the utility to have some generators that will be on call for only a few hours a day. But of course, different types of generators cost different amounts to build and to run, which brings us back to the clash between capital costs and operating costs. The owner of a small business can afford perhaps to be blasé about operating costs, and consider only capital costs in building his plant, but the utility, with its immense appetite for energy, cannot ignore the costs of operation.

These cost tradeoffs can best be represented for the utility by a *screening curve*. Figure 9.9 shows a screening curve based on costs as of 1972. The utility uses such a curve to decide which sort of source is most cost-effective for each purpose. Gas turbines provide the cheapest

peak energy, nuclear reactors the cheapest base load. A utility would not construct a nuclear facility for peaking power nor use gas turbines more than sporadically in times of high demand. Of course, costs have changed drastically since that time for both construction and for fuel, but the tradeoffs indicated remain relatively the same. Table 9.3 illustrates how costs for nuclear energy have changed over the decades of the 1970s and 1980s (8). Nuclear energy is still the most expensive alternative and getting more so (9), and the United States is just about the most expensive place to build a reactor (see Table 9.4) (10). As a result, no orders for nuclear plants since 1974 have resulted in completed plants. Nuclear energy has borne a heavy burden of regulation and public disenchantment because of perceived safety issues and because of large cost overruns

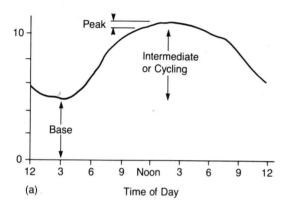

(a) Time of Day

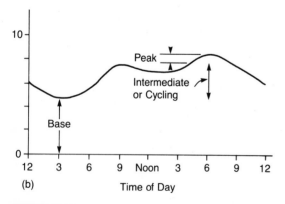

(b) Time of Day

FIGURE 9.8
(a) Summer demand in southern California (30 August 1976). (Data courtesy of Southern California Edison.) (b) Winter demand in southern California (6 January 1976). (Data courtesy of Southern California Edison.)

TABLE 9.3
Capital cost per kilowatt of coal-fired versus nuclear.

	Coal	Nuclear
1971	912	964
1978	983	1510
1988	977	1709

SOURCE: Adapted by permission from reference 8, corrected to December 1984 dollars. Copyright 1981 by the New York Times Corporation.

TABLE 9.4
Comparative cost of nuclear energy in various countries ($/kW installed).

Country	Cost
France	680
Italy	812
Belgium	876
Netherlands	1148
West Germany	1213
United Kingdom	1298
United States	1434
Japan	1438

SOURCE: Reference 10. Copyright © 1983 by The New York Times Company. Reprinted by permission.

ECONOMIES OF SCALE IN TANKER TRANSPORT OF CRUDE OIL (6)

The costs involved in transport of crude oil by tanker are partly capital expenses—that is, shipbuilding expenses—and partly costs of ship operation. Let us consider two tankers of different size built at the same time, all other things being equal. We shall take the small tanker to carry 100,000 barrels of oil, and the large tanker to have a capacity of 800,000 barrels (eight times as much as the smaller tanker).

In order for one tanker to have eight times the capacity of another, it must be approximately twice as long, twice as wide, and twice as deep. The larger ship requires only four times the volume of steel as the smaller ship for a hull of the same thickness because the amount of steel necessary is determined by the ship's surface area. Actually, the thickness of the hull must be increased, but not by a factor of 2, so that there is a savings in cost of steel per barrel of capacity.

Also, the power plant necessary to run the ship is not as expensive per barrel of capacity because the bigger ship's system weighs and costs less per barrel of capacity, and because the larger ship needs only about three times the power to go as fast as the smaller one (6).

Operating expenses are smaller also. Propulsive power per barrel is smaller for the large ship, so less fuel is consumed per barrel of capacity. The crew sizes of the two ships are about the same—the type and number of sea duties have little to do with size. Hull and deck maintenance on the larger ship use only half as much work per barrel of capacity.

In all, there is about a threefold reduction in transport cost per barrel of oil carried by the larger ship (6). This is the *economy of scale*.

FIGURE 9.9
Screening curves for production of energy from gas turbine, fossil steam, and nuclear steam plants is shown, for 1972 costs.

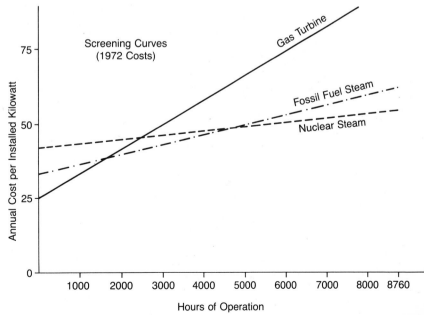

in the construction of nuclear plants. Costs as of late 1984 and early 1985 for construction of various facilities are presented in Table 9.5 (11,12).

Note that the installation cost in 1972 (Figure 9.9) of a nuclear power station is the greatest cost of the alternatives shown ($42/installed kW), but its operating cost is low ($1.50/kWh$_e$). The installation cost is a yearly cost amortized over the life of the installation (usually assumed to be fifteen years). If the facility is used for more than 5000 hours a year, the figure would indicate that the nuclear facility is the cheapest available way to supply the demand. The utilities would tend to construct nuclear power stations for the base load, at least as long as the relative relationships shown in Figure 9.9 continue to hold. One of the biggest problems facing the nuclear industry is its reliability. Nuclear plants are often shut down for various reasons. The reliability is increasing (13), but this is a problem that must be overcome to assure the utilities that nuclear energy is the appropriate choice for base load.

The installation cost of the gas turbine is the lowest ($18/installed kW), and turbines can be turned on and off easily, but the fuel is the most expensive ($9.10/kWh$_e$). This energy is thus useful for peak load; it may be used up to about 2200 hours per year and remains the cheapest way to generate the energy. For operations requiring more than 2200 hours per year but fewer than 5000 hours per year, coal-fired or oil-fired generators would give the lowest overall cost. It is for these reasons that utilities mix their modes for generating energy.

Other schemes are available for meeting peak demand. In Chapter 18, we discuss many alternatives for storage of energy to be used at peak demand.

Generating Units Become Larger

Whatever type of generating facility is built today, it is likely to be large. The history of the largest size unit on order is shown in Figure 9.10 (14). The largest units are now at about 1000 MW$_e$, so the logistic curve is coming into play yet again. The historical increase in size is a reflection of both the effect of technology in keeping electricity prices low and the incentive to realize economies of scale. As the producers of energy learn more from their experience in running a facility, they are better able to innovate the next time they build a plant to produce energy. The same holds true in other endeavors; for an ex-

TABLE 9.5

Capital costs of various generating units ($/kW installed).

	Southern California Edison (1000 MW units)	Edison Electric Institute (1000 MW units)
Nuclear	3730	1750 (pre-TMI completion)
Coal	2730	2540 (pre-TMI completion)
		1100 (southern state)
Coal-gas combined cycle	2820	
Biomass	4200	
Geothermal	1330	
Wind	1550	
Fuel cells	1350	
Solar thermal	2550–3650	
Solar photovoltaic	3440	
Hydroelectricity	1500	
Compressed air storage	750	
Combustion turbine	550	

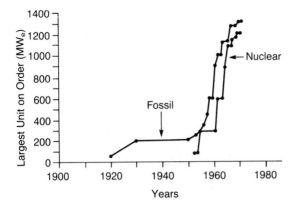

FIGURE 9.10
The trend in size of U.S. steam generating facilities between 1920 and 1969 is indicated in data supplied by the U.S. Atomic Energy Commission.

ample from another industry, the price of hand calculators has declined as the total number produced has increased. This experience has been summarized by Fisher (15):

Each time there is a doubling of the total cumulative number of units manufactured, or transported, or processed in some economic activity, there results a fixed percentage decline in cost.

If we assume a demand of 16 TkWh$_e$ in the year 2000, we would have to build a new 1000 MW plant every week until the end of the century (14). Of course, as we have discussed, the more realistic estimate is 1 to 3 TkWh$_e$, which would require far less construction. Still, a lot of new facilities will have to be built.

"Hard" and "Soft" Energy Paths and Cogeneration

In addition to economies of scale, a larger power plant can recover efficiency by operating at higher temperatures than are possible for a smaller facility. However, in this case, there is a price to be paid—increased pollution. More nitrogen oxides are emitted when the combustion temperature is higher (16). These factors have led to

ever-increasing centralization of electric generating facilities. Amory Lovins (17) has called the continuation of the trend toward more centralized energy generation the *hard energy path*. Lovins favors development of renewable, decentralized energy sources, his so-called *soft energy path,* as the rational and fruitful way to approach the future. He views the strategy of "more of the same" with great disaffection and sees disincentives to change as structural; he believes that market forces can induce changes that will lead to increased conservation efforts. The price rise in gasoline throughout the 1970s and early 1980s has proved this to be true for energy efficiency in cars.

Lovins sees "soft" technologies as able to provide all energy use after about 2025. These soft paths are characterized by:

reliance on renewable energy sources,

diversity of sources of supply, with total supply aggregated from many modest inputs,

flexibility and use of equipment of low technological level,

match in scale and geographic distribution of supply to end use, and

match in energy quality.

As we have earlier emphasized, we lose two-thirds of the energy value in producing electricity from fossil fuels. If we want to heat a room, it is usually better to burn the fuel oil or gas directly. This is what Lovins refers to when he speaks of energy quality. Smaller installations would minimize transmission losses; roughly half the amount of our electric bills in the United States goes to pay transmission and distribution costs, not generation costs (17). Lovins favors *cogeneration,* a time-honored method by which, for example, a paper mill could generate steam for making paper and incidentally get electricity as a byproduct. The Public Utilities Regulatory Policies Act (PURPA), passed by Congress in 1978, allowed cogenerators or producers to sell their extra electricity (up to 80 MW) to utilities at the marginal (peak, or "avoided") cost of generating energy. Avoided cost is the most expensive energy that the utility buys or generates. PURPA has encouraged many producers and cogenerators (Figure 9.11)(18), even including the Bronx Zoo (18), in a sense turning the clock back to the situation a century ago, when generators were widely scattered. Utilities claim that the law costs their customers more for energy because utilities are forced to pay producers at the utilities' own highest cost of production under PURPA.

In *district heating,* a generating facility supplies waste heat for space heat in large districts of cities. The Europeans have used the district heating concept extensively (19), and conditions favoring district heating exist in some American cities. Baltimore Steam Company in Baltimore, Consolidated Edison of New York, Detroit Edison in Detroit, and Eugene Water and Electric Board in Eugene, Oregon, among others, have significant district heating programs; and the Trenton, New Jersey (20), and Cedar Rapids, Iowa, business districts are heated by cogeneration.

Other similar ideas have been suggested that would lead to smaller generating facilities. For example, in *total energy systems,* small generators run on the burning of trash from a small housing or apartment complex, producing electricity. Transmission losses are small, and waste heat from the generator is used to supply space heat in the homes. This is a promising idea and should certainly be pursued. At present, it suffers from problems in maintenance and efficiency (trash combustion temperature is low). Maintenance costs at least ten times as much per kWh_e for a small power installation as for a utility. It would be difficult for such a small operation to afford more than the cost of a full-time operator to run the facility; it would not be enough to hire a full-time technician. The utilities spend less than 0.1 cents per kWh_e on maintenance, and have qualified personnel on hand to deal with emergencies (7), again demonstrating economies of scale.

It is probably more feasible to set up such a total energy system for a large town or part of a city. This would assure a larger supply of trash and waste to be burned, and would increase the number of customers for the electricity generated and the waste heat to be used for space heat. If the usage base is large enough, it should be possible to afford proper maintenance.

The idea of coordinating electricity generation with the supply of low-grade waste heat for space heating or for process steam (used in industrial applications) is not new. As previously mentioned, Consolidated Edison of New York sells much of its reject heat. As another typical example: Dow Chemical was to have bought reject heat from a nuclear plant built by Consumers Power Company of Midland, Michigan. Construc-

FIGURE 9.11
The number of kilowatts of cogeneration capacity available in the New York Consolidated Edison service area have increased substantially, especially since the passage of the Public Utilities Regulatory Policies Act. (Data courtesy of Consolidated Edison.)

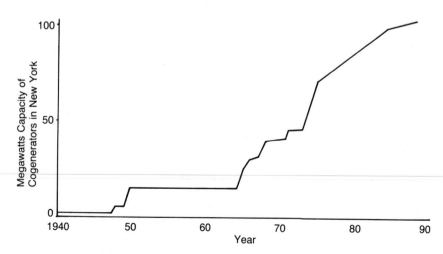

tion delays caused Dow to reconsider and reduce the amount of steam it would buy in 1974; as of 1985, the plant was still uncompleted, one among the many nuclear plants cancelled in the 1980s (21). Yet another example is that of paper companies, which must maintain boilers to obtain steam for processing of paper. Before the state governments stopped them from competing with the electric companies by setting up electric utility monopolies, paper companies sold much of the electricity they were able to generate in this incidental way.

Some utilities are following the suggestion of researchers like Lovin and are getting involved with renewable energy. California utilities in particular have embraced various renewable energy sources: Southern California Edison is using windmills near Palm Springs, as well as "Solar One," a large-scale solar generating facility developed in cooperation with the Department of Energy. These alternative sources are discussed in more detail in Chapter 15. A 1984 survey of sixty-five utilities conducted by the Electric Power Research Institute (22) revealed that twenty-four see future hydroelectric generation as important to them, twenty are doing research on biomass alternatives (see Chapter 16), and fourteen are working on wind energy.

Other factors may be pushing the utilities in Lovins' direction. It appears that the largest plants are proving much less reliable than smaller plants (23), casting into some doubt the gains possible from economy of scale. On the other hand, smaller plants may cause more deleterious effects per kilowatt than larger ones (24). The Tennessee Valley Authority, one of the successes of federal involvement in the Depression, committed a great deal of its resources to the development of large-scale coal-fired and nuclear plants. The TVA has since responded to criticism by returning to the ways of its past (25), encouraging conservation and dropping plans to increase its commitment to nuclear energy.

This trend is reflected in the industry as a whole. Orders for new plant capacity fell from 94 GW$_e$ in 1974 to an average of 3.5 GW$_e$ in 1979

to 1981 (26,27) to an average of 0.9 GW$_e$ for 1982 to 1984 (27). As of the mid-1980s, new orders are essentially absent; in 1984, utilities cancelled 23 GW$_e$ on order (27). This downward trend causes industry analysts some concern. Most generation facilities take roughly a decade from order to first use. If more energy is needed a decade hence, the plants must be started now. Peak demand is expected to increase from 465 GW$_e$ in 1985 to 546 GW$_e$ in 1994 (27). The projected capacity in 1994 is 704 GW$_e$, bringing the reserve margin below 20 percent (27). A healthy reserve margin is necessary in order to deal with weather-induced demand, unanticipated shutdowns, and necessary maintenance (27). The utilities are planning to meet the demand by purchasing hydroelectricity and nuclear electricity from Canada. These power imports are projected to rise from 196 MW$_e$ in 1985 to 571 MW$_e$ by 2000 (27). Demand, which rose to an annual rate of 5.7 percent from about 2 percent as the recovery from the 1981 to 1983 recession began, fell back to an annual rate of 3.4 percent in 1984 and is still falling (27). With all the uncertainty, it appears that the best way for the utilities to proceed in the present poor investment climate (26) is to continue to encourage conservation and to plan to build modular units that can be erected quickly and added to as necessary (see, for example, the discussion of the Cool Water Coal Gasification Plant in Chapter 13). In this instance, we see again, as we saw in the previous chapter, how important reliable projections are, and how difficult they are to make. Electricity is vital to the functioning of the country, but we have little idea of what even the near future will bring. This uncertainty may well be the most persuasive cause of the change to a softer energy path if Lovins' projections prove true.

The director of the Italian Nuclear and Alternative Energy Agency has argued that electricity could be useful in development in the Third World (28): It is still coupled to growth in Gross Domestic Product (see Chapter 8); it is convenient; rural electrification could stem a rush to the cities; and it could be small-scale and based locally on renewable resources. These arguments follow Lovins' arguments for the soft approach. Europeans have been somewhat more receptive to these ideas than Americans (29).

ENERGY USE BY APPLIANCES AND THE PHENOMENON OF SATURATION

Of particular interest is the distribution of energy use by small appliances. These appliances, as shown in Table 9.6 (2), consumed over 47 billion kWh$_e$ in 1969, but this amounted to a tiny 0.8 percent share of total energy use. While it is useful to practice economies by halting gratuitous use of these appliances, their effect on total national consumption is small indeed. Table 9.7

presents a more balanced picture of household appliance energy use.

By this time, over a century past the first large-scale commercial use of electricity, we may reasonably hope that we have reached the end of the road in these small ingenious devices. In any case, they all take such a small share of demand that they are of no practical importance. While the U.S. population was growing at a rate of only 1.3 percent per year between 1960 and 1968, basic energy consumption increased an average of 4.1 percent per year (2). It is clear that faddish devices such as electric toothbrushes and electric carving knives could not be responsible for this great growth; the increase is attributable mostly to space heating and water heating, frost-free re-

TABLE 9.6
Electrical consumption of selected small appliances.

	Annual Number kWh per Item*	Number of Items (millions)†	Consumption (billion kWh)
Bed coverings	147	27.0	3.97
Blenders	1	16.0	0.02
Broilers	85	14.0	1.19
Clocks	17	55.0	0.94
Coffee makers, automatic	140	50.0	7.00
Dehumidifiers	377	3.8	1.45
Fans (circulating)	43	75.0	3.21
Food disposers	30	13.5	0.41
Hair dryers	25	22.5	0.56
Humidifiers	163	4.0	0.65
Frypan skillets	100	33.0	3.30
Heaters (portable)	176	17.0	2.98
Hot plates	90	15.0	1.35
Irons	60	57.0	3.42
Knives (carving)	8	13.0	0.15
Mixers	2	49.0	0.10
Radios	86	57.0	4.90
Shavers	0.5	24.0	0.01
Toasters	39	54.0	2.11
Toothbrushes	1	15.0	0.01
Vacuum cleaners	46	53.0	2.44
Total			40.17

*Source is Edison Electric Institute, 1987.

†Estimated number in households as of mid-1969. Sources are *Merchandising Week*, 23 Feb. 1970, and Stanford Research Institute estimates.

TABLE 9.7
Electricity consumption
of appliances, 1987.

	Wattage(W)*	Average Number of Hours Used per Year	Annual Use (kWh)*
Food			
Dishwasher	1200	30	36
Drip coffee maker	1500	60	90
Freezer			
15 ft³ manual	320	—	1512
15 ft³ automatic	440	—	1824
Hot plate	2350	24	30
Knife	100	6	0.6
Microwave oven	1500	360	540
Range			
Small surface unit	1300	163	212
Large surface unit	2400	82	197
Oven	3200	240	768
Self-cleaning oven	3200	240	614
Toaster oven	1400	24	34
Trash compactor	400	3	1.2
Clothes			
Dryer	5000	Regular	684
		Permanent press	360
Iron	1000	60	60
Washer	500	90	45
Water heating	2500	1600	4000
Household			
Blanket	175	1200	210
Clock	2	8760	18
Lamp	75	200	15
Shaver	14	60	0.8
Stereo	100	120	12
TV			
Black and white	55	2160	119
Color	200	2160	432
Toothbrush	7	60	0.4
Vacuum cleaner	650	24	16

*Estimated.
SOURCE: Ohio Edison, 1987.

frigeration, and growth in air conditioning. The distribution of energy to the consumer changed very little between 1960 and 1968, except for heating. This was probably a result of an over-selling of total electric homes during the 1960s; the miscalculation became unhappily apparent when the price of electricity skyrocketed in 1973. Older uses of energy, such as for clothes washing and cooking, have not really increased much since the 1950s.

The relative stagnation of the older uses of energy is a reflection of the fact that there is a limit to how much hot water, how many stoves, and how many washing machines we need. When every family has a stove, then the rate of sale of stoves will simply reflect the growth rate of the population and the retirement of old stoves. This phenomenon is called *saturation,* which is defined as the proportion of households having a particular appliance to the total number of households. Refrigerators and stoves are in just about all American homes. Washing machines are in three-quarters of homes (apparently, this is about the limit because some apartment-dwellers must use, and others prefer to use, laundromats). Black-and-white TVs are a supersaturated commodity, but the market for color sets is still alive and well. Clothes dryers and air conditioners still have some growth potential, but water heaters are in practically every home already. At some future time, the market for newer uses of energy will become just as saturated as that for stoves or water heaters. Figure 9.12 shows a typical saturation curve from its beginning to total saturation. We have seen such a logistic curve before, in Chapters 4 and 8. Figure 9.12 gives us a basis for estimating how long it will take an appliance to reach saturation. The growth rate of all large appliances will eventually decrease to the growth rate of the American population.

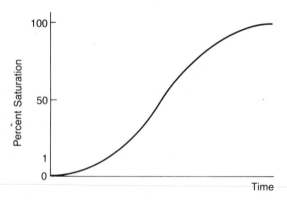

FIGURE 9.12
Saturation of appliances with time follows a logistic curve.

In Appendix 3, some additional background material for this section is presented, as well as a discussion of how graphs can be misleading. Included in this material are explanations of how the tables and charts of this section follow from the tables of Reference 2.

TRANSPORTATION OF ELECTRICITY

In Chapter 5, we discussed the reasons that high-voltage transmission lines are used to transport electricity. You will recall that power generated by a power plant is the product of the current and voltage produced, and that, the greater the current, the greater the power lost to Joule heating. It follows that the power plant's output is stepped up to high voltage so that the energy loss is minimized.

Long-distance transmission has changed in character over the years. Originally, "long distance" referred to interlinkages between adjoining utilities. The depression of the 1930s brought massive federal involvement in energy in the United States: in the South and Midwest, the TVA; in the West, the massive Columbia River and Colorado River projects. With these projects far away from urban concentrations, it became necessary to transport energy over long distances. Those first transmission lines were 138 kV; today the lines are 765 kV or greater (30). There is still no agreement as to whether there are significant dangers to people living near high-voltage transmission lines (30,31). The USSR has set exposure limits for their industrial workers, and people have been known to experience painful skin sensations when they are in regions where the electric field is greater than 15 kV/m (30).

Despite these possible risks, transmission lines interconnect, and since 1976, the entire country has been joined into one interconnected transmission grid, which also reaches into Canada (31). These interconnections are pictured in Figure 9.13 (32). They help utilities smooth out hour-

FIGURE 9.13
The interconnected system of electrical distribution in the U.S. and Canada. (Courtesy of the National Electric Reliability Council.)

to-hour variations among users (for example, near adjoining time zones, where peak usages occur an hour apart) and allow the United States to import much surplus hydroelectricity from Canada (4). The huge hydroelectric project at James Bay in Canada is partly based on the assumption that the excess electricity may be sold in the United States (33).

There have been some decided problems due to interconnections. The blackout in most of the northeast states on November 9, 1965, was traced to a problem in the transfer point at Niagara Falls. A blackout in the New York metropolitan area on July 13, 1977, was traced to three lightning strikes on transmission lines in upstate New York

(34). Surge protectors and automatic cutoffs prevented this blackout from spreading as far as the one in 1965, although the 1977 blackout lasted more than a day.

Long-distance transmission of electricity costs about 8 percent of what local transmission costs per kilometer (35). Nevertheless, the total transmission cost can loom large—over 50 percent of the cost of energy (17). As a result of this high cost, research is proceeding on methods of electric transmission for which Joule heating is much less of a problem than in the current methods. One promising method involves use of superconducting transmission lines (see Chapter 18 and the box following).

SUPERCONDUCTIVITY

Kamerlingh Onnes discovered in 1911 that very cold lead ($\cong 11$ K) lost altogether its resistance to the flow of electric current. He soon discovered other materials for which this happened at low temperatures. This phenomenon is called *superconductivity*. In a superconducting current loop, the current will circulate without changing for a very long time (an experiment carried out for over a year verified this aspect of the phenomenon).

Many materials are superconductors. Each material exhibits a transition to the superconducting state at a different temperature. Most transition temperatures are very close to absolute zero—liquid helium at 4.2 K is used to cool down most samples. Since the discovery of superconductivity at very low temperatures, there has been research aimed at finding superconductors that work at higher temperatures (36). As time passed, materials were discovered for which the transition occurred at higher tempertures. While no room-temperature superconductors have yet been found, compounds made of yttrium, barium, copper, and oxygen have exhibited superconductivity in the region up to about 100 K, with hints of superconducting behavior at higher temperatures.

Materials with high transition temperatures would be ideal to use for electricity transmission if they are found. Even so, the building of superconducting transmission lines at present is problematic. Materials for transmission lines remain to be fabricated. Estimates are that it will be economical to install such superconducting transmission facilities only for transmitted powers in the gigawatt range unless true room-temperature superconducting materials can be discovered. Even though in theory there are no losses, eddy currents (see Chapter 5) will still cause losses. The eddy current Joule heating losses must be kept to below 0.1 μW/mm^2 (36). There are unresolved problems with current surges (three to ten times the rated current in AC lines), and doubts about the ability of the relatively high-temperature superconductive materials to withstand large temperature changes repeatedly. Research continues on superconductor transmission because it will make possible the cutting of losses by over 50 percent. Even if the highest transition temperatures remain at about 100 K, this is above the temperature at which nitrogen liquefies, and it is cheaper to use a nitrogen liquefier to supply refrigerant than to continue to accept Joule heating losses in conventional transmission lines.

SUMMARY

Transportation and space heat uses account for about half of the yearly energy use in the United States. In both these uses, liquid fuel or natural gas provides the major energy source.

Electric utilities, originally small local facilities competing for business, became publicly regulated monopolies around the turn of the twentieth century. The major reason for this is the difficulty of storing electricity, since demand varies from hour to hour and month to month. Another reason has to do with economies of scale, which make energy generation cheaper as the facilities are made larger.

In order for a utility to decide what sort of facility it wishes to use to generate energy, the capital cost of the facility must be known and the operating cost of the facility must be projected. The utility's screening curve is then used in conjunction with its load curve to determine the lowest cost alternative.

The law known as PURPA has caused the energy utilities' monopoly grip to loosen a bit. Small installations can supply energy by cogeneration or alternative energy sources and sell the excess, up to a maximum, to the utilities, which must buy it at the "avoided cost."

Because of advances in the technology of electricity transmission over long distances, almost all of North America is interconnected into a grid. This allows shuffling of cheap hydroelectricity and nuclear electricity from Canada into the United States and among domestic utilities and helps utilities smooth out peaks and valleys in demand. At some time in the future, superconducting transmission lines may lessen transmission losses still further.

REFERENCES

1. Department of Commerce, *Statistical Abstract of the United States 1984* (Washington D.C.: GPO, 1984).
2. Stanford Research Institute, *Patterns of Energy Consumption in the United States* (Washington, D.C.: GPO, 1972).
3. S. Novick, *Environment* 17, no. 8 (1975): 7 and 35.
4. Economic Council of Canada, *Connections* (Ottawa: Canadian Government Publishing Centre, 1985).
5. C. A. Berg, *Science* 184 (1974): 264.
6. J. A. Fisher, *Energy Crises in Perspective* (New York: John Wiley & Sons, 1974).
7. R. Stuart, *New York Times,* 10 Aug. 1975; also data from Consolidated Edison of New York, quoted in M. Leebaw and H. Heyman, *New York Times,* 12 Sept. 1976.
8. R. D. Hershey, Jr., *New York Times,* 8 March 1981.
9. M. L. Wald, *New York Times,* 1 Jan. 1984 and again 20 Jan. 1985.
10. P. Lewis, *New York Times,* 11 Dec. 1983.
11. D. S. Johnson, Senior Planning Engineer, Southern California Edison, private communication, June 1985.
12. C. F. Fisher, Jr. and W. R. Schriver, *Power Plant Cost Study* (Washington, D.C.: Edison Electric Institute, October 1984).
13. D. McInnis, *New York Times,* 4 Oct. 1981; J. M. Fowler, R. L. Goble, and C. Hohenemser, *Environment* 20, no. 3 (1978): 25; and R. L. Goble and C. Hohenemser, *Environment* 21, no. 8 (1979): 32.
14. Atomic Energy Commission, *The Safety of Nuclear Power Reactors and Related Facilities,* WASH-1250 (Washington, D.C.: AEC, 1973).
15. J. A. Fisher, *Energy Crises in Perspective* (New York: John Wiley & Sons, 1974).
16. C. A. Berg, *Science* 181 (1973): 128. In 1973, cost of maintenance was 0.03 cents/kWh$_e$. This is 0.07 cents/kWh$_e$ in 1984 dollars.
17. A. Lovins, *Foreign Affairs* 55 (1976): 65.
18. S. Diamond, *New York Times,* 24 June 1984.
19. J. Karkheck, J. Powell, and E. Beardsworth, *Science* 195 (1977): 948.
20. C. Stern, *ASHRAE J.* 27, no. 2 (1985): 36.
21. W. D. Metz, *Science* 188 (1975): 820.
22. J. S. Feher, F. S. Young, and R. W. Zeren, *Environment* 26, no. 10 (1984): 12.
23. J. M. Fowler, R. L. Goble, and C. Hohenemser, *Environment* 20, no. 3 (1978): 25; and R. L. Goble and C. Hohenemser, *Environment* 21, no. 8 (1979): 32.

24. H. W. Lorber, *Environment* 22, no. 4 (1980): 25.
25. S. D. Freeman, *Environment* 27, no. 3 (1985): 6.
26. P. Navarro, *New York Times,* 7 July 1985.
27. M. Crawford, *Science* 229 (1985): 248.
28. U. Columbo, *Science* 217 (1982): 705.
29. *Environment Policies for the 1980s* (Paris: Organization for Economic Cooperation and Development, 1980).
30. L. B. Young, *Environment* 20, no. 4 (1978): 16.
31. M. W. Miller and G. E. Kaufman, *Environment* 20, no 1 (1978): 6; and A. A. Marino and R. O. Becker, *Environment* 20, no. 9 (1978): 6.
32. R. Stuart, *New York Times,* 25 May 1976; P. Khiss, *New York Times,* 18 Feb. 1979; and M. L. Wald, *New York Times,* 11 Nov. 1984.
33. G. Laroque, *Bull. Am. Phys. Soc.* 30 (1985): 32.
34. A. McGowan, *Environment* 19, no. 6 (1977): 48.
35. T. H. Maugh II, *Science* 178 (1972): 849.
36. M. K. Wu et al., *Phys. Rev. Lett.* 58 (1987): 908; S. R. Ovshinsky et al., *Phys. Rev. Lett.* 58 (1987): 2579. See also T. H. Geballe and J. K. Hulm, *Sci. Am.* 243, no. 5 (1980): 138; and J. K. Hulm and B. T. Matthias, *Science* 208 (1980): 881.

PROBLEMS AND QUESTIONS

True or False

1. A product is fully saturated only when it reaches 100 percent distribution.
2. Replacement sales are a large component of the air-conditioning industry's sales.
3. Novelty items will eventually become *major* drains on our energy supply.
4. Superconductors offer major immediate savings in costs of energy transmission.
5. The U.S. electricity grid allows transmission of electric energy nationwide.
6. It is usually less expensive to build something that is partly prefabricated than to build everything onsite.
7. Cogeneration of electric energy has occurred ever since the beginnings of the electric distribution system a century ago.
8. PURPA has made a great difference in the economy of buying cogenerated energy for the utility.
9. It is usual for base-load energy to be the cheapest to produce per kWh.
10. It is advantageous for saving energy for each family to have its own electric generating system.
11. As the utilities build more generating facilities, the cost per kWh_e should decline.

Multiple Choice

12. As a manufacturer, for your best future sales, you should probably produce a
    a. refrigerator.
    b. black-and-white TV.
    c. iron.
    d. home computer.
    e. food processor.

13. Which of the following is *not* an economy of scale? Savings due to:
    a. more capacious oil tankers.
    b. larger petroleum refineries.
    c. higher temperature in boiler vessels in power plants.
    d. increased size of power shovels used in strip mining.

14. Which method of transmitting energy probably costs the least?
    a. Transmission of electric energy underground
    b. Transmission of chemical energy using a bucket brigade
    c. Transmission of chemical energy using tanker trucks
    d. Transmission of chemical energy using a gas pipeline
    e. Transmission of electric energy on 765 kV high-tension wires

15. In terms of human safety near high-voltage transmission lines, it is
    a. clear that it is safe for people to be nearby.
    b. clear that it is dangerous for people to be nearby.
    c. never going to be possible to decide whether it is safe for people to be nearby.
    d. not yet clear whether megavolt transmission lines have any effect on people.

16. The largest energy consumer in the average electric house (outside of space heat) is probably
    a. cooking.
    b. refrigeration.
    c. cleaning.
    d. entertainment.
    e. water heating.

17. Which of the following methods could lead us toward "softer" energy paths?
    a. Adoption of more individual solar energy units for electricity
    b. Use of cogeneration
    c. Using the sun to heat water
    d. Adoption of total energy systems
    e. All of the above

18. Which of the characteristics listed below are likely to be the least descriptive of the hard energy path?
    a. Monopolistic control of the energy distribution systems
    b. Use of depletable resources
    c. Many different sources of supply
    d. Large-scale generation of electricity
    e. Centralization of decision making

19. For the customer demand as indicated in Figure 9.8, how would you mix your generating facilities if you were a utilities manager?
    a. Build all-nuclear generating facilities.
    b. Build all-fossil-fuel generating facilities.
    c. Build nuclear facilities to supply about 50 percent of maximum load, fossil fuel to supply about 40 percent, and gas turbines to supply the remaining 10 percent.
    d. Build mostly nuclear facilities, with coal or oil reserved for peak demand.
    e. Build mostly coal or oil generators, with gas turbines reserved for peak demand.

20. The screening curve of Figure 9.9 indicates that the cheapest option for producing energy needed throughout the year is
    a. gas turbine.
    b. coal- or oil-fired steam.
    c. nuclear steam.
    d. impossible to determine from the curve.
21. The screening curve of Figure 9.9 indicates that the cheapest option for producing energy needed for less than 2000 hours per year is
    a. gas turbine.
    b. coal- or oil-fired steam.
    c. nuclear steam.
    d. impossible to determine from the curve.

Discussion Questions
22. Should electric utilities be monopolies?
23. List the advantages and disadvantages of "total energy" systems.
24. It is already possible to build sun-powered water heaters. Would this be a useful energy-saving option?
25. Discuss advantages and disadvantages of the nationwide electric grid.
26. Why should Lovins prefer the "soft" energy path? What arguments against this view can you think of?
27. Discuss some possible disadvantages of "economies of scale" (e.g., superconductors, very large concentrations of energy generation facilities).
28. Why should annual load peaks be different in different locations?
29. Explain how the details of summer and winter daily demand arise (see Figure 9.3).
30. Identify advantages and disadvantages of district heating in North America, Scandinavia, and Jamaica.
31. Discuss the difference between the "corrected" and "uncorrected" energy figures referred to in Figures 9.1 and 9.2. How does one fairly compare a kWh_e from hydro sources and a kWh_e from thermal sources?
32. How could we reduce the break-even time for a generating facility?

10

ENVIRONMENTAL EFFECTS OF UTILITY GENERATING FACILITIES

In the preceding two chapters, we have discussed issues relating to the historical development and construction of generating facilities. Some of the consequences of choosing a particular energy option include effects on the land, the air, and the water, as well as upon plants and animals dependent upon these resources. Such consequences are explored in this chapter.

KEY TERMS   *waste heat*
biochemical oxygen demand
hypolimnion
epilimnion
stratification
wet cooling
dry cooling

We consider some general effects following from the method of generation by the utilities and some features common to both fossil-fuel and nuclear energy generation. These include the problem of waste heat production, the various negative effects of constructing a generating facility, and the ultimate gain in energy of a generating facility. Issues peculiar to one particular mode of power production are considered in later separate chapters.

EFFECTS OF WASTE HEAT ON AQUATIC SYSTEMS

Consider a power plant burning fuel energy at a rate of 3000 MW to produce electricity for sale. Of the 3000 million joules per second produced, 1000 million constitute electrical energy transmitted to the consumer, and the other 2000 million joules are exhausted into the environment as *waste heat*. In practice, this means that an amount of water is heated in the power plant and then returned at a higher temperature to a river or lake, where it gradually cools by heat exchange with other water and the air, or the waste heat is exhausted directly into the atmosphere by cooling towers. The waste heat exhausted into water is a particularly serious problem for two reasons: its effect on the dissolved oxygen content of the water, and the effect of heat on aquatic systems. Both fossil-fuel and nuclear plants generate waste heat and use large quantities of water. The availability of water, as well as other uses of water in the energy system, are discussed in Chapter 11.

Almost any pollution in a river or lake demands oxygen in its neutralization. If the pollutants are organic, the amount of oxygen necessary to break them down is called the *biological oxygen demand*. If the pollutants are chemical, the amount of oxygen necessary to neutralize them is called the *chemical oxygen demand* (1). We shall classify these together and simply refer to *biochemical oxygen demand—* BOD for short. Cold water contains more dissolved oxygen than warm water. Consequently, waste heat imposes an additional stress on the "cleaning power" of a river or lake and reduces the amount of oxygen available. This has a debilitating effect on aquatic life.

The other effect of heat on aquatic life has to do with the rate of chemical reactions. It is a rule of thumb that rates of reaction double for each 10°C rise in temperature. Among other effects, this causes increased plant and algal growth (2). For fish, the problem is most serious. A slight temperature change can result in the replacement of previously dominant species by others (trout and salmon, for example, prefer relatively cool, well-aerated water, and are especially sensitive). There are adverse thermal effects on most physiological processes, including reproduction, at temperatures several degrees below lethal temperatures (3); fish under constant heat stress are more susceptible to disease.

Many electric facilities are situated on lakes (or bays) rather than on a river. A temperate-zone lake generally exhibits thermal stratification in the late spring and summer, and the addition of waste heat increases and prolongs the stratification. The lower, denser layer of the lake is called the *hypolimnion;* this is much colder than the *epilimnion,* or upper, less dense, layer of the

lake. During the summer, the epilimnion gradually becomes thicker and the amount of dissolved oxygen in the lake—in the cooler hypolimnion—gradually decreases as a result of natural BOD. In the fall, the cooling of the epilimnion, which increases its density, allows the layers to mix (or turn over), replenishing hypolimnion oxygen levels. Valuable fish, such as lake trout, live in the hypolimnion and need this oxygen to stay alive and healthy. Thus, the longer the stratification exists, the less oxygen there is for the animals living there.

WASTE HEAT

Many power facilities use "once-through" cooling to eliminate waste heat. If a lake is used as a source of cooling water (or as a low temperature reservoir) for a generating facility, water is taken from the lake and pumped through condensers. The steam or other working fluid is cooled in the condensers after flowing through the turbines. It is then recycled to the boilers to be made into steam again. The cooling water is exhausted to the lake 5–10°C warmer than it was when it was pumped from the lake. Most likely, water will be taken from the cool hypolimnion and exhausted to the warmer epilimnion, increasing the difference in density. The exact temperature change of the water depends on the rate at which water is pumped through the condensers. To produce a given amount of energy requires a plant to exhaust a given amount of waste heat. This amount is fixed by the energy requirements, not the pumping rate, of the cooling water. If we pump water through faster, we simply take away a smaller amount of waste heat for each liter of water pumped through.

An Environmental Defense Fund study for a proposed nuclear facility on Cayuga Lake in New York (4) found effects that *could* seriously affect the lake and its inhabitants:

Stratification would begin earlier in spring and extend later into fall.

The length of the growing season for plants and animals would be increased in the epilimnion.

Water from the hypolimnion that would be brought to the surface would contain nutrients that were previously unavailable in the epilimnion.

There would be a greater capacity for biological production in the epilimnion.

Prolonged stratification extends the period of oxygen depletion. The effect of the waste heat burden and the environmental stress on the lake could cause breeding problems for the fish in the lake. The increased biological production is on a primary level—algae growth is encouraged; then, when the algae decay, the process uses up oxygen, to the detriment of fish and invertebrate life. This process, called *eutrophication,* often occurs gradually in the natural aging of lakes, but it is accelerated by waste heat exhaust. Waste heat exhausted into rivers is generally less of a problem than it is in lakes, but river-dwelling fish can still suffer from the effects of heat.

Once-through cooling is rarely chosen to dissipate waste heat in the 1980s (5), because most prime sites (that is, sites with a sufficient quantity of water available) are already taken and because of public concern over temperature rises. Use of manmade cooling ponds or canals for once-through cooling is possible, the heat subsequently being dissipated by evaporation in place of some of the hot water. The pond option is not often chosen because sufficient land is lacking to develop ponds in wetter climates, and sufficient water is lacking in drier climates. Occasionally, it is possible to build a facility on the site of an abandoned water-filled quarry.

There is another strategy for dissipation of waste heat: cooling towers. Currently, this is the most popular method for dissipating waste heat. In "wet" cooling towers, the heat is dissipated by evaporation. This evaporation can cause fog to form in areas around the generating plants, and has even been responsible for increased local

rainfall and snowfall levels (6). These "wet" methods are less effective when the relative humidity of the area is high.

In "dry" cooling towers, the heat is taken away directly by the atmosphere. Air passes across pipes containing hot water and carries the heat away by convection (that is, actual mass motion of the air does the cooling). The warm air will rise until its temperature is the same as the air around it; to do this, it will expand, since the air normally is cooler as the altitude increases. The rate at which dry air cools with increase in height is called the adiabatic lapse rate (0.01°C/m, see Chapter 22), because the process takes place adiabatically (rapidly), without exchange of energy to the surroundings. Dry towers are less expensive to maintain than evaporative towers, and can be used even in areas having little water. However, dry towers are less thermally efficient than wet towers (5).

These alternative cooling plants are expensive to build and run compared to those that use once-through cooling. In addition, once-through cooling can usually provide lower water temperatures than towers (6). A comparison of the continuing expenses of competing strategies is shown in Table 10.1 (7). Despite the costs, the cooling pond or tower method has been adopted because of the growing public perception expressed succinctly by Walter Heller in public testimony (8): "The public is subsidizing these industries at least twice—once by rich tax bounties and once by cost-free or below-cost discharge of waste and heat."

In the economic climate of the 1980s, upgrading of older towers is more important than new construction, because there is very little new construction (5). Natural-draft hyperbolic towers (the outer shell is in the shape of a segment of a hyperbola), more energy-efficient than forced-draft units, were a more popular option in the 1970s. Such units have performed well in North America and Europe. New units being designed are 130 m in diameter and 200 m high (5). Europeans have favored smaller towers, with me-

TABLE 10.1
Power requirements for cooling systems (percentage of output of 800 MW plant).

Cooling System	Type of Fuel	
	Fossil	Nuclear
Once through: discharge to lake, river, ocean, or cooling pond	0.4	0.6
Wet cooling tower		
Natural draft	0.9	1.3
Mechanical draft	1.0	1.7
Dry cooling tower		
Natural draft*	0.9	1.5
Mechanical draft	3.0	4.8

*Construction of natural-draft dry cooling towers is quite expensive.

SOURCE: Compiled from cooling data in M. Eisenbud and G. Gleason, eds., *Electric Power and Thermal Discharges* (New York: Gordon and Breach, 1969), 372.

chanical-draft assist units, because of lack of space. In America, most plants in operation have cooling towers of the forced-draft variety. More towers are being built of concrete rather than wood because of substantially lower maintenance costs (5).

CONSIDERATIONS IN BUILDING A POWER PLANT

In this section, specific real-life examples are given illustrating issues that were of concern to the utilities and the public during the preparatory environmental impact study for the Kaiparowits coal-fired plant. Kaiparowits was proposed by California utilities, and was to be built in remote southern Utah. Partly as a result of public response to the impact statement, this plant was abandoned in 1976.

Construction

Many large nuclear power plant installations as well as conventional installations are built on the site of the plant rather than in a factory. Factory fabrication is advantageous, if it is feasible, because it is cheaper to build units in a factory and then transport them to the site. This follows from our discussion of economy of scale in Chapter 9. Building a power plant at the site essentially means that one builds a factory whose output is a power plant; by using prefabricated units, the factory cost is amortized over all units produced (9). Often such onsite construction is necessary because the facilities are one of a kind or because of the difficulty of shipping already-constructed modules to the site.

In order to illustrate the savings associated with factory production, suppose the cost per installed kilowatt drops by a quarter each time production is doubled. We may assume a large factory to produce 1000 megawatts of installed capacity per year, and we suppose that we wish to build a 1000 MW generating facility. Typical construction time is five to ten years; therefore, we shall choose a construction time of eight years (i.e., 125 MW capacity installed per year). A comparison of the construction cost for differing amounts of onsite construction is shown in Figure 10.1. It is clear that the more construction that can be done in a factory, the better. The startup of onsite construction introduces the greatest increase in construction cost. The cost has increased over 50 percent if only 10 percent of construction is in the field; full onsite con-

FIGURE 10.1
Due to the learning curve, the unit cost of power plant capacity decreases as the produced amount increases (here, we assume a 25% decrease when production doubles). For a 1000 MW plant built over 8 years, the minimum cost of field construction as compared to that in a factory producing 1000 MW of installed capacity per year is 2.3 as great.

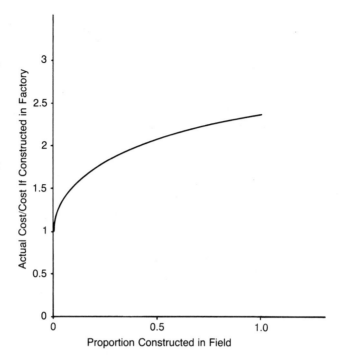

struction is only about 60 percent more expensive than the first 10 percent. The last 84 percent of construction introduces only as much extra cost as the first 16 percent of construction onsite.

Construction in remote underpopulated areas can result in even greater cost if worker housing must be supplied and labor brought in from outside. In addition, field labor is often of lower quality than that employed in factories, causing expensive maintenance or repair problems after the plant is in operation.

Boom-Town Syndrome

Undeveloped areas having a supply of natural resources may be suddenly transformed by plans for rapid use of the resource (the coal lands of Wyoming and Montana are recent examples of this phenomenon). The problem here is that a small indigenous population is pressed to provide greatly expanded local service to large numbers of new residents. Because of the lack of school space, recreational facilities, and other infrastructure necessities, it is difficult to attract the service personnel needed, and high turnover is thus much more likely. This leads to a drop in productivity, and thence profits, and further, to a drop in public services as tax revenues drop. This in turn causes even greater strain on the abilities of public servants, and coping becomes harder, leading the cycle further downhill (10).

While local government is stalled in its effort to minimize the impact of development because of widespread local sentiment against any interference with property rights, state and federal government can help by assisting the locality in the institutional changes necessary to cope with development.

The "boom-town" problem is now part of the overall environmental impact considered in siting new energy facilities. The now-abandoned plan for a large coal-fired power facility near Utah's Kaiparowits plateau required preparation of an environmental impact statement that dealt in part with where the incoming population would be housed and serviced (11). We might regard this part as dealing with "social pollution."

Other social problems include a high rate of divorce (at least a third higher than in comparable counties, as reported in a study of Wyoming), greatly increased use of alcohol and consequent drunkenness, and a higher rate of school dropouts than in comparable counties (12).

The Kaiparowits plan envisioned a scheme in which 25 percent of the plant's support staff would be housed 80 kilometers away in Page, Arizona, where facilities were available to decrease the boom-town pressure. (Of course, this would greatly increase the net fossil-fuel cost of plant operation, since that 25 percent of the workforce would commute each day by private car or contractor bus.)

The plan was ultimately dropped because of lower-than-expected increases in electric consumption in Los Angeles (13), a casualty of the first 1970s energy crisis.

Environmental Effects

Like other proposed generating facilities, the Kaiparowits plant was to be located in an area of great natural beauty. Kaiparowits had been sited close to Glen Canyon, Capitol Reef, Bryce Canyon, Zion Canyon, Lake Powell, and several national forests. A smoke plume would be out of place in this natural setting. The plant itself would have been visible over great distances, and the transmission lines also would have impaired the splendid natural vistas.

There would have been "unavoidable deterioration of air quality" (11) were the Kaiparowits plant to be built, including "periodic yellow-brown atmospheric discoloration," (11) due to nitrogen oxide emissions (perhaps as much as 10.4 tons/hour). The air quality would also have been lowered by the emission of particulates (0.58 tons/hour) and sulfur dioxide (2.2 tons/hour), and by the minor estimated emissions of 980 tons/year of particulate matter and 700 tons/year

of sulfur dioxide added by the new residents brought in to work on the plant. Air pollution is discussed in detail in Chapter 22.

The proposed removal of 420 megatons of coal from a nearby strip mine over the lifetime of the plant would have affected 63 square miles, with concomitant noise and contamination of aquifers (formations for storage of ground water) from mined-out coal beds. Also, another 420 megatons of coal would have been rendered unrecoverable. The proposal would also have involved excavation of 1.6 million cubic yards of sand and gravel for construction and mining of 6.8 megatons of limestone from the region for concrete.

Other damage from implementing the Kaiparowits project would have included pollution of Lake Powell from the final estimated 50 million cubic yards of fly ash scrubber residue (containing arsenic, barium, fluorine, lead, mercury, and other materials); increased sediment yield; and increased salinity in local streams as well as Lake Powell. Erosion losses would have been serious because the local soils have such a poor capacity for recovery (almost three-fourths of the local soils would not recover during the twenty-year project span). Salinity would have increased for several reasons: pumping-induced rock fracture between salt and fresh water aquifers, leading to mixing; evaporative salination of the evaporation ponds, liquid from which would have then leaked into ground water; and salt drift from the cooling towers onto the nearby vegetation, thence into the waters. The Colorado River would have increased in salinity by 2 milligrams/liter (see also Chapter 11).

Thus, to build Kaiparowits would have meant a loss of vegetation, impairment of habitat for wildlife, a decrease in amount and aesthetic value of recreational land, and general degradation of the environment. While some of the particular effects cited are unique to the Kaiparowits plan, most of the problems would be encountered in the attempt to build a large coal-fired facility anywhere.

Energy Gain

Given the environmental impacts of Kaiparowits, it is interesting to consider the net gain the plant would have contributed in the energy supply system. By *gain,* we refer to the high-quality energy available to people (mostly as electricity) as a result of construction of a facility. In a universal sense, we merely transform some energy into other forms—energy is conserved. It is high-quality energy, however, that is of direct use to us. Somewhat earlier in this section, we referred to one of the "invisible" energy costs of Kaiparowits: the amount of fuel that would have been used by workers in commuting. Other such "invisible" costs are found in this sort of a project, including the energy cost of construction materials and machines and the energy value of the fuel used in the construction and operation of the facility.

It is not possible to evaluate comprehensively these hidden costs of building an energy generating facility. The losses that can be measured are staggering enough. Using the projections in the Kaiparowits study (14), only 50 percent of the coal in the underground formation would actually have been mined (the remainder, left in place to support the excavation, would have been lost forever), and 25 percent of the coal mined would have been wasted. Each year on the average, mining would have cost 710 GWh_e, and transporting of the coal to the plant would have required 730 GWh_e. Assuming the plant to be 38 percent efficient in converting kWh_t to kWh_e, its overall efficiency in terms of the energy in the coal mined would have been only 25 percent; if we include in the energy available the other half of the coal resources that are rendered permanently useless, the overall efficiency would have been 12.5 percent. (Reference 12, which does not properly allocate kWh_e, obtains an overall efficiency of 15.5 percent.) It must be noted that this efficiency is further reduced if one takes into account the energy cost of producing the cement and steel used in the plant, the energy cost of

TABLE 10.2
Ratio of energy out/energy in.

	0.3% Ore	0.007% Ore
Pressurized water reactor*	8–19	0.5–4.8
CANDU*	5–13	5–7.4
Heavy water reactor*	9–13	2.1–2.9

*See Chapter 14, "Conventional Reactor Types."

transporting the material, the requisite earth-moving, and so forth. This still does not count the energy cost of moving people to the site or the energy cost of daily commuting.

To be fair, this cost is shared by any such large undertaking and is not restricted to coal-burning facilities. Analyses of the energy cost of a 1000 MW_e nuclear installation used for twenty-five years (15) show that it would provide a net energy loss during the first eight to nine years after it is placed on line for reasons similar to those cited above. Only after this time does the high-quality energy produced by the plant exceed the energy cost of its construction. (If one used a thermal energy criterion, the break-even time is reduced to about seven years.) Ultimately, of course, there is a substantial net energy gain, as is indicated in Table 10.2 (15). This is also true of solar energy (see Chapter 15). Solar facilities built in the 1980s may not become net producers until after 2000 (16).

The point of these remarks is that we must be careful, if we build any sort of generating facilities, to be sure that the pace of construction results in a rather slow increase in generating capacity. If we were to try to build all these facilities all at once, we could cause a net energy *consumption* in the short run.

SUMMARY

Utilities build generating facilities, which decision causes effects on local land, air, and inhabitants. In making a decision about which type of generating facility to build, the utility must assure that sufficient water is available for cooling, so that the effects of waste heat rejection on the local biota are minimized, including suppression of BOD; must decide how much of the facility should be prefabricated as opposed to constructed onsite; must consider the effect of building a facility on the social structure of the community servicing it; and must deal with any emissions of pollutants into the atmosphere and water. Utilities must also consider the rate at which they build facilities, in order to spread out demands on capital as well as to assure that some *net* output of energy is available.

REFERENCES

1. An exhaustive analysis of the effects of pollution on a river is found in J. M. Fallows, *The Water Lords* (New York: Bantam, 1971).
2. J. W. Gibbons and R. R. Sharitz, *Am. Sci.* 62 (1974): 660.
3. W. A. Brungs, in H. Foreman, ed., *Nuclear Power and the Public* (New York: Anchor, 1972), 68ff.
4. A. W. Eipper, in *Patient Earth,* ed. J. Harte and R. Socolow (New York: Holt, Rinehart & Winston, 1971).
5. T. C. Elliot, *Power* (December 1985), special report.
6. M. L. Kramer et al., *Science* 193 (1976): 1239.
7. J. S. Steinhart and C. E. Steinhart, *Fires of Culture* (N. Scituate, Mass.: Duxbury, 1974), Table 8-2; or *Energy Sources* (also Duxbury, 1974), Table 10-3. These are based on data taken from *Electric Power and Thermal Discharges,* ed. M. Eisenbud and G. Gleason (New York: Gordon and Breach, 1969), 372.
8. W. Heller, Testimony before the Committee on Interior and Insular Affairs (U.S. House of Representatives) on conservation of energy.
9. S. Rattner, *New York Times,* 26 November 1976.
10. J. S. Gilmore, *Science* 191 (1976): 535.
11. U.S. Department of the Interior, *Kaiparowits, Environmental Impact Statement,* (Washington, D.C.: U.S.D.I., 1975).
12. *Kaiparowits EIS,* Figure III-68, page III-323.
13. G. Lichtenstein, *New York Times,* 3 August 1975; G. Hill, *New York Times,* 15 April 1976; G. Lichtenstein, *New York Times,* 16 April 1976.
14. *Kaiparowits EIS,* pages VI-32,33.
15. P. Chapman, *New Scientist* 64 (1974): 866; and J. Wright and J. Syrett, *New Scientist* 65 (1975): 66.
16. C. Whipple, *Science* 208 (1980): 342.

PROBLEMS AND QUESTIONS

True or False

1. Slight changes in temperature can greatly affect fish speciation in lakes or rivers.
2. Fertilizer runoff encourages explosive growth of algae. This decreases the BOD in the water.
3. For a given level of available nutrients in a lake, a 10°C temperature rise will probably cause a doubling of the algal biomass in the lake.
4. There is a possibility of heavy metal contamination of water in the vicinity of coal-fired generating facilities.
5. Large-scale excavation for generating facilities can lead to large-scale erosion.
6. Dry cooling towers would be more appropriate than wet cooling towers for a power plant constructed in the arid Southwest.

7. Significant social dislocations can occur in the wake of any large construction project.
8. Field construction of energy-generating facilities would be less of a problem if there were a common utility industry-wide design standard for generating units.
9. The most significant reason for the setback for the U.S. nuclear industry in the decade 1975–1985 was massive public antinuclear demonstrations.
10. Cost overruns have become common in construction projects for public utilities.

Multiple Choice

11. Waste heat from a generating facility
   a. is more harmful when released into a lake than when released into a river.
   b. is more harmful when released into a river than when released into a lake.
   c. should always be released through cooling towers.
   d. should never be released through cooling towers.

12. The most important reason that the Kaiparowits facility drew especially heavy criticism from environmentalists was that
   a. Los Angeles did not really need the energy.
   b. it was located in Utah, too great a distance to send the electrical energy.
   c. it was located close to so many scenic national parks and monuments.
   d. it would not have broken even on energy before 1990.
   e. it would cause many social problems for the local area.

13. Dry cooling towers are
   a. more efficient at cooling than wet cooling towers.
   b. more efficient for nuclear plant cooling than for fossil-fuel plant cooling.
   c. an environmentally acceptable way to dissipate waste heat from a generating facility.
   d. more expensive to operate than wet cooling towers.
   e. likely to cause increased local snowfall.

14. Cooling towers are used instead of once-through cooling because of
   a. lack of acceptable sites.
   b. concern over the effects on river and lake biota.
   c. concern over the pollution of drinking water by contaminants.
   d. all of the above reasons.
   e. none of the above reasons.

15. Water quality depends most on the presence of
   a. carcinogens.
   b. PCBs.
   c. kepone.
   d. high BOD.
   e. dissolved oxygen.

16. Water is often used for cooling of power plants because
    a. it is readily available in most locations.
    b. it has a large specific heat.
    c. engineers are familiar with the properties of water.
    d. water piping can be bought "off the shelf" (it is available almost everywhere in standard sizes).
    e. of all the reasons above.

17. During the early years of a power facility's operations
    a. it represents a net gain in electrical energy supply.
    b. it represents a net loss in energy because of the energy cost of its construction.
    c. it has practically no environmental impacts.
    d. None of the above statements is true.
    e. Only statements a and b are true.

Discussion Questions

18. Since nuclear facilities must be constructed mainly in the field, and this is expensive, how does the utility justify building the nuclear facility?

19. About half the waste heat generated in fossil-fuel plants is exhausted into the atmosphere directly; the other half is exhausted into cooling water. All the waste heat of a nuclear plant is exhausted into the cooling water. Compare the effects of waste heat of the two sorts of plants, both for direct exhaust of heat into the water and for use of cooling towers.

20. Explain why air cooling is more complicated than water cooling for transporting waste heat away from a generating facility.

11
WATER AND ENERGY

In Chapter 10, we learned that construction of generating facilities can affect water quality. No matter what type of facility we choose to produce energy, water is necessary, whether as a resource for hydroelectricity (solar energy of a sort) or as a necessity to provide a low temperature reservoir for a heat engine. In this chapter, we discuss some methods of generating energy from water; in this context, we also discuss water resources. The theme of solar energy continues in Chapters 15, 16, and 17.

KEY TERMS
precipitation
evapotranspiration
undershot wheel
overshot wheel
schistosomiasis
bilharzia
latent heat of vaporization
pumped storage
aquifer
oil shale

WATER AND POLITICS

A student of water politics has said in regard to water and politics: "Water is energy, and in arid lands it rearranges humans and human ways and human appetites around its flow" (1). Ground water is a nonrenewable source of such energy.

We saw in Chapter 10 that water is necessary for cooling most types of power plants. Water is a necessity for making energy from heat. The heat engine would not work if there were just one temperature reservoir. The water is the low temperature reservoir, whether it is used as a once-through system, or whether it circulates in a closed-cycle heat exchanger, as in a cooling tower.

Water can also make energy directly: Hydroelectricity is energy won from water as it falls. Water from oceans and lakes absorbs energy from the sun to supply *latent heat of vaporization* (see Chapter 7), thereby changing liquid water to water vapor. The water vapor forms drops, and collections of drops make clouds. Clouds eventually release their water as rain. The rain falls on land, runs off into streams or rivers, and may be caught behind a dam. The gravitational potential energy of the water behind the dam is simply stored solar energy.

While water may exist on Earth in staggering quantities, only a tiny amount is usable for drinking, agriculture, and industry. Americans used 400 million m³ of water per day in 1980 (2) (see Figure 11.1a), but only 1.3 percent of this was water used for generating energy and not returned (3). Another 7.7 percent is used for manufacturing and mineral extraction (3). Steam electric utilities account for almost half of total use, and the balance, about 8 percent, is used by public water utilities (3).

To get a feeling for how much water there is, imagine Earth to be a smooth sphere coated with all the Earth's water (4). The estimated 1.5 × 10⁹ cubic kilometers is distributed as described in Table 11.1 (2,3,6). Note that only two water molecules in every ten thousand are fresh surface water.

Precipitation, averaged over the world, amounts to about 1 m per year. Over the United States, the precipitation averages about 750 mm/ yr (7), which is marginally better than the world average of 710 mm/yr for land precipitation. For normal years, the precipitation in the continental United States ranges between 2.06 m/yr in the Pacific Northwest to less than 100 mm/yr in parts of the Pacific Southwest (7). Figure 11.1 shows annual precipitation in the United States; regions with rainfall levels under 24 inches are short of water in most years. The variability of wet and dry years can make the interregional disparity even greater.

Agricultural irrigation is the major water consumer in the United States (3). Irrigation is very

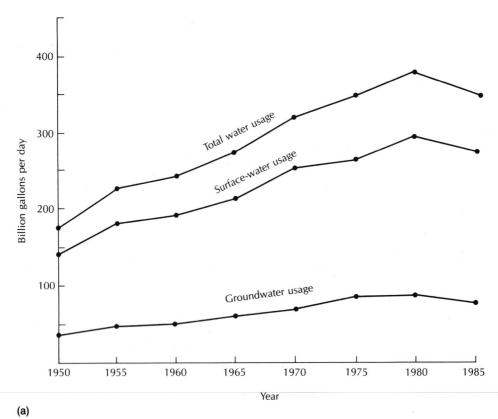

(a)

FIGURE 11.1
(a) Fresh water consumption in the United States. (U.S. Department of the Interior.) (b) Precipitation in the U.S. in inches per year. (National Weather Service.)

TABLE 11.1
Water inventory on a
smooth Earth.

Smooth	Percent of All Water	Height above Smooth Spherical Earth
Salt water	97	2700 m
Ice, snow	2.25	120 m
Antarctic ice cap	1.93	
Greenland ice sheet	0.20	
All other glaciers	0.12	
Fresh ground water	0.73	45 m
Surface (water in soil, rest below soil)	0.006	
Above ≈1 km (below ground level)	0.36	
Below ≈1 km (below ground level)	0.36	
Fresh water	0.22	1 m
Water vapor	—	0.03 m

SOURCE: References 2, 5, and 6.

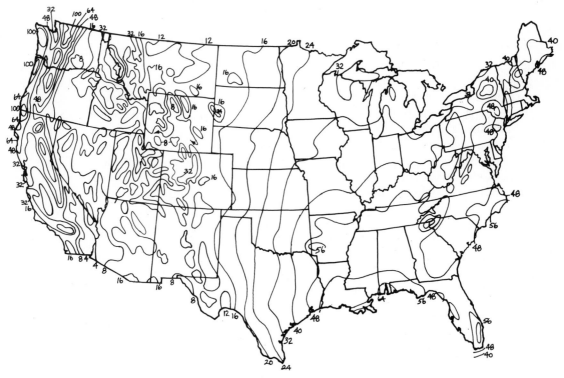

(b)

FIGURE 11.1 (continued)

useful because it increases the cultivated area (raising yields); it increases crop yield; it allows for multiple cropping; and it provides farmers with some independence from the vagaries of weather (8). It costs energy to irrigate; the major portion of energy used in irrigated agriculture is for supplying the water.

There are sometimes unexpected consequences of damming and irrigation. Vegetation loses water through its surfaces, a process called *transpiration*. Since most precipitation eventually evaporates from open-water surfaces, building dams or irrigating dry land increases evaporative losses. Over a large part of the North American land mass, the loss exceeds the precipitation. *Evapotranspiration* in the United States averages 630 mm/yr (leaving about 120 mm/yr for river flow from the 750 mm/yr of precipitation). Both of these effects are present in the salt loading of the Colorado River, which has caused friction between the United States and Mexico. The loss from open water in the seventeen western states amounts to about 8 percent of capacity (9).

Irrigation can have political ramifications as well. For example, American farmers in the Wellton-Mohawk district using the Colorado River for irrigation have made the lower Colorado so saline that the United States has built a 100 million gallon per day desalination plant in order to fulfill its water treaty obligations to Mexico (10–12). This deficit of water has led to proposals such as the "Parsons plan," which would pump water from Canada and the American Pacific Northwest to Colorado, New Mexico, and Arizona (12,13).

ENERGY FROM WATER

One of the important uses of water is to produce energy. It is not clear exactly when water power for purposes other than transportation was first used, but it was certainly by Roman times. The water wheel preceded the windmill, and the water wheel created wealth for the millers who owned them. In the milling of grain, the miller earned a portion of the grain in exchange for his services in grinding it, and sometimes this share was large. Millers were often the butt of ribald stories and songs (see, for example, Chaucer's *Canterbury Tales*) because people were jealous of them and their economic success.

The industrialization of America began in New England because New England was rich in water resources that could be used to make a water wheel turn. New England and the Middle Atlantic states boasted 23,000 water wheels in the nineteenth century (14).

Water Wheels

The world's first water wheels were probably the vertical-axis type, in which an influx of water drives into the wheel's blades, causing rotation. Such wheels may be used in a streambed itself, with the minor modification that streamflow be directed onto the blades. The horizontal-axis water wheel, in use since Roman times, requires a good deal more engineering. There are three general types of horizontal-axis water wheels. The undershot wheel can be run on streamflow like the vertical-axis wheel: The flowing water pushes at the partially submerged blades, causing the wheel itself to rotate. The breastshot wheel has water brought in part way up the wheel so that both the impact of the water and the weight of the water cause rotary motion. In the last of the three, the overshot wheel, water is brought to the top of the wheel by a flume, or penstock. Here again, the wheel rotates because of the water impact and the weight of the water. Both the breastshot and overshot wheels involve nontrivial engineering (14).

The overshot wheel is most efficient, delivering about 65 percent of the available power; the undershot wheel is the least efficient, delivering about 25 percent of the available power (14). Modern turbines are considerably more efficient than these, at approximately 90 percent efficiency.

FIGURE 11.2
A rehabilitated 2.25 MW dam on the Battenkill River in Greenwich, New York, supplies an industrial site and the local utility with energy. (Photo courtesy of New York State Energy Research and Development Administration.)

Dams

Most modern ways of tapping the stored solar energy of water (see box on evaporation) involve dam building. Dams increase the available *head* over the (generally small) head available from stream diversion. Of course, the silt (alluvium) carried along by the stream in its flow is lost, as the silt settles out in the calm reservoir. In the case of the Tarbela Dam in Pakistan, the silt flow is so intense that the deep reservoir will be entirely filled with silt forty years after its construction was completed (the late 1970s) (15). The silt that settles in the reservoir cannot fertilize land further downstream.

These siltation processes call into question the rationale behind such projects as the Aswan Dam in Egypt. Since construction of the dam, the precious alluvial matter (silt carried down from upriver) settles at the upper end of Lake Nasser, where current flow slows. Erosion in Ethiopia is reducing the life of the Aswan Dam (16). Meanwhile, some land now irrigated in Egypt is becoming salt flat as capillary action brings up salts from the ground underneath (10,17).

This problem has led to a multibillion-dollar World Bank-financed installation of drainage tiles on a million acres in Egypt along the model of the tiles installed in the Imperial Valley (17), where 14 percent of the land now in drainage tiles has

FIGURE 11.3
A recently refurbished
dam in upper New York
supplies local energy.
(Photo courtesy of New
York State Energy Re-
search and Development
Administration.)

become productive. Even in Egypt, the gigantic cost of land reclamation has stopped the increase in cultivated area (it has shrunk by 200,000 acres since the mid-1960s) (17).

The early twentieth century Aswan Dam, constructed in 1905, led to many similar complications. It appears that the experience garnered from the smaller dam was ignored by Egyptian planners during construction of the new dam in 1969, in their eagerness to obtain the energy necessary to industrialize.

Other effects of the dam have included erosion in the Nile Delta because of loss of sediment from upriver, a loss of fisheries along the Egyptian coast, riverbed erosion, and of course, the increased salinity of the Nile River water from its impoundment in Lake Nasser (16). The increase in stagnant canal waters due to the dam set off an epidemic in Egypt of schistosomiasis, or bilharzia, which causes blindness. The disease is of parasitic origin, and the stagnation of the water allows the snails harboring the parasite to breed

(17). A similar epidemic of schistosomiasis struck Brazil in the wake of dam construction there (18).

It must by now be obvious that hydroelectric power is not problem-free. We might mention here several other byproducts of dam construction that are not at first apparent.

1. Drowning of productive land in the reservoir is often the case. Farmers fight to keep their rich farmlands from being inundated by dams. A project that became something of a political "hot potato" in the late 1970s and early 1980s, the Garrison District Reclamation in North Dakota, would have resulted in a net reduction of 8150 acres of cropland and 34,000 acres of grassland if the dam had been built as originally planned (19).

2. Changes in insect speciation often follow reservoir construction. Most mosquito species prefer to breed in stagnant water. Reservoirs are as attractive as discarded water-filled tires are to breeding mosquitoes (20).

3. Changes in fish speciation usually follow reservoir construction. Some fish live in swift-moving water. These fish disappear when a lake replaces a river. Their places are taken by other species of fish not necessarily so interesting to the angler.

4. Changes in the ground water level can ruin good farmland by making marsh out of pasture, allowing upward movement of salts, and so on.

5. In the tropics, decomposition of flooded trees in reservoirs can cause a penetrating stench and produces hydrogen sulfide as well as increased BOD. The resultant acid water may corrode the turbines in the dam used to produce electricity (18).

6. Especially in the tropics, still waters are ideal environments for water plants. These can cover or fill a reservoir, fouling water intakes and cutting the rest of the water off from essential sunlight (18).

7. The cost of relocation of displaced people is not usually counted as a cost by those making the decision to build a dam: 80,000 had to move for Aswan; 75,000 for Lake Volta in Ghana; 57,000 for Lake Kariba in Zimbabwe and Zambia (16).

8. Stratification in the lake behind the dam (see Chapter 10) can change the temperature of the water released downstream, affecting the amount of dissolved oxygen in the water.

9. Loss of fisheries may result from dam construction. For example, salmon runs in the Columbia River have fallen off greatly since the dams were built.

While this is by no means a complete list, it should impart the idea that the building of a dam should proceed with due caution. This is not to say that all such efforts are disastrous: In Israel, the application of technology to water has allowed the Israelis to get the utmost from their limited fresh water supplies (21). Israeli dryland farming produces large yields with very low water input. The Israelis now market their drip irrigation technology, which supplies water in very small amounts directly to the root system of each plant, all over the world.

The Dead Sea Project

Another project may be nearing reality in Israel. The Dead Sea is 400 m lower than the Mediterranean. Since the water that once flowed to the Dead Sea from the Jordan River is now being utilized for irrigation, the Dead Sea has shrunk. Water levels are now 8 m or so below those of a half-century ago. There is a proposal to construct a canal from the Mediterranean running south of Beersheba into the Dead Sea (22). In the course of refilling the Dead Sea to 1930s levels over ten to twenty years, the 1.6 km³ of water per year could generate some 1600 MkWh (1.6 TWh). Storage facilities would exist so that electricity demand fluctuations would be met. Even after refilling, about two-thirds as much water will still have to be brought in every year to replace evaporative losses, generating about 1 TWh (22).

Pumped Storage

Another study in water use is provided by the Storm King Mountain pumped storage facility proposed by Consolidated Edison of New York City in 1962 and finally abandoned in 1981. In pumped storage, a reservoir is constructed some distance above a source of water. In times of slack demand, electricity is used to pump water from the supply below to the reservoir above. When demand for electricity is high, the water in the higher reservoir returns to the lower level through turbines, adding to the supply of energy available. The idea of pumped storage is that it is most efficient for the utility to increase its base load, evening out the daily fluctuations. The utility should consequently try to store the extra energy it produced when demand was low to recover it at times of peak demand. Such pumped storage returns about three-quarters of the energy used to pump the water uphill—not bad by comparison with technologically competing possibilities.

The Environmental Defense Fund of New York City had calculated that a conventional system would be a cheaper alternative than Storm King, but despite its own economic crisis and the prospective cost of 750 million dollars (23), the utility was convinced that pumped storage was its best alternative.

Proponents expected that the mountaintop reservoir could be used for recreation, but one wonders what happens when a small lake is filled

or drained at the design pump rate of 8 million gallons per minute from or into the Hudson River (23). Some of the most effective opponents found that the huge fluctuations in river flow rate would destroy 25 to 75 percent of the striped bass hatch in the Hudson River. Consolidated Edison finally agreed to give up Storm King in return for the dropping of demands by the Natural Resources Defense Council that cooling towers be built onto existing Con Ed thermal plants (24).

Low-Head Hydroelectricity

Already-existing small dams may be used to produce electricity. Such low-head facilities were often built for electricity but abandoned as utilities formed monopolies to gain from economies of scale. More than 770 hydroelectric plants have been abandoned since 1940 (25). Of over 2800 dams in New England, only 200 now produce electricity (26). Towns and individuals are rehabilitating old sites to increase local energy supplies, causing a growing market for industrial upgrading and rebuilding efforts (27). While such efforts do make sense and make use of equipment already in place, the total amount of energy produced is not large. In Michigan, eighty hydro sites produce 2.7 percent of the state's power. If the next best thirty sites were added, the total would still be under 3 percent (25). In New York State, as of 1981, small hydroelectric plant sites produced about 4.2 MW, another 5.4 MW of capacity was under installation, and 3.4 MW of capacity was in design. Further sites that could produce a total of 12.4 MW were under study (28). As of 1984, New York had thirty-five small hydroelectric plants operational (29). The developments in New York came about partly because of a 1979 state law setting the price of energy from independently owned plants at 6 cents/kWh (29). The estimated total potential of the reasonable hydroelectric sites in New York is 1000 MW (29). New equipment is being designed and installed to take advantage of this new market (27).

China has made use of some of its small hydro possibilities. It now has 90,000 small dams to generate electricity and a booming export market in small turbines and generators (30). These turbines can generate energy from streamflow, just as tethered water wheels once did. Many developing countries, such as Burundi, Papua-New Guinea, Liberia, and Nepal are developing low-head hydroelectric resources (31).

WATER QUALITY

One often hears that in the year 1990, or 2000, or 2010, or ..., the nation's power plants will require the total national water supply for cooling purposes. This is some exaggeration, since water used for cooling upstream can at a later time be used for cooling downstream. However, any use degrades the water to *some* extent and renders some unfit for consumption. Recent tests of municipal water supply systems (32–34) have shown substantial quantities of carcinogenic material. This has also been true in Britain (35), where information on pollutants has not in general been available to the public. The inadvertent poisoning of the Hudson River (New York) by General Electric with the noxious polychlorinated biphenyls (PCBs), and of the James River (Virginia) by a subsidiary of Allied Chemical with kepone also stand out as water-degrading uses.

Water is required for many uses. In many parts of the country, water use is already perilously close to supply. Making separate projections by region, only New England, Ohio, and the southern Atlantic-Gulf states will have an assured supply that exceeds commitment (7). This is indicative that the "pollution dilution solution" is simply not feasible over broad areas of our country (36).

A striking example of commitments exceeding supply is the salination of parts of the aquifer underlying Florida (32). Salination occurs when water is rapidly pumped out of the porous limestone formation. The decrease in pressure pulls in salt water from the ocean. A similar occurrence on Long Island, New York, in the 1960s was due to population pressure on the limited water resources. The aquifer underlying Long Island is now totally contaminated by salt water. It may be many years before the salt water intrusion is repulsed.

When salt water intrudes because of surface pumping, it can be removed only by supplying a head of fresh water. This is the reason that Long Island has such a long-term problem. Water distribution in Florida is becoming more and more of a problem as the population grows. As in California, water is being exported from unpopulated regions to more populated regions. Unlike California but like Long Island, much of the water comes from underground. The east coast of Florida has the largest current salt water intrusion problem, but the relentless pace of development on the west coast spells trouble ahead for the entire peninsula. Florida still has remnants of Big Cypress swamp and Everglades National Park, which supply tremendous quantities of water to the west coast aquifer system. Development of the west coast of Florida, with the diversion of water into canals and thence out of the "river" of water flowing south through the swamps, could ruin the aquifer system by encouraging salt water intrusion. The canals and impounds of one development near Naples, Florida, flushes an awesome 150 billion gallons a year out of the swamp system (37)!

The concern voiced about our water supplies has generated new ideas for water purification. One of the most interesting proposals is the use of bulrushes (*Scriptus lacustris*) in manmade marshes as filters for manmade sewage (36). Another possibility involves spraying wastes into sandy soil. The water percolates through the sand filter to arrive purified at ground water level, while the solid wastes eventually become humus. Of course, the place is not a very pleasant spot to visit while the process is at work (34).

Water and Energy in Semi-arid Regions

Another area of new perception involves water use in obtaining energy in the semi-arid West. The states of Montana, Wyoming, and the Dakotas are problem areas. The push for energy will have a mammoth effect there, since there are very thick near-surface coal beds (in contrast to experience in the East). With the coal beds near the surface, extraction will be by strip mining—removing overburden and piling it, thereby despoiling many hectares of land. In addition, these states do not get much rain, and thus the weathering of the piles is very slow. There are 128 million acres of this coal and lignite (38).

The North Central Plan, a joint endeavor of twenty-five utilities, proposed development of water and coal resources in over 250,000 square miles centering on the Gillette-Coalstrip Wyoming–Montana area. This plan, now temporarily shelved, proposed forty-two sites in five states for coal-fired steam generating plants to produce an additional 50 GW_e by the year 2000. (The thirty-nine existing coal plants in the western states now generate 9.3 GW_e.) Thousands of miles of 765 kV transmission lines would have linked Medicine Bow, Wyoming, to St. Louis, Missouri. This may yet happen if the plan is revived.

Three new pumped storage facilities would provide another 3 GW_e. The project would consume 855,000 acre feet/yr (half the New York City consumption; an acre-foot is the amount of water required to cover an acre of land a foot deep in water). If gasification plants are added, the demands rise to 2.6 million acre-feet/year (38,39). This is one-third the average flow of the Yellowstone River.

In a wide swath of the semi-arid Western states, surface and near-surface rock is permeated with an oil-like mineral. Such rock is called oil shale (see Chapter 13). Some attempts have been made to exploit the energy locked up in the oil shale. Prototype treatment facilities have been designed.

Oil shale treatment would require large amounts of water: three barrels of water for each barrel of oil produced (38). In addition, there is the problem of disposing of the spent shale in the arid climate. Runoff from a shale oil project could increase the salinity of the Hoover Dam reservoir by 50 percent (40).

There are other water problems that can follow upon exploitation of such energy minerals as coal and oil shale. A National Academy of Science study (40) warns that:

1. Only areas with rainfall greater than 250 mm/yr seem to have high potential for rehabilitation (given good management). These make up about 60 percent of strippable western coal lands.
2. *Restoration* is "not possible anywhere."
3. Near-surface coal seams are also ground water aquifers, serving livestock and domestic wells. Mining could "dewater" a large number of wells.
4. There is the problem of destruction of ephemeral streams (dry gullies and arroyos) that carry off thunderstorm or snowmelt. These are vital features of the arid bioscape.

With the water shortages already mentioned in this region, it behooves us to decide rationally whether we wish to use the water in the place to create another Four Corners (an enormous coal-fired plant near the point where Utah, Colorado, New Mexico, and Arizona meet; its smoke plume is visible from orbit) in the beautiful landscape of Wyoming, Montana, or North Dakota. For water-short regions near oceans, there are always desalination plants (Saudi Arabia, for example, currently desalts 480 million gallons per day) (41), or the possibility of transporting icebergs from the Arctic or Antarctic seas. For inland regions, there are no alternative supplies except perhaps for pipelines (for example, as in the Parsons plan).

SUMMARY

Water is necessary for almost all forms of energy generation, from production of coal-fired electricity to production of food. The building of dams for water and for electricity is usually considered benign. However, dams can cause increased soil salination; increased disease carried by mosquitoes and water-borne human parasites;

drowning of productive land; replacement of fish and insect species; changes in ground water levels; dislocation of people; and fouling by water plants.

Pumped storage facilities also have environmental side-effects. Fewer side-effects accompany refurbishing of old dam sites, as is being done in New England and the upper Midwest.

Water is already in short supply in the southwestern United States and in arid regions the world over; it may become so in other areas of the country and the world in the future.

REFERENCES

1. C. Bowden, *Killing the Hidden Waters* (Austin: Univ. Texas Press, 1977).
2. R. Reinhold, *New York Times,* 9 August 1981.
3. Department of Commerce, *Statistical Abstract of the United States, 1984* (Washington, D.C.: GPO, 1984).
4. H. L. Penman, Chap. 4 in *The Biosphere* (San Francisco: Freeman, 1970).
5. R. L. Nace, U.S. Geological Survey, Circular 536, 1967.
6. F. H. Forrester, *Weatherwise* 38 (1985): 83.
7. A. M. Piper, U.S. Geological Survey Water Supply Paper 1797 (Washington, D.C.: GPO, 1973).
8. R. P. Ambroggi, *Sci. Am.* 243, no. 3 (1980): 100.
9. J. S. Meyers, U. S. Geological Survey Prof. Paper 272D, as quoted in Reference 2.
10. E. Groth III, *Environment* 17, no. 1 (1975): 28.
11. D. Sheridan, *Environment* 23, no. 3 (1981): 6.
12. A. F. Pillsbury, *Sci. Am.* 239, no. 1 (1981): 55.
13. A. L. Dellon, in *Alternative Energy Sources II,* ed. T. N. Veziroglu (Washington, D.C.: Hemisphere Publ. Corp., 1984), 3715ff; see also W. Robbins, *New York Times,* 25 August 1974.
14. D. McGuigan, *Harnessing Water Power for Home Energy* (Charlotte, Vermont: Garden Way, 1978); and R. Wolfe and P. Clegg, *Home Energy for the 'Eighties* (Charlotte, Vermont: Garden Way, 1979), 167–81.
15. R. Revelle, *Science* 209 (1980): 164.
16. H. Tanner, *New York Times,* 4 May 1975.
17. D. Deudney, *Environment* 23, no. 7 (1981): 16.
18. C. Canfield, *Nat. Hist.* 92, no. 7 (1983): 60.
19. A. M. Josephy, Jr., *Audubon* 77, no. 2 (1975): 77.
20. D. Zimmerman, *Smithsonian* 14, no. 3 (1983): 28.
21. A. Weiner, *Am. Sci.* 60 (1972): 466.
22. I. Steinhorn and J. R. Gat, *Sci. Am.* 249, no. 4 (1983): 102; and Y. Ne'eman and I. Schul, *Ann. Rev. Energy* 8 (1983): 113.
23. L. J. Carter, *Science* 184 (1974): 1353.
24. P. Hine, *NRDC News* 1, no. 2 (1981): 1.
25. P. Kakela, G. Chilson, and W. Patric, *Environment* 27, no. 1 (1985): 31.
26. B. P. Smith, *Environment* 20, no. 9 (1978): 16.
27. C. A. Neumann, *Alt. Sources Energy* 82 (1986): 24; and D. Marier, same issue, p. 10.
28. E. J. Dionne, Jr., *New York Times,* 4 Oct. 1981.
29. E. A. Gargan, *New York Times,* 11 March 1984.
30. *Futurist* 19, no. 3 (1985): 50.
31. M. Johnson, *Alt. Sources Energy* 78 (1986): 23; and A. R. Inversin, *Alt. Sources Energy* 82 (1986): 33.

32. J. L. Marx, *Science* 186 (1974): 809; and R. Redd, *New York Times,* 20 Nov. 1974.
33. J. M. Fallows, *The Water Lords* (New York: Bantam, 1971).
34. D. Zwick and M. Benstock, *Water Wasteland* (New York: Bantam, 1972).
35. J. Tinker, *New Scientist* 65 (1975): 551.
36. G. V. Jarnes, *Intern. J. Environmental Studies* 1 (1970): 47.
37. B. Webster, *New York Times,* 7 March 1975.
38. R. Gillette, *Science* 182 (1973): 456.
39. B. A. Franklin, *New York Times Magazine,* 29 Sept. 1974.
40. W. D. Metz, *Science* 184 (1974): 1271.
41. E. Sciolino, *New York Times,* 21 April 1985.

PROBLEMS AND QUESTIONS

True or False

1. Since the federal government is now involved in pollution control, polluting by government installations has ceased.
2. There are few good reasons for building dams.
3. The fall of the river valley civilizations implies that all land will eventually wear out from constant irrigation.
4. About half of the water used in the United States is used for agricultural purposes.
5. More water falls over a given area of land every year than over the ocean (on average).
6. There are few regions where the water demand exceeds runoff.
7. Dams are environmentally safe.
8. Water loss through evaporation exceeds input for much of the land area of the United States.
9. The building of the Aswan Dam brought difficulties as well as electricity to Egypt.
10. Industrial development began historically with exploitation of water power resources.

Multiple Choice

11. Which site for a proposed dam would probably be worse from the point of view of producing increased water salinity? A dam in
    a. the Pacific Northwest west of the Cascades.
    b. Southern California.
    c. New England.
    d. Labrador.
12. Which site for a proposed dam would probably cause the worst smell to be produced?
    a. Labrador
    b. Siberia
    c. Tierra del Fuego
    d. Chilean Alps
    e. Columbian lowlands
13. Which of the following is *not* an environmental cost of dam building?
    a. Decomposition of plant matter, causing acidification and increased BOD
    b. Relocation of displaced people
    c. Drowning of productive land by the reservoir
    d. Heavy metal pollution of the reservoir
    e. Changes in insect speciation; for example, increased populations of mosquitoes

14. Which of the following is the most efficient for utilizing water energy?
    a. An overshot wheel
    b. An undershot wheel
    c. A breastshot wheel
    d. A modern turbine
    e. A water-screw

15. Fresh water available for human use constitutes
    a. a tiny fraction of all water on Earth.
    b. a large minority of all water on Earth.
    c. about half of all water on Earth.
    d. a majority of all water on Earth.
    e. most of the water on Earth.

16. Which use of water is most responsible for the draining of the Ogalalla Aquifer (which underlies Nebraska, Kansas, Oklahoma, and Texas)?
    a. For irrigation
    b. For human consumption
    c. For animal consumption
    d. To dilute saline rivers
    e. For manufacturing

17. Salinity damage affects mostly
    a. soils near the sea.
    b. arid soils.
    c. waterlogged soils.
    d. tundra in the Arctic.
    e. soils near dams.

Discussion Questions

18. Can there be international ramifications of dam building and consequent water pollution? Think of cases for which this could be true.

19. From the information on water use entailed in the exploitation of shale oil, examine the utility of such schemes for energy self-sufficiency.

20. Evaporation losses from impoundments could be countered by release of millions of white ping pong balls to float on the water's surface. The white surface of the balls would reflect most of the incident sunlight, and reduce evaporative losses substantially. Assess the feasibility of such a measure in the "real world."

12

MINERAL RESOURCES

Energy and resource exploitation are closely re-lated. Many fuels are mineral resources—coal, oil, gas, oil shale—and this chapter explains the nature of such resources, how they are produced, and the ultimate limitations and costs of their production.

In this chapter we discuss the distribution of minerals in the Earth's crust and the intercon-nection of the availability of materials, standard of living, energy use, and growth of world pop-ulation.

KEY TERMS   *plate*
                mantle
                magma
                subduction zone
                plate tectonics
                Curie temperature
                resources
                triage
                reserves
                manganese nodules

The distribution of minerals is intimately con-nected to the process of energy. Energy is nec-essary to obtain the mineral. Energy is required to refine the raw material. Without energy, the final products could not be made. Coal and pe-troleum are minerals themselves as well as en-ergy sources, and will be discussed in the next chapter. Fertilizer minerals (see Chapter 17) are necessary for crop production.

Without raw materials to work with, we could not support the economy. In some products, the raw material is an insignificant part of the cost, but absolutely essential to create product value (for example, consider the germanium necessary to dope a transistor element in an integrated circuit). The phenomenon of great value added for small amounts of materials used will probably be a characteristic development of modern econ-omies (1).

Production of goods and services is the basis of value added. Without any material production, we could not afford to produce energy. To dis-cuss energy without talking about raw materials is like discussing cooking without mentioning the ingredients. Conversely, without the necessary materials, we would not be able to exploit energy resources. It is becoming more difficult to extract energy as the most easily obtainable resources are used first (see Chapter 13).

THE SOURCE OF MINERALS

The Earth's surface is constantly being reshaped by sedimentation, weathering of exposed rock, movement of the continents, volcanism, and the activities of its living inhabitants. Some 200 million years ago, there was just one continent on Earth. It existed for an eyeblink in geologic time. In 1912, the irregularities of the continents' outlines, which seemed almost to match one another, led the German geologist Alfred Wegener to postulate that some of the continents were once adjacent, and had subsequently moved apart somehow. In his time, he met with general ridicule, but by the end of the 1960s, his ideas had been vindicated. The continents rest on *plates,* and the plates move relative to one another. At one point, the motion of the plates brought the continents together, and then the motion of the underlying plates broke them apart again.

Earth has a liquid *core,* surrounded by a *mantle,* and overlain by a *crust.* The crustal plates are patches of light rock that cover the Earth's mantle like a patchwork quilt. The internal heat of the Earth drives up molten material, *magma,* between the places where adjacent plates abut, causing them to be pushed apart. This generally occurs on the seabed and is referred to as *seafloor spreading.* At their other edges, the plates collide, and one is pulled downward under the other plate, in regions called *subduction zones,* by the cooler, descending parts of the plates. Many geologists have spent the last two decades (since Wegener's vindication) studying the plates and their motions as they move about over the semiliquid mantle. They call their area of study *plate tectonics;* the word *tectonics* comes from the Greek word for "builder." In architecture, tectonic refers to the construction of buildings, and in geology, it refers to the study of the structural deformation of the Earth's crust.

Through this mechanism and direct volcanic action, material is fed into the mantle rock and back to the Earth's crust in the largest recycling project ever conceived. The Himalayas rise out of the collision between the plate bearing India and the Asian land mass. The plate moving across the Pacific had traces of its movement written on it by a hot spot; the movement produced chains of parallel islands or seamounts. The Hawaiian Islands–Emperor Seamounts are one visible remnant of this movement. The Andes and the Rocky Mountains are pushed up by the collision of the South and North American plates, causing subduction of the Pacific plate under the mountains (the rocks are in plastic flow under isostatic equilibrium, so that, wherever a mountain rises from the surface, there is an upside-down mountain underneath it).

Many elements are found in the magma, or liquid material, that oozes from the plate boundaries or is ejected from volcanoes as lava. Iron is one of these elements. Iron and other materials that can form magnets are called *ferromagnetic.* Magnets form in nature when hot rock containing iron cools in the presence of an external magnetic field. Above a temperature known as the *Curie temperature,* the magnetic domains (see Chapter 4) are free to move about and line up pointing along the direction of the external field. As the material cools below the Curie temperature, the domains become frozen, unable to jiggle about. If there is an external magnetic field on the material as it cools below the Curie temperature, the result is the formation of magnetic rock.

Geologists have been able to verify seafloor spreading because of this physical principle and because the Earth's magnetic field occasionally flips direction from north to south. Seafloor made up of cooled magma spreading from a midocean ridge will have a record of the Earth's magnetic field at the time of cooling locked into the rock. The flip-flops of the Earth's magnetic field cause symmetrical patterns of "magnets" in the rock of the seafloor—a sort of geological "tape recording." Such symmetrical patterns have been observed near the Strait of Juan de Fuca in the Pacific Northwest and on the seafloor along the Atlantic rift.

All these motions and collisions give rise to geologic processes that generate ores. Ores seem to have been deposited in spurts (most ores of

FIGURE 12.1
Selected mineral imports as a proportion of supply are shown for 1982. (U.S. Bureau of Mines)

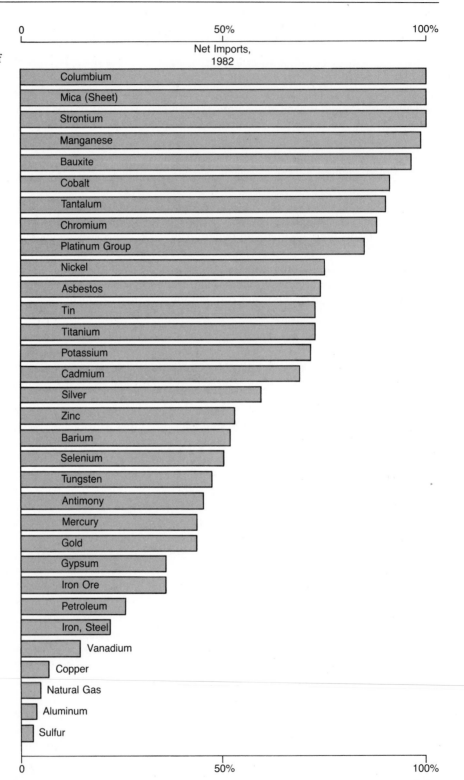

Net Imports, 1982

one kind are found in restricted geographic provinces created at roughly the same time) (2,3). The spurts are correlated with tectonic and magmatic activity in the American cordillera (3), and may have something to do with the interaction of Earth's biomass on surface chemistry and tectonic activity (2).

Mineral Distribution

The United States has been a net importer of minerals since the 1920s. Of the key metals, chromium, aluminum, nickel, and zinc, America gets more than half of its supply from abroad. A recent survey of stockpiles of minerals that would be necessary in case of a national emergency (a war of three years' duration) showed that for twenty-four of the forty-two most critical minerals, the nation is dependent on sources overseas (4). The five most critical minerals are cobalt (93 percent imported from Zaire and Zambia), chromium (91 percent imported from South Africa and Zimbabwe), manganese (all imported), platinum metals (87 percent imported from South Africa

FIGURE 12.2
Supplies of aluminum, copper, zinc, and lead for the United States are shown for the period between 1950 and 1971. (U.S. Department of the Interior)

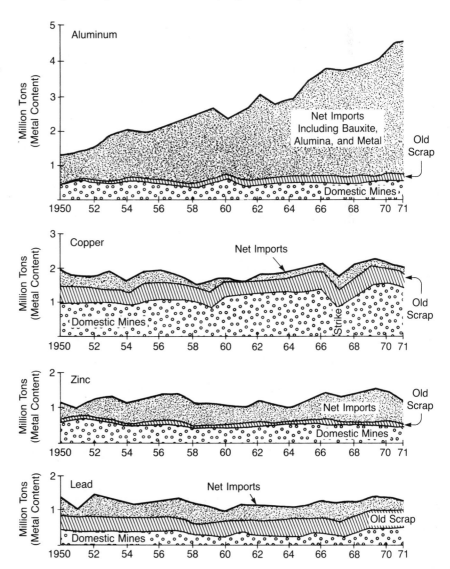

and the Soviet Union), and titanium (4). Titanium is mostly imported from Australia, which produces three times as much as the United States; Canada produces twice as much as the United States (5). Our reliance on foreign supply is illustrated in Figure 12.1, on page 172 (6) and Figure 12.2, on page 173 (7). In 1970, with 5 percent of the world population, the United States imported 27 percent of the world's raw materials exports (5).

We are not unique in finding it necessary to import raw materials. Virtually every nation must import material needs, because no part of the Earth is self-sufficient in minerals. North America is rich in molybdenum but poor in tin, tungsten, and manganese; the situation is the reverse in Asia. South Africa and the Soviet Union have much of the world's gold and platinum. Cuba and New Caledonia supply half the world's nickel (5,8). Our supplier nations are distributed worldwide; about a third of the world's countries are major suppliers for some minerals.

Thus, there is a politics of materials. One of the reasons for our diplomatic involvement with so many countries, and why worldwide strategic interests are a reality, is our vulnerability to interruptions in the immense flow of materials into our country. We will need even more minerals in the future, since our rate of raw materials consumption tends to increase at *twice* the rate of population growth.

This is the point at which population growth, standard of living, and questions of international morality intersect. Each additional American born or immigrating exerts a disproportionate leverage on materials use (see box). Increasing our share of use of the world's materials will become more difficult as the rest of the world tries to catch up to our standard of living. The pressure on resources will be intense—recall that per capita materials use and standard of living are related. The dilemma is that the high population growth rates of the undeveloped world are a virtual promise that their standard of living cannot become "American" (Figure 12.3); yet the high population growth rate is, in a sense, a re-

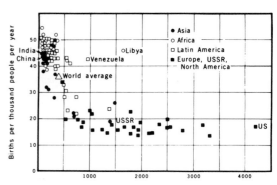

FIGURE 12.3
World births are plotted by country in terms of countries' GNP. (Chancellor and Goss, copyright 1974 by AAAS)

sponse to the low standard of living (children provide security in illness or old age). Like the Red Queen in *Through the Looking Glass,* the world must keep running simply to stay in the same place.

In addition to pondering the morality of our continuing to amass riches in contrast to most of the world, we must also consider the ramifications of our minerals acquisitions policy on the supplier countries. There is some truth to the claim that supplier countries benefit monetarily, but resource development seldom initiates other industrial development. Bolivian tin mines have not boosted Bolivian industrialization. Neither the copper of Chile nor that of Zambia have spread benefits of industry broadly, or caused the growth of industrial development. Small nations also leave themselves open to manipulation by powerful nations, which generally find ways to get resources when they need them (9). The question of what is to be done is still open.

MINERAL SUPPLY ADEQUACY

Our previous discussion has made it obvious that our domestic supplies of many materials are insufficient. We must turn to world supplies. Are world supplies *adequate* for sustained world production? Adequacy is a slippery concept. It might

MINERALS, CONQUEST, AND WAR

Until the time of the industrial revolution, wars were generally fought for land or religion, as well as for access to resources such as gold. Spanish and Portuguese colonization of the New World was for the ostensible purpose of converting heathen Indians to Catholicism. Actually, it allowed poor but capable men who were unafraid to tackle it a chance to make their fortunes. Later colonization in the seventeenth century also served to relieve population pressure and often involved transplantation of religious minority groups (Puritans to New England, Quakers to Pennsylvania, Catholics to Maryland, Presbyterians to North Carolina, Huguenots to South Africa) to regions having little to offer but land and few aboriginal inhabitants.

These colonies eventually sent back agricultural raw materials (and even some minerals such as gold and silver) to the mother countries and took manufactured goods in return. In some countries, most notably Great Britain, the new agricultural materials served to improve the nutritional condition of the population and also encouraged the rise of light manufacturing industries such as refining and textiles, which led to the eventual development of heavy industry. Raw materials for industry, in short supply in Europe, were abundant in the "New World."

During the nineteenth century, with increased population pressures and growing demand for raw materials, the emphasis of colonization shifted from agricultural colonies such as India, Indonesia, and the West Indies to settler colonies such as Victoria, South Australia, Natal, New Zealand, and Canada. Countries began being colonized for raw materials or for the prestige of colonial ownership. In this group are all the colonies of Asia and Africa not already mentioned, though many of the African colonies developed around various slave trade stations on the West African coast.

Britain had many more colonies than Germany, which was partly because Germany was not united under the Kaiser until the 1860s. Since the German industrial revolution was later than the British, their equipment was more modern. The Germans attempted to wrest Alsace and Lorraine from the French, not only because the territory was German-speaking but also because of the iron deposits in the area. The German need for raw materials and the disinclination of the British to give materials to the Germans was one of the factors contributing to the outbreak of World War I.

The late entry of the Japanese into the colonial scramble was met by a lack of available colonies; the Japanese were able to colonize only Korea for raw materials. The Japanese needed greater access to raw materials and could not get it from the British, Dutch, or Americans. Eventually, they developed the Greater East Asia Co-prosperity Sphere, which led to their invasion of China and ultimately to the existence of a Pacific Theater in World War II.

Some see the conflict between the United States and the Soviet Union as one involving the safeguarding of raw materials from rivals. Possession of raw materials has caused and will continue to cause international rivalry.

OF TRIAGE AND LIFEBOATS

In World War I, the French were terribly short of medical supplies and doctors to treat their troops. They decided that they had to use the medical resources they had in the most efficient way. Some of the wounded soldiers would survive without treatment of any kind; some of them would die no matter what the treatment; and some would die *unless* they were treated. They decided that the best way to use their meager resources was to separate the wounded into these three categories, and attempt to treat immediately those who would die without treatment or medicine. Because of the division into three groups, the practice became known as *triage* (10).

Socioeconomist Garrett Hardin has made the controversial suggestion that we practice a form of triage among nations (11) that he characterizes as the "lifeboat" analogy. Hardin pictures the world as an ocean, and the various countries as survivors of a downed ship. The world's food and resources constitute the lifeboat. If all nations share resources equally, it will overload the lifeboat and all will perish. He suggests that the weakest should be jettisoned in the interest of saving those who can be saved.

mean that the required amounts of the particular material should be available at present prices; or, that the rise in prices be no steeper than some specified rate; or, that, with a certain percentage of scrap being recycled, materials can meet domestic demand.

Let us make an attempt to estimate the adequacy of presently known supplies by considering the connection between supply and rate of use. The use rate is historical information, expressed as number of tons of a material used in a year (or any other convenient measure of time). Known supplies, given in tons of material, divided by the use rate gives an estimate of supply lifetime:

$$\text{lifetime of supply of a material} = \frac{\text{known supply}}{\text{rate of supply use}} \cdot$$

Table 12.1 was prepared using data from the Bureau of Mines of the U.S. Department of the Interior (4). There are three ways in which this estimate is deficient:

1. The actual rate of use of a supply of a certain material changes with time; the general trend in the use of materials is that the rate rises. The Great Depression of the 1930s is an exception to this rule, as was the recession of the early 1980s.

2. The amount of known supply increases as new resources are discovered. For example, tremendous amounts of iron ore and bauxite were discovered in Australia in the 1960s.

3. The market for a material could become saturated.

This table does illustrate one basic point: that it is difficult to sustain exponential growth in anything.

There is an indication that a decline in per capita steel use sets in after a per capita income of $2000 is attained (12). If all materials are subject to such a use peak, perhaps there is yet hope that we do not need an infinite resource supply to support a reasonable living standard.

Total consumption of any mineral resource is all but impossible. One cubic kilometer of "average" crustal rock contains 200 Mtonnes of aluminum, 140 Mtonnes of iron, 0.8 Mtonnes of zinc,

TABLE 12.1
Mineral supplies, use
rates, and projected
lifetimes.

Material	Supplies (Mtons)	Rate of Use in 1968 (ktons/yr)	Lifetime at 1968 Rate (years)
Cobalt	2.4	22.1	109
Iron	96,720	423,000	226
Manganese	736.5	3250	96.7
Molybdenum	5.4	33.575	161
Nickel	75	233.5	322
Aluminum	1170	10,288	114
Magnesium	2600	5274	493
Titanium	137.4	1424	96.5
Copper	307.9	7290	42.3
Fluorine	38.8	1800	22
Phosphorus	21,800	11,475	1900
Potassium	109,800	14,250	7600

SOURCE: U. S. Bureau of Mines.

TABLE 12.2
Comparison of clarke
and ore for selected
materials.

	Element	Clarke	Typical Ore	Cutoff Grade
Abundant	Aluminum	8.3×10^{-2}	2×10^{-1}	1.85×10^{-1}
	Iron	5.8×10^{-2}	3×10^{-1}	2×10^{-1}
	Manganese	1.3×10^{-3}		2.5×10^{-1}
Intermediate	Nickel	8.9×10^{-5}		9×10^{-3}
	Copper	8.7×10^{-5}	6×10^{-3}	3.5×10^{-3}
	Chromium	1.1×10^{-4}		2.3×10^{-2}
Rare	Mercury	9.0×10^{-8}	3×10^{-3}	1×10^{-3}
	Tungsten	1.1×10^{-6}		4.5×10^{-3}
	Gold	3.5×10^{-9}		3.5×10^{-6}

and 0.2 Mtonnes of copper. Even though some see advantages in mining "average" crustal rock rather than ore (the "mine" can be adjusted to the level of technology available rather than vice versa) (13), dilute minerals in rock will probably never be mined (14). Some minerals may be extractable from the seawater of which they are minor constituents.

In order to get some measure of how much of an element exists, and to see what we mean by an ore, we define the *clarke*. The clarke of an element is its average crustal abundance. Table

12.2 (15,16) compares ores to the respective clarkes of several elements. We presently mine high-grade ores, those for which the abundance is much higher (by two to four orders of magnitude) than the clarke. The grade of an ore is its concentration in the ore body.

For some materials, the relatively high concentrations of ore gradually change to the clarke as we move outward from the center of the ore body. An analysis by Lasky (17) holds for these ores. Lasky found that there was a relation between ore grade and amount mined:

WHY LOWERING GRADE SOMETIMES DOES
NOT INCREASE RESERVES BY MUCH

Mercury (Hg)

An analysis of forty-three mercury deposits by the Bureau of Mines and the U.S. Geological Survey (15) showed

0.37 Mtons of grade 0.008 ore (2960 tons of Hg)
1.22 Mtons of grade 0.00125 ore (1525 tons of Hg)
0.29 Mtons of grade 0.0008 ore (228 tons of Hg).

Decreasing the grade by a factor of 6 involves handling four times as much ore, but provides only a 50 percent increase in the amount of mercury obtained.

Helium (He)

Helium is used as a coolant, for pressurizing liquid fuel for rockets, for arc welding, and so on. The atmospheric He concentration is 5×10^{-6}. Natural gas fields sometimes contain appreciable He concentrations: One such field is an astounding 8.2 percent He; others are around 5.5 percent He. Larger fields with concentrations ranging from 5×10^{-4} to 5×10^{-2} exist, but the fields do not vary continuously in concentration. (We discussed the now-abandoned helium conservation program in Chapter 7.) (8,18).

Copper (Cu)

An analysis of Peruvian copper mines (and this analysis is corroborated by North American experience) (16) suggests that ore deposits do not show a compensating increase in reserve tonnage as grade decreases:

20 Mtons of grade 0.0132 ore (0.26 Mtons of Cu)
430 Mtons of grade 0.0099 ore (4.26 Mtons of Cu)
102 Mtons of grade 0.0032 ore (0.33 Mtons of Cu)
1057 Mtons of grade less than 0.0020 ore (0.067 Mtons of Cu).

Also, when the concentration of a metal is high, it tends to be present in lumps; when it is low, it is present in crystal form in silicates. Such compounds require much more energy to process.

---

average grade of ore
= $c - c' \log_{10}$ (cumulative tonnage mined).

Here, c and c' are constants for a given body of ore. The same principle may be stated alternatively as

volume of ore of a particular grade $\approx 10^{-(\text{grade})}$.

This relation means, roughly, that resources increase in quantity as the grade mined declines. Lasky's work has been interpreted to mean that raising the price increases the volume of mined rock, and hence the amount of material available. In such an interpretation, the world will not run out of its inexhaustible resources; those resources will simply increase in a monetary and energy cost. Since the dollar value of minerals has averaged only about 2 percent of the gross national product (GNP) over the last five decades, it is probably possible to increase mineral prices several-fold (8) if we are willing to ignore the cost in energy efficiency of extraction.

For most materials, however, the ore bodies are in geochemically isolated outcrops. Different sorts of rock are separated by sharp boundaries. Limestone carrying metal of grade 2 to 3×10^{-5} is extremely close to a body of ore of grade 10^4 times the clarke. Lowering the grade from 20 percent metal to 10 percent metal in this case does not greatly increase reserves (15,16). Thus, the solution of simply mining greater volumes of rock for our materials is probably not practicable, except for materials such as iron and aluminum, for which the Lasky analysis holds (8).

We could always return to the idea of mining average crustal rock. This shares the same problem just described in the mining of lower grade ores in immense quantities: an increase in energy consumption per unit of output.

Cheaper energy, in fact, would do little to reduce total costs (chiefly capital and labor) required for mining and processing rock. The enormous quantities of unusable waste produced for each unit of metal in ordinary granite (in a ratio of at least 2000 to 1) are more easily disposed of on a blueprint than in the field (12).

The difficulty in obtaining energy for exploitation of resources was part of the resource catastrophe evidenced in the "limits to growth" analysis (19). The history of the energy cost of mineral production is shown in Figure 12.4 (15).

In addition, there are mountains of rubble produced in mining large volumes of rock. The

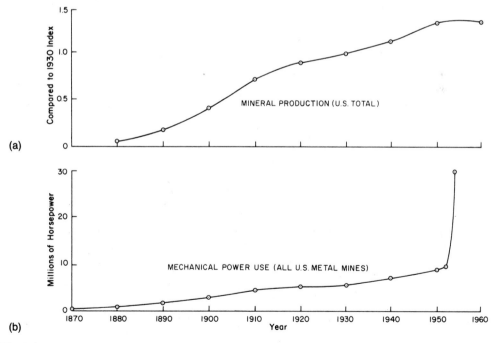

FIGURE 12.4

(a) U.S. mineral production from 1880 to 1960 is shown. (b) Mechanical power is used in ever-greater amounts to extract minerals, even though production has not grown commensurately. (Figure from *Resources and Man,* used by permission of W. H. Freeman.)

ground water would probably leach out contaminants and become polluted. The decrease in effluent per unit of production must more than overcome an increase in production. The inevitable accident is made more likely the longer the production, as is its impact. Most pollutants cause problems even with chronic low-level exposure. There is the possibility of pollution on a truly global scale.

Estimating Resources

Most mineral resources are hidden under the Earth. It is useful to distinguish *reserves* (economically recoverable material in identified deposits), from *resources* (which include reserves, identified deposits not presently recoverable, and deposits not yet discovered) (12). It is difficult to estimate even reserves with a high degree of accuracy until they are largely mined out. Estimates made in advance of production are generally 25 percent too low. The error in estimating incompletely explored deposits is much greater (12). As of this writing, copper is in oversupply.

We may classify resources as proved, probable, possible, and undiscovered. The distinction

between some resources and reserves is economic (Figure 12.5). This distinction can be made by applying one of three descriptors (12): *recoverable* (at less than or equal to present cost), *paramarginal* (recoverable at a cost less than 1.5 times present cost), and *submarginal* (recoverable at some price above 1.5 times present cost). History shows that there is a movement from submarginal to paramarginal to recoverable, mostly attributable to technological advances. The cutoff grade of copper ore has been reduced by a factor of 10 since 1900, and by a factor of 250 over the history of mining (12).

With these definitions, it is possible to relate reserves to the abundance of minerals in the geologic environment. The tonnage of minable reserves of well-explored elements in the United States is approximately related to the crustal abundance (reserves are 10^9 or 10^{10} times the clarke) (12) (Figure 12.6). This provides us with a tool for estimating, with a given level of technology, what total resources amount to.

Many argue that resources can be extended by substitution of one material for another (1,20). For example, transistors replaced vacuum tubes, and germanium, selenium, and other once-use-

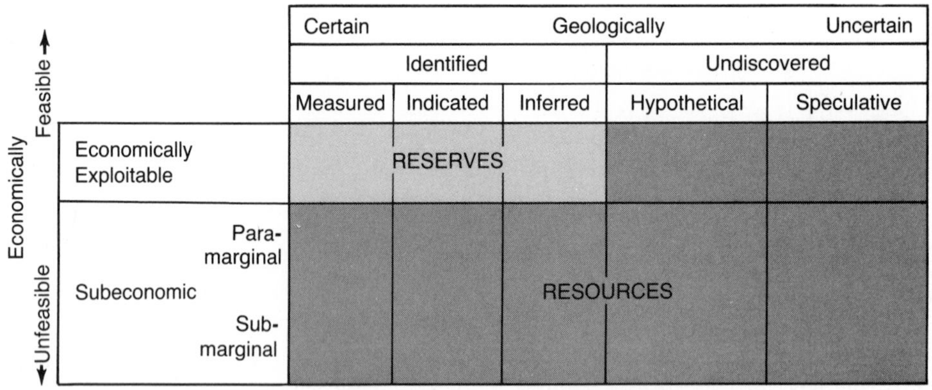

FIGURE 12.5
Reserves may differ from resources in two ways. First, they may not be available at the current price (ie., they are subeconomic). Second, they may not exist. Hypothetical resources are those that should be there on the basis of analogy with known geologic formations. Speculative resources are just estimated with very little evidence.

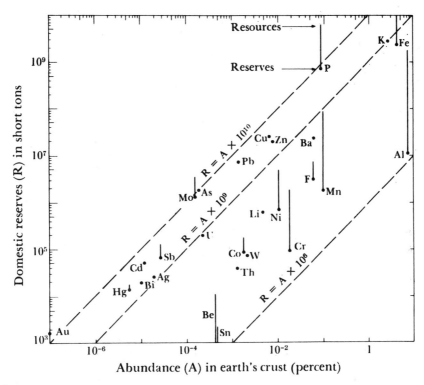

FIGURE 12.6

A comparison of domestic reserves to their abundance in Earth's crust. The dots represent tonnage of minable ore at present; tonnage exploitable at higher prices is indicated by the bars. (V. E. McKelvey, Copyright 1972 by the Society of the Sigma Xi.)

less materials became resources (21). Old, abundant materials may be used in new ways (see, for example, Hillig's discussion, in Reference 22). Magnesium or aluminum may be used instead of steel. Table 12.3 (23) presents contrasting use "lifetimes" of various metals as given in Meadows et al. (18) and as estimated by the German government. The difference is partly due to replacement of metals by plastics, and by the reduction in weight of cars since the early 1970s, as well as the effect of the general world economic malaise in the 1980s, which caused a huge drop in demand (23). Table 12.4 (23) may hint that the peak use rate of some common metals may have passed (with caveats as indicated above, a recognition that ten years is too few to establish a

trend, and due recognition that exponentially growing populations will eventually force use rates to rise).

Another source of materials is common seawater and the seabed. Table 12.5 lists some common elements found in seawater and their value (20,24). Manganese nodules are found in certain regions of the ocean in amounts up to 10 kg/m², assaying up to 27–30 percent manganese, 1.1–1.4 percent nickel, 1.0–1.3 percent copper, and 0.2–0.4 percent cobalt (25). These nodules grow at a speed of from 1 mm to 10 mm per million years by several different mechanisms (26,27). Some areas of the seafloor, such as in the Red Sea, contain new deposits of minerals of virtually inestimable value. The worth of these seabed

TABLE 12.3
Estimated lifetimes* of selected resources at current use rates.

Resource	Meadows et al.[†] 1972		German Bureau for Geological Sciences and Raw Materials 1984
	(static)	(exponential)	(static)
Aluminum	100	31	253
Chrome	420	95	364
Cobalt	110	60	112
Copper	36	21	70
Iron	240	93	184
Lead	26	21	43
Mercury	13	13	28
Nickel	150	53	110
Tin	17	15	41
Tungsten	40	28	49
Zinc	23	18	39

*There are two ways to interpret the lifetime: lifetime at current use rate, with use of the current year's demand, for which there is no future growth in use assumed (static assumption); and lifetime at current use rate, assuming that the demand for the quantity has been growing and will continue to grow exponentially (exponential). We use both in this table. Note that the economy was near the top of a business cycle in 1972 and near the bottom of a business cycle in 1984.

[†]THE LIMITS TO GROWTH: *A Report for the Club of Rome's Project on the Predicament of Mankind,* by Donella H. Meadows, Dennis L. Meadows, Jørgen Randers, William W. Behrens, III. A Potomac Associates book published by Universe Books, N.Y., 1972. Graphics by Potomac Associates.

TABLE 12.4
Metals use (in kilotonnes) in 1973, 1979, and 1983.

	1973[a]	1979[a]	1983[b]
Aluminum	11,187	12,587	11,760
Copper	6927	7536	6765
Lead	4052	4197	3795
Nickel	514	584	472
Tin	204	179	154
Zinc	4822	4682	4317

[a]Top of business cycle
[b]Bottom of business cycle

resources has made agreement on rules for exploitation of the ocean basins difficult and has exacerbated arguments about territorial limits that were once centered on disagreements over fishing rights.

Bartlett (28) has proposed a different sort of solution to the problem of extracting depletable resources, which he calls "sustained availability." He proposes that resource exploiters voluntarily arrange that the rate of extraction will decline

TABLE 12.5

Concentration and 1985 value of selected elements in seawater (not including water).

Element	Concentration (kg/m³)	Per Mgallon Value (1985$)
Chlorine	0.4859	924
Sodium	0.4180	378
Magnesium	0.485	4130
Sulfur	0.0255	101
Calcium	0.0091	150
Potassium	0.0088	91
Carbon	0.0021	0.0001
Bromine	0.0007	190

exponentially with time. Such a program would assure that a resource would be available forever (in declining amounts, to be sure). He admits that the plan would seem too simple and naive to those advocating immediate plunder for financial gain, but its simplicity is a powerful asset. He believes that we should adopt a more responsible attitude toward exploitation of resources and leave to our descendants some modicum of resources.

SUMMARY

Minerals are extracted using energy, and some minerals supply energy. Minerals are deposited in place by chemical and geological processes, especially processes involving molten rock and the subsequent interaction of cool rock and water.

Minerals are concentrated at various grades in various parts of the world. Seldom does one country or even continent contain all or most minerals. The United States is dependent on the rest of the world for most of its minerals. Many minerals appear to be in short supply; the supply of most materials is measured in decades.

Reserves are economically recoverable material in identified deposits. Resources include reserves, unrecoverable deposits, and deposits not yet discovered.

REFERENCES

1. E. D. Larson, M. H. Ross, and R. H. Williams, *Sci. Am.* 254, no. 6 (1986): 34.
2. C. Mayer, *Science* 227 (1985): 1421.
3. G. P. Easton, *Am. Sci.* 72 (1984): 368.
4. C. Holden, *Science* 212 (1981): 305; see also *Mining and Minerals Policy,* a report by the Secretary of the Interior to the Congress, 1973.
5. Bureau of Mines, *Mineral Facts and Problems* Bulletin 617 (Washington, D.C.: GPO, 1980).
6. Department of Commerce, *Statistical Abstract of the United States 1984* (Washington, D.C.: GPO, 1984). See also R. C. Kirby and A. S. Prokopovitsh, *Science* 191 (1976): 713.
7. Department of the Interior, *First Annual Report of the Secretary of the Interior under the Mining and Minerals Policy Act of 1970;* this report to Congress was delivered in March 1972.
8. P. E. Cloud, Jr., *Texas Quarterly* 11 (1968).
9. D. B. Brooks and P. W. Andrews, *Science* 185 (1974): 13.
10. W. Greene, *New York Times Magazine,* 5 Jan. 1975, 9.
11. G. Hardin, *BioScience* 24 (1974): 561; and *Psychology Today,* Sept. 1974.
12. V. E. McKelvey, *Am. Sci.* 60 (1972): 32.
13. P. Flawn, *Mineral Resources* (New York: Rand McNally & Co., 1966), 14.
14. C. F. Westoff, *Sci. Am.* 231, no. 3 (1974): 108.
15. T. S. Lovering, in *Resources and Man* (San Francisco: Freeman, 1969), 109ff.
16. E. Cook, *Science* 191 (1976): 677.
17. S. G. Lasky, *Eng. Mining J.* 151, no. 4 (1950): 81.

18. D. H. Meadows et al., *The Limits to Growth* (New York: Universe, 1972).
19. J. Harte and R. Socolow, in *Patient Earth,* ed. J. Harte and R. Socolow (New York: Holt, Rinehart & Winston, 1971); W. D. Metz, *Science* 183, (1974): 59. Metz points out that each billion cubic feet of helium retrieved from air costs 10 percent of the U.S. annual energy use. For a more recent analysis, see E. F. Hammel, M. C. Krupka, and K. D. Williamson, Jr., *Science* 223 (1984): 789.
20. H. E. Goeller and A. M. Weinberg, *Science* 191 (1976): 683.
21. A. G. Chynoweth, *Science* 191 (1976): 725.
22. W. B. Hillig, *Science* 191 (1976): 733.
23. I. Mayer-List, *die Zeit,* 13 August 1984.
24. P. E. Cloud, Jr., in *Resources and Man* (San Francisco: Freeman, 1969), 135.
25. A. L. Hammond, *Science* 183 (1974): 502, 644.
26. R. A. Kerr, *Science* 223 (1984): 576.
27. F. T. Manheim, *Science* 232 (1986): 600.
28. A. A. Bartlett, *Am. J. Phys.* 54 (1986): 398.

PROBLEMS AND QUESTIONS

True or False

1. All mineral resources seem to be distributed equally among the continents of the globe.
2. The calculation of resource lifetime gives the most important information about that resource.
3. According to the Lasky formula, the higher the grade, the smaller the amount of ore of that grade.
4. The Lasky formula is universally applicable in describing mineral resources.
5. The light rock floating on the mantle is thickest underneath the great mountain ranges such as the Alps and Rockies.
6. The idea of continental drift was accepted as self-evidently correct from the time it was proposed.
7. A half century ago, the United States was dependent on imports for very little of its mineral needs.
8. Resources have been used as a means of maintaining international peace.
9. Supplies of most key minerals exceed a century in lifetime.
10. Many mineral supplies are in overabundance as compared to the 1970s because of the huge drop in demand during the depression of the early 1980s.

Multiple Choice

11. The amount of aluminum imported to the United States generally
    a. increases with time.
    b. decreases with time.
    c. varies erratically from year to year.
    d. None of the above

12. A basic reason to support a helium conservation program is to
    a. thwart the second law of thermodynamics.
    b. save energy in the long run.
    c. save money in the short run.
    d. All of the above
    e. None of the above

13. Under an international *triage* policy, which nation is most likely to be allowed to succumb?
    a. India
    b. Iraq
    c. Columbia
    d. Bangladesh
    e. China

14. It has been suggested that we could solve the resource problem by mining "average rock." What is the largest drawback of such an approach?

a. Too much energy is necessary, and energy is expensive.

b. The quantities of waste are very large.

c. Ground water could become contaminated.

d. The amounts of valuable minerals are too small to make it worthwhile.

e. Machinery for such a task is not available.

15. Agreement on a seabed resources treaty has been impossible to reach. Which of the reasons given below is *not* probably a reason for this?

a. Landlocked countries want a share in seabed resources along with countries with coastlines.

b. Manganese is in short supply in most areas of the Earth, and substantial manganese resources are available in the manganese nodules on the seafloor.

c. There are many minerals found in seawater, and no nation wants to give up the opportunity to obtain these minerals.

d. Countries with boundaries on oceans or seas cannot agree on a common territorial limit.

e. Some countries see the arguments over resources availability as part of "cold war" contention between the superpowers.

16. The cost of mining has increased spectacularly in recent times. Which factor listed below is most probably responsible for this?

a. The cost of energy has been dropping in real terms.

b. The cost of mining machinery has declined recently.

c. More machinery has been used because miners' salaries are too high, so owners replace people with machines.

d. The grade of ore now being mined is lower than in times past, so much more material must be handled per unit of material production.

e. The distance between the mine and ore processing facilities has been increasing.

17. The rush to colonization by European countries in the eighteenth century was prompted mostly by the

a. need for more workers to run machinery.

b. need for new sources of raw materials.

c. belief in the God-given right of Europe to rule the world.

d. need for agricultural products.

e. population explosion in Europe.

18. Which agencies are important for production of minable minerals on

a. Sedimentation

b. Volcanism

c. Upwelling of material between separating plates (seafloor spreading)

d. All of the above

e. None of the ab

19. Evidence for seafloor spreading comes from
    a. matching of rock types on both sides of the upwelling.
    b. symmetry in the shapes of mountains on both sides of the upwelling.
    c. study of marine fossils on both sides of the upwelling.
    d. the symmetrical pattern of the magnetic field captured in rock on both sides of the upwelling.
    e. the easily measured movement of the seafloor.

20. Cobalt is not mined in the United States, but comes from sometimes unstable regions of the world. Where could the United States get an assured alternative supply?
    a. Mine manganese nodules from the ocean floor.
    b. Recover cobalt from seawater (less than 1 gram per cubic meter).
    c. Take over supplier countries as colonies.
    d. Buy off politicians in supplier countries.
    e. Mine average crustal rock.

Discussion Questions

21. What sort of ramifications are there to a "lifeboat" philososphy? Is our expectation that resource-rich nations act in their own interest in accordance with the "lifeboat" view?

22. What is the purpose and morality of the "triage" view?

23. Is it likely that the United States could function as a self-contained isolated unit?

24. Could the problems envisioned in *The Limits to Growth* (22) generally be solved by the acquisition of a cheap energy supply?

13

FOSSIL-FUEL RESOURCES

Continuing the examination of resources available for production of energy, begun in Chapter 11, this chapter considers the long-term possibilities for fossil fuels: petroleum, coal, natural gas, and alternatives such as shale oil. All fossil fuels are present in finite amounts. How do we predict how long a supply will last? When does it make sense to use coal in preference to oil? This chapter attempts to address some of these issues.

KEY TERMS   *fossil fuels*
steam vessel
turbine
condenser
hydrocarbons
production rate
normal curve
delay time
logistic curve
finding factor
primary recovery
secondary recovery
tertiary recovery
fractionating tower
catalytic cracker
kerogen
fluidized-bed combustion
power gas
synthetic natural gas

Since ancient times, surface oil and asphalt deposits have been used for medicinal purposes. Then, in the late Middle Ages, the British began to use their coal resources for heating—the first large-scale use of fossil-fuel energy. They were forced to begin using coal because the profligate use of wood as fuel for heating and for cottage industry had reduced the once-magnificent forests of the British Isles to pitiful remnants. However, despite these examples, it is safe to say that the total use of fossil-fuel resources was negligible prior to 1800.

In Figure 13.1, we see how the use of coal and oil has grown over the past 200 years. This world use history may be compared to use in the United States, shown in Figure 13.2. Coal use doubled worldwide about every 16 years (4.4 percent per year growth rate) in the period from 1860 to 1913, doubled about every 93 years (0.75 percent per year growth rate) in the period from 1913 to 1945—two major wars and a depression occurred during this time—and doubled every 20 years (3.6 percent growth rate) from 1945 onward (1). Note that U.S. coal use lagged somewhat behind the total world use; this happened because the wood economy was able to persist longer than in developing countries elsewhere. U.S. oil use led the world in development because, by the late 19th century, the United States was already an industrial power, and industry was developing rapidly in the coal-poor northeast. World crude oil use has doubled every decade (7 percent per year growth rate) (2).

These fossil fuels are used in small power plants for transportation of goods and people, and in large power facilities to supply industrial steam and electric energy.

THE FOSSIL-FUEL PROCESS

A simplified schematic of a fossil-fuel facility is shown in Figure 13.3. Fuel is introduced and burned in the boiler. There are various ways to burn fuel in the boiler, but the most common are similar in operation to those found in home

FIGURE 13.1
(a) World production of coal and lignite as it varied between 1800 and 1982 (M. K. Hubbert, Fig. 3, in *U.S. energy resources as of 1972*, U.S. Senate Committee on Interior and Insular Affairs, document 93–40, GPO, Washington, DC 1974 and U.S. Department of Energy, *Coal Data: A Reference*, GPO, Washington, DC, 1985, Fig. 19). (b) World production of crude oil as it varied between 1880 and 1977 (M. K. Hubbert, Fig. 3, Am. J. Phys. *49*, 1007 (1981). Reprinted with permission.)

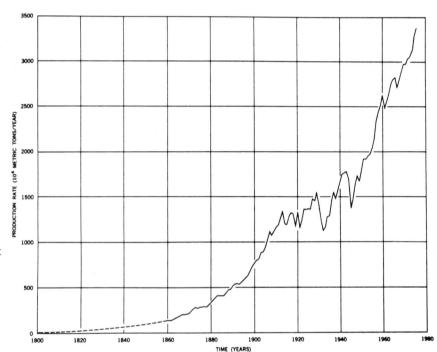

(a)

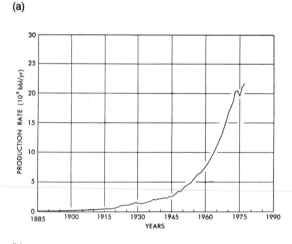

(b)

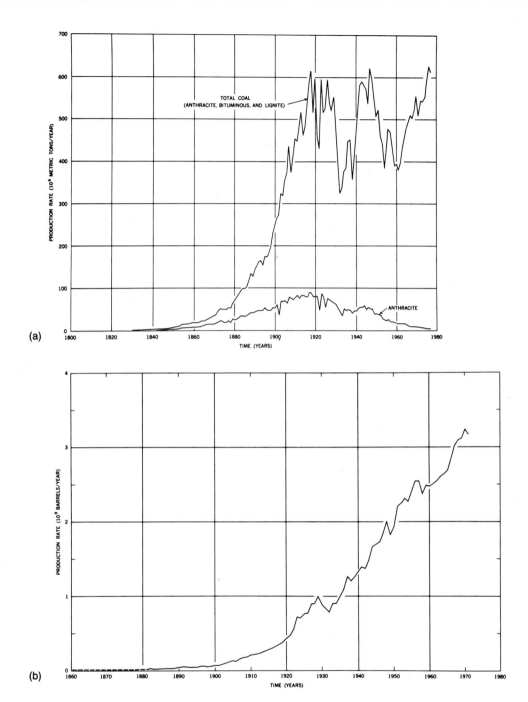

FIGURE 13.2
(a) U.S. production of coal and lignite as it varied between 1800 and 1971 (M. K. Hubbert, Fig. 5, in *U.S. energy resources as of 1972*, U.S. Senate Committee on Interior and Insular Affairs, document 93–40, GPO, Washington, DC 1974). (b) U.S. production of crude oil as it varied between 1880 and 1971 (M. K. Hubbert, Fig. 6, in *U.S. energy resources as of 1972*, U.S. Senate Committee on Interior and Insular Affairs, document 93–40, GPO, Washington, DC 1974. Reprinted with permission.)

FIGURE 13.3
Schematic of a steam
electric generating facility.

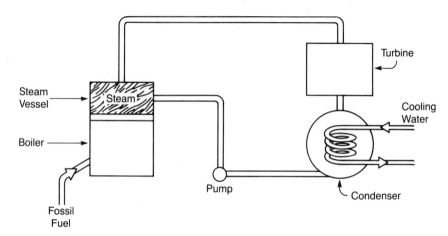

furnaces. The gas or fuel oil burns in an open flame fed by pipes. Lump coal is carried to a grate by a conveyer, dumped onto the grate, and burned, with the ash falling through the grate. Coal is sometimes piped in as slurry, or small lumps of coal in water. Heat in the furnace evaporates the water, the coal then burns, and the ash falls as before. Most often, coal is pulverized into dust-like powder and blown into the furnace.

The open flame heats the feed water in the *steam vessel,* which produces steam under pressure. This gives an increase in the overall efficiency of the process because it raises the temperature at which water boils and thus enhances the temperature difference between high and low temperature reservoirs. The high pressure is also useful in driving the steam along its path. The steam is then piped to the *turbine.* In the turbine are blades that intercept the steam as it flows. The blades are canted at an angle (like a propeller), and the steam bounces off the blades, pushing at them. This causes the turbine blade assembly to rotate about its center. Since the blades are connected to a shaft at the center, the shaft can be connected to a coil outside the turbine. The coil then rotates in a region where there is a magnetic field, as described in Chapter 4. This produces the alternating current (AC) we ordinarily use.

The steam, which has continued through the turbine, flows on to a *condenser,* where the spent (cooling) steam condenses again to water. This is the lowest temperature point in the cycle. To keep the condenser at ambient temperature, cooling water from a lake, river, or cooling tower is employed (Chapter 11). This cooling water merely circulates through the system, and is later exhausted at a slightly warmer temperature. The feed water in the condenser is generally recycled back into the steam vessel to close the cycle.

A recent development, called the *Kalina cycle,* holds the promise of raising the efficiency of conversion to electricity from ≈33 percent to 45 percent (3). Instead of using water alone in the closed cycle as the working fluid, in the Kalina cycle, water is mixed with ammonia to lower the boiling point and consequently to liberate more steam than water alone at the high temperature. The ammonia is diluted further with water to raise the dew point; condensation then occurs at a reasonable temperature, and the water is distilled out to reconstitute the original ammonia–water mixture.

In the automobile, hydrocarbon fuels refined from fossil fuels provide energy for transportation. A fuel (gasoline or diesel fuel) is mixed with air, introduced into a cylinder, and then ignited by a spark device or its own heating by compres-

Within my living memory, gasoline cost 19.9 cents per gallon. Anyone looking at a pump in a service station realizes that those days are no more. The price of fossil-fuel energy was unrealistically low, leading to profligate use and wastage. We humans were squandering the heritage of the last several hundred million years of geologic deposition, and doing it over a very short period of time. From this perspective, the price rises of the 1970s were a welcome dose of reality.

The year 1969 saw the United States produce the most oil in the history of oil production. The subsequent slow decline in production and in reserves went almost unnoticed. The world was awash in imported oil. The new era began in November 1973 with the so-called Yom Kippur war between Egypt and Israel in the Mideast. The Arab nations had control of the Organization of Petroleum Exporting

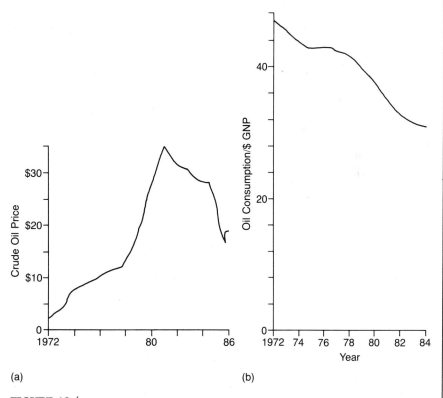

(a) (b)

FIGURE 13.4

(a) Oil prices ($ per barrel) are given from 1972 to 1986. (U.S. Department of Energy). (b) Oil consumption per GNP dollar is given from 1972 to 1984. (U.S. Department of Energy)

Countries (OPEC), the producers of the major fraction of exported crude oil, and embargoed exports to the United States and other Western countries. Their goal was to put indirect pressure on Israel to settle territorial disputes in the Sinai and to resolve the Palestinian refugee problem. In the wake of the cutbacks, the price of oil roughly quadrupled (see Figure 13.4). R. D. Hershey, in a retrospective article on the energy crisis of the 1970s, writes (4) that "it was no coincidence, analysts agree, that huge price increases and the Arab embargo came not long after United States oil production peaked, at 9.6 million barrels a day, in 1970."

Prices were high enough and, because of the long lines of cars waiting for gas and a general crisis atmosphere, nerves of politicians were raw enough, that President Nixon (1969–1974) felt obliged to proclaim "Project Independence," an attempt to assure a domestic supply of liquid fossil-fuel energy. This project was carried on by President Ford (1974–1977). These developments promised an evanescent independence and a rapid depletion of domestic resources, with attendant environmental effects (5).

President Carter (1977–1981) presided over still another administration that discovered how vulnerable America was to foreign intervention. The revolution in Iran in 1979 led to another quadrupling in oil prices, and to gasoline rationing in parts of the country: Cars with odd-numbered license plates could buy gas only on odd-numbered days. The political response this time was to encourage development of alternative energy sources (see Chapters 15, 16, 17, and 18) and of domestic coal for energy, an emphasis begun with "Project Independence." President Reagan (1981–1989) presided over an era of lowered energy prices, mostly the result of conservation efforts launched earlier, and allowed the initiatives proposed by his three immediate predecessors to wither away. Alternative energy research of all kinds was subject to funding cuts, and big cars once more came into vogue.

All of these national policy developments have hinged on presidential perceptions or misperceptions. The point of this discussion is that presidential energy policies and ideas, such as "Project Independence," arise from political considerations. It is not possible to discuss the utilization of fossil fuels without discussing human influence and activity.

sion. The resultant explosion increases the pressure of the gas in the cylinder as the chemical energy in the fuel is transformed into kinetic energy of the gas molecules. Work can be done, and the piston inside the cylinder is driven, pushing on a cam device that ultimately couples to the wheels.

The Energy Crises of the 1970s

The United States (and the world at large) had gotten used to cheap crude oil. Such low cost led to the importation of foreign oil, even in the oil-rich United States. Other oil-importing countries recognized their vulnerability to a cutoff of supply and taxed imported oil to dampen consumption. The United States government took a more *laissez-faire* attitude, and consumers and contractors took this as an invitation to waste oil.

The steep price rises of the mid-1970s caused a slowing in the appetite for oil consumption; the price rise of 1979 on top of that sent consumption down almost 7 percent over the next three years (6). Economic growth in the late 1970s and early 1980s (when it occurred) took place at about twice the rate of energy growth. In earlier times, the two rates had been essentially identical (6). Much of the effect of the price rise on home oil use is still to be felt in Europe, because of the relatively new oil heating equipment (dating from the early to mid-1970s) now in place (7). Even though prices have dipped, the impetus to conservation by industry and in the home will probably not falter unless the price of oil repeats the collapse of 1986, during which the price of a barrel of crude oil dropped from about $30 in November 1985 to as low as $10 by July 1986. Thereafter, prices rose again to a level of $18 to $20 per barrel. Of course, oil prices influenced the producing countries, causing domestic price rises, lower production for export, and grandiose spending schemes based on the enormous fund transfer to the oil-producing countries (8). As mentioned earlier, the price rises of the 1970s were triggered by an oil scarcity in the United States. The price rises caused lowered demand (4,8,9), which in turn impelled conservation measures (9). According to this view, the price had

to fall again, because a glut had built up (9) (see Figure 13.5 for a droll comment on this change).

Cheaper oil is a boon to the air travel industry, and also influences consumers to buy larger cars. Since the United States imports so much oil, economists expected the lower oil bills to decrease the deficit in the balance of payments, since each $5 per barrel drop in price translates to an annual savings of $8 billion dollars (10). Instead, the balance of payments continued to register the worst deficits in history. The spectacular drop of early 1986 disrupted the economies of the major oil-producing states, causing layoffs, bankruptcy, and bank failures.

Of course, any gains from low oil prices are short-term, because the price of oil cannot long remain so low at our volume of exploitation. The economic fallout of the 1986 price drop did not result in outright disaster, but it was very unpleasant for the economy as a whole. The lower oil prices led to increased U.S. oil use after some years of relatively flat consumption, pushing up U.S. imports once again (11). By the mid-1990s, with imports predicted to exceed 50 percent,

FIGURE 13.5

A 1985 cartoon comments on the political demise of OPEC in the mid-1980s. (Dana Summers, copyright 1986, Washington Post Writers Group, reprinted with permission.)

OPEC may again be able to control oil prices. The United States appears to be heading toward another oil crisis.

OIL AND THE PREDICTION OF DEPLETABLE RESOURCES

There appears to be an adequate predictive measure of how much energy it will ultimately be possible to extract; that is, of how much fossil fuel there is in the world. For example, it is estimated that U.S. natural gas and crude oil resources total only 10 percent of coal energy resources (167 gigatonnes coal, 7.5 Tm^3 natural gas, 48 billion barrels (Gbbl) of crude oil). The estimating method used in these predictions was derived by geologist M. K. Hubbert, who worked at various times for Mobil Oil and for the U.S. Geological Survey. Hubbert notes (1,2) that the rate of production must be zero at two times: before exploitation of a resource, and after exhaustion of the resource. In between, a certain amount is produced, with the time over which it is produced depending on the *rate* of production (the amount produced per unit time). This follows because the amount produced in a time interval Δt is (see Figure 13.6):

rate of production $\times \Delta t$.

The shaded area in Figure 13.6 is just this amount. By adding up areas like those in Figure 13.6 for

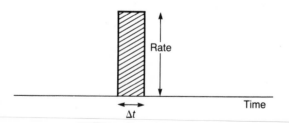

FIGURE 13.6
Production equals rate times time interval. The shaded area is equal to ΔP, the amount produced.

all the changing rates and their corresponding Δts, we find the total area beneath the production curve. The total area under the production curve is the total amount of energy resource produced.

Hubbert begins with this (perhaps obvious) point and then extrapolates, or projects ahead, observations related to the production rate of the energy resource in order to predict the total area under the curve—the total energy resource that will have been produced by the time the resource is exhausted. This method could also apply to rates of commodities use, or rates of industrial activity, or other kinds of activity. This method's weakness is that the phenomena it analyzes strongly reflect human activity, which is influenced by economic and political factors. McKelvey (12) in particular has criticized this method, preferring to extrapolate observations relating to the abundance of a mineral in its geologic environment. One such method used in dealing with oil exploration (the Zapp hypothesis) is discussed later in this chapter. These methods are purportedly more capable of allowing for major breakthroughs in technology that would lead to increased recovery.

Let us denote the rate of production of a resource by P, and the amount produced by Q. Thus, we can say

$$\text{or} \quad \begin{aligned} P &= Q/\Delta t \\ Q &= P\Delta t \end{aligned}$$

as we found earlier. The ultimate amount produced, we will call Q_∞, will be greater than (or equal to) the estimate of Q from geologic information. The production curve may assume any number of shapes. However, our latitude of choice is reduced because the technology of production mandates an exponential growth of production at the start of exploitation. One curve with these properties is the *normal curve,* illustrated in Figure 13.7. We will assume here that this is the production curve.

Figure 13.8 shows the actual U.S. production data for crude oil (Figure 13.2b) along with the corresponding normal curve as predicted in 1967.

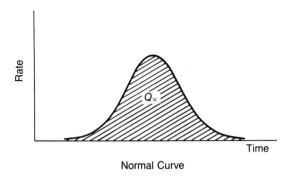

FIGURE 13.7
A normal curve.

The maximum rate of production was predicted to occur in 1969 or 1970 for the continental United States. In fact, 1969 *was* the year of maximum production from domestic wells (12). Domestic output, exclusive of Alaska, has been dropping ever since.

In order for there to be production in a given year, there must have been a discovery of oil in a particular area some time earlier. After some time has passed for installation of equipment to tap the oil, production can begin. This interval is called the *delay time*. Thus, we expect the quantities of crude oil lying in the ground of which we have knowledge (proven reserves) first to increase as more discoveries are made and then to decrease as these discoveries are used up and we run out of crude oil altogether. Hence, the total accumulated amount of crude oil produced rises exponentially at first, slows, and then gets slowly closer to its final value—the total amount of recoverable crude oil in the crust. Such behavior has already been seen: The cumulative production follows a *logistic curve*. Since production follows discovery by the delay time, the cumulative discoveries must resemble the cumulative production, but these are separated in time by the delay time, Δt. This is illustrated

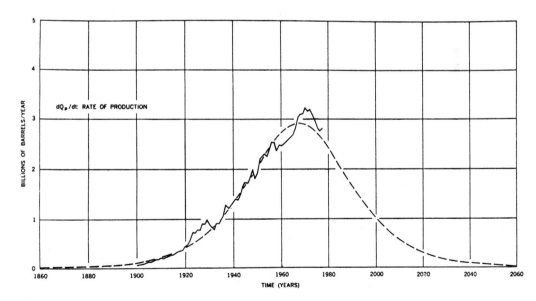

FIGURE 13.8
Data on U.S. oil production (excluding Alaska) fall approximately on a normal curve (M. K. Hubbert, Fig. 14, in L. E. St. Pierre, ed., *Resources of Organic Matter for the Future: Perspectives and Recommendations,* Multiscience, Montreal, Canada, 1978). Q_∞ is about 165 billion barrels.

in Figure 13.9. More realistically, we expect the actual production curve to be skewed somewhat to the left (the start of exploitation) and to collapse abruptly toward the end of exploitation.

These data fix the curve that allows us to predict the total U. S. crude oil production. Taking 1969 as the peak production year and using a delay time of twelve years, we estimate $Q_\infty = 171$ Gbbl of crude oil for the continental United States.

This method obviously may be generalized to other fossil-fuel resources, and has been accepted by the government as its method of estimation.

Is There Such a Thing as Energy Independence?

Before the energy crisis, developing countries as well as the developed countries were hooked on cheap energy. They were, however, much less able to adjust to the price rise than the more developed countries, and we had quite a time even in the United States in coping with the gigantic reallocation in capital from oil consumers to oil producers. The oil price rises were re-

sponsible for a tenfold jump in the trade deficit of the developing countries in the decade 1973–1982, as well as for the sixfold increase in national debt (32). Because the systems of all countries, developed and undeveloped, came to depend on oil as an essential commodity, it has been extraordinarily difficult to shift people's unconscious expectations. People want the status of car ownership, and automobiles need gasoline for fuel.

All these considerations taken together indicate that "Project Independence" would not have been able to end American dependence on imported oil and (eventually) gas. As was pointed out by President Carter (1977–1981), only conservation of our scarce reserves and resources has a chance of protecting the United States and Canada from total dependence on imports at some future time. Carter's plan would have emphasized coal production, because of the vast quantities of coal that the United States still has, as we shall see in subsequent discussion.

No country that is developed or developing in the world is insulated from oil addiction, and it is just as impossible to go "cold turkey" from

FIGURE 13.9
Curves for cumulative discoveries (Q_d), cumulative production (Q_p), and proven reserves (Q_r), during a cycle of production. The data are through the end of 1977. The time lapse between discovery and production is 12 years (M. K. Hubbert, Fig. 11, in L. E. St. Pierre, ed., *Resources of Organic Matter for the Future: Perspectives and Recommendations*, Multiscience, Montreal, Canada, 1978).

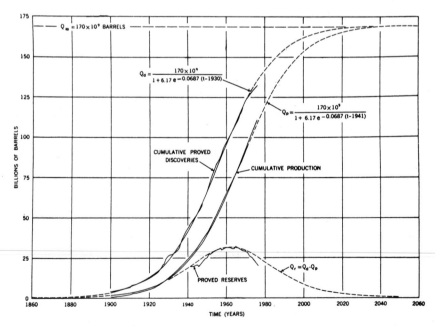

oil as from an addictive drug. All countries must find their own alternative energy sources, be it coal, or geothermal, or energy from biomass, or alternate means of transportation. Whatever one's estimates of total oil reserves, it is absolutely clear that oil will run out, for all practical purposes, by 2050 at the latest unless civilization itself collapses. A Congressional Research Service expert, Joseph Riva, predicts a 17 percent decline in world oil production by 2000 as his most optimistic estimate—a 29 percent drop is more likely (33). The alternatives, whatever they are, will have to be found.

The United States established a strategic oil reserve in salt caverns in Louisiana, which is designed to hold between 750 Mbbl and 1 Gbbl of oil eventually. The reserve has continued to be filled, in spite of opposition from oil-producing countries, at a rather slow rate. By 1984, only about 330 Mbbl had been put in the reserve—at 1.7 Mbbl per day, that supply would last for about 180 days (34). This reserve is of some help in delaying interference in internal American politics by the oil producers. Some observers have advocated a national tax on oil consumption similar to what has been done in other countries as a method to encourage substitution of other fuels where possible, preventing an undesirable shift back toward gas-guzzling cars and making an attempt to have consumers pay the full cost of fuels (35). Such a tax will not be instituted, despite the prudence of such a course, because of the current ascendence of the "free market" concept in American political thought. The decline in consumption as a result of the price rises of the 1970s proves that eventually, as the price moves high enough, demand drops.

Shale-Oil, Oil Mining, and Tar Sands

Another alternative for increasing the reserves of oil is the utilization of the hydrocarbon compounds (called *kerogen*) in oil shale. Kerogen, a solid hydrocarbon, is found in the process of oil formation; it is partially cooked petroleum. The oil shale in Colorado was reportedly dis-

covered to contain fuel resources by a local settler who used the shale to construct a fireplace and chimney. This rock's oil caught fire; the settler's first fireplace fire was his last in *that* house.

Unlocking the oil from the kerogen in the oil shale requires a tremendous amount of water: three barrels of water for each barrel of oil produced. Another problem is that there is a yield of only 50 liters or so of oil per tonne of rock, so that an incredible volume of waste rock is produced in the process of removing the kerogen from it (36).

A few formations have a relatively high yield. The Green River basin of Colorado, Wyoming, and Utah has as much as 15 gallons of oil per tonne of rock—30 gallons per tonne in one formation, the Piceance Creek Formation (37). Some shale-oil was mined in Rifle, Colorado, in 1978 that produced 25 gallons per tonne (38).

In the first flush of the 1973 energy crisis, several companies examined the possibility of oil production from shale. The Colony Development Corporation had a plan to mix hot ceramic balls with finely crushed shale, recovering all the oil. In 1973, Colony estimated the capital cost at $250 M. By October 1974, the estimated cost had spiraled to $850 M, which would have resulted in a cost of over $12 per barrel for shale crude (still higher than 1974 oil prices), over twice the original National Petroleum Council estimate (39). Colony thereupon suspended plans for its 50 kbbl/day plant.

Another approach was developed by Occidental Petroleum. Occidental's plan was to blast a room in the rock, build an exit channel, and then ignite the rock in the closed-off room. This method was estimated to be capable of recovering about 80 percent of the oil (37).

Among the problems of shale-oil production, there is the fact that to get 1 Mbbl of oil per day, 1.2 Mtonne of residue is produced (37). Visual pollution is created, and revegetation is difficult since the shale-oil–bearing area of the West gets only 37 cm of rain or less a year. Large-scale production would also disturb the habitats of local wildlife and would release dust, sulfur diox-

CRITICISMS OF THE HUBBERT ESTIMATE

As we indicated earlier, there are some criticisms of the Hubbert method. Zapp (1,12) argues that the best way to express the rate of discovery is in terms of the cumulative footage of exploratory drilling; that is, that the historical average experience of about 120 bbl per foot drilled will continue into the future. This argument sounds reasonable, and would predict that approximately five times as much oil would ultimately be discovered as had been until 1959: That amount was 60 Gbbl. Thus, Zapp would estimate Q_∞ = 300 Gbbl, approximately doubling Hubbert's estimate.

Zapp's argument fails on two counts. One follows from a study done comparing production from wells drilled using a random method of site selection (and utilizing the relatively well-known distribution of domestic oil to predict production) to production from the wells as they were found historically (13). The study found that drilling randomly was *more* likely to produce oil than was historically the case up to at least 1940. A similar study (14) was done for the oil in the Permian Basin in West Texas; this study documents a decline in barrels of oil found per well drilled. More than 60 percent of the oil and gas came from 2 percent of all fields, and these giant fields were among the first discovered.

The second argument against the Zapp view has been made by Hubbert, who examined the data on crude oil discoveries per foot of exploratory drilling. If the Zapp hypothesis is correct, the curve of discoveries per foot drilled versus time should be a straight line. A comparison between the Zapp hypothesis prediction and the actual discovery rate is shown in Figure 13.10 (1, 14, 15) along with Hubbert's projection of the declining discovery rate. Using the dotted-line projection in Figure 13.10, we derive the value of Q_∞ as 165 Gbbl, in good agreement with Hubbert's previous estimate. Note from the figure that production per foot drilled is declining despite the advances in petroleum exploration and production techniques mentioned previously. According to Zapp's hypothesis, the oil industry has been 130 times less successful than if it had drilled at random (16). Presently, only one in ten exploratory wells finds sufficient oil to justify production, and only one

ide, high levels of arsenic, and complex organic substances into the Colorado River, as well as increase its salinity (a problem similar to that detailed in Chapter 11) (40). One problem with the Piceance Creek Formation in particular is that it holds one aquifer above the shale and another aquifer below the shale. Shale mining, in drilling into the shale formation, could open communication between the aquifers and cause large-scale contamination.

When the energy costs for machinery, plants, electricity, and other necessities are taken into account, some observers estimate that little or no net energy would be gained in the exploitation of oil shale (39).

There are also heavy oils in the sands of various locations around the world. The amount of oil in oil sands is staggering. Estimated reserves in Venezuela are somewhere in the neighborhood of up to 3000 Gbbl; in Canada and the USSR, over 1000 Gbbl; and in the United States about 175 Gbbl (41). For the sake of comparison, the Prudhoe Bay oil find contained about 15 Gbbl. The Athabasca oil sands in Canada alone contain 869 Gbbl of oil, of which only 74 Mbbl are readily available in an area of surface strip-mining (42).

well in fifty repays its total cost (17). This is at a time when the oil industry is using sophisticated geologic information to drill in the most promising formations. However, the discovery rate does not seem to be truly exponential; rather, it appears to be falling modestly over the last few million kilometers drilled (18).

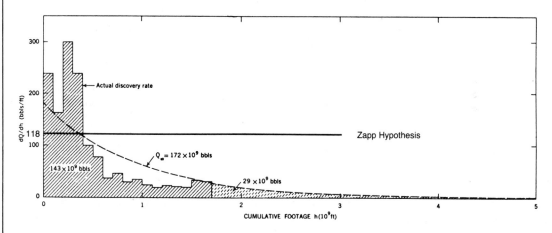

FIGURE 13.10
Comparison between the Zapp hypothesis for the oil finding rate in the U.S. (predicting 118 barrels/ft) and actual U.S. experience. (Adapted from M. K. Hubbert, Fig. 50, in *U.S. energy resources as of 1972*, U.S. Senate Committee on Interior and Insular Affairs, document 93–40, GPO, Washington, DC 1974).

At present, about 45 kbbl/day is being produced from 100 ktonne of rock processed (38). The main problems encountered in Canada are abrasiveness of the sand (mostly quartz) and the ferocity of the winters. Large chunks of steel can shatter as machines try to loosen and transport oil sands from frozen windrows to the refinery.

Still another way to increase fossil-fuel supplies is to mine oil. We have seen that oil wells typically leave two-thirds of all oil in place, while coal mines can recover up to 95 percent of the coal in a formation by stripping. Exploration of this connection led to construction of several working oil mines. The oldest appears to have been a gravity flow of oil to collection sites through holes drilled into an oil formation in Ventura, California, in 1866 (38). During World War I, Pechelbronn in Alsace was the site of an oil mine of the same sort as that in Ventura. By the end of World War II, 5.4 Gbbl had been recovered (38,43). The Germans also mined oil in Wietze, near Hannover, for a cumulative production by 1950 of 5 Gbbl.

The Russians mine oil in Yarega, 185 meters down. The Yarega mine holds viscous heavy oil, so in addition to the holes drilled into the rock

OTHER ESTIMATES OF RESERVES

There have been more recent estimates of oil reserves. Depending on whose estimates are believed, the United States could run out of oil sometime between 1998 (lowest estimate) and 2075 (highest estimate) at 1975 rates of consumption (19). The Workshop on Alternative Energy Strategies predicted that world production would fail to meet demand sometime between 1990 and 1997 (20).

A recent National Academy of Sciences (NAS) study (19) predicted remaining reserves as of 1975 to be 113 Gbbl of oil and 15 Tm³ (530 Tft³) of gas. The U.S. Geological Survey predicted 200 to 400 Gbbl of oil and 28 to 56 Tm³ (1000 to 2000 Tft³) of gas (21). To put this in perspective, British probable reserves as of this same time are about 13.6 Gbbl from its North Sea oil finds (22). The U.S. Geological Survey has recently given estimates more in line with Hubbert's projections.

The major disagreement involves the estimate of the *finding factor*—the amount of oil ultimately available as a result of drilling in barrels per foot drilled (the Zapp hypothesis is that it is 1; Hubbert's, that it is about 0.1). The argument against the Hubbert–NAS methods is that they are essentially *economic* rather than *geologic*. In fact, some critics have claimed that as much as 485 Gbbl are yet recoverable (23). A sampling of estimates is shown in Table 13.1 (17,24,25). Figure 13.11 shows the area containing oil resources in North America. Predictions of total American reserves include oil and gas from the continental shelf. In 1981, the U.S.

TABLE 13.1
Estimates of ultimate oil recovery in the United States.

Reference	Oil in Place (billion barrels)	Ultimate Recoverable Reserves (billion barrels)	Undiscovered Recoverable Reserves (billion barrels)
Hubbert (1,2)		190	62
National Petroleum Council	727	228	90
American Association of Petroleum Geologists	824	257	120
U.S. Geological Survey (24)		1718	550
Doescher (24)	460	148	
Halbouty and Moody (25a)		987	
U.S. Department of Energy (25b)	28.4		82.6
British Petroleum (25c)	32.5		

NOTE: Undiscovered recoverable reserves have been estimated by subtracting the present cumulative production of 100×10^9 barrels and the expected reserves of 38×10^9 barrels from ultimate recoverable reserves.

Department of the Interior estimated that about 27 Gbbl of oil would be found offshore (about 40 percent of all undiscovered oil) (26). By 1985, 100 dry wells had been sunk offshore, and the Interior Department had reduced its estimate to 12.2 Gbbl of oil (27). R. Nehring of RAND Corporation estimates that offshore wells will produce only 3.5 Gbbl of oil (15). Alaska has between 14 Gbbl and 19 Gbbl of oil (15).

Because of our (almost profligate) increases in demand from 2.6 to 4.3 Gbbl/year between 1955 and 1975 (17), there has been unprecedented exploration. In

FIGURE 13.11
The geographic distribution of North American oil reserves is shown schematically.

those twenty years, U.S. oil reserves dropped from 14 percent to 7 percent of world reserves. U.S. production dropped over those same twenty years from 44 percent to 19 percent of world production; furthermore, the number of wells fell from 600,000 to 500,000, the number drilled per year dropped by a factor of 2, and the distance drilled dropped from 71,000 km to 42,000 km, in spite of improved technology for drilling deeper wells.

Since the United States is the best-explored region in the world, and the region with the most experience in extrapolating production, there is less certainty in applying this experience to estimation of world energy resources. Therefore, estimates of world oil resources are even less trustworthy than estimates of American discoveries. Even though the ground in the United States has been extensively studied and drilled, there are great disparities among estimated U.S. reserves, as we have seen.

More clues can be found in the history of oil in the United States. The peak in discoveries in the 1930s coincides with the great number of giant fields discovered. Reserves peaked in the 1940s, and began to decline exponentially following the exponential decline in discoveries (14,16) by the delay time. Peak use occurred in 1969.

If we refer to the previous figures, using Hubbert's projections for the United States, we see that the first 10 percent of oil production occurred before 1934, and the middle 80 percent will be gone by 1999. Actually, the effect of the energy crises and the political decisions made will change the profile a bit from that of a normal distribution. But there is no doubt that oil production has peaked and will inevitably drop off in the United States (the third-largest producer of oil in the mid-1980s) as well as in the USSR (the largest producer of oil in the mid-1980s) and, by implication, in the world.

World reserves have stopped growing, and it is unlikely that the finding rate of the early 1970s (15 Gbbl/yr) can ever be attained again (13). Worldwide reserves are beginning to fall; the U.S. Department of Energy estimates 697 billion barrels of proved reserves worldwide (25b). The USSR, another large oil producer, saw its production of oil peak in 1985 (28). The lack of growth is a problem because it is not possible to recover more than 10 percent of reserves in any year without reducing the amount ultimately recoverable (18). Cumulative crude oil production for the world will range from 1350 to 2100 billion barrels, based on estimates of L. G. Weeks and W. P. Ryman (quoted in Reference 1).

(at ≈30 m intervals), other holes are drilled to inject steam into the formation. The oil that drips out is collected and then pumped to the surface. The Russians have achieved about 50 to 60 percent recovery (38).

Another method called *block-carving* undermines weak oil-bearing rock, which fractures and can be removed to the surface for processing. Finally, there is open-pit surface mining. This method is the least expensive of the alternatives discussed. Costs at the terrace mine in Kern River, California, are approximately $11/bbl, and at the terrace mine in Sunnyside, Utah, about $21/bbl (38). An open-pit mine in Edna, California, produces oil at $18/bbl. A strip mine in Santa Cruz, California, produces oil at a cost of $18/bbl (38). Since this is at or below the world oil price, these oil resources are economically exploitable. The

INCREASING RECOVERY

About 70 percent of all oil found is being left in the ground. Many factors (including government regulation) contribute to this. One problem is that the amount of oil produced is sensitive to the rate of extraction. The rate assuring optimal production is seldom the rate of optimal economic return (13,17,27). These rates are set by careful attention to the heterogeneity of the reservoir, the efficiency of displacement of oil by water in the rock pores, and the sweep efficiency, which measures how much oil is bypassed in the reservoir because of the geometry of the well. It is for this reason that yearly production should be no greater than 10 percent of reserves. *Primary recovery,* recovery under natural pressure, allows about 20 percent of the oil to be taken from a petroleum resource (13,30). *Secondary recovery* takes place when water or gas is pumped into a secondary well to push the oil toward the producing well; it can recover from 20 to 50 percent (average: 40 percent) of the oil in light oil reservoirs (13,31). About half of the U.S. production is due to waterflooding (13,30,31). The difficulty with waterflooding is that a lot of oil gets entrained in the water and is lost forever unless surfactants (chemicals allowing oil and water to mix) are added to the water—and surfactants are very expensive (31). *Tertiary recovery* consists of injecting steam or chemicals into the well. Steam injection is economically justified only when the reservoir contains more than 1000 barrels per acre-foot and is more than 15 meters thick (31). In the most favorable reservoirs, the energy cost of crude recovery by steam injection is 25 percent of the total recoverable; in California, it costs about 35 percent of the total energy recovered (31).

A method now being used to generate steam without an *external* energy penalty is to set the oil burning underground by injection of air; then, after burning is established, injection of air is alternated with injection of water. The air keeps the fire burning, and the water generates steam (31). Another method of tertiary recovery is flooding with carbon dioxide. High-pressure CO_2 mixes well with most oils (pressures used are about 300 times atmospheric pressure). A high volume of CO_2, about 270 m^3, seems necessary to recover each barrel of oil (28,31). Oil would need to cost an extra $15 to $50 per barrel if this method were adopted. The net result of secondary and tertiary recovery has been to increase the well recovery rate to about 40 percent (31).

other methods are more capital-intensive, and the price of oil will have to rise again before it would be economically worthwhile to use them to produce oil.

NATURAL GAS

Natural gas, which is perceived as a "clean" fuel because it is almost pure methane (CH_4), has experienced a surge in use followed by a drop. From 1961 to 1971, consumption jumped 73 percent (13). The U.S. Department of Energy puts U.S. proven reserves at 5.5 Tm^3 (25b). We estimate the Q_∞ for natural gas using historical experience as a guide; 1700 cubic meters (m^3) of gas is discovered for each barrel of oil discovered. This gives a rough estimate of 28.3 Tm^3 for Q_∞. The Potential Gas Committee estimated that

REFINERIES

The crude oil that is discovered must first be extracted from the ground and then processed in order to be useful as a fuel. The instrument for this transformation is the refinery. The refinery works by several means. Since crude oil is a mixture of different organic compounds, which have a continuum of properties such as boiling points, viscosities, and so on, it is common to speak of fractions.

Crude oil is said to be made up of various fractions. The main characteristic of the fraction is its boiling point (or rather, its range of boiling points). The fractions are separated out in a fractionating tower, in which they undergo a process of vacuum distillation. In vacuum distillation, the pressure is lowered, which allows the liquid to boil at a lower temperature. (It is because of the reduced pressure phenomenon that coffee made in La Paz, Bolivia, high in the Andes, is lukewarm.)

In the fractionating tower, the oil is sent in as a hot vapor (except for the heaviest fraction, known as residual oil). In the tower, trays are stacked upward. The trays are at different temperatures. The heavy fractions condense in the lower trays, which are at the highest temperature. The lighter fractions bubble through and rise until they are captured in trays at successively lower temperatures. The lightest fraction, captured in the coolest tray, is called *straight-run gasoline*. The next fraction is called *kerosene;* succeeding fractions are heavier and heavier oils.

The residual oil may be mixed with rock to make asphalt, or it may be used in the next stage of the refinery process, catalytic cracking and reforming. The straight-run gasoline is low in octane levels. The octane level measures the energy content of the gasoline. Engines running on low-octane gas may knock when the car accelerates or when it is climbing hills. To make the fuel higher in energy content, the heavier fractions can be broken up into smaller constituents (reformed) that are then added to the straight-run gasoline to enrich it. Thus, the output of the refinery can be made to match almost any spectrum of demand.

The catalytic cracker takes heavy fractions and mixes them at very high temperature with a fluidized catalyst. The heavy fractions break into pieces and are reformed. The catalyst provides a surface area on which the breakup and reformation can occur. Most of the demand from refineries is for gasoline (auto and truck fuel) and kerosene (jet fuel), the lightest fractions.

Q_∞ is 36.5 Tm^3. The current estimate is 71.2 Tm^3 (44), another twofold discrepancy.

Figure 13.12 (44) shows how the proven world reserves of 71,210 Gm^3 (102,200 Gm^3 is a more recent estimate [25b, 25c]) and probable world reserves of gas of 158,360 Gm^3 are distributed among the continents and subcontinents. While crude oil will last about forty-eight years at current use rates, natural gas will last about sixty-nine years (44). Natural gas deposits in North

America occur in much the same areas shown in Figure 13.10, although they are more broadly distributed. Consumption of natural gas fell after the initiation of government price deregulation, but use began climbing again in 1983 (45). Deregulation has increased gas supplies by stimulating exploration, but the end of these supplies is also in sight. If one applies the Hubbert analysis to U.S. resources of natural gas, as was done earlier in this chapter for oil, it is found that the

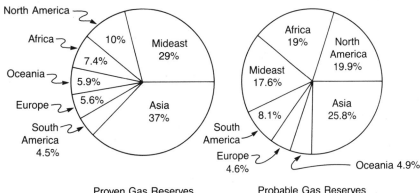

FIGURE 13.12 Estimates of (a) proven and (b) probable gas reserves according to German sources (1980) (Ref. 40).

Proven Gas Reserves   Probable Gas Reserves

middle 80 percent of natural gas should be produced between 1950 and 2015, with peak production in the mid-1980s.

COAL

The emphasis on coal is a recent development. Prior to 1973, coal production had been almost stable, because of the interest in the cheaper forms of energy, and because much of domestic coal is "dirty" (has a high sulfur content) and requires sophisticated equipment to control emissions.

Estimates of domestic coal production are about 1500 billion tonnes, and of world production, about 10^{14} tonnes (1,2). It would be difficult to place too much reliance on predictions for coal production since we are only *beginning* the cycle of coal use. Averitt estimated world coal reserves (1,46) totaling 7623 Gtonne. The USSR has total estimated reserves of 4310 Gtonne, and the United States, of 1486 Gtonne; together, the two countries account for three-quarters of total world coal resources. The Workshop on Alternative Energy Sources estimates a total of 11,500 Gtonne of coal available, of which only 1300 Gtonne are *known* reserves and only 740 Gtonne of that are economically recoverable (47). (Estimates by British Petroleum [25c] of 1018 Gtonne and of the U.S. Department of Energy [25b] of 897 Gtonne for known reserves are in general agreement with this estimate.) North American

reserves are found in large areas (Figure 13.13): a band stretching along the Rockies from Albuquerque to north of Edmonton; a band running from an Omaha–St. Louis axis south toward Dallas; a large area south of Chicago; and a band stretching along the Appalachians.

Coal comes in various grades: anthracite, bituminous, subbituminous, and lignite in descending order of energy content and hardness. Anthracite is found in very small quantities. An average bituminous coal might have an energy content of 30 MJ/kg, while lignite may have an energy content of 16 MJ/kg. As with oil, some coal is left in place when it is mined. In deep mines, anywhere from 30 to 75 percent of the coal is left to support the overlying rock. Surface, or strip, mines generally yield 85 to 95 percent of the coal present (47). Coal has, roughly speaking, a chemical formula $CH_{0.8}$ as compared with diesel fuel with a formula of roughly CH_2 or to natural gas, methane, which has a chemical formula CH_4 (48), as discussed in Chapter 5. It is clear that there is more carbon per unit mass in coal than in methane, so more carbon dioxide is emitted per unit of energy released for coal than for alternative fossil fuels (see Chapter 24).

The wanton wastage of the environment by strip miners led Congress to pass the Surface Mining Act of 1977, which mandates environmental safeguards. The Act has added $1 to $5 per ton extra cost for surface coal to pay for reclamation (49). This reclamation cost may or

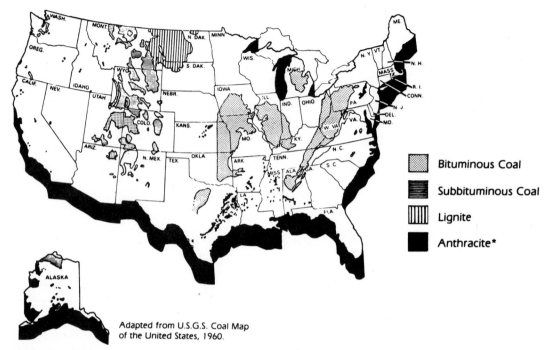

Adapted from U.S.G.S. Coal Map
of the United States, 1960.

▨	Bituminous Coal
▤	Subbituminous Coal
⦀	Lignite
■	Anthracite*

*Principal anthracite deposits are in Pennsylvania. Small deposits occur in Alaska, Arkansas, Colorado, New Mexico, Virginia, Washington, and West Virginia.

Note: Coal Rank not distinguished in Alaska

FIGURE 13.13

The geographic distribution of North American coal reserves is shown schematically.

may not be a factor in actual sales, depending on the cost of transportation of the coal to its point of combustion. Transport cost may be two to three times the actual cost of the coal itself.

Slurry pipelines—pipelines carrying ground-up coal suspended in water—are in competition with unit coal trains to transport coal. The most important problems hampering construction of more pipelines are the lack of rights-of-way and the difficulty of obtaining water in the dry West (49,50).

Burning of coal produces many pollutants: waste heat (see Chapter 11), acid rain from nitrogen and sulfur oxides (see Chapter 22), and carbon dioxide (see Chapter 24). Local emission problems have been tackled by the addition of tall smokestacks, which aid dispersal, and by ap-

TABLE 13.2

Average yearly emission estimates from a coal-fired 1000 MW$_e$ plant.

Emissions	Amount (1000 tonnes)
Air	
Carbon monoxide	2.5
Hydrocarbons	2.6
Nitrogen oxides	40.0
Particulates	6.3
Sulfur oxides	32.0
Water	
Organics	0.4
Dissolved solids	95.0
Suspended solids	1.6
Solid wastes	3500.0

THE ORIGIN OF COAL AND OIL

No matter which of the fossil fuels we concentrate on, we find we are using it at a rate that means it will last mere centuries at most, while it takes roughly 500 million years to develop new fossil-fuel resources. The future is clear: The supply of fossil fuels is small and will soon disappear.

Coal results from geologic processes involving burying of vegetation under anaerobic conditions in swamps. There, it becomes peat and is overlain gradually by rock, raising the temperature and pressure of the partially oxidized organic matter. Petroleum forms from more dispersed organic matter, such as organic sediment on a continental shelf that is buried by geological processes and, as with coal, subjected to high-pressure cooking in the earth. After about a hundred million years in either case, the material is what we know as coal and oil.

All nongaseous fossil fuels are made up of complicated molecules with backbones of many carbon atoms. Over five hundred different chemicals have been identified in crude oils (52). Analysis of samples from many different parts of the world led researchers to a surprise–all samples contained similar profiles of complicated carbon molecules containing from twenty-seven to thirty-five carbon atoms each (52). These compounds, characteristic of all fossil fuels, seem to have been ubiquitous among precursors of modern fossil fuels. Compounds of mass ≈191 and thirty-five or thirty-six carbons, called *hopanoids,* were implicated somehow in the formation process. Further research isolated such compounds in primitive bacteria and blue-green algae living today (52). The compounds are probably early ancestors of the cholesterol produced in animals (which has twenty-seven carbons), and their presence means that all fossil fuels owe their existence to bacterial and algal biological processes.

---

plication of the best available control technology (BACT). Scrubbers, which sweep particulates from the smoke electrostatically and use a limestone slurry to trap sulfur-bearing molecules, have had many problems, not the least of which is the generation of great volumes of throwaway material—sludge from the limestone–gas mixture (49–51). This material is still an undesirable waste but is in a concentrated form. It is usually stored in "ponds" and allowed to weather indefinitely. Typical emissions from coal-fired generating facilities are presented in Table 13.2.

Coal-Fired Power Plants

The fossil-fuel combustion process described at the beginning of the chapter applies to coal as well as oil. In earlier days, generating facilities burned lump coal. In more modern times, the industry standard has been the pulverized coal boiler. The next step appears to be atmospheric pressure fluid-bed combustion (53); many units are already in service worldwide.

There are several reasons for this development: Many different fuels, of different characteristics, may be burned; fuels of varying heats of combustion may be burned equally satisfactorily; and many designs incorporating two-stage combustion allow lowering of emissions of nitrogen oxides (53). A reactor that has a bed of combustion through which the fuel circulates may well become the newest standard for large-scale (>60 MW$_t$) combustion (53). These reactors began to be sold commercially in the early 1980s. As we see in the next section, fluidized-bed reactors are very important in gasification of coal.

Coal Gasification

A widely broached possibility is gasification or liquefaction of coal or high-sulfur residual oil. The main advantage of these techniques is that it is relatively cheap to remove sulfur from gas. Squires (54) has estimated that the H_2S from "power gas" costs only $20 per kW to remove, while the equivalent cost of removal from stack gas is $70 per kW (in 1984–85 dollars). Fluids are also easily moved by pipeline (see Chapter 21), so there may be an economic reason to gasify coal to transport it.

These alternatives look more attractive in light of persistent delays in construction of nuclear plants and concomitant cost increases. The newer oil plants (built during the period from 1974 through 1978) need a daily ration of 1.25 Mbbl of oil (54). With oil costs low, this is attractive. However, as oil prices rise again, this may become uneconomic. Most energy experts look to coal to fill the energy needs of the near future. The current mix of fuel sources for electricity

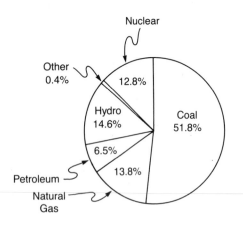

FIGURE 13.14
U.S. sources of energy for generating electricity as of 1982.

generation is shown in Figure 13.14 (55). While coal has been available all along, it has not been used for several reasons according to Osborn (56): the low cost, convenience, and availability of natural gas; the convenience and availability of low-sulfur oil; the inability to obtain long-term, low-sulfur coal supplies; the uncertainty as to practicality of sulfur removal from stack gases; the more stringent environmental, health, and safety regulations in coal mining; and recurrent transportation problems.

The first known coal gasification took place in 1792 when a Scotsman, William Murdock, used gas to light his home (57). Two types of gas are usually produced. *Power gas* is a mixture of carbon monoxide, hydrogen, and nitrogen. The simple technology for producing power gas dates to the 1880s (58). It has about one-sixth the heating value of natural gas, about 5.6 MJ/m³, which is too low a heating value to be worth transporting over any substantial distance. *Synthetic natural gas* (SNG) is made from power gas by shifting the ratio of hydrogen to carbon monoxide (to 3 : 1), cleaning, and then transforming to methane (59). It can then become ammonia, methanol, or synthetic gasoline, or can be fed into a natural gas pipeline as methane.

The proven Lurgi (medium-pressure) and Koppers-Totzek (low-pressure) processes convert only about 55 percent of the coal to synthetic natural gas. These processes are also expensive; the gas costs two to four times as much as natural gas from the wellhead (57,58,60–63).

The Cool Water Coal Gasification Plant   At the Cool Water plant, the initial gasification takes place in oxygen so that no nitrogen oxides are formed in the gasification stage (67–69). A form of power gas is generated that is low in methane. Methane is not desired since the gas is used *in situ,* thus wasting the extra energy cost required to produce methane. The plant, a 100 MW$_e$ modular demonstration facility, was completed in under two years and under budget (a feat not often seen today!) (69). In its first year of operation,

June 1984 to June 1985, the plant generated 413 MkWh of electricity with an on-line factor of ≈60 percent (68–70).

A slurry of ground coal and water is fed to the gasifier. The gas from the gasifier, which is preheated to 1260°C and brought to forty times atmospheric pressure, consists (using Utah coal) of 42.5 percent carbon monoxide (CO), 38.2 percent hydrogen (H_2), 18.6 percent carbon dioxide (CO_2), 0.3 percent methane (CH_4), 0.4 percent argon (A) and nitrogen (N_2), and about 50 parts per million (ppm) of hydrogen sulfide (H_2S) and carboxylsulfate (COS) (69). The gas has an energy content of 10.7 MJ/m³. The sulfur is scrubbed out from the synthesis gas using commercial processes developed for oil refining (Selexal, SCOT, and Claus processes) (69).

At such high temperatures, over 1200°C, no tars or phenols are produced. The heat produced in the gasifier is used to generate steam, which is mixed with the steam generated in the gas combustion to run the turbine producing electricity (this gives the facility a "bottoming cycle"). Therefore, the facility runs two generators—one operating on the "normal" generator fired on the coal gas (with the generator modified to take larger air flow rates), and the other, on steam from both combustion and gasification. Steam is injected with the coal gas in the main generator to reduce combustion temperature and thereby cut nitrogen oxide emissions. Essentially all the carbon in the coal is burned. The levels of pollution are compared to EPA standards in Table 13.3 (69). Such a combined-cycle plant is a very good bet for clean combustion of coal.

The Cool Water plant is the only coal-fired plant ever to be licensed in California, and it has exceeded the requirements of the California clean air standards during its operation. It uses less land and 60 percent less water than conventional plants, as well as consuming 10 percent less coal to generate the same amount of energy. No scrubbers are necessary on such combined-cycle plants. The effluents are, as Table 13.3 shows, carbon dioxide, coal clinker, and sulfur. The clinker is nonhazardous and will be sold for use in road fill. The sulfur is sold. The Cool Water plant has demonstrated that coal plants can be constructed to meet very stringent effluent standards.

The Cool Water plant is still experimental, and as such, costs more than the proposed production modular units. An advantage of the modular units is that they can be installed in phases. It has been suggested that the first purchase for the utility be gas turbines, at first burning natural gas or oil. As prices rise or load requirements change, the turbines could be integrated into a coal-gasification facility (71).

The Cool Water plant saves importation of 4300 bbl of oil per day (69,72). Nevertheless, the project could be completed only because of 130 million dollars in guarantees from the government-owned Synthetic Fuels Corporation; the plant cost $238 million to build (69,70). The owners are guaranteed $12.50/MBtu for the first 255 Gm³ and $9.75/MBtu thereafter (73). This subsidy will have been justified if the plants that are developed on the basis of this prototype can replace dirty older plants as they are retired.

The other commercial-sized gasification and liquefaction project, the Great Plains Coal Gasification Project, at Beulah, North Dakota, is currently undergoing testing. Product price supports ($800 M) have been requested from the Synthetic

		EPA New Source
Pollutant	Plant Data	Performance Standard
NO_x	0.061 lb/MBtu	0.6 lb/MBtu
SO_2	97% removal (0.034 lb/MBtu)	90% removal (1.2 lb/MBtu)
Particulates	0.0013 lb/MBtu	0.03 lb/MBtu

TABLE 13.3
Cool water plant pollutant levels.

NEWER GASIFICATION AND LIQUEFACTION TECHNOLOGY

There were difficulties with proven processes (old technology)—for example, agglomeration of pulverized coal and large volumes of waste. These difficulties led the U. S. Bureau of Mines as well as private corporations to research newer types of gasifiers: Hygas, Bi-Gas, Synthane, the carbon dioxide acceptor process (60,61), the Exxon donor solvent (64) and potassium catalyst (57) processes, and the Texaco process (65).

Gasifiers may generally be classified as gravitating-bed, fluidized-bed, or suspension, depending on whether they work with lump coal, crushed coal, or pulverized coal, respectively.

In the gravitating-bed gasifier, new coal enters from the top into a region of humidified (steamy) air that rises from the bottom. Oxygen is removed from the air in a combustion zone in the coal column, the residue of which is a powdery carbon-free ash that falls into a grate. The degree of humidity controls the rate of reaction of steam with carbon, and of carbon with carbon dioxide. The Lurgi process, mentioned previously, is a gravitating-bed gasifier in which steam and coal react at twenty times atmospheric pressure.

In fluidized-bed gasifiers, rising gases buoy the granular coal as it burns, assuring uniform temperatures. In the Ignifluid process (developed in France during the 1920s), the crushed coal travels in on a moving bed that carries off the self-adhering ash (60,66). Such gasifiers generally supply gas at a much higher rate than gravitating-bed gasifiers, but do not burn the coal so thoroughly.

In suspension gasifiers, the finely pulverized coal reacts in a dilute suspension with oxygen and steam at atmospheric pressure, and the ash falls as clinkers (glassy cinders). The Bi-Gas process is a two-stage suspension gasifier, in which the reaction of coal and steam takes place at about 900°C and the synthesis gas from the lower chamber is gasified further with oxygen and steam at about 1500°C (55,58,59,61). Less carbon dioxide is produced in the latter process.

After the coal has produced some gas, the subsequent processes involve one of several alternate methods: pyrolysis, or destructive distillation; gasification followed by synthesis; hydroliquefaction; and solvent extraction (40). Examples of synthesis are the Fischer-Tropsch process for synthetic fuel liquid (one of the oldest processes known) and the Mobil process for making methanol from synthesis gas by passage over an acidic catalyst (40). An example of hydroliquefaction is the Ashland H-coal process, in which coal is liquefied by treatment with hydrogen at ≈450°C and seventy to two hundred times atmospheric pressure (40). The hydrogen is not especially cheap, and it takes additional energy to increase the pressure so much

Fuels Corporation. The current price of the Great Plains facility is only about 60 percent of the amount needed to keep it open beyond 1990. Even if The Department of Energy does not wish to continue to operate the plant, political pressures are at work to subsidize it through the Department of Defense (DOD). Legislation was written to direct the DOD to purchase 10,000 barrels per day of coal-based jet fuel from the plant (about half the plant's capacity) (74).

above atmospheric. Examples of solvent extraction are the Exxon donor solvent and the Gulf-SRC-II processes (40), in which coal is mixed in a solvent to form a slurry, hydrogen is added, and the coal is then gasified. The relative advantages and disadvantages of these various processes are still under study.

The largest installation in the world for production of liquid fuel from coal is operated by Sasol, the South African oil company. Since South Africa has become nearly an international pariah, not looked on favorably by OPEC producers nor by independents such as Britain, there was an internal political reason to use an uneconomic process to produce gasoline. Sasol uses Lurgi dry-ash gasification to produce synthesis gas, which is reformed with oxygen and steam over a nickel catalyst and then liquefied by the Fischer-Tropsch process involving powdered iron as a catalyst (48). This process was first used by the Germans in World War II to make up for shortfalls in supplies of petroleum, since Germany has (as does South Africa) large domestic coal reserves. The Sasol process produces gas that is composed of 60 percent carbon monoxide (CO) and hydrogen (H_2), 29 percent carbon dioxide (CO_2), 9 percent methane (CH_4), 1 percent argon and nitrogen (A, N_2), and 0.5 percent hydrogen sulfide (H_2S) (48). This gas composition is not atypical (67). The Lurgi process, in addition to its thermal inefficiency, has the drawback that it produces tars as byproducts and that it wastes much of the carbon.

Other gasification processes promise great possibilities for reduction of pollution. Contamination may be controlled at the source before dilution in the atmosphere. In processes using air, improved rates allow combination at lower temperatures, lowering the nitrogen oxide emissions; sulfur combustion is suppressed, and sulfur captured readily, before the gas is burned (65,66).

The Texaco process has led to two commercial ventures: a process by which Tennessee Eastman is making acetic anhydride from syngas instead of oil (65), and the Cool Water Coal Gasification Plant developed by Southern California Edison to supply electricity. Acetic anhydride is used in large quantities in the manufacture of many organic acetates, such as vinyl acetate. The plastics are used for many purposes, including use as fibers. The Eastman plant alone is designed to produce about 250 ktonne of acetic anhydride per year. Acetic anhydride is made from ketene (formed by pyrolysis of acetic acid) and acetic acid. The process, based on organic compounds, is expensive, time-consuming, and produces unwanted byproducts. The Texaco process uses coal to produce synthesis gas (CO and H_2); water is added to shift the proportions of CO and H_2, and methanol is formed; the methanol is reacted with CO to make methylacetate; and then the methylacetate is reacted with CO to make acetic anhydride. This process also may be used to yield vinyl acetate through hydrocarbonylation of the methylacetate and thermal cracking (65).

Onsite Gasification Another proven technique involves onsite production of gas at the coal field. The U.S. Bureau of Mines ran an experiment near Hanna, Wyoming, in which sixteen wells were drilled over a 10-hectare area into a 10 m-thick coal seam, which was hydraulically fractured. Air was injected far from the center, and gas rose at the center installation. In this experiment, about 25 tonnes of coal a day were gasified; and the well produced 70,800 m³ (2.5

Mft³) per day of 5.27 to 5.64 MJ/m³ gas. The gas produced was 5 percent methane, 10 percent carbon monoxide, 15 percent hydrogen, 15 percent carbon dioxide, and 54 percent nitrogen (56). Of course, the latter two gases cannot burn.

This technique seems feasible for areas where there are deep coal seams but where underground mining is hazardous, allowing recovery of only about a quarter of the coal, or for coal seams that are impracticable to mine for various reasons. We might note here that some deep beds being mined produce $0.34 \, Mm^3$ ($12 \, Mft^3$) of methane per day ($5 m^3$/tonne coal) (56).

The town of Centralia, Pennsylvania, has had to be abandoned because of a two-decade-long fire in the coal seam underlying the town. This gives evidence to the feasibility of extended burning *in situ*.

Alternative Strategies   There are additional opportunities for savings at the burning end. Other methods and strategies could be technologically competitive right now, especially in the area of increasing overall combustion efficiency. One such strategy involves construction of a centralized station, a "fuelplex" (54). The "fuelplex" would address energy generation from a combination of methods, using and/or producing low-sulfur coke, synthetic natural gas, clean liquid fuel, electrochemical generation, power gas, electricity, and so on. This concentration could increase efficiency and allow a wider range of options.

SUMMARY

Within the last decade and a half, the United States has seen two energy crises. Each was caused by forces outside government control. The vulnerability we have experienced has led to increased domestic oil exploration at the same time as reserves are decreasing. This has occurred because U.S. oil is fast on its way to exhaustion, and world oil is sure to follow soon. The model of M. K. Hubbert, which asserts that production follows a normal curve, seems verified by domestic experience. The largest oil fields were the most quickly discovered, and it has become progressively more difficult to find oil.

Research has increased the amount of oil ultimately recoverable from an oil-bearing formation from about 20 percent to about 32 percent of the oil in place. The world seems to be "hooked" on fossil liquid fuel. However, experience with the price rises of the 1970s verifies that if the price rises enough, the demand decreases.

Shale oil and oil sands, which are found worldwide, contain staggering amounts of oil. However, exploitation of most of these resources involves large-scale environmental disruption.

Natural gas will last somewhat longer than oil, but coal reserves are so huge that coal may be able to supply energy for the next several centuries, while oil and natural gas will be gone within a century. All fossil energy has environmental problems associated with its exploitation. Coal-burning by traditional methods produces the most pollution. Research has proceeded on ways to reduce pollutants from coal, and it appears that some gasification processes are inherently cleaner than combustion modes now in general use. The Synthetic Fuels Corporation has provided backing for alternatives involving synthetic gases. Many such processes have been developed around the world.

REFERENCES

1. M. K. Hubbert, *U.S. Energy Resources, a Review as of 1972,* U.S. Senate Committee on Interior and Insular Affairs Report (Washington, D.C.: GPO, 1974); M. K. Hubbert, *Am. J. Phys.* 49 (1981): 1007; M. K. Hubbert, Chap. 8 in *Resources and Man* (San Francisco: Freeman, 1969).
2. M. K. Hubbert, Chap. 3 in *Energy and Power* (San Francisco: Freeman, 1971).
3. C. Sims, *New York Times,* 4 Feb. 1986.
4. R. D. Hershey, Jr., *New York Times,* 25 Sept. 1983.

5. Federal Energy Administration, *Project Independence Report* (Washington D.C.: GPO, 1974), as quoted in R. Gillette, *Science* 186 (1974): 718. See also R. Stuart, *New York Times,* 1 Sept. 1974.

6. D. Martin, *New York Times,* 29 August 1985.

7. L. Schipper and A. N. Ketoff, *Science* 230 (1985): 1118.

8. H. Herberg, *Reports of the DFG* 2–3 (1984): 35.

9. S. F. Singer, *Ann. Rev. Energy* 8 (1983): 451.

10. N. D. Kristof, *New York Times,* 8 April 1986; R. W. Stevenson et al., *New York Times,* 22 Jan. 1986.

11. R. L. Hirsch, *Science* 235 (1987): 1467. See also M. Crawford, *Science* 354 (1987): 626.

12. V. E. McKelvey, *Am. Sci.* 60 (1972): 32.

13. H. W. Menard and G. Sharman, *Science* 190 (1975): 337.

14. D. H. Root and C. J. Drew, *Am. Sci.* 67 (1979): 648.

15. R. A. Kerr, *Science* 212 (1981): 427.

16. H. W. Menard, *Sci. Am.* 244, no. 1 (1981): 55.

17. R. R. Berg, J. C. Calhoun, Jr., and R. L. Whiting, *Science* 184 (1974): 331.

18. W. L. Fisher, *Science* 236 (1987): 1631.

19. D. Shapley, *Science* 187 (1975): 1064.

20. A. R. Fowler, *Sci. Am.* 238, no. 3 (1978): 42.

21. R. Gillette, *Science* 187 (1975): 723.

22. M. Kenward, *New Scientist* 65 (1975): 140.

23. S. F. Singer, *Science* 188 (1975): 401.

24. M. K. Hubbert, Reference 1; National Petroleum Council, *U.S. Energy Outlook* (Washington, D.C., 1972); R. H. Cram, ed., *Future Petroleum Provinces of the United States—Their Geological Potential,* vol. 1, A.A.P.G. Memoir 15; P. K. Theobald, S. P. Schweinfurth, and D. C. Duncan, U.S. Geological Survey Circular 650, 1972; T. M. Doesher, *Am. Sci.* 69 (1981): 193.

25. a. R. A. Kerr, *Science* 223 (1984): 382.

    b. Department of Energy, *Annual Energy Review 1986* (Washington, D.C.: GPO, 1987).

    c. British Petroleum, *BP Statistical Review of World Energy* (London, 1987).

26. *Energy,* National Geographic Special Report, Feb. 1981.

27. C. Norman, *Science* 228 (1985): 974.

28. T. Scanlon, *Science* 217 (1982): 325.

29. D. R. Bohi and M. A. Toman, *Science* 219 (1983): 927.

30. F. M. Orr and J. J. Taber, *Science* 224 (1984): 563.

31. T. M. Doesher, Reference 24.

32. J. Dunkerley and W. Ramsay, *Science* 216 (1982): 590.

33. R. A. Kerr, *Science* 226 (1984): 426.

34. J. H. Lichtblau, *New York Times,* 1 April 1984. The reserve total is from Table 993 in *Statistical Abstract of the United States, 1984* (Washington, D.C.: GPO, 1984).

35. See, for example, P. W. McAvoy, *New York Times,* 14 March 1982; and E. J. Oppenheimer, *New York Times,* 1 April 1984.

36. G. Szegö, *The U.S. Energy Problem,* RANN Report, Appendix S (Washington, D.C.: GPO, 1972).

37. W. D. Metz, *Science* 184 (1974): 1271.

38. R. A. Dick and S. P. Wimpfen, *Sci. Am.* 243, no. 4 (1980): 182.

39. J. P. Sterba, *New York Times,* 3 Nov. 1974.

40. A. Schriesheim and I. Kirschenbaum, *Am. Sci.* 69 (1981): 536.
41. E. Marshall, *Science* 204 (1979): 1283.
42. G. D. Mossop, *Science* 207 (1980): 145.
43. T. H. Maugh, *Science* 207 (1980): 1334.
44. Verbandes der Gas- und Wasserwerke, Baden-Württemberg e.V., *Haushalt mit Energie* (1980).
45. W. D. Smith, *New York Times,* 17 Oct. 1984; T. Shabad, *New York Times,* 13 Oct. 1974.
46. J. Walsh, *Science* 184 (1974): 336.
47. E. D. Griffeth and A. W. Clarke, *Sci. Am.* 240, no. 1 (1979): 38.
48. M. E. Dry, *Endeavour* 8 (1984): 2.
49. H. Perry, *Science* 222 (1983): 377.
50. R. L. Gordon, *Science* 200 (1978): 153.
51. S. C. Morris et al., *Science* 206 (1979): 654.
52. G. Ourisson, P. Albrecht, and M. Rohmer, *Sci. Am.* 251, no. 2 (1984): 44.
53. A. M. Squires, M. Kwank, and A. A. Avidan, *Science* 230 (1985): 1329.
54. A. M. Squires, as quoted by W. D. Metz, *Science* 179 (1973): 54.
55. Department of Commerce, *Statistical Abstract of the United States, 1984* (Washington, D.C.: GPO, 1984), Table 981.
56. E. F. Osborn, *Science* 183 (1974): 477.
57. J. T. Dunham, C. Rampacek, and T. A. Henrie, *Science* 184 (1974): 346.
58. A. M. Squires, "The Fossil Fuel Development Gap."
59. R. L. Hirsch et al., *Science* 215 (1982): 121.
60. A. L. Hammond, *Science* 193 (1976): 750.
61. A. M. Squires, *Science* 184 (1974): 340.
62. H. R. Linden et al., *Ann. Rev. Energy* 1 (1976): 65.
63. H. C. Hottel and J. B. Howard, *New Energy Technologies* (Cambridge, Mass: MIT Press, 1971).
64. G. K. Vick and W. R. Epperly, *Science* 217 (1982): 311.
65. H. E. Swift, *Am. Sci.* 71 (1983): 616.
66. P. F. Fennelly, *Am. Sci.* 72 (1984): 254.
67. D. F. Spencer, M. J. Gluckman, and S. B. Alpert, *Science* 215 (1982): 1571.
68. D. F. Spencer, S. B. Alpert, and H. H. Gilman, *Science* 232 (1986): 609.
69. V. Shorter, Head Engineer, Cool Water Coal Gasification Plant, private communication.
70. W. N. Clark, Project Manager, Cool Water Coal Gasification Plant, news release, 25 June 1985.
71. R. E. Balzhiser and K. E. Yeager, *Sci. Am.* 257, no. 3 (1987): 100.
72. M. Crawford, *Science* 228 (1985): 565.
73. M. Crawford, *Science* 228 (1985): 1410.
74. C. Holden, *Science* 230 (1985): 1022.

PROBLEMS AND QUESTIONS

True or False

1. The disagreement as to the exact value of Q_∞ means that there is probably enough fossil fuel to last several thousand years at present consumption rates.
2. The energy available in oil shale is about the same amount as the oil reserve of Saudi Arabia.
3. The United States has a larger share of world coal than would be expected on the basis of its land area.

4. No matter what sort of curve you use to draw oil consumption, it must return to the axis after the lapse of a sufficient amount of time.

5. Using oil as an energy source might be fruitfully compared to living off one's capital rather than investing the capital and living off its income.

6. The British have more oil reserves per unit area of their country than the United States.

7. There is no reason to expect that the Q_∞ for natural gas should be able to be predicted in the same manner that the Q_∞ for oil was.

8. Coal reserves are sufficient for several centuries worth of energy production at current rates of consumption.

9. The presence of hopanoids in all fossil fuels implies that the material making up what later became the fossil fuel was derived from living things.

10. In gasifying coal, almost all the energy in the original coal is present in the end-product gas.

Multiple Choice

11. Why is there a delay time between discovery of a resource and production from it?

   a. Since there is so much of each resource available, it takes some time to get around to using what is found.

   b. The price of resources is rising, and developers are thus loath to produce.

   c. It takes a long time for information on resource availability to come to the attention of companies involved in resource exploitation.

   d. It takes time to acquire the capital and equipment to exploit the resource.

   e. Most people involved in resource exploitation live in the tropics, and the pace of life is slower there.

12. How can different experts arrive at such disparate predictions for American oil resources as described in this chapter?

   a. No two groups can agree on the facts.

   b. The only way for an expert to make a reputation in the business is to produce something different.

   c. Slightly different expectations for the percentage of oil eventually recovered lead to large differences in estimates of oil resources.

   d. Each expert takes as a model a different country's experience. Since many countries are better explored than the United States, the experts expect to have their predictions borne out.

   e. Different hypotheses for the expectation of finding oil per meter drilled can lead to great disagreements in predictions.

13. The world price for oil depends mainly on

   a. political considerations.

   b. supply and demand.

   c. the cost of extraction of the oil.

   d. the strength of the OPEC cartel's solidarity.

   e. an amalgam of all the reasons listed.

14. M. K. Hubbert argues that the production history (amount produced) of an energy commodity such as oil is best described by a
    a. logistic curve.
    b. normal curve.
    c. quadratic curve.
    d. exponentially rising curve.
    e. exponentially falling curve.

15. In order to remove as much as 40 percent of the oil in place in any oil deposit, one must use
    a. only primary recovery (pumping under natural pressure).
    b. secondary recovery (water or gas pumped down an adjacent well).
    c. tertiary recovery (steam injection).
    d. oil-mining techniques.
    e. a technique that has yet to be developed.

16. Which of the following is *not* a method for producing synthesis gas from coal?
    a. Pyrolysis
    b. Deep fracture combustion
    c. *In situ* burning
    d. Solvent extraction
    e. Hydroliquefaction

17. What chemical compounds might be found among products of coal combustion?
    a. CO
    b. CO_2
    c. H_2S
    d. CH_4
    e. All the above compounds are found.

18. Commercial gasification processes
    a. have yet to prove themselves to be profitable.
    b. produce "dirty" exhausts.
    c. involve low-temperature combustion.
    d. have problems of contamination by nitrogen oxides.
    e. evolve gases, all of which are combustible.

19. The value of Q_∞ for coal is
    a. well-predicted by a Hubbert-type analysis.
    b. not much larger than that for oil.
    c. uncertain because exploitation of coal reserves lags far behind that of oil reserves.
    d. unlikely to be modified by new survey information.
    e. independent of the amount of coal predicted to be on the continental shelf.

20. The time required to form new reserves of fossil fuels is measured in the
    a. hundreds of years.
    b. thousands of years.
    c. millions of years.
    d. hundreds of millions of years.
    e. billions of years.

Discussion Questions

21. Is it possible that the 1973–1974 or the 1979–1980 energy shortages could have been "manufactured" by the oil companies, rather than being genuine?

22. Discuss the reliability of Hubbert's projections.

23. If we experience an average increase of a certain number of barrels of oil per foot drilled over all our oil history, do you think that means we shall also find that increased number of barrels per foot drilled in the future?

24. Do Hubbert's ideas depend only on geologic data and hard scientific facts?
25. M. K. Hubbert has been quoted (1) as saying that the "children born in the mid-1960s will see most of the world's oil consumed in their lifetime." Comment on the basis for this statement, and give evidence for or against it.
26. Hubbert has also said (1) that the effect of oil is "very, very brief in terms of human civilization." This is also true for coal. Explain how Hubbert could support such a view. (How long has human civilization existed?)
27. It has been suggested that demand for oil fell only because of price, and that people were less comfortable as a result of using less oil. Is there any truth to this view?
28. How important are political considerations to the price of energy? Support your opinion with facts gleaned from current events.
29. Does it seem likely that further advances in combustion technology will occur? On what do you base your answer?
30. A punster in the utility industry has referred to the burning of coal in the Cool Water plant of Southern California Edison as "immaculate combustion." Explain how serious he could have been by reference to Tables 13.3 and 13.4.
31. "Sting me once, shame on you. Sting me twice, shame on me." Apply the moral of this folk saying to energy resource price fluctuations over the last thirty years.

14

ENERGY FROM NUCLEAR REACTIONS

Chapters 13 through 16 discuss various facets of energy generation. The focus of this chapter is generation of energy through nuclear processes. In Chapter 6, we considered the basic nuclear physics involved in fission and fusion of nuclei. This physics is utilized here to explain in some detail the operation of reactors.

In this chapter, we try to dispel some of the myths surrounding the operation of nuclear fission reactors and the fuel cycles of these reactors. We discuss the breeder reactor; compare American and foreign experience with nuclear energy; and discuss the possibility of controlled nuclear fusion energy (which has already become surrounded by myths of its own, even though no such power plants will be operating until at least twenty or thirty years from now).

chain reaction
moderator
prompt neutrons
delayed neutrons
control rod
enrichment
gaseous diffusion
yellowcake
breeder reactor
fusion
Lawson criterion
tokomak
magnetic mirror
plasma
toroid

KEY TERMS   *nuclear*
            reactor
            boiling water reactor
            pressurized water reactor
            CANDU
            high-temperature gas-cooled
                reactor
            graphite-water reactor
            binding energy
            fission

The most easily remembered symbol of the nuclear age is the mushroom-shaped cloud caused by the release of nuclear energy in a bomb. Some believe that the specter of such a destructive force unleashed from the nucleus has poisoned nuclear energy in the mind of the public. Many of the scientists involved in the project to build the bomb during World War II have worked hard to build *peaceful* uses of nuclear energy in the time since then.

All current and proposed forms of nuclear energy involve production of steam to drive turbines. The delivery system, from turbine on, is identical for all facilities producing electricity, so we concentrate on the parts of the facilities that are different for nuclear systems as opposed to fossil fuel systems. The reader is referred to References 1 and 2 for more detail than it is possible to give in this chapter.

CONVENTIONAL REACTOR TYPES

There are several different types of nuclear reactor used in generating electricity on a commercial basis. The most common types are briefly discussed before the physics behind them is explained. Three types of reactor use water in their cores: the boiling water reactor (BWR), the pressurized water reactor (PWR), and the Canadian-designed pressurized water reactor (CANDU). In addition, some energy is generated in high-temperature gas-cooled reactors (HTGR), and in the USSR in graphite-water reactors.

In the BWR, the fuel elements are loaded in the reactor core, surrounded by water. As fissions occur, fragments are produced that move through the elements and the water, transferring their energy to other materials. This kinetic energy thus gets "spread out" in many collisions to become thermal energy. The thermal energy boils the water in the core, producing steam at about seventy times atmospheric pressure (6.75 MPa,

or 1000 lb/in²). The steam drives the turbine and arrives in the condenser where it collects as liquid feed water and is pumped back to the core. The maximum thermodynamic efficiency for this process is about 45 percent and the actual efficiency is about 30 percent, substantially less than for the newest fossil-fuel boilers. The BWR is shown schematically in Figure 14.1. Thirty-eight BWRs are operating in the United States.

In the PWR, the fuel elements are in a closed pressurized container. The high-temperature water ($\approx$320°C) at about 150 times atmospheric pressure (15.2 MPa, or 2250 lb/in²) is circulated through a "steam generator" (a heat exchanger for the higher temperature steam). The water in the heat exchanger vessel boils as a result of the heat provided by the reactor, and this steam runs through a turbine as in a conventional fossil fuel plant. The maximum thermodynamic efficiency for this process is 49 percent, but the actual efficiency is again about 30 percent. The PWR is shown schematically in Figure 14.2. Sixty-seven PWRs are operating in the United States.

The CANDU reactor uses heavy water (deuterium oxide, D_2O) as a moderator and can use natural uranium in its core, rather than enriched uranium, which has a higher concentration of fissile uranium. Because of its simplicity and efficiency, it is used in eight different countries around the globe (3).

The HTGR has been used mainly in Europe but may now be more attractive in the United

FIGURE 14.1

A schematic representation of a boiling water reactor (BWR).

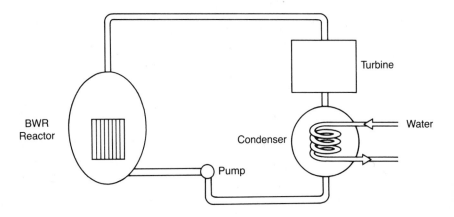

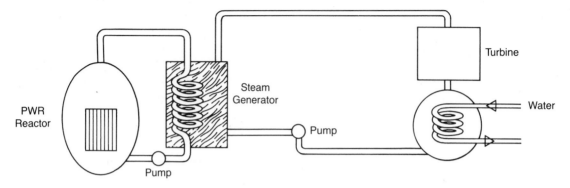

FIGURE 14.2

A schematic representation of a pressurized water reactor (PWR).

States, partly because of the recent concern over reactor safety, and because it operates at a higher temperature than conventional BWRs and PWRs. Because of the higher operating temperature, it has a greater thermodynamic efficiency. The 330 MW$_e$ prototype reactor at Ft. St. Vrain, Colorado, has operated at 39.2 percent efficiency since 1978 (4–6). The gas in this HTGR is helium, and the reactor operates at about seventy times atmospheric pressure and at a temperature of nearly 700°C. The fuel used is highly enriched uranium wrapped into 50 mm diameter balls of graphite and ceramic (Figure 14.3). Because this material has such a high melting point, it is possible that a total loss of helium, which would raise the temperature to 1500°C or higher, could be withstood without long-term reactor loss. However, there would still be problems in case of such an accident.

The graphite-water reactors (RBMK) in use in the USSR are known as pressure-tube reactors (7). They operate on scantily enriched uranium

FIGURE 14.3

An enlarged view of a fuel pellet for a high temperature gas-cooled reactor. The photo is courtesy of the HBK project at the nuclear research facility in Jülich, West Germany.

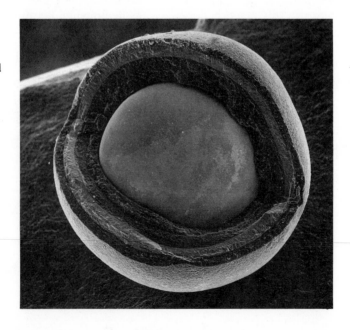

fuel, and differ from BWRs and PWRs in that it is possible to refuel as the reactor runs. The 100 MW_e reactor is made of 2488 columns of graphite blocks, through which pass 1661 pressure tubes containing zirconium–niobium fuel elements and water (Figure 14.4). The water system for this reactor has been characterized as a "plumber's nightmare."

The rate of reaction is controlled by 211 boron carbide control rods, which are cooled separately from the fuel rods. The reactor is heavily monitored and computer controlled because of the fact that it has a "positive reactivity coefficient." This means that if the temperature rises (as it would if power output increased), the core becomes *more* reactive, because steam bubbles absorb fewer neutrons than the water they replace. BWRs and PWRs have negative reactivity coefficients, and so are less prone to and less dangerous during accidental failure. This type of reactor, the RBMK-1000, was involved in the 1986 disaster at Chernobyl in the Ukraine (discussed in Chapter 25).

THE FISSION PROCESS IN REACTORS

The heat provided by a fission reactor comes from the kinetic energy of the fragments of the broken-up nucleus. The nucleus is made up of protons and neutrons, and in Chapter 6 we described the fission process using the curve of the average binding energy per nucleon versus A, the mass number.

Because protons and neutrons have practically the same mass, the mass of a nucleus containing a number A of nucleons has a mass roughly A times the average proton–neutron mass. It is slightly less than the sum of all proton and neutron masses involved because some mass energy is converted into kinetic energy during the binding process.

The nucleons are bound together by what we call the *strong force,* which is greater than the electrical force of repulsion, but only for separation distances less than 10^{-15} m. Because the strong attractive force between nucleons exists, nucleons can give up energy by binding together

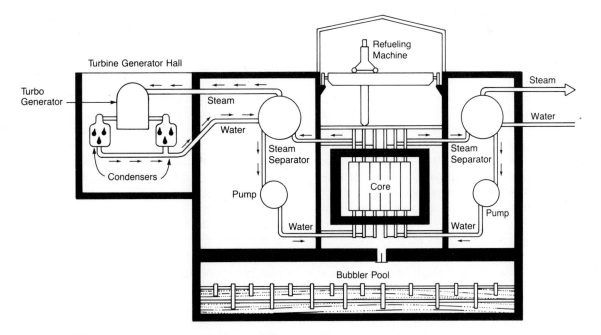

FIGURE 14.4

A schematic diagram of the RBMK reactor of the type used in Chernobyl.

to form a nucleus. They become more stable by doing so, just as two balls rolling into a trough will end up sitting at the bottom of the trough, or two chemicals making bonds will release chemical energy and make themselves more stable. They can then be released from a nucleus, or be taken out of the trough, or reconstitute the original chemicals only if a substantial energy price is paid. In the case of the bound nucleus, by conservation of energy, this price is exactly the energy given off in the fusion process forming the nucleus. The amount of energy the nucleons give up to remain together is called the *binding energy*.

As the number of nucleons increases, there is a rapid increase in the average binding energy per nucleon. At high mass number (high A) the average binding energy per nucleon is approximately constant. The curve of the actual average binding energy per nucleon as a function of A, the mass number, was presented in Figure 6.2.

Fission reactions involve the breakup of a high-A nucleus into two smaller A nuclei, in each of which the "average" nucleon is more deeply bound than in the more massive nucleus. The process produces this mass energy difference as kinetic energy of the products as well as that of the two or three extra neutrons (recall that higher A nuclei have a greater neutron-to-proton ratio than lower A nuclei). These neutrons are ejected during formation of the fission products; they can then move away and interact with other fissile nuclei, causing additional fissions. This is the essence of the idea of the chain reaction (note that each neutron going in produces two or three new neutrons in causing a fission).

A tabletop model of an uncontrolled chain reaction has been made using mousetraps and ping-pong balls. If all the mousetraps are set, and several ping-pong balls are placed on the front of the trap, one ping-pong ball can cause a mousetrap to snap, sending three ping-pong balls flying. Each of these springs a trap, and so on. Balls fill the air with explosive suddenness.

When the uranium fuel rods are put into water in BWRs or PWRs, and carbon rods are inter-mingled with the uranium fuel elements, the neutrons produced in fission are absorbed in water molecules and slowed down by collisions with the atoms in the water and the carbon rods. These carbon rods are called *moderators*. Moderators slow neutrons from fission in reactors. It is essential that the neutrons be slowed, because slow neutrons are most effective at being captured by the ^{235}U nucleus and causing it to fission. In a sense, one could say that slow neutrons dawdle long enough inside the nucleus to cause the internal nuclear excitation leading to fission, whereas fast neutrons just zip by without having time to interact with the protons and neutrons in the ^{235}U nucleus. The nonfissile ^{238}U can actually be made to fission by fast neutrons, but its "effective size" is much smaller than that of the ^{235}U for slow neutrons.

Some of the neutrons released in fission emerge immediately upon fission (these are called *prompt neutrons*). In some fissions, however, neutrons are held in the nucleus of the fission product for a time before release (these are called *delayed neutrons*). The presence of some delayed neutrons among the average net 2.43 neutrons per fission allows greater operational safety in reactors by giving the operators leeway in mechanically initiating shutdown if even a hint of danger is detected. The reactor functions in a steady state when exactly one neutron from each fission is allowed to cause a subsequent fission.

In use in the reactor, enough material is present so that there is a net of less than one *prompt* neutron per fission. It is the delayed neutrons that make up the difference and allow the reactor to continue operation. This means that there is some extra time interval during which safety devices can work. About 0.65 percent of the neutrons from the decay of fissile uranium are delayed; only 0.3 percent of those from fissile plutonium are delayed (8). This gives an extra margin of safety to the uranium reactor as compared to the plutonium reactor. Delay times are so short that shutdown must depend on mechanical devices. *Control rods,* which absorb neutrons and can cause the reactor to cease op-

THE OKLO REACTOR (9)

A French scientist analyzing ore from Oklo, Gabon, in 1972 stumbled on an anomaly (9). While in normal uranium ore 0.7202 percent is $^{235}_{92}$U, in his sample the percentage was only 0.7171. Further detective work turned up a sample containing only 0.44 percent $^{235}_{92}$U.

The only acceptable hypothesis is that the ore body had been a reactor at some point in the distant past. This is possible because, in the past, not so many of the $^{235}_{92}$U atoms present when the Earth formed had decayed. With higher concentrations of $^{235}_{92}$U, a natural reactor could have formed (the concentration in a modern PWR or BWR is about 4 percent).

Further studies have estimated that the natural reactor released about 15,000 MWyr, approximately the energy release of a 1000 MW reactor in 15 years. The reactor probably ran for at least 150,000 years, about 2 billion years ago.

---

eration, can be inserted mechanically at any time on seconds' notice. The mousetrap and ping-pong ball model can be extended to controlled reactions by use of flypaper hanging above the table as a model for control rods. Some ping-pong balls can stick to the strips of flypaper; the number of strips can be adjusted until only one ping-pong ball from each "fission" escapes to produce a subsequent "fission."

Enrichment of Uranium

In order for the BWR or PWR to work, there must be a suitable mixture of ^{235}U with ^{238}U. In naturally occurring uranium ore, ^{235}U constitutes only 0.7 percent of the uranium. To make the uranium suitable for BWR or PWR reactor use, the proportion of ^{235}U to ^{238}U must be increased to about 4 percent. The HTGR uses fuel with between 20 and 93.5 percent ^{235}U. The process by which the proportion of ^{235}U is increased is called *enrichment*.

The uranium presently used for reactors is enriched by gaseous diffusion. The uranium is made into uranium hexafluoride, a gas. The gas molecule made with ^{235}U has a slightly higher chance of penetrating a thin membrane placed in its way than the gas molecule made with ^{238}U.

If the gas is made to pass through many membranes, the proportion of ^{235}U in the mixture increases, at a considerable energy cost. The gaseous diffusion plant in Oak Ridge, Tennessee, uses about 2 percent of the total electric energy generated in the United States. It is not possible to charge all of this energy used to the energy cost of producing nuclear fuel, since much of the enriched uranium is very enriched indeed and goes into missile and bomb warheads. However, it is certainly energy-expensive to enrich uranium.

Recent advances in separation technology have led to discovery of several processes using much less energy per kilogram of enriched uranium than gaseous diffusion: gas centrifuge enrichment and laser enrichment. Laser enrichment involves shining light of a very precisely determined frequency on a uranium mixture. The frequency causes one of the electrons of $^{235}_{92}$U to be stripped away, ionizing the uranium atom, while no electrons from $^{238}_{92}$U are stripped away. This occurs because the energies of the electron states (Chapter 6) in the two nuclei differ slightly; if the light is tuned to the precise frequency that supplies enough energy to ionize ^{235}U, only this species will be moved in an external electric potential. Ionized $^{235}_{92}$U can then be separated from the other

URANIUM RESOURCES

Relatively large deposits of uranium are found in the United States, Australia, Canada, Gabon, South Africa, and the USSR. The uranium compound (U_3O_8) produced from the ore is called *yellowcake* (1 kg of yellowcake contains 0.85 kg of uranium).

The history of uranium resources in the last twenty years resembles a ride on a roller coaster. In the early 1970s, many nuclear reactors were being built, here and elsewhere. In the growth atmosphere of that time, projections made by drawing straight lines on semilogarithmic paper (see Chapter 8) indicated a demand for yellowcake in the United States of 60 ktonne and for the world, of 114 ktonne, by the year 2000 (12). Prices rose until the mid-1970s, even to the extent of causing a major supplier to renege on its contracted price (13).

Since that time, few nuclear plants have been ordered, and many previously ordered have been cancelled (due to cost overruns sometimes exceeding 1000 percent [14].) No plant ordered between 1974 and 1978 is still under construction (15), and no new plants have been ordered at all in the United States since 1978 (5). As a result, projected demand for enriched uranium has dropped tenfold since 1975 (10), and the price has fallen by a factor of 3. Furthermore, detailed studies of uranium ore deposits led to the conclusion that, for every decrease in ore grade of a factor of 10, there is a 300-fold increase in actual mass of recoverable uranium (16), implying that supplies would be adequate far into the future.

Hubbert (17) points out that the Gassaway member of the Chattanooga shale outcrop is hundreds of square kilometers of 5-meter-thick 0.006 percent uranium. Similarly, the New Hampshire Conway Granite is 750 km² (330 mi²) and a few km deep. It contains 56 grams of thorium per tonne of rock. The density of granite is about 3 tonnes/m³; thus 1 m³ of this rock contains about 150 grams of thorium. If *breeder reactors* are in use, this thorium is usable and equivalent to 1.5×10^{14} barrels of crude oil per cubic meter of rock. The exploitation of this resource would naturally cause large-scale visual pollution and ground water contamination, among its other consequences. At any rate, we may conclude that, with sufficiently large energy input, the resources could sustain the nuclear industry for quite a while.

The reasonably assured supply of cheap yellowcake (less than $100/lb) is estimated at about 4.8 Mtonne, which corresponds to 4.1 Mtonne of available uranium. There is of course a much larger supply than was originally estimated if the acceptable grade of ore is reduced (16), as was noted above.

isotopes of uranium (10). This technology is still under development. A gas centrifuge plant, which was under construction in Portsmouth, Ohio, despite the collapse of the commercial market (discussed later in this chapter) for enriched uranium, was finally scrapped in mid-1985 (11). It would have produced $^{235}_{92}U$ for about 5 percent of the energy cost of the gaseous diffusion enrichment (10).

Why Reactors Cannot Explode

There are internal safeguards in nuclear reactors in addition to moderators and control rods. As

TABLE 14.1
Dilution of fissile
materials in reactors.

	Commercial* Reactors	Research† or Test Reactors	Weapons
Fuel enrichment (%)	1–4	90–95	90–100
Fissile concentration in fuel (grams fissile/cm³ fuel)	0.1–0.4	1–1.5	18
Assembled density‡ (grams fissile/cm³ solid, including clad and structure)	0.1–0.3	0.2–0.4	18
Fissile concentration in core (grams fissile/cm³ core volume)	0.05–0.15	0.1–0.2	18

*Thermal neutron power reactors, excluding those fueled with natural uranium.
†Thermal neutron type.
‡Without transient compaction.

temperature increases, the ^{238}U nuclei absorb more neutrons. Because the moderator expands, there are more open spaces, and it does not slow down as many neutrons as at a lower temperature. Since only the slow neutrons can initiate the fission reaction, the reactor must then cool down. There is really no possibility of a reactor undergoing a nuclear explosion because the density of fissile material in the reactor's core is much too low, as Table 14.1 (8) shows. We shall discuss some actual dangers of reactor operation in Chapter 25.

BREEDER REACTORS

Reactors such as the BWR and the PWR are called *converters*. They produce some ^{233}Th and ^{239}Pu, which could in turn be used to run reactors from the reaction of fast neutrons with ^{232}Th and ^{238}U, but they extract only 1 to 2 percent of the energy theoretically available in uranium (8). The typical BWR or PWR produces 0.55 atoms of ^{239}Pu (17) per atom of uranium spent. This number, called the *conversion coefficient,* is used to classify reactors. If the conversion coefficient is zero, the reactor is called a *burner;* if it is between zero

and one, it is called a *converter;* and if the conversion coefficient is greater than one, the reactor is classified as a *breeder*. In a breeder, extra ^{238}U is put into the reactor core in such a way as to encourage fast neutrons to make ^{239}Pu. The ^{239}Pu is highly fissile, and can be used as fuel for a converter. Since the conversion coefficient is greater than one, the breeder produces more fuel than it burns. The higher the conversion coefficient, the more fuel bred.

The liquid metal fast breeder reactor (LMFBR) is a breeder that uses liquid sodium metal as the working fluid rather than water. Such reactors operate at a higher temperature than those conventional reactors using water as the working fluid. France, West Germany, and the United States considered building prototype breeders. Environmental concerns halted the projects in West Germany and the United States in the early 1980s. The breeder utilizes reactions such as

$$^{238}_{92}\text{U} + n \rightarrow {}^{239}_{92}\text{U} \rightarrow {}^{239}_{93}\text{Np (plus electron and neutrino)}$$

and

$$^{238}_{92}\text{U} + n \rightarrow {}^{239}_{92}\text{U} \rightarrow {}^{239}_{94}\text{Pu (plus electrons and neutrinos)}$$

Thus, the uranium atoms capture fast neutrons, and the resultant nucleus undergoes one or several neutron beta decays to produce the more stable, but fissile, isotopes of neptunium and plutonium. Plutonium has a much smaller critical mass than uranium, so reprocessing must be done continuously to prevent the breeder from reaching a state of self-criticality (becoming a "bomb").

In the LMFBR, the uranium dioxide or plutonium oxide fuel rods are sealed in stainless steel tubing. The reactor core is cooled by a loop of liquid sodium (much as in the PWR), with the heat exchanged to another liquid sodium loop and finally to a water steam loop. This can give a high actual efficiency, estimated to be about 41 percent, because the boiling point of liquid sodium is 892°C. Proponents of the breeder reactor point out that it promises essentially unlimited nuclear energy to rescue us from a possible future resource bind. In the breeder, 70 to 80 percent of fertile material produced is available for energy production (1).

While the United States has dropped its breeder program because of doubts about safety, the French government is proceeding with the development of the Phenix and Superphenix breeder reactor. The French are heavily committed to nuclear energy (currently, $\approx$30 percent of the electricity in France is generated in nuclear facilities) and are worried about cutoff of foreign supply, while most other nuclear countries have access to potentially large amounts of uranium from domestic sources. For this reason, they are willing to spend $2800 per installed kilowatt for Superphenix at a time when the normal French reactor costs about $1200 per installed kilowatt (18).

NUCLEAR FUSION TO PRODUCE ENERGY

The average binding energy curve of Figure 6.5 indicates that fusion reactions release energy as long as the product has a mass number less than

56. Fusion reactions are responsible for the production of all elements besides hydrogen. Our entire world was built of material forged in the centers of suns. Of the many possible reactions shown in Chapter 6, the fusion of deuterium ^2_1H to produce helium (^4_2He) is worth further study.

We may estimate the extent of the energy resource using this fusion reaction by finding out how much energy is available in common seawater if all the deuterium were to be used in fusion devices. Since 1/6500 of hydrogen atoms are deuterium atoms (19), 1 m^3 of water contains 34.4 grams of deuterium (1.03×10^{25} atoms). Taking the volume of all the oceans to be about 1.5×10^9 km^3, the oceans contain 34.4 grams/$m^3 \times 1.5 \times 10^{18}$ $m^3 = 5.16 \times 10^{19}$ grams of deuterium. Using the first reaction listed, the oceans can supply 1.135×10^{24} kWh. At the U.S. 1985 use rate, the deuterium will last

1.135×10^{24} kWh/$(1.88 \times 10^{12}$ kWh/year)
$= 6 \times 10^{11}$ years.

The age of our solar system and of the universe is only about 10^{10} years, so our calculation indicates that we could supply the United States at its 1985 consumption rate for twenty to fifty ages of the universe. Even if we use only 1 percent of the ocean's deuterium, the energy would last the United States for over 5 billion years. If we further assume that the rest of the world uses energy at the 1985 U.S. rate (with present populations), the resource would last, at 1 percent utilization, almost 300 million years.

If the reaction $^2_1H + ^2_1H \rightarrow ^3_2He + n$ is used, the entire United States could be supplied with energy at the present rate for input of 10 kg of deuterium per hour (20). Since ^2_1H costs 3×10^{-6}/kWh (21), and demand is about 350 MkW, we find a cost of $1050 per hour, if deuterium supplied all U.S. power needs.

Fusion looks as if it is a cornucopia—unlimited power for all. The catch is that small-scale controlled fusion has not yet been developed. We do know that it is possible, since all normal stars in our universe run on controlled fusion reactions. The hydrogen bomb is an uncontrolled fusion device. The question is whether fusion will be able to operate economically on a scale much smaller than that found in a star.

The two key problems for fusion on Earth are *heating* and *containment*. The reactants must be at sufficiently high temperature ($\approx 10^7$ K) so that they collide with enough energy to fuse. This heat must not be allowed to dissipate before more energy emerges than we had to put in. If 10^7 K particles were to hit any material obstacles, they would quickly give up their energy, and all possibility of fusion occurring would be lost. How long the nuclei must be confined depends on the density. The greater the density, the more collisions, and the more energy is given out in a given time. This is expressed by the *Lawson criterion*:

(density) $\times$ (time) $= 6 \times 10^{13}$ cm^{-3} s.

The Lawson criterion must be satisfied if fusion is to occur. The first case of a device exceeding the Lawson criterion occurred in late 1982 at the Alcator, a doughnut-shaped fusion reactor, at the Massachusetts Institute of Technology. The (density) $\times$ (time) was about 6 to 8×10^{13} cm^{-3} s (22).

Research Fusion Reactors

There are three main strategies by which fusion is being pursued: magnetic confinement (nonmaterial container); laser implosion (reaction before the possibility of escape to the container walls); and inertial confinement (particle beam fusion). The implosion and confinement options are more speculative. What reactor engineering research has been done has focused on *tokamaks* and *tandem mirrors* (23).

The magnetic confinement idea arose because, at 10^7 K, all particles are charged. The magnetic field, if it is intense, can keep the charged particles confined inside itself—it forms a "bottle" or "mirror." The most promising (and best-funded) approaches thus far are the tokamak figure-8 toroid, the doughnut toroid geometry, and the tandem mirror. In the figure-8 configuration, the *plasma* (the charged particles at high temperature are said to be a plasma) is trapped inside the figure-8 magnetic field. Recent developments in tokamak research have brought the plasma extremely close to the Lawson criterion at temperatures in excess of 75 MK (24). One of these involves changing the shape of the plasma trapped in the figure-8 doughnut from a circular to an oval cross section (23). In addition, early tokamaks operated in a pulsed mode. The most recent development of radiofrequency waves for driving currents holds out the promise of continuous operation (23).

The simple toroid geometry will not confine plasma inside, because the magnetic field strength that causes confinement drops away from the central axis. The charged particles tend to move out to the periphery of the doughnut. In practice, a poloidal magnetic field, to keep the particles in the doughnut, and a vertical field, to push plasma to the doughnut's center, are necessary (23). Research outside the United States has involved greater emphasis on torus geometry machines other than tokamaks (23).

The new mirror reactors differ substantially from older versions, in both the magnets used (minimum-B magnets, which are shaped like a baseball's seam) and the use of electrostatic bottle ends (23). The ends of the old mirrors were rather leaky; the electrostatic plugs are not. However, power losses from flow of plasma to the plugs is still quite large (23).

The laser implosion idea might work because the deuterium is held in place in a glass or plastic balloon 50 micrometers in diameter at pressures of 50 to 100 atmospheres. The laser beams are split, amplified, and then focused by lenses and mirrors onto a small spot occupied by the fuel pellet to vaporize the outer part of the pellet and cause implosion of the fuel at temperatures higher than 10^7 K (25). The resulting microexplosion should give out more energy than was put in.

The original laser beam is divided into many beams. These beams are directed so as to hit the pellet from many separate directions to allow more uniform irradiation of the fuel pellets. (The usual number of beams is thirty-two.) The advent of frequency-converted lasers to raise the energy of the laser light has been essential to recent advances (25). The fusion device would run rather like an internal combustion engine. However, this work is proceeding mostly because of its potential for weapons research applications rather than for its use in a practical energy device.

In 1986, a nuclear weapon was detonated underground in this research program. It produced high-energy x rays in its blast. The x rays bombarded the deuterium–tritium mixture inside the fuel pellet, succeeding in causing fusion. Work is proceeding on both weapons and fusion power plant designs utilizing this "microfusion" process (26).

There is some hope that accelerators designed for high-energy physics research could be used to focus beams of particles onto one point and cause the microballoons to implode. This version is known as *inertial confinement fusion*. Some researchers are optimistic that these focused particle beams can bring workable fusion technology more quickly and cheaply than other methods (27), but the decade-old idea will not be subjected to a practical test until the completion of the particle beam fusion accelerator (PBFA II) at Sandia Labs in New Mexico. PBFA II will use lithium ions at high energy to cause pellet implosion (28).

Fusion Reactor Engineering

Methods of energy conversion are still under discussion. Some argue for direct conversion of the current of charged particles from the reaction. This is more easily achieved with the linear tandem (magnetic) mirror than with toroids. Most designs advocate capturing the neutrons liberated from the fusion reaction as heat in a liquid lithium blanket surrounding the reactor. Heat is then converted to electricity in the conventional way.

Capital costs for fusion reactors are within a factor of 2 of being competitive in today's market (23), and the cost of fuel is very cheap. The designs to be used must be for a reliable, steadily available machine that will not produce large volumes of pollutants. The major contaminant to be released to the environment from the fusion reaction is tritium, radioactive hydrogen, $_1^3H$. Wastes from the generating facility are projected to be amenable to near-surface burial (23).

In stars, in contrast to reactors, both the high temperature and the confinement are produced by the gravitational compression of the stellar material. Because of the high densities in our sun, its interior temperature can be rather low ($\approx$10 MK).

The timetable for fusion of any sort is not certain, but most authorities do not believe it is feasible to develop fusion for energy any time before the next century. Part of the fusion problem may be that fusion research for weapons is easier and has a greater priority than fusion research for energy. As to the status of fusion as a possible energy source, one scientist was quoted in 1976 as saying "We see smoke but not fire" (29). This remains true over a decade later.

SUMMARY

There are four major types of reactor currently in use: the boiling water reactor; the pressurized water reactor; the Canadian deuterium reactor; and the breeder reactor (still experimental, and only in Europe). The high-temperature gas-cooled reactor has several working prototypes, but is still not being used for everyday generation of electricity. Pressure-tube reactors have been built only in the USSR.

Fission reactors are identical to other sorts of reactors from the turbine on. The difference is found in the reactor vessel, where nuclei split and give the energy released in the fission to the products of the decay, which further transfer energy to the material in the reactor vessel. Eventually, the temperature gets high enough to produce steam for the turbine.

All working reactors are fission reactors. In fission, the many nucleons in the large nucleus can gain stability by dropping deeper into each of the daughter nuclei. In going into the nuclear energy well more deeply, each nucleon gives up some energy. As a result the larger nuclei have masses smaller than the sum of the masses of its constituent nucleons. The difference is the binding energy of the nucleons in the nucleus.

BWRs and PWRs cannot undergo a nuclear explosion because the proportion of ^{235}U is too low, only about 4 percent. Bombs contain over 90 percent ^{235}U. Breeder reactors, such as the French-built Phenix and Superphenix, are reactors blanketed in ^{238}U that then produce ^{239}Pu. Such reactors are more unstable than conventional reactors, but do hold the promise of an infinite supply of reactor fuel.

Nuclear fusion research has demonstrated in the 1980s that fusion is feasible, but any working fusion reactors are probably more than thirty years in the future. Fuel for fusion reactors is virtually inexhaustible. Best estimates of pollution emission for fusion are that the levels should be very low, but this must be the subject of engineering research as the fusion reactor is developed.

REFERENCES

1. S. Glasstone, *Sourcebook on Atomic Energy,* 3d ed. (New York: Van Nostrand Reinhold Co., 1967).
2. S. Glasstone and W. H. Jordan, *Nuclear Power and Its Environmental Effects* (Chicago: American Nuclear Society, 1980).
3. Economic Council of Canada, *Connections* (Ottawa: Canadian Government Publishing Centre, 1985).
4. H. M. Agnew, *Sci. Am.* 224, no. 6 (1981): 55.
5. E. Marshall, *Science* 219 (1983): 265.
6. E. Marshall, *Science* 224 (1984): 699.

7. N. A. Dollezhal', *Nucl. Energy* 20 (1981): 385. Annex 2, Report of the Commission on the Causes of the Accident at the Fourth Unit of the Chernobyl Nuclear Power Plant, 1986. See also B. G. Levi, *Phys. Today* 39, no. 7 (1986): 17.

8. AEC, *The Safety of Nuclear Power Reactors and Related Facilities,* WASH-1250 (Washington, D.C.: GPO, 1973).

9. G. A. Cowan, *Sci. Am.* 239, no. 1 (1978): 36.

10. C. Norman, *Science* 221 (1983): 730.

11. C. Norman, *Science* 228 (1985): 1407.

12. G. Szegö, *The U.S. Energy Problem,* RANN report, Appendix E (Washington, D.C.: GPO 1972); 1973 AEC report to Congress (Washington, D.C.: GPO, 1974).

13. E. Cowan, *New York Times,* 18 July 1976; D. Burnham, *New York Times,* 16 Nov. 1975.

14. M. W. Wald, *New York Times,* 24 Feb. 1984.

15. M. W. Wald, *New York Times,* 20 Jan. 1985.

16. K. S. Deffeyes and I. D. MacGregor, *Sci. Am.* 242, no. 1 (1980): 66.

17. F. J. Shore, *Phys. Teacher* 12 (1974): 327.

18. A. M. Weinberg, *Science* 232 (1986): 695.

19. M. K. Hubbert, Chap. 8 in *Resources and Man,* (San Francisco: Freeman, 1969).

20. R. F. Post and P. L. Ribe, *Science* 186 (1974): 397.

21. W. D. Metz, *Science* 178 (1972): 291; R. F. Post and F. L. Ribe, *Science* 187 (1975): 215.

22. M. M. Waldrop, *Science* 222 (1983): 1002.

23. K. I. Thomassen, *Ann. Rev. Energy* 9 (1984): 281.

24. H. P. Furth, *Sci. Am.* 241, no. 2 (1979): 51.

25. R. S. Craxton, R. L. McCrory, and J. M. Soures, *Sci. Am.* 255, no. 2 (1986): 68. See also W. D. Metz, *Science* 186 (1974): 519 and 1194.

26. W. J. Broad, *New York Times,* 21 March 1988.

27. G. Yonas, *Sci. Am.* 239, no. 5 (1978): 50.

28. J. P. VanDevender and D. L. Cook, *Science* 232 (1986): 831.

29. W. D. Metz, *Science* 194 (1976): 307.

PROBLEMS AND QUESTIONS

True or False

1. The supply of uranium for reactors depends on the price consumers are willing to pay.
2. Most of the world uranium supply is in North America.
3. A converter is a reactor that produces more fuel than it consumes.
4. The PWR has a higher thermodynamic efficiency than the BWR.
5. Future reactor development must include breeders in order to provide fuel for burners and converters.
6. There is a gigantic supply of cheap deuterium available for fusion reactors.
7. There are several ways to achieve controlled fusion currently being investigated.
8. The Lawson criterion tells us when fission is possible.
9. One reason for the early success of toroid-geometry fusion reactors is that they have no ends through which to lose plasma.
10. No fusion reactions have yet been able to produce net energy on Earth.

11. Uranium enrichment is, both in principle and in practice, extremely energy-expensive.
12. A fission reactor is always in danger of exploding like a bomb.

Multiple Choice
13. In bombs, the concentrations of $^{235}_{92}U$ is over 90 percent. In light water reactors, the concentration of uranium is about 4 percent. In natural uranium as mined, the concentration of uranium is about 0.7 percent. French scientists have found evidence that a uranium deposit in Gabon functioned as a reactor about 2 billion years ago. What does this imply about $^{235}_{92}U$?
    a. The deposit was made of artificially enriched uranium.
    b. Aliens from outer space visited Earth about 2 billion years ago.
    c. Some $^{235}_{92}U$ has been spontaneously changed to $^{238}_{92}U$.
    d. The uranium had much more plutonium in it 2 billion years ago.
    e. Uranium-235 must undergo decay with a half-life of about a half-billion years.

14. What is the function of the control rods in a nuclear fission reactor such as a boiling water reactor or a pressurized water reactor?
    a. To delay neutrons from fissions of ^{235}U a preset amount
    b. To slow down neutrons from fissions of ^{235}U
    c. To turn the reactor on and off
    d. To reflect neutrons from dense regions of the core
    e. To prevent radioactive material from leaking into the environment

15. Which of the following fusion test reactors would probably be able to provide energy by direct conversion from the current of charged particles?
    a. Joint European Torus
    b. Tokamak Fusion Test Reactor
    c. Stellerator (torus)
    d. Alcator-C (torus)
    e. Tandem mirror reactor

16. Supplies of cheap uranium ore
    a. are available for the foreseeable future.
    b. are available mostly in the USSR.
    c. are unavailable to the French.
    d. are less readily available in Canada than in the United States.
    e. contain a smaller proportion of ^{235}U than supplies of more expensive uranium ore.

17. The conversion coefficient (ratio of ^{239}Pu created to ^{235}U burnt) in a reactor is 0.9. The reactor would be classified as a
    a. burner.
    b. converter.
    c. transformer.
    d. breeder.
    e. fusion device.

18. Which of the following is *not* a method of uranium enrichment?
    a. Use of a laser to ionize only the ^{235}U in a mixture of isotopes
    b. Use of a centrifuge to separate the isotopes
    c. Sinking of the heavier isotope by gravity in a mixture of uranium hexafluoride gas
    d. Progressive separation by diffusion through membranes of uranium hexafluoride gas

19. Weapons-grade uranium might be used as fuel in which of the following types of reactor?
    a. Pressurized water reactor
    b. Tokamak fusion test reactor
    c. Breeder reactor
    d. Boiling water reactor
    e. High-temperature gas-cooled reactor
20. Which of the following is *not* a safety feature in fission reactors?
    a. Use of delayed neutrons in designing the steady-state reactor operation
    b. Special ventilating fans to keep the airborne concentration of radioisotopes low
    c. Use of moderators such as water in the reactor
    d. Use of control rods to absorb neutrons
    e. Design of an emergency core cooling system in case of accident

Discussion Questions

21. Why are delayed neutrons so important?
22. Are breeders superior to converters in resource utilization? If so, how?
23. In what ways does the nuclear generator differ from the fossil fuel generator in the steam vessel–turbine–condenser system?
24. Since only 4 percent of the electric energy generated in 1970 was from nuclear reactors, and 2 percent of the electric energy generated has been used for gaseous diffusion, is nuclear energy a bad bargain?
25. Will fusion power plants alter the Earth's heat balance?

15

SOLAR ENERGY

Continuing the examination of energy resources begun in Chapter 13, this chapter examines several forms of solar energy. The chapter begins with an examination of the sun (fueled by nuclear fusion—see Chapter 14) and its output of energy. Stored solar energy was discussed in Chapter 13; wind energy and tidal energy are among the unconventional solar energy sources examined here. Solar cells and large-scale solar energy projects are discussed in detail. In the next two chapters, solar energy resources stored in biomass are examined. In Chapter 19, the last chapter of this group, passive solar energy is the major focus.

KEY TERMS   electromagnetic radiation
frequency
period
wavelength
spectrum
blackbody
Stefan-Boltzmann law
emissivity
tidal range
ocean thermal gradients
salt-gradient solar pond
active solar
passive solar
photodissociation
photovoltaic cells
photoelectric effect
photon
lattice
conductor
insulator
band gap
valence band
conduction band
semiconductor (n-type, p-type)
amorphous silicon
heat island
heat balance
heliostat
load-shifting

The surface of the sun, which is what we see if we look at the sun, is hot enough to cause the electrons in atoms to be excited and to emit energy in the form of what is called *electromagnetic radiation*. The sun delivers all of the energy that makes objects visible, keeps the Earth warm, and drives the winds and the ocean currents.

ELECTROMAGNETIC WAVES

There are many ways to produce electromagnetic radiation; one way is to wiggle a charged particle (such as an electron). For example, our electric generating facilities make electrons wiggle back and forth sixty times per second. As they wiggle, they cause electric and magnetic fields to be generated that travel off in all directions. The electric and magnetic fields, taken together, are the entirety we refer to as electromagnetic radiation. Other familiar forms of electromagnetic radiation include radio waves, microwaves, infrared radiation light, ultraviolet radiation, x rays, and gamma rays. We distinguish forms of electromagnetic radiation from one another by their *frequencies* (number of repetitions per second, measured in hertz, abbreviated Hz). Hence, electric power produces electromagnetic radiation at 60 Hz.

If we were on a boat in a lake in which there is a steady periodic water wave, we could measure the period of time it takes our boat to go through a complete cycle of motion up and down (see Figure 15.1). This is called the *period* of the wave. If we know the period, we also know how many repetitions there are in a second; that is, we know the frequency, the inverse of the period. A water wave with a period of two seconds will complete half its cycle in one second (½ Hz); a wave with a period of ½ second will complete

two cycles in a second (2 Hz); and so on. This relation is summarized in the equation

$$\text{period} = \frac{1}{\text{frequency}}.$$

If someone were to take a snapshot of the water surface as in Figure 15.1, we would see that there is a repetitive behavior of the wave, one that repeats itself after a certain distance. The distance necessary for the pattern to repeat is called the *wavelength*. If the person in the boat timed the wave for one period, it is easy to see from the diagram of Figure 15.2 that the wave would have traveled one wavelength. Thus its wave speed, which we denote c, is (see Chapter 3) wavelength/period, or alternatively,

$$c = (\text{wavelength}) \times (\text{frequency}).$$

Knowing wavelength, we also know frequency— and vice versa *if* we know the wave speed c.

Electromagnetic radiation, which includes light, travels through empty space at a speed of 3×10^8 m/s. This speed is usually called the speed of light. The equation relating wavelength, frequency, and wave speed holds for electromagnetic radiation as well. Our 60 Hz radiation from power plants corresponds to a wavelength of

300,000,000 m/s/60 Hz = 5,000,000 m.

FIGURE 15.1
Water waves illustrate the concept of wavelength.

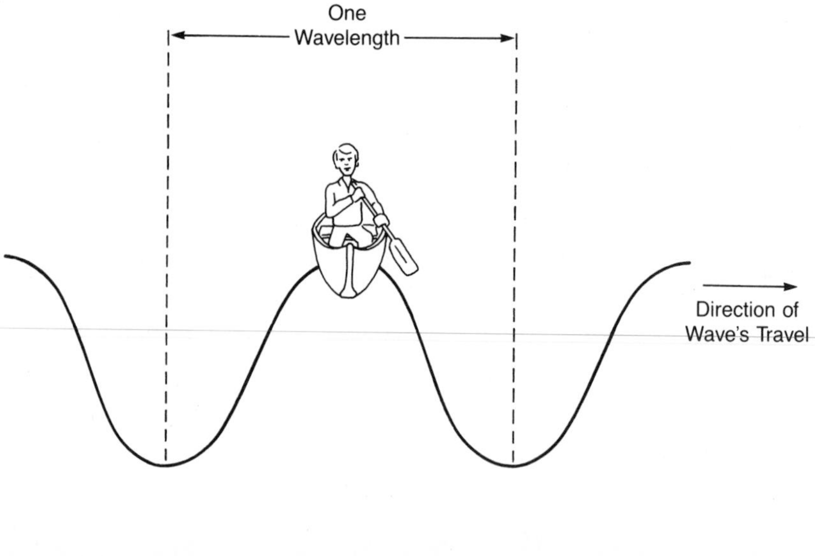

FIGURE 15.2
Water waves move past a boat. (a) Start. (b) 1/4 wavelength after start. (c) 1/2 wavelength after start. (d) 3/4 wavelength after start. (e) 1 wavelength after start.

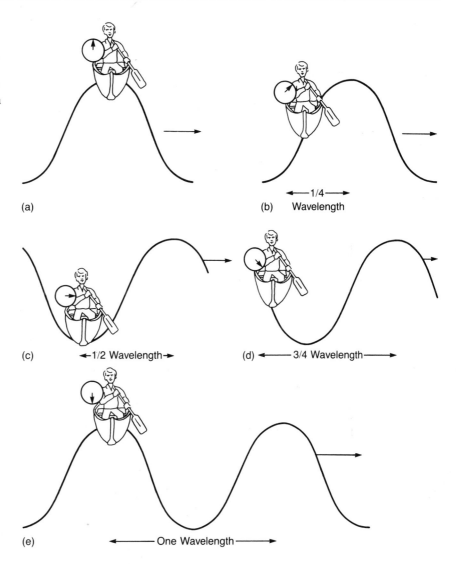

(a)

(b)   ←—1/4—→   Wavelength

(c)   ←1/2 Wavelength→

(d)   ←—— 3/4 Wavelength——→

(e)   ←——— One Wavelength ———→

The visible white light from the sun reaching the Earth can be made to pass through a slit and then onto the face of a prism. The light is spread out by the prism into a continuum of color that changes uniformly from red (lower frequency) to deep blue (higher frequency), as shown in Figure 15.3. This spread-out light is called the *visible spectrum*. The wavelengths corresponding to red and deep blue are about 650 nm and 450 nm, respectively. In fact, we receive other wavelengths of electromagnetic radiation from the sun,

but these are not visible to us. The heat we feel from the sun from wavelengths longer than that of visible red is known as *infrared radiation*. The radiation causing sunburn comes from regions of wavelengths smaller than visible blue and is called *ultraviolet*. We extend the meaning of the word spectrum, originally applied only to visible light, to the entire set of electromagnetic radiation. Figure 15.4 presents the spectrum on a logarithmic scale. Note that visible light constitutes only a tiny region of the electromagnetic spec-

FIGURE 15.3
An incident light beam
is refracted (bent) and
dispersed (separated by
color) by a prism.

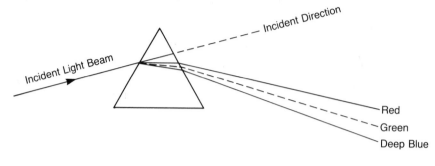

FIGURE 15.4
The electromagnetic
spectrum.

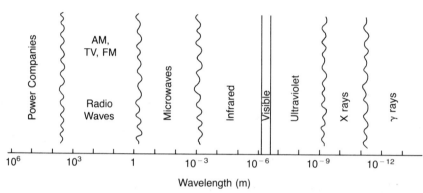

trum. The boundaries between regions on the spectrum are rather vaguely defined (indicated in Figure 15.4 by a wavy line).

The Sun

The surface of the sun that we can see is called the *photosphere*. The temperature of the photosphere is about 6000 K. The gases at that temperature emit a characteristic blackbody radiation spectrum (see box).

The amount of power radiated by the sun is 3.9×10^{26} W. This awesome amount of power can scarcely be comprehended. The total energy produced in the world each year by all the world's energy sources is equivalent to the amount the sun produces in about 5 billionths of a second.

The sun's energy is derived from fusion reactions taking place inside the sun, where the temperatures can be measured in megakelvins. The minimum temperature at which fusion reactions can occur is about 10 MK (10^7 K), and the sun's core is slightly hotter. The net result of

fusion is to transform hydrogen into helium. The curve of the average binding energy per nucleon vs. nucleon number, shown on Figure 6.2 in Chapter 6, reveals that this reaction liberates about 27 GWh/kg. This large energy release reflects the incredible stability of the helium nucleus. The heat travels through the gases making up the sun (mostly hydrogen and helium) by conduction and convection to the limits of the solar atmosphere. The various layers of material in the sun are somewhat insulated from one another, and, as we previously mentioned, the photosphere temperature is only about 6000 K.

The sun's energy radiates out into space because of the temperature difference between the sun's surface and that of interplanetary space (about 3 K). Solar energy is in the form of electromagnetic radiation, which can travel in a vacuum, so that no material medium is necessary to enable heat to be transferred. This is most definitely not the case for conduction and convection (Chapter 7).

Despite the fact that we cannot see it, infrared and ultraviolet radiation cause a radiometer (Figure 15.5) to rotate just as visible light does. The fact that infrared radiation is invisible, but acts in the way that light does, allows instruments to be designed that will send out infrared light and detect and amplify the reflected infrared light. The device uses an image intensifier to make infrared light visible to the observer. Such a device, called a *sniperscope,* was attached to a rifle and used by U.S. soldiers in Vietnam to detect warm objects (people) in the dark. A more modern version now in use is the night vision goggle, which is small and easy to use. The 0.68 kg goggle has a quick-release mechanism and is compatible with U.S. and British gas masks and helmets.

FIGURE 15.5
A radiometer, a device whose vanes turn under the influence of light.

Solar energy streams off from the sun into space in all directions. In most directions, there is nothing much to intercept the light. As the energy gets further and further away from the sun, it is spread out uniformly over a larger and larger spherical area. A very small fraction of the sun's radiated power falls on the Earth because each square meter of surface at the Earth's orbital distance, about 150 million km from the center of the sun, receives 1382 watts (1). Not all of this energy reaches the Earth's surface because it is reflected by clouds, ice, and shiny surfaces. The energy that does reach the surface and gets absorbed by vegetation is responsible for the con-

BLACKBODIES

In our everyday experience, we notice that black objects can get very hot, while white objects stay cooler. White is overwhelmingly used in tropical cities because white reflects sunlight. Black objects are good absorbers of radiation.

Consider a body that is a total absorber of radiation. Such a body is called a *blackbody*. Although such a body would seem to be "ideal" rather than "real," it *is* possible to make a real blackbody. A solid object with a cavity inside will trap all radiation inside that cavity. By this we mean that all radiation absorbed by the walls is re-emitted. If the object is heated to a temperature T, then the spectrum of radiation trapped inside is characteristic of a blackbody at temperature T. While much of the radiation is in the visible region, much is also in the infrared, and some is in the ultraviolet. The sun's spectrum is shown in Figure 15.6. The spectrum of the sun is characteristic of a 5800 K blackbody. An object at a different temperature would have a different spectrum.

The radiant power emitted by a blackbody depends on its area. The greater the area, the greater the radiated power. It is therefore usual to discuss the radiated power per unit area for blackbodies. By taking real objects with spherical cavities and drilling small holes in them, we can observe the power spectrum of a black-

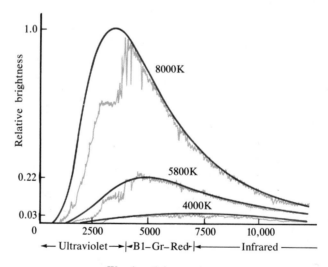

FIGURE 15.6

Blackbody spectrum for the sun [Fig. 6–17, from Protheroe, Capriotti, and Newsom, *Exploring the Universe*, 4th ed. (Columbus, OH: Merrill, 1989).

body. The properties of blackbodies were investigated by the nineteenth century physicist Josef Stefan (1835–1893). He found that the amount of power per unit area increased as the absolute temperature increased:

$$P/A = (5.67 \times 10^{-8} \text{ W/m}^2\text{K}^4) \, T^4.$$

This result is called the *Stefan-Boltzmann law,* and the constant given is known as the *Stefan–Boltzmann constant*. Hence, for the sun, the amount of radiated power per unit area is 6.4×10^7 W/m². Since the sun's radius is 6.96×10^8 m, the surface area is 6.1×10^{18} m², and the radiated power is 3.9×10^{26} W.

Blackbodies are not common in the universe outside of stars themselves. Thus, we must take into account other properties of bodies that describe how well they absorb (or emit) radiation. A perfect absorber is a perfect radiator. A perfect reflector does not radiate. This is the reason that thermos bottles have silvered surfaces inside—to trap the radiation. The *emissivity* describes how close to being a blackbody an object is. A perfect emitter or absorber has an emissivity of 1, while a perfect reflector has an emissivity of 0. For example, the Earth has an emissivity of about 0.5. Since the Earth is at an average temperature of about 290 K, we find its average radiated power per unit area by

$$P/A = 0.5 \, (5.67 \times 10^8 \text{ W/m}^2\text{K}^4) \, (290 \text{ K})^4 = 229.6 \text{ W/m}^2.$$

Interestingly enough, the Earth's temperature can be predicted using radiation balance—that is, the Earth emits as much radiation as it absorbs. If this were not true, then the Earth would either heat up or cool down, neither of which has been observed. The Earth's average temperature is remarkably stable. The mean solar irradiance at the Earth's distance from the sun is 1382 W/m² (1). The total power absorbed by the Earth is then

$$P_A = (0.5) \, (1382 \text{ W/m}^2) \, (\pi \, R_E^2),$$

since the Earth looks like a disk to the light rays from the sun. The power radiated by the Earth is

$$P_R = (0.5) \, (5.67 \times 10^{-8} \text{ W/m}^2 \text{ K}^4) \, T^4 \, (4\pi R_E^2),$$

where the last term is the Earth's total surface area. Since $P_A = P_R$, we conclude that

$$T^4 = 1382 \text{ W/m}^2/(5.67 \times 10^{-8} \text{ W/m}^2 \text{ K}^4),$$

which gives $T = 279$ K. This predicts a mean Earth temperature of 6.2°C, not far off from the actual mean temperature of about 17°C. The reason for this difference is discussed in the chapter on the Earth's climate (Chapter 24).

tinued lives of all living beings, and for all the coal, oil, and gas deposits found on Earth (see Chapter 13).

Electromagnetic Radiation Reaching the Earth

As we shall see in Chapter 24 on the Earth's climate, the atmosphere prevents much of the radiation from the sun from reaching Earth's surface. Most of the infrared and ultraviolet radiation is reflected or absorbed in the upper atmosphere by water vapor, ozone, methane, and other trace gases. Visible light makes it through practically unscathed, which is why Earth animals have light detectors (eyes) sensitive to these frequencies (colors) of light and why plants absorb radiation at these frequencies (see the discussion in the next chapter).

Since the sun's radiation diminishes rapidly away from the visible region, few x rays and gamma rays reach us from the sun (the short wavelength region), and few microwaves and radio waves reach us from the sun (the long wavelength region). Most forms of radiation found on Earth other than infrared light, visible light, and ultraviolet light are artificially produced. Radio waves transmit radio and television signals. Microwaves are used in ovens to cook materials containing water and for communications such as telephones and satellite relays. X rays are useful because flesh is reasonably transparent to them, while the denser bone is not. As a result, x rays are used for diagnostic purposes—to examine patients for broken bones and decayed teeth, for example.

USES OF THE SOLAR ENERGY REACHING EARTH

There are many different ways to use the solar energy that reaches Earth. Water is evaporated, clouds form, the clouds produce rain, and the runoff from the rain may be captured by dams and used to run water wheels or turbines. This is an "old technology," in common use, as dis-

cussed in Chapter 11. In this chapter, we discuss the more direct ways to use this energy: wind power, tidal power, solar energy from sea temperature differentials, photolysis for use in fuel cells, direct solar "farms," and photovoltaic solar cells. An NSF/NASA study in 1976 indicated that by 2000, solar energy could account for 35 percent of heating and cooling, 30 percent of U.S. gaseous fuels, 10 percent of U.S. liquid fuels, and 20 percent of the nation's electric needs (2). The study's projection now seems wildly optimistic. Later realities have dampened enthusiasm for solar power except, perhaps, in California. As of 1983, about 5 percent of California's energy was from nonconventional sources; officials predict that this will be up to 10 percent by 1990 (3). California wind energy costs 5 to 7 cents/kWh (4) but may cost as much as 9 to 12 cents/kWh after tax credits expire (5); this is somewhat higher than the cost of energy from conventional sources. By 1985, wind energy alone was providing California with 600 MkWh$_e$ annually (5).

Wind Energy

About 2 percent of solar energy is fed into the winds, in such a way that, at any one time, about one day's worth of solar energy is stored in the kinetic energy of the atmosphere. It has been estimated that about 10,000 TkWh per year could be captured from the winds, about twenty times the current world electricity use (6).

The windmill is a proven technology. Windmills have been used for a thousand years, mostly to pump water and grind grain. In the United States, the old farm windmill has been used for decades to pump water and generate electricity. It is still used for electricity in areas too remote for rural electrification. Table 15.1 (6) shows how wind power compares to other forms of "renewable" power.

This great potential untapped source has fired the imaginations of some engineers. One proposal calls for a chain of 333,000 windmills 240 meters tall across the great plains from Texas to Canada. It is believed that such a system could

POWER FROM THE WINDS

Windmills produce power at a rate that depends on the effective area swept by the windmill blades and on the speed of the wind. Local wind speed is the most important parameter, since the power transferred to the windmill grows as the cube of the wind speed. Since the kinetic energy of any air parcel of mass m is $\frac{1}{2} mv^2$, the rate of mass flow through the blade's effective area, A, determines the power transferred to the blades. The rate of mass flow is given by ρAv, where ρ is the air density, since the mass flowing through area A in a small time Δt is the density times the volume $A(v\Delta t)$ of air having that mass. As a result, the wind's power is

$$P = \frac{1}{2} (\rho Av^3).$$

Windmills cannot extract all the power in the wind, both because air must be able to move away from behind the blade, and because conversion of mechanical energy involves losses. The maximum fraction of wind power extractable is only 16/27 (≈ 0.6) (11,12). The limitation arises because, if the blades slow the wind too much, most wind will just flow around the blades, but if the blades do not slow the wind enough, the energy will be lost (13). Sites with mean wind speeds in excess of 7 m/s at 25 m height are prime wind energy resources.

TABLE 15.1

Power in renewable resources.

Wind	130,000 GW
Tidal energy	60 GW
Geothermal energy	130 GW
Hydropower	2900 GW

SOURCE: Reference 6, copyright 1979 by AAAS.

be built in less than thirty years and would ultimately generate 1.5 TkWh (7). It had been estimated in 1970 that such huge windmills could be constructed at a cost comparable to that of coal-fired power plants (7). The same person who proposed the chain of windmills, W. E. Heronemus, also proposed floating offshore windmills (8,9). The winds are three to five times as strong just offshore of New England as on land (9). Heronemus proposed using the wind energy to dissociate water and then piping the hydrogen ashore. Such offshore windmills were considered in great detail in the report of the EURO-CEAN group (10).

Denmark, which has rejected nuclear energy and has wind in abundance, leads the world in windmill manufacture. The Danes plan to get 10 percent of their energy from wind by 2000 (14). China is planning to use windmills where it lacks other energy sources (15). However, the largest current market for windmills is in the United States. At some places in the world, the flow of wind through a square meter (at 25 m height) can produce 500 W (11). Efficient windmills can produce 175 W/m² of area swept by the propeller (11). This compares favorably to the total solar annual energy flux. Someone planning to erect a wind turbine would look for a region with steady, relatively fast winds, because the turbine's delivered power grows as the cube of the speed.

Hawaiian Electric brought thirty 2.5 MW machines in 1980 to operate on the local trade winds (about 25 km/h), and Southern California Edison (SCE) buys energy from "windmill farms" in the

San Gorgonio Pass (Figure 15.7) and in the Te-hachapi Mountains. SCE is also testing various devices at its test site in the San Gorgonio Pass (Figure 15.8) near Palm Springs (9). A private company put more than seven hundred wind machines in a mountain pass near San Francisco to sell to Pacific Gas and Electric (5). The temperature difference between the cool ocean and the warm interior valley of California causes an air exchange, with cool air coming in from the coast and warm air rising. Passes in the mountains funnel the winds, which blow steadily through them.

Because of the immense amount of energy that could be produced by tapping the winds, even at 35 percent efficiency, the 1970s and 1980s have seen many companies get into the wind turbine business (16). Private companies are building machines to tap the winds and then selling electricity to the power companies. This is in large part due to the Public Utilities Regulatory Policies Act (PURPA) passed by Congress in 1978. Under terms of PURPA, companies may sell energy to utilities at their marginal or avoided cost (the cost of their most expensive energy), not their average cost. Southern California Edison is currently paying 5.7 cents/kWh as its avoided cost (17). (Marginal, or avoided, cost is the cost of adding capacity; under PURPA, this has been interpreted to mean the cost of the most expensive additional energy, usually from gas turbine generation.) Southern California Edison, as of April 1985, had contracts to purchase about 4700 MW from renewable energy sources (14). The utilities are protected because the private companies are prohibited from supplying more than

FIGURE 15.7
Windmills in the San Gorgonio Pass near Palm Springs, California. (U.S. Windpower, Inc.)

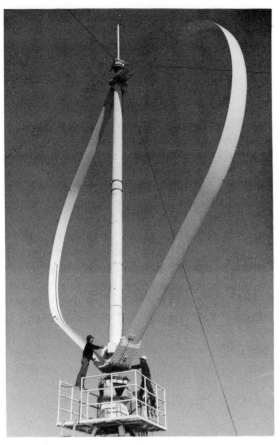

FIGURE 15.8
A windmill on the test site of Southern California Edison.

FIGURE 15.9
An old windmill sits on the prairie. Wind energy has been used in the U.S. for almost a century on the farm. (U.S. Department of Energy)

80 MW to the utilities. Because the utilities in California were being oversubscribed under PURPA, the California Public Utility Commission suspended sign-ups in April 1985.

Solar Farms

One interesting proposal envisions a solar energy "farm" (18) that would be able to provide energy continuously (night as well as day) by using sunlight to heat liquid sodium to about 500°C. Liquid sodium has a high specific heat and thus can store a lot of energy for each degree rise in temper-

ature. Heat exchangers allow the liquid sodium to cool, giving off the energy absorbed to water. The resulting steam would drive conventional turbines.

The "farm" would have to be located in a sunny region, such as a desert. It would consist of long rows of energy absorbers painted black to absorb heat, alternating with rows painted white to reflect heat. If this alternation were not done, the thermal balance of the large area would change, which could change the weather. By alternating reflectors and absorbers, the average amount of heat absorbed by the plant would be the same as if it had still been desert. If the solar energy could be captured at 100 percent efficiency, a land area of only 1300 km² could supply all the energy currently used in the United States. At 10 percent efficiency—a more realistic figure—such a farm would need a net area of only 13,000 km² (5000 mi²).

Solar Energy in Water

Energy in the Tides   Tidal energy arises from the interaction of the sun (and the moon) with Earth. The oscillatory flow of water in filling and emptying a partly enclosed region along a coast can be tapped to produce electricity. This energy may be made available by cutting off the basin with dams to create a water level difference. Of course, the total amount of energy available depends strongly on the difference between the high and low tide levels, since it uses the potential energy generated by the rise and fall of the tides to generate hydroelectricity. The total world tidal energy available is estimated to be 2.9 TkWh$_e$ and is only about 1 percent of the world's available water power (7,12). The largest facility in operation, at 240 MW installed capacity, is at La Rance, France. It has a tidal range (difference between high and low tide levels) of ≈ 8 m and produces about half a billion kWh$_e$ yearly (about 20 percent of the maximum available energy).

In the Bay of Fundy, which has a 15 m average tidal range, a plant is proposed for the Minas Basin; if built, it would provide about 15 GkWh$_e$ yearly (19). The total potential electric generation capability of the entire Bay of Fundy system is between 30 and 60 GWh/year (12). The scheme would consist in 128 turbines with an installed capacity of 5 GW$_e$, and with the intermittent character of the tidal rise and fall, the power output would average ≈ 1.75 GW$_e$ (19). The Bay of Fundy has the largest tidal range in the world, which is due to the resonance between the tidal forcing and the Gulf of Maine, of which Fundy is a part. (*Resonance* refers to what happens to the response amplitude at some particular frequency of forcing—for example, if your tires are out of

balance, the imbalance provides a periodic jar; the jarring can cause the car to shake as if it would fall apart at one particular speed. If you speed up or slow down, the shaking vanishes. You have just found a resonant frequency for your car. Bodies of water can resonate, too. Ask any child who used a washcloth to push on water in the bathtub with just the right frequency to make a huge wave go over the edge of the bathtub onto the floor!)

The Gulf of Maine has a natural period of about 13.3 hours, while the moon-caused tidal period is 12.4 hours (19). This near-match causes the huge water flow: about 1/40 (km)³/s. There is a possible problem with the Minas Basin project. Models indicate that closing off the Minas Basin would decrease the natural period of the Gulf of Maine, causing an increase in the tidal range as a result of the closer match (19). Therefore, tides would be higher as far south as Cape Cod. This project might even change tides in Britain, although by only a few millimeters (19). The effect on the Georges Bank, a major world fishery, is still to be determined.

OTEC   Temperature gradients in the sea can be used to produce power: This technology is known as *OTEC* (Ocean Thermal Energy Conversion). It has been estimated that the thermal gradients of the Gulf Stream could produce energy at a rate of 182 TkWh$_e$ per day, about seventy-five times the 1980 U.S. energy use rate (20). One kilogram of water changing its temperature from 25°C to 5°C gives as much energy (≈ 8 kJ) as it could by falling about 8.5 meters. Of course, the efficiency of such a plant is less than the maximum thermodynamic efficiency (see Chapter 7) of 20/298 = 0.067, or 6.7 percent. A plant producing 7.8 MW and 24 million liters of fresh water daily might be built for as little as \$18.4 M; the energy would cost about \$0.006/kWh, and the water, \$1/1000 gallons (10,21). The usual option is a heat exchanger using a fluid—perhaps ammonia—that operates much like the fluid in the home refrigerator or air conditioner (10). The low temperature reservoir is the deep ocean's 4°C water; the high temperature reservoir is the

TAPPING TIDAL ENERGY

Energy from the tides is generated in much the same way as hydroelectric energy from dams. Tidal basins (partly enclosed river mouths, bays, etc.) empty and fill with the tides. If the basins are closed off by dams, water level differences will exist between the basin and the ocean that could be used to run turbines.

The total amount of energy available from such an enclosed basin is Mgh, where M is the mass of extra water held by the basin at high tide as compared to low tide, g is the acceleration of gravity (9.82 m/s²), and h (the tidal range) is the difference in heights between high and low tide. The mass M is proportional to the surface area of the basin and to h. Thus the mass of captured water is given by the density of water (1000 kg/m³) multiplied by the volume of water captured by the dam; the volume is M = (density of water) × (area) × h. This means that the stored energy grows as the square of h. For this reason, regions with high tidal ranges—such as the Bay of Fundy—have the greatest potential.

The energy actually available is generally only a fraction of the maximum potential. The Rance, one of only two working tidal power installations, produces between 20 and 25 percent of the theoretical maximum. Still, although tidal energy is only a minor *world* energy resource, locally it could have a large impact indeed. The closing of the San Jose Bay in Argentina; Mont Saint-Michel in France; Severn in England; Cook Inlet in Alaska; and the White Sea, the Sea of Okhotsk, and the Penzhinsk Gulf in Russia could produce average power in the gigawatt range. Other potential sources are relatively minor compared to these.

upper ocean. The cold fluid is pumped to the surface, expands at the high temperature, and drives a turbine. The first working OTEC facility operated in Cuba's Havana harbor in the 1920s. A facility for OTEC research has been established in Hawaii (16) with involvement of the Department of Energy's (DOE) Solar Energy Research Institute.

The upwelling of cold water, which contains nutrients, through an OTEC installation could cause large increases in fish populations, adding a valuable food resource. One could even build farms at the upwelling to minimize fishing costs. Also, the upwelling cold water could alter the entire world climate if such plants were built on a large enough scale. The ability to alter climate by building large numbers of OTEC installations could help humanity prevent another ice age if we can gain a better understanding of the inter-

action between the oceans and the atmosphere (21).

Wave Energy Just as the sun drives the winds, the winds drive the waves. In some regions, wave power can reach a megawatt per meter (21). This energy could be used to run a water pump, controlled by a check valve, up to a point at which the water would be allowed to fall through a turbine. Japan has built ten turbines in the Sea of Japan; they generate a total of 1.25 MW$_e$ utilizing oscillating water columns (22).

In 1986, two wave energy demonstration projects were completed in Norway that generate a combined total of 0.85 MW$_e$. The country is planning a 10 MW$_e$ wave energy station for 1990 (22). In one of the projects, wave action pushes water up a tower, driving air upward through a turbine; as the water falls, air is pulled downward through the turbine. It is designed so that both effects cause rotation in the same direction. The demonstration apparatus has one serious drawback: It sounds like a siren and can be heard for long distances (22).

A device known as Salter's duck, which consists of a series of wing-like objects connected together, appears to absorb most of the energy in waves (23). Prototype elements are 15 m by 30 m and have a mass greater than 400 tonnes (10). Use of Salter's duck could serve two purposes: energy generation, and use as a breakwater to protect harbors, since wave energy extracted by the duck cannot cause waves in a harbor. The 1979 EUROCEAN study (10) indicated that it was one of the least expensive wave energy conversion devices available (estimated price 7.4 cents/kWh$_e$).

Salt Water and Solar Energy   A recently recognized energy source is available from the mixing of salt water and fresh water as rivers flow into oceans. In fact, the mixing energy in a seawater–fresh water interface is equivalent to a dam 240 meters high (24,25). Several possible ways to get useful energy from the mixing of salt water and fresh water have been suggested. Most methods use membranes to separate the two fluids, but this involves expensive pumping and short membrane life; vapor exchange between two solutions at high temperatures could produce power without the use of membranes (25). Because the fresh water is constantly renewed by the sun and because the resource potential is so large, such a resource can be used even if the process has a very low efficiency.

Salt water is also involved in another type of solar energy storage. Ponds are constructed in which salt water sits covered by fresh water. The water naturally separates into regions of different density (or salinity). This salt density gradient provides zones in which heat is trapped and stored. The water at the top is fairly fresh. There is a middle region that is very stable, and that lets light through into the bottom but not out (it acts something like a window). The middle zone

also prevents the very salty hot water at the bottom from mixing with the rest of the water.

Such ponds can use the stored heat to heat water or air somewhere else. Professor C. Nielsen of Ohio State University has built a pond that demonstrates the economical use of salt-gradient solar ponds for heating barns and swimming pools in the northern United States (26).

In most temperate zones, solar ponds can provide 250 kWh$_e$m^2 of heating in the winter, for an efficiency of 18 percent (26). Work on such ponds is proceeding in areas with lots of sunlight, especially Israel. Measurements in Beer Sheva, Israel, indicate that an average of 18.4 MJ of energy reaches each square meter of surface per day (27). Salt-gradient solar ponds in Israel are used for production of electricity. The two reservoirs for the heat engine are the high temperature reservoir at the bottom of the pond and the low temperature reservoir at the top. A turbine using a low-boiling-point fluid operates between the two temperatures. A 1250 m^2 pond built in the early 1970s is still producing 6 kW; a 7000 m^2 pond built in 1979 is producing 35 kW (and can produce a peak of 150 kW) (28). The Israelis are planning to produce around 2000 MW by the year 2000 from such solar ponds.

Active Solar Water Heating and Space Heating

Solar energy has been used for decades to produce cheap hot water for home use in Israel, Florida, and Australia. Many such systems were being marketed across the United States because of the tax credit available for investment in solar energy. Some solar water and space heating systems are competitive even without tax incentives, but only in the "sunbelt" (29). Mixed systems of solar plus conventional energy supplements are feasible everywhere in the United States. Some argue that the technological match of solar systems with the utilities is poor, and that most of the savings from solar energy come from the energy storage rather than the solar component (30). Others argue that the match is appropriate, especially when conservation is taken into account (31).

Since solar energy can be intermittent, it is always necessary to have backup systems as well as storage, or to disperse widely the components of the system (31). Flat plate collectors, as shown in Figure 15.10, oriented toward the sun can be up to 70 percent efficient in warm weather but are only about 10 percent efficient in colder weather because of conductive losses (32). Water is circulated by pumps through the collectors and then brought into a storage tank (generally in or adjacent to the basement). Performance can be improved, at additional cost, by circulating water (or some other fluid) in tubes inside evacuated glass pipes (32).

Such systems are known as *active solar systems*. Since these systems cost energy to build, it is reasonable to ask if a widespread switch to solar energy will involve so much new manufacture that there may be a cumulative energy debt, and if so, how long it will take to break even. There are indications that this payback period could extend beyond the year 2000 (33). Of course, new homes should last much longer than that.

Another system that uses energy but does not actively distribute it is called *passive solar*. Masonry walls exposed to the sun through a window during the day (Trombe walls) radiate the heat later, as the temperature drops, cutting costs for space heating. This works well, since people feel warmer at a lower temperature if they are in an environment in which the surroundings radiate heat. Some studies in West Germany have led to development of heating systems that heat interior masonry walls or concrete floors during the heating season (34). While it may be somewhat expensive to raise the temperature of the walls at the start of the heating season (stone has a substantial heat capacity), the temperature of the interior walls could be maintained at a modest cost. Thick-walled structures could absorb substantial amounts of heat during the daytime that would be released after sundown. Perhaps the medieval castles were more comfortable, when drafts were closed off by draperies, than we have traditionally supposed.

The distinction between passive solar and "conservation measures" is fading away as both become more common. The use of passive solar energy in new homes complements "conservation measures" in construction. Passive solar systems are described in much more detail in Chapter 19, which is concerned with energy "conservation."

Photoelectricity

Light-Caused Chemical Reactions   In the methods discussed so far, solar energy is converted to electricity through intermediaries or is converted to "lower quality" energy. One reaction that is direct is the *photoelectrochemical reaction*. Such reactions are chemical reactions in which light causes production of a fuel, such as hydrogen, which may subsequently be burned to produce energy. The most attractive such reaction is photodissociation of water into hydrogen and oxygen. Unfortunately, direct photolysis in the laboratory is not presently possible (35,36). To this point, there has not been much success in satisfying all the constraints of such a fuel cycle. There are, however, glimmerings of possibility (37): It has been found that small particles suspended in electrolytes liberate H_2 and O_2 under certain circumstances (37), and with high effi-

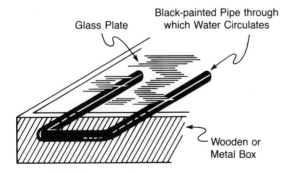

Glass Plate

Black-painted Pipe through which Water Circulates

Wooden or Metal Box

FIGURE 15.10
Flat plate solar collectors.

ciency if the "catalyst islands" cover only a small fraction of the surface (38). It also appears possible to use catalysts for the production of methanol from carbon dioxide. If any one of these possibilities is realized, it would mean that fuel could be produced by the sun and either piped to a central location or used locally to generate energy as needed.

Solar Cells   Direct conversion of sunlight to electricity is accomplished in *photovoltaic cells*. Photovoltaic (solar cell) research cost about $700 M between 1973 and 1984 (39). Partly as a result of this, the cost per watt dropped from $1000 before the space program in 1960 to about $10, which is still too high to compete with conventional energy sources where they are available. A cost of about $2 per watt (in 1984 dollars) appears necessary for there to be real competition with these established sources (40). However, for the 55 percent of the world population not connected to an electric grid in remote areas, the price now is competitive (41).

The physical basis for solar cells is the *photoelectric effect* (it was this for which Einstein actually won his Nobel Prize). It is the photoelectric effect that allows construction of the automatic door openers that work when you walk through a light beam. Experiments had been done in which both light frequency and light intensity were varied. It was observed that, if the frequency was below a certain characteristic value (a different frequency for each material), nothing was observed to happen no matter how intense the light was made. For frequencies above that critical value, electrons were observed near the surface of the material. If the intensity of the light was increased, many more electrons came away from the material. If the frequency was increased, the electrons moved away from the material with greater kinetic energy. This is shown in Figure 15.11. Einstein argued that this was evidence that light has particle nature (that is, for the existence of *photons*). These bundles of light carry energy proportional to their frequency. It takes a certain minimum energy (the ionization energy) to kick

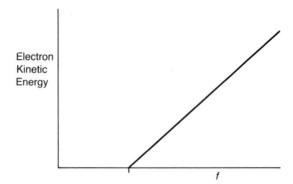

FIGURE 15.11
In the photoelectric effect, the electron kinetic energy increases as the incident photon frequency increases for the frequency above some critical value. If the photon frequency is below this value (i.e., the photon energy is too low), no electrons are released from the material.

an electron completely out of an atom. When atoms are bound together, electrons become associated with the whole material. An energy lower than the atomic ionization energy, known as the work function, is necessary to free an electron. If the energy of the photon is too low (i.e., the frequency is too low), then no matter how many photons hit the atom, it cannot remove the electron. If the energy of the photon is above the minimum value, an increase in the intensity increases the number of photons, which in turn increases the number of electrons that can leave atoms.

A useful analogy is that of a person at the bottom of a well. To get water out of the well, the person must raise the water at least as high as the top of well. Water that does not reach the top falls back in. If water *does* make it over the top, then the harder it was thrown, the further out it will go.

A simple experiment demonstrates this effect for ultraviolet light. You need a black light and an electroscope (as described in Chapter 4). If you charge the electroscope by induction with a piece of zinc having a freshly sanded surface and placed on top of the electroscope, then exposure to ultraviolet light will cause the leaves of the electroscope to fall gradually. The electrons that leave carry the extra negative charge away.

In the solar cell, the photoelectric effect can act in a similar way. Of course, in most materials, recombination of the electron with the ion is almost immediate. Only materials or devices in which it is possible to prevent recombination are candidates for solar cell building materials. Materials called *semiconductors* are used to make solar cells.

When atoms are stuck together in a solid, they often end up being separated from one another with regular spacings in all three directions, forming a *lattice*. Such a regular array of atoms is called a *crystal structure*. Because the atoms in the lattice influence one another, the electron energy levels of the atoms combine together in such a way that the possible energies of the electrons get spread out into "bands." Some bands are partially or totally full, and some are empty. In metallic materials, such as silver, the band is not totally full, and the existence of the band allows electrons to wander about wherever they are pushed within the material. Such materials are *conductors*. In other materials, known as *insulators,* when the bands are formed, all available levels in the band are filled. Since other, unfilled, bands are at a somewhat higher energy, no electrons are free to move unless they receive energy equal to the gap in energy between bands, which is known as the *band gap* (refer to Figure 15.12).

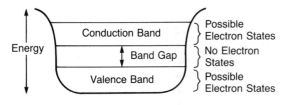

FIGURE 15.12
Electron bands in a solid. The valence band lies below the band gap and the conduction band lies above the band gap. Acceptor states would lie just above the valence band. Donor states would lie just below the conduction band.

In semiconductors, the filled band (*valence band*) and the band in which electrons are free to move (*conduction band*) are separated by a potential difference of about 1 volt. Hence, light coming in can push an electron from the valence band into the conduction band if it has an energy of about 1 electron volt. The electron in the conduction band is free to move. If it is kept from recombining, it can give up its energy in an external circuit before coming back to the material.

To prevent recombination, two different types of *doped semiconductor* are grown together to make the solar cell. Pure silicon is grown in a furnace in the presence of silicon vapor. The silicon vapor is doped with acceptors or donors (*p*-type and *n*-type semiconductors) to deposit layers of *p*-type or *n*-type material. On occasional lattice sites in *p*-type semiconductors are atoms or compounds with *fewer* electrons in the outermost shell than in the rest of the atom. Thus, there are occasionally "vacancies," or holes for electrons, within the lattice of a *p*-type semiconductor. An *n*-type semiconductor has occasional lattice sites occupied by an atom or compound with *more* electrons in the outermost shells than in the rest of the atoms. There are occasionally excess electrons in the lattice in *n*-type semiconductors. The extra electrons can move around in the *n*-type material in response to an external potential; in the *p*-type material, the holes move around in response to an external potential. Such materials therefore conduct electricity better than would be guessed from the band gap.

When two different types of semiconductor are bonded together, current will flow in response to an external potential in only one direction. Such devices are called *diodes*. Solar cells act in a way similar to the diode, so that current flows in only one direction when the cell is exposed to light. A typical cell cross section is shown in Figure 15.13a–c.

Energy in excess of the band gap is delivered as heat to the solar cell (42). It is necessary to match the band gap to visible light in order to minimize this heating. It is possible to use ma-

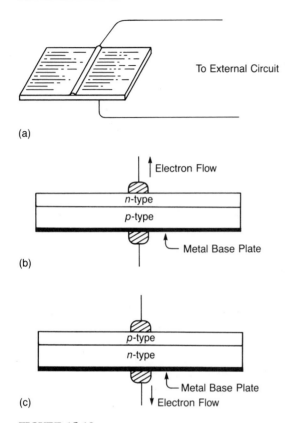

FIGURE 15.13
(a) Drawing of a solar cell. (b) Connections for a *p*-type solar cell. (c) Connections for a *n*-type solar cell.

terials with band gaps of 1 to 2 electron volts (corresponding to the size of the visible spectrum), but a band gap of about 1.5 electron volts is optimum for a solar cell (35).

The most popular choice for solar cells is silicon, with a band gap of 1.1 electron volts, production cell efficiencies of about 12 percent (43), and a maximum efficiency of about 15 percent, and gallium arsenide, with a band gap of 1.4 electron volts and a maximum efficiency of about 22 percent. Both of these materials must be grown as single crystals under very precisely controlled conditions to minimize imperfections, which can cause recombination (42). Both types are easily eaten away by chemical reactions with the holes (41). ARCO (Atlantic Richfield) is one of the com-

panies involved in silicon photocell development. Its subsidiary built a 1 MW solar facility at the Lugo substation near Hesperia, California, which sells electricity to the local utility (Figure 15.14).

Materials that have no crystal structure are classed as *amorphous,* from the Greek, meaning "lack of structure." Amorphous silicon has no crystal structure, and its atoms are ordered over only a very short distance; small pieces of silicon crystal abut one another at random orientations in such a way that no long-distance structure exists. Amorphous silicon solar cells in thin films exhibit better absorption than pure silicon, but because of the many structural defects, they are only about 11 percent efficient at maximum, and most cells are about 4 to 8 percent efficient (43). Amorphous silicon is much easier to make than grown silicon crystals, and by using several layers, each set for a different band gap, a greater part of the visible spectrum can be used (43). For this reason, it is the material of choice. Japanese companies are mounting a challenge to U.S. manufacturers with plans to produce amorphous silicon solar cells at $3 to $4 per watt. In response, many U.S. companies have abandoned the field (40,44). As the experience of the last decade leads to ever-cheaper solar cells, we can expect to see mass market penetration by the 1990s. Solar energy strikes a resonant chord in many people because it promises independence from the power companies and the government. When it does become cheap, there may be an explosion of demand even if tax incentives are lacking. Unfortunately, U.S. industry may not reap the benefits of its investment because of Japanese penetration of the U.S. market (45).

We should note that all the aforementioned solar energy devices (as well as ones not explicitly mentioned, such as using the sun to dissociate water into hydrogen and oxygen) involve no *net* addition of heat to the atmosphere. There is, rather, a redistribution of heat from one place to another. Most conventional energy generation involves net additions of heat, which can affect climate in local areas, raising the possibility that we may be tampering with the Earth's overall

FIGURE 15.14
The ARCO solar installation at the Southern California Edison substation at Hesperia, California.

heat balance (see Chapter 24). This could of course happen even with redistribution, as mentioned in the discussion of ocean thermal gradient power.

Local climatological effects include increased rainfall and snowfall near generating facilities (46). Cities, with their high energy use levels, are consistently warmer (they are referred to by some as "heat islands") than the surrounding countryside the year around. The Los Angeles basin ultimately bears a local heat burden that is over 5 percent of the total solar energy absorbed by the basin! This heat burden will continue to grow.

LARGE-SCALE SOLAR ENERGY PROJECTS

A technology that could change the world's heat balance if it were widely adopted is the use of large fields of tracking mirrors (*heliostats*) to focus sunlight onto central towers. A pilot plant at Daggett, near Barstow, California, built by Southern California Edison, with the support of Sandia Labs, the Department of Energy, the Los Angeles Department of Water and Power, and the California Energy Commission (47–51) has been named "Solar One." Solar One does not attempt to compensate for the additional heat absorbed

by the system. Such a project is clearly not meant to appeal to an individual user, but rather to mesh with the present character of the publicly regulated electric utility industry.

Solar One has been generating electricity since 1982 (46–50). The best performance to date for a single day was a net of 80 MWh during June 1985. Figure 15.15 (48) shows the monthly operating experience through the end of 1984. The heliostats (Figure 15.16) reflect and focus sunlight onto the nearly 100 m tall tower (Figure 15.17). The tower's absorber panels, although they appear to be white in Figure 15.17, are painted black and absorb 88 percent (96 percent when newly painted) of the incident light (17). The heliostat field (Figure 15.18) is oriented mostly toward the north to capture sun from the south, and the heliostats to the south (to the right on the photo) are focused on six panels to preheat the water that goes to the eighteen superheat panels, which receive most of the sunlight. Water leaves the superheat panels at 510°C and is at a hundred times atmospheric pressure (17,46,47).

The hot water goes either to turbines, where it generates electricity at 35 percent thermodynamic efficiency, or it is sent to a heat exchanger, where it heats oil that is then sent to a thermal storage tank. The tank has a volume of about

FIGURE 15.15
Monthly energy production of Solar One (1984–85). (Data courtesy of Southern California Edison)

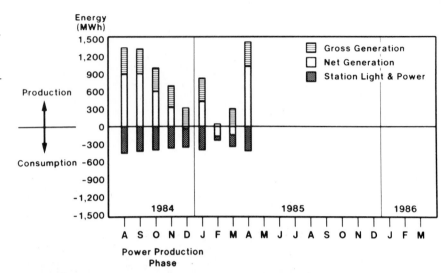

FIGURE 15.16
Heliostats at Solar One in Daggett, California.

4230 m³, and contains 4120 tonnes of crushed granite and 2060 tonnes of sand (49) through which the oil circulates. There is a temperature gradient in the tank (the idea is from Rocketdyne, and is proprietary) from the low temperature (244°C) to the high (entry) temperature (304°C). The heat from the tank can be drawn back through the heat exchanger to produce steam for the turbine at 274°C (49). The thermal storage system

is important and cost-effective because it allows load-shifting, thus increasing profits. Off-peak electricity is worth 5.7 cents/kWh, while peak electricity is worth 12 cents/kWh in winter and 14 cents/kWh in summer to Southern California Edison. In addition, the thermal storage allows buffering for passing clouds and so can keep the plant operating through changes in weather conditions—the plant needs to be up at least two

FIGURE 15.17
The tower at Solar One
in Daggett, California.

FIGURE 15.18
A view of Solar One
from the air. (U.S. De-
partment of Energy)

hours per day to make it cost-effective to start up the pumps (47).

Solar One has proved that the water/steam cycle is reliable, that the system can meet expectations, and that thermal storage is cost-effective. The computer control system, partly developed in use on Solar One, is now used in other utility control rooms as well (47). Future versions may incorporate provisions to keep the heat balance over the collector area the same as

it would have been without the concentrating mirrors.

Adjacent to the Solar One site in Daggett is a facility, the $62 M Solar Electric Generating System I, or SEGS I, built by Luz International, Ltd. SEGS I is the largest commercial solar energy project in the world and produces 13.8 MW_e. The system has 72,000 m^2 of parabolic trough collectors (Figure 15.19). The black collector lines (they appear bright in the photos) carry oil that is heated

FIGURE 15.19
The Luz solar facility (SEGS-I) at Daggett, California.

FIGURE 15.20
Storage tanks for hot oil at the Luz solar energy facility in Daggett, California.

and runs into storage tanks (Figure 15.20). The hot oil is used to produce steam, which is raised to a temperature of 415°C with a gas superheater to drive a turbine generator. The investors have added SEGS II, with another 165,000 m² of collectors, at a cost of $93 M. SEGS II produces 30 MW$_e$.

In the next two chapters, we consider the one important form of solar energy not addressed in this chapter: biomass. Nature has solved the problem of creating convenient fuels from sunlight.

SUMMARY

Light energy is emitted in the form of electromagnetic waves, which travel at the speed of light, 300,000 km/s. The product of the wavelength and frequency of the wave is the speed of the wave. From longer to shorter wavelengths, we speak of radio waves, microwaves, infrared light, visible light (from ≈700 nm to ≈400 nm), ultraviolet light, x rays, and gamma rays.

The sun is the source of almost all the world's energy, and the source of the sun's energy is fusion inside the sun. The sun radiates energy from its surface characteristic of a blackbody at a temperature of 5800 K, the sun's surface temperature. Some of this radiation impinges on the Earth and is absorbed. Because the Earth is at a nonzero temperature, the amount of energy it absorbs from the sun is exactly balanced by the amount of energy it radiates away.

Solar energy that reaches the Earth drives the winds and ocean currents or is used by plants. The winds and the ocean can be tapped for energy. Hydroelectricity is stored solar energy, as is the fresh water–salt water interface where rivers flow into the ocean. Solar energy can be used to heat water to be used in the home, to fall on photovoltaic cells to produce electricity, to fall on chemical solutions to produce electricity, or to be concentrated to make ultrahot water to generate steam (as in Solar One).

REFERENCES

1. S. Wieder and E. Jacobi, *Am. J. Phys.* 45 (1977): 981.
2. A. L. Hammond, *Science* 179 (1973): 1116.
3. S. Blakeslee, *New York Times,* 4 Dec. 1983.
4. S. Bronstein, *New York Times,* 23 June 1985.
5. L. Fisher, *New York Times,* 26 Oct. 1985.
6. M. R. Gustavson, *Science* 204 (1979): 13.
7. S. N. Paleocrassas, *Solar Energy* 16 (1974): 45.
8. W. E. Heronemus, in *Perspectives on Energy,* ed. L. C. Ruedisili and M. W. Firebaugh (New York: Oxford Univ. Press, 1975): 364.
9. D. R. Inglis, *Environment* 20, no. 8 (1978): 17.
10. J. Constans, *Marine Sources of Energy* (New York: Pergamon Press, 1979). This is the report of the EUROCEAN group, presented to the United Nations Office of Science and Technology. It investigates the engineering feasibility of alternative sources of energy.
11. B. Sorensen, *Am. Sci.* 69 (1981): 500.
12. M. F. Merriam, *Ann. Rev. Energy* 3 (1978): 29.
13. P. M. Moretti and L. V. Divone, *Sci. Am.* 254, no. 6 (1986): 110.
14. R. Kahn, *Alt. Sources Energy* 81 (1986): 44.
15. G. G. Piepers, *Alt. Sources Energy* 81 (1986): 40.
16. R. J. Smith, *Science* 207 (1980): 739. See also J. W. Shupe, *Science* 216 (1982): 1193. Since 1984, wind energy has supplied about 10 percent of Hawaii's energy needs.

17. J. Reeves, Alternative Energy Project Manager, Southern California Edison, private communication.

18. C. Garrett, *Endeavour* 8 (1984): 58.

19. A. B. Meinel and M. P. Meinel, *Phys. Today* 25, no. 1 (1972): 44.

20. J. H. Anderson and J. H. Anderson, *Mech. Engrs.,* April 1966, p. 41.

21. D. F. Othmer and O. A. Roals, *Science* 182 (1973): 121; and C. Zener, *Phys. Today* 26, no. 1 (1973): 48.

22. M. W. Browne, *New York Times,* 10 Feb. 1987.

23. J. Lear, *Saturday Review,* 5 Dec. 1970, p. 53.

24. R. S. Norman, *Science* 186 (1974): 350.

25. J. D. Isaacs and W. R. Schmitt, *Science* 207 (1980): 265.

26. C. Nielsen, private communication. Prof. Nielsen was a consultant for the first commercial use of a salt-gradient solar pond: for heating a public swimming pool in Miamisburg, Ohio.

27. A. I. Kudish, D. Wolf, and Y. Machlav, *Solar Energy* 30 (1983): 33.

28. T. H. Maugh II, *Science* 216 (1982): 1213.

29. R. H. Bedzak, A. S. Hirschburg, and W. H. Babcock, *Science* 203 (1979): 1214.

30. J. G. Asbury and R. O. Mueller, *Science* 195 (1977): 445.

31. E. Kahn, *Ann. Rev. Energy* 4 (1979): 313.

32. A. L. Hammond and W. D. Metz, *Science* 201 (1978): 36.

33. C. Whipple, *Science* 208 (1980): 342.

34. H. H. Wiechmann and Z. Varsek, *Ziegelindustrie International,* May 1982, p.307.

35. V. Balzani et al., *Science* 189 (1975): 852.

36. J. R. Bolton, *Science* 202 (1978): 705.

37. T. H. Maugh II, *Science* 222 (1983): 151.

38. A. Heller, *Science* 223 (1984): 1141.

39. E. A. DeMeo and R. W. Taylor, *Science* 224 (1984): 245.

40. C. Norman, *Science* 226 (1984): 319.

41. T. H. Maugh II, *Science* 221 (1983): 1358.

42. A. Wilson, *Alt. Sources Energy* 81 (1986): 8.

43. E. A. Perez-Albuerne and Y.-S. Tan, *Science* 208 (1980): 902.

44. T. J. Lueck, *New York Times,* 16 Oct. 1983.

45. M. L. Kramer et al. *Science* 193 (1976): 1239.

46. S. Bronstein, *New York Times,* 23 June 1985.

47. C. Lopes, Research and Development site manager, Solar One, private communication.

48. I. R. Straughan, *Middle Management Perspectives Program,* Southern California Edison, 1985.

49. J. J. Bartel, "Solar 10 MW$_e$ Pilot Plant Fact Sheet," (New Mexico: Sandia Labs, 1984).

50. F. Kreith and R. T. Meyer, *Am. Sci.* 71 (1983): 598.

51. J. Thornton, presentation delivered at the Southern Ohio AAPT Section Meeting, Columbus, Ohio, 4 May 1984. A good source for general information on this and all the solar energy alternatives discussed in this chapter is L. Hodges, ed., *Solar Energy, Book II* (College Park, Md.: AAPT, 1986).

PROBLEMS AND QUESTIONS

True or False

1. The wind can supply a large proportion of the energy used in the United States.
2. Photovoltaic cells can be up to 50 percent efficient.
3. Sea temperature gradient energy, because of its low efficiency, can never be a major source of energy.
4. Solar ponds trap solar thermal energy in a layer of fresh water underneath a layer of salt water.
5. A tidal energy installation with a tidal range of 4 m can produce twice the energy of an installation with a tidal range of 2 m.
6. Wind energy is currently competitive with conventional energy sources.
7. Energy from the tides could supply a substantial proportion of the world's energy needs if it were all developed.
8. The longer the wavelength of a wave, the faster the wave travels in space.
9. The mean wavelength of energy emitted by Earth is longer than that emitted by the sun.
10. The Public Utilities Regulatory Policies Act of 1978 has spurred development of alternative energy sources.

Multiple Choice

11. Which of the following statements is *false*?
    a. The sun's interior experiences temperatures in excess of 10 million °C.
    b. As one of nine planets, the Earth absorbs about a ninth of the total energy emitted by the sun.
    c. The sun produces light in the ultraviolet and infrared regions of the electromagnetic spectrum in addition to visible light.
    d. The Earth radiates roughly as much energy to space as it absorbs from the sun.
    e. An object placed in the sun will not be able to have its temperature raised indefinitely because it will begin to radiate as it is heated.

12. Which of the following is *not* a form of radiant energy?
    a. Ultraviolet radiation
    b. Gamma radiation
    c. X radiation
    d. Alpha radiation
    e. Microwave radiation

13. Which of the following orders the electromagnetic spectrum in terms of increasing wavelength?
    a. Radio waves, ultaviolet, microwaves
    b. Visible, ultraviolet, gamma rays
    c. Infrared, microwaves, x rays
    d. X rays, visible, microwaves
    e. Gamma rays, ultraviolet, x rays

14. Two blackbodies are at temperatures of 300 K and 600 K. What is the ratio of the power emitted by the hotter body to that emitted by the cooler body?
    a. 2
    b. 4
    c. 6
    d. 8
    e. 16

15. Salt-gradient solar ponds
    a. can be used for energy only in the tropics.
    b. can supply space heat for residences.
    c. have water layers separated by sheets of plastic or glass.
    d. must have the bottom surface painted black in order to be able to absorb sunlight.
    e. get energy from the mixing of salt water and fresh water.

16. Solar cells can be made to work
    a. on any electromagnetic radiation.
    b. only with green light.
    c. with materials in which all electrons are in the conduction band.
    d. with materials in which all electrons are in the valence band.
    e. with materials for which the band gap is about 1 to 2 eV.

17. In the photoelectric effect,
    a. electrons are emitted for all intensities of incident light.
    b. electrons are emitted for all frequencies of incident light.
    c. for some high frequencies and low intensities, no electrons are emitted.
    d. for some low frequencies and high intensities, no electrons are emitted.
    e. electrons are absorbed to produce very intense beams of monochromatic light.

18. Which of the following applications of solar energy is not subject to commercial exploitation?
    a. Solar energy reflected by a field of space-borne mirrors onto Earth's surface
    b. Solar energy absorbed in solar cells and fed into the power grid
    c. Solar energy captured in parabolic troughs, heating material carried at the focal point to drive turbines
    d. Solar energy reflected by a field of heliostats onto a central tower, heating a fluid to drive turbines
    e. Solar energy used in residences for water heating

19. The 1985 cost of amorphous silicon solar cells is $4 per installed watt. This is a factor of roughly 100 times cheaper than the first solar cells of the 1950s.
    a. Such a cost makes solar cells available only for essential uses, such as in the space program.
    b. Solar cells are currently competitive as energy sources for uses in remote areas.
    c. Solar cells will be commercially competitive only when the cost falls to less than 10 cents per installed watt.
    d. Since solar cells provide DC power, energy from solar cells cannot be absorbed by the current energy distribution system.
    e. Development of solar cells to provide AC power directly is under way.

20. A solar cell does not respond to red light, but does respond to green light. The solar cell will be able to respond to
    a. microwaves.
    b. infrared light.
    c. ultraviolet light.
    d. all of the sources listed above.
    e. none of the sources listed above.

21. In water, the speed of light is about 200 million m/s. Electromagnetic radiation (radiant energy) of frequency 5×10^{14} Hz travels through water. The wavelength of this electromagnetic radiation is
    a. 400 nm.
    b. 2000 nm.
    c. 10^{23} m.
    d. not determinable from the information given.
    e. not given by any of the above choices.

22. A sniperscope is able to detect some objects it illuminates with infrared light by detecting the reflected light. It is also possible for an observer to see objects that are not being illuminated because of the
    a. chemiluminescent character of the objects.
    b. very low temperature of these objects (a few kelvin).
    c. emission of photoelectrons by the objects.
    d. high amount of emission by the objects in the infrared.
    e. high amount of emission by the objects in the ultraviolet.

Discussion Questions

23. Why should anyone be worried about the Earth's heat balance?

24. What are the characteristics of areas in which solar water heating is economically feasible?

25. Why haven't sea wave power installations such as Salter's duck already been constructed? Would you expect development to happen soon? Why or why not?

26. If the internal temperature of the star is 10 MK, why is the surface temperature only 5800 K? How is heat transferred to the stellar surface from the interior?

27. The Meinels' "solar farm" takes up a lot of space. Get a feeling for how much energy is produced by comparing this area (5000 mi²) to our area in farmland and its productivity (see Chapter 17). The Meinels' farm would produce 10^6 MW_e of continuous power, i.e., about 8 TkWh$_e$ per year.

16

BIOMASS ENERGY

In the last chapter, we saw several different methods for utilizing solar energy. This chapter and the next focus on solar energy and the production of biological organisms (biomass). Chapter 16 deals mainly with the energy and chemical resources in biomass.

KEY TERMS
solar window
photosynthesis
chlorophyll
chloroplast
reaction center
adenosine triphosphate (ATP)
C_3 plants
C_4 plants
ethanol
anaerobic digestion
bagasse
feedstock
jojoba
guayule
cassava

The sun has shone on the Earth for about 5 billion years. The original atmosphere was composed of hydrogen, helium, nitrogen, methane, ammonia, and some carbon dioxide. By 3 billion years ago, the hydrogen and helium had gone, and life was beginning in the oceans. It was only because there was available solar energy to keep the Earth warm and to supply energy for chemical processes that life was able to develop.

It should not be surprising that plants have succeeded at using and storing solar energy over the course of 3 billion years, while humanity has not yet been successful at capturing and storing solar energy for use at later times. Without plants, animals could not exist in their present forms. The original plants were able to use the relatively abundant carbon dioxide (CO_2), methane (CH_4), water (H_2O), and sunlight to produce sugar-like compounds and oxygen. The sugars are food for animals, and without oxygen, animals could not metabolize the sugars. The oxygen atmosphere, with its protective 2 billion-year-old ozone layer, is an artifact of life on Earth. Since oxygen is a byproduct of plant photosynthesis, new forms of plants had to arise as adaptations to new conditions. Many of the old bacteria are still with us. These remnants of the early Earth are responsible for *anaerobic processes* (processes that take place in the absence of oxygen). Bacteria were also responsible for a decrease in atmospheric carbon dioxide, a great amount of which was processed through the oceans and deposited as limestone.

SOLAR WINDOW

The atmosphere, as we shall see in Chapter 23, has a "window" allowing solar radiation in a restricted range to arrive on Earth: About 300 nm to 700 nm is clear for transmission of all solar energy (which is why this is the visible part of the spectrum); there are also some regions in the range from 700 nm to ≈3000 nm that allow some near-infrared from the sun through the atmosphere. If you were going to design a solar energy receptor, you would probably just want to use the wavelength range 300 nm–700 nm, since longer wavelength infrared radiation would cause heating, which could interfere with the chemical reactions you would like to use in photosynthesis. A leaf would evolve to avoid wavelengths from 700 nm to 3000 nm. It is interesting that real leaves do exactly this (1): Figure 16.1 shows the absorptance of the leaf of the eastern cottonwood (*Populus deltoides*). The absorptance peaks in the infrared are caused by

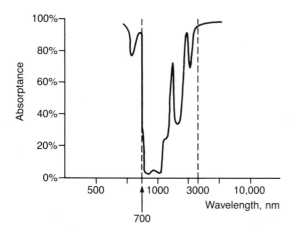

FIGURE 16.1

Absorptance (percentage of incident radiation absorbed) of *Populus deltoides* as a function of wavelength of the incident light. (C. G. Granqvist, copyright 1981 by Optical Society of America)

water in the leaf and correspond to gaps in transmitted solar radiation. Little heat is therefore absorbed in leaves. The broad peak in the visible region is due to absorption by the cells of the leaf.

PHOTOSYNTHESIS

Photosynthesis involves the chemical reaction of water and carbon dioxide to produce oxygen and organic materials such as sugars. As emphasized in Chapter 5, photosynthesis still holds mysteries for us. Nevertheless, during the course of the activity, three processes are known to occur: Membranes called *chloroplasts,* containing chlorophyll, absorb light; there is a separation of charges across the membrane; and electrons are donated or accepted on the sides of the membrane (2). Photosynthesis takes place in bacteria as well as in plants and algae, but plants and algae release oxygen and fix carbon, while bacteria do not (3).

Chlorophyll molecules of two different sorts exist in the chloroplast in plants and algae. The chlorophyll molecules can absorb the blue and the red light from incident sunlight and so appear to be green (because we may think of white light as the sum of blue, green, and red light, and because the color of nonluminous objects depends on the color of the light reflected from them). The energy of the incoming photons is collected in a reaction, or "trapping" center, in which water is split, oxygen is released, and energy is stored in intermediate compounds. The energy stored in the compound adenosine triphosphate (ATP) by the one sort of chlorophyll is used to make sugars; animals also use ATP in their internal energy economy (2,3). The energy stored in the compound nicotinamide ademine dinucleotide (NAD) by the other sort of chlorophyll is used to store hydrogen (2,3). Energy is exchanged between the two sorts of chlorophyll, as well as electron flow. It is this "light" part of the reaction that is the most mysterious, especially the one involving the splitting of water and the synthesis of ATP.

The rest of photosynthesis consists of a "dark" chemical reaction, one not using light energy, in which CO_2 is combined with water into sugars, using the energy from the ATP already produced.

Photosynthesis is made difficult because of the presence of oxygen, which inhibits photosynthesis, but it uses about 42 percent of the incident sunlight at about 700 nm. Only about two-thirds of the sunlight is of the appropriate wavelength, that is, the wavelength corresponding to half the energy necessary to drive the reaction. Sunlight of lower wavelength (higher energy) is absorbed as well. The excess energy is released as heat. As a result, the efficiency drops to about 27 percent averaged across the visible region, which in turn is only 44 percent of available sunlight (3). Hence, plant photosynthesis is only 12 percent efficient per se in transforming radiant energy to chemical energy, not counting losses due to plant metabolism (the taking in of carbon dioxide and the emission of oxygen). Assuming 10 percent loss to reflection, the overall efficiency cannot exceed 10 percent for conversion of solar energy to chemical energy (3).

Photosynthesis rates are limited by the small proportion of CO_2 in the atmosphere (3). Increased CO_2 concentration causes more rapid plant growth. Efficiency varies from species to species, being greater for corn or sugar cane than for wheat or soybeans (these are representative of species of plants utilizing different types of photosynthesis, called C_4 and C_3, respectively). A study of a complete ecosystem in Hubbard Brook, New Hampshire, Experimental Forest reveals that the actual year-round efficiency is only about 0.8 percent (4). Of course, plants grow only in the summer months, so that actual efficiency is somewhat higher. Sixty years after it was logged over, the forest is still accumulating about 628 kJ/m² each year (4). When the forest finally matures, the energy flow should be in balance.

For most human purposes, production should be out of balance in order to enable us to produce food for people and animals. About 1.5 Gtonne of grain is produced each year, of which 218 Mtonne is traded (about half is from the United

States alone) (5). As we saw in Chapter 2 and will investigate further in the next chapter, this energy is vital to our well-being. This chapter focuses on biomass as used for purposes other than feeds.

BIOMASS FOR FUEL

Production of biomass is a desirable way to use land. It is one of the few methods of energy production that does not arouse much public opposition. Biomass is renewable. Agricultural fields produce every year, and even trees can grow over periods of thirty to a hundred years. Biomass also is well placed to take advantage of the ambient conditions. Plants grow almost everywhere, despite the fact that they spend half their lives in darkness. Plants are automatic energy storage devices, transforming solar energy to plant matter, and are well suited to the intermittent availability of sunlight. Some algae can convert nearly 10 percent of available light energy to biomass. Sugar cane is about 2 percent efficient, and corn about 1 percent efficient. Most other cultivated plants are less efficient than the Hubbard Brook forest in utilization of solar energy. Plants are not as efficient as the most efficient solar cells.

Given all of these advantages, it is natural to ask whether we could use some of the stored solar energy as a fuel for transportation, heat, or electricity. The answer is clearly yes. Many of us buy gasoline with ethanol mixed in. Wood provides about half as much energy as is generated in nuclear plants (6), perhaps as much as 5 percent of the total U.S. energy supply (7) and may be cheaper than alternatives. A small Kentucky college was able to save some $221,000 a year by burning wood, instead of gas, to heat dormitories (8). In lumbering areas, such as Oregon, power plants burn "hog fuel," which is wood residue, to produce energy. The Eugene Water and Electric Board, a municipal utility in Eugene, Oregon, can generate up to 34 MW or produce about 1000 tonnes of steam per hour to be used by local industry or for local heating (9). The

McNeil Generating Station, in Burlington, Vermont, produces 50 MW (7,10). Bagasse (woody residue) from sugar cane supplies about 7 percent of Hawaii's energy needs (11).

Sugar cane is also involved in an ambitious Brazilian project to produce ethanol in order to reduce dependence on imported oil for automobiles and trucks. Brazil has few oil reserves and, in response to the oil price rises of 1973 and 1979, implemented more stringent measures than were taken in the United States. In addition, research had determined that less than 2 percent of Brazil's land area grown in sugar cane could produce enough fuel to replace all imported petroleum (12). All gas sold in Brazil now is 15 percent ethanol; the country is going forward with a program involving research on producing ethanol from sugar cane, cassava, and sorghum (13). Table 16.1 presents their results, showing that cassava is probably not going to be useful in the Brazilian context.

The energy question is more open in the United States because the higher cost of labor invites the substitution of energy for labor (14). Production of ethanol from wastes has worked in the United States. The notorious Al Capone distilled bootleg liquor from Chicago garbage during Prohibition (15). One analysis shows that distillation of ethanol from corn (or other grains) can produce disparate results. At best, it can produce a *gain* equivalent to one gallon of gasoline if coal, biomass, or solar energy is used in the distillation process; the ethanol is added to gasoline; and the distiller's grain (left-over mash) is used as animal feed. At worst, it can produce a *loss* of as much as a quarter of a gallon of gasoline when oil or gas is used in distillation; the ethanol is used straight; and the distiller's grain is not used (16).

New technologies for distillation are under active investigation. It is of little importance whether distillation shows a net energy gain or loss when the resulting ethanol is marketed as a drink for people, but it is of crucial importance in considering the use of ethanol as a replacement for all or part of our gasoline use. Techniques showing promise include using cellulose to absorb water, using permeable membranes to separate alcohol and water, using CO_2 at high pressure to take up water, and using molecular sieves (zeolites, working by ion exchange) to absorb water from alcohol (17).

Given that it is possible to make alcohol from agricultural products, should we do it on a large scale? We need to study the feasibility of moving to a totally solar-derived energy economy. Actually, such studies have been done by the Review Panel on Biomass Energy, chaired by Professor D. Pimentel of Cornell University (5), by Professor Pimentel and collaborators (18), and by others (16,19). They conclude that there is scant chance of biomass replacing total energy as solid, liquid, or gaseous fuel in the United States.

In 1979, energy from biomass totaled about 2 percent of U.S. energy use, roughly the same amount as in 1850, when biomass produced over 90 percent of our energy (5). The most important categories of source are shown in Table 16.2 (5).

TABLE 16.1
Analysis of Brazilian net energy gain for selected crops.

Crop	Produced/Consumed	Gross Gain	Net Gain
Sugar cane	4.53	2.43	1.43
Cassava	1.71	1.16	0.16
Sorghum			
Grain and stems	3.39	1.89	0.89
Stems only	2.27	2.01	1.01

SOURCE: Reference 13, copyright 1978 by AAAS.

TABLE 16.2
Net biomass energy resources.

| Source | Potential | | Current Use | |
	Biomass Total (Mtonne)	Total Available (Mtonne)	Amount (Mtonne)	Heat Energy (GkWh$_t$)
Wood (total)	651	261		
Mill residues (forest)	135	118	58–75	207–214
Logging residues	164	45	3.3	12
Thinnings	43	43	2.2	9
Residential fuel wood	27	27	27	59
Mortality, excess over harvest	282	27	—	—
Forage	682	118	—	—
Animal wastes	159	45	1	0.3 (biogas)
Grains	321	9–20	1	2.4 (liquid)
Bagasse	3.6	3.6	3.6	12
Food-processing waste	14	14	—	—
Industrial wastes	90	21	7.1	15
Municipal solid wastes	124	60	—	—
Municipal sewage	12	9	—	5.7 (biogas)*
Aquatic plants	18	3–12	—	—
Crop residues	430	73	—	—

*Energy is used in sewage digestion, so is not net energy. Reprinted with permission from *Solar Energy* 30 (1983): 1, Review Panel on Biomass Energy, "Biomass Energy," copyright 1983, Pergamon Journals, Ltd.

We consider sources in the order listed in the table.

Wood Costs of wood collection could exceed energy gained (5). Biomass growth in forests and growth in commercial forests exceeds 2×10^{12} kWh$_t$ per year (19). Sustainable yields could be increased to about 1.1 percent of current fossil fuel production by 2000, but such increases might be accompanied by severe erosion on steep hillsides when residues are taken (9,18). It would take 634,800 hectares of forest to supply a 100 MW$_e$ plant (18). Since wood is burned at low temperature, emissions of sulfur and nitrogen oxides are low (9,16). However, wood stoves have caused local pollution problems, especially when many are used in valleys, and a natural inversion layer traps the smoke (see Chapter 22).

Forage Some land now used for forage could be used for crops instead, or the forage itself

could be used to produce energy by burning, gasification, or conversion to alcohol. It would be necessary to increase the price of forage, which would increase the cost of meat. It would also be necessary to fertilize fields to a greater extent than is now done (15,18), but the ground cover after harvest will prevent erosion.

Animal Wastes Much waste (about 60 percent) is dropped on the range, where it is not readily retrievable for agricultural fertilizer. However, feedlots are concentrated sources of olfactory pollution and water pollution from runoff. Fermentation of manure to methane and fertilizer by use of anaerobic bacteria could help clean up the environment. A plant in Guymon, Oklahoma, sends methane into the gas pipeline (20). A proposal has been made to set up a closed-system dairy farm to use gas and produce cattle feed from the digested waste. A herd of a hundred cattle could get about 66 percent of their needs recycled in summer (52 percent in winter) (5). A jokester once proposed tapping the flatulence of cattle; a cow can produce 100 to 500 liters of methane per day (16).

Grains   In 1980, a million tons of corn supplied 100 million gallons of ethanol. It would take 10 percent of the yearly corn yield to supply only ≈1 percent of gasoline consumption (5). Pure replacement seems doubtful. If much corn were used to produce ethanol, the demand would rise, causing corn prices to rise (and the cost of meat as well, since much corn is used as animal feed). This has caused controversy among people concerned about world food shortages. Grain that fuels cars cannot feed people. Historically, however, the United States has had an oversupply of food, and much rots in storage. Fermentation of spoiled stored grains would seem a feasible and useful alternative.

Bagasse   Bagasse is the woody residue of sugar cane. This resource is being used in Hawaii, Brazil, and the Caribbean. The amount is limited, and this is not expected to be a major source of energy elsewhere.

Food Processing Waste   Very little waste is now being processed. Such processing would be of environmental benefit, since it would reduce the volume of waste products to be disposed of. Treatment is similar to that of animal wastes. A plant in Florida to produce gas from food waste proved uneconomic (21).

Industrial Waste   Obtaining energy from industrial wastes is often problematic because of contamination in the waste by toxic substances, especially heavy metals. Nevertheless, some energy (mostly in the form of paper) is currently being recovered.

Municipal Solid Wastes   The organic material in garbage is capable of generating methane of heat content from 7450 to 22,350 kJ/m³ (21). There are currently about twenty-five landfills in the United States tapping this gas (20). Some municipal power facilities burn wastes (see Chapter 18).

Municipal Sewage   Sewage treatment plants yield about 1 m³ of digester gas per kilogram of volatile solids. Many plants utilize this gas *in situ,* so it makes no net contribution (Table 16.2), though it is an alternative energy source for the treatment plants.

Aquatic Plants   Plants are mostly water, and as such can produce energy when fermented (see box on duckweed). Such plants can be used to concentrate the heavy metals in waste (5).

Crop Residues   If too much ground cover is gathered, problems can result from fertilizer replacement, increased erosion of soil, and probable net energy cost rather than gain. Research has indicated that some wastes can be gathered with the harvest and used to dry the grain with no adverse effects. In Illinois and Indiana, delivered corn residue is more expensive than coal before processing (5). The Review Panel on Biomass Energy (5) concluded that potentially adverse environmental consequences of current crop management practices, even under ideal cropping conditions, are so serious as to draw into question the use of residues for biomass energy production until crop production technologies are improved.

Biomass For Feedstocks

Large quantities of the petroleum and natural gas used in the world are consumed to make chemicals (that is, they are feedstocks). There are many reasons to believe that biomass is currently better suited to production of chemicals than to fuels (23,24). Fuel production is very large, but the chemical production scale matches biomass availability rather well. The problems of fuels are not those of chemicals, and the oil price rises of the 1970s have engendered a shift in costs. In 1970, ethylene cost half as much per kilogram as

DUCKWEED (22)

Rushes and reeds have been used to clean water for some time. Other plants, such as water hyacinth and duckweed, can remove organic and inorganic materials from treated water. They also concentrate heavy metals. Duckweed can grow very fast. It sends out fronds, which produce more duckweed, producing more fronds, and so on (22). Duckweed is a good bet because it can grow even in brackish or polluted water, as long as the water is still (22).

Duckweed flows easily in water, and the plants are small enough to be pumped through pipes (22). They do not need to be chopped to be fermented as hyacinths do. Duckweed can be mixed with other materials or fermented by itself.

If one used duckweed with a one-hundred-head dairy farm, fermenting the methane for energy and feeding the duckweed on the holding lagoons directly to the cattle, 60 percent of the protein requirement could be met in summer (15 percent in winter).

cornstarch; it presently costs twice as much (23). Also, biomass can react to market changes by changes in composition, while petroleum composition is set by geology. Biomass is flexible as well as being adaptable to market changes (23). In the fuel market, biomass will certainly have to adapt to conditions set by the greater volume petroleum and natural gas markets. In chemical production, it is possible for biomass to be changed into new chemicals which could build markets for themselves. One example is levulinic acid, which comes from biomass, and can be used to make biodegradable strong fibers and transparent fibers (23).

Industrial processes are often proprietary. However, most processes for fuels from biomass necessitate an investment of energy greater than the energy values of the resultant fuel. One that does not was developed at MIT, using specially selected strains of *Clostridium* bacteria (24). Cellulose is anaerobically transformed into sugars, then fermented into ethanol and residues that can be burned to get steam or electricity. Other promising industrial processes involve production of long cellulose chains for paper-making, alcohol, and mixed sugars suitable for cattle feed (24).

NEW RESOURCES FROM BIOMASS

Several plants can make contributions to the chemical or fuels industry. Some plants are especially suited to increasing total production because they grow well in dry climates. The bladderpod (*Lesquerella fendleri*) produces an inedible oil, about 20 to 30 percent by weight, that is a source of chemicals. It can be used to make a very strong plastic and grows in areas getting only 30 to 40 mm of rain per year (25). The buffalo gourd (*Cucurbita foetidissima*) is another dry-climate plant. It produces edible oil as well as protein from seed ($\approx$65 percent). The roots can produce industrial starch (food additives), and the vines yield forage (25–27). The yield is about 2 tonnes of seed and 8 tonnes of roots per hectare (25). Jojoba (*Simmondsia chinensis*) is nearest to commercial status of any of these plants. Its oil ($\approx$50 percent by weight) is very similar to that produced by the sperm whale. It resists biological degradation, so is non-rancid and can be used as a replacement oil in foods. It can also be used in cosmetics, hair oil, and transmission fluid (25–27). The price of jojoba had declined to about $10/kg by 1984, and more uses are expected to be found as the price continues to fall. The United States produced 150 tonnes of jojoba oil in 1983, which exceeded demand. Over 11,000 hectares are planted in jojoba, and production is expected to reach 130 ktonne by 1990 (25). The meal residue currently has no commercial use—the meal contains materials making it toxic and unpalatable as feed, and it is not a good fertilizer (27).

Guayule (*Parthenium argentatum*) produces rubber almost identical to that from the rubber tree. During World War II, guayule was planted in the Southwest as a hedge on loss of Malayan rubber. It is still uncertain whether crops can be grown economically in the United States (25,26).

The gumweed (*Grindelia camporum*) produces a rosin-like material. Rosin is a very important naval stores item with a market in excess of 700 ktonne per year. Rosins are used for adhesives, tackifiers, paper sizings, and other industrial products. At present, gum rosin is extracted by the slashing of pine trees, which is very labor-intensive; wood rosin is obtained from 300-year-old pine stumps, and an estimated ten to fifteen years' supply is left (25). The residues of gumweed are high-quality animal feed, and the plant could become a chemical feedstock of importance for pharmaceuticals and for agricultural products.

Crambe (*Crambe abyssinica*) grew originally in countries bordering the Mediterranean, and has been grown in many American states (27). It is a source of erucic acid to replace rapeseed, from which amounts of erucic acid are declining. Erucic acid is used in the production of plasticizers and lubricants (27).

Kenaf (*Hibiscus camabinus*) is a bushy plant having fibers ideally suited for making high-quality paper and paperboard (27,28). Kenaf grows naturally in Africa, Asia, and Central America (27,28). It could be a valuable resource in regions such as Scandanavia, which currently must import wood chips (27).

The gopher plant (*Euphorbia lathyrus*) grows wild in northern California. Its latex is an emulsion of hydrocarbons in water, and it can produce about twenty-five barrels of crude oil per hectare in California (26,29). It grows well in both wet and dry regions.

A tree discovered in 1979 in Brazil produces what seems to be pure diesel fuel (30). Local residents use it for medicinal purposes. They drill a 5 cm hole into the trunk, into which a bung is placed. About twice a year they remove the bung to collect 15 to 20 liters of hydrocarbon. A hundred trees could produce twenty-five barrels of diesel fuel per year and cover only an acre (30). The millions of barrels used per year would of course require immense plantations, and readily available corn oil may also be used directly to fuel diesel engines.

Politics of Biomass Energy

As an illustration of the political effect of benign appropriate technology, consider the situation in India as described by J. B. Tucker (36): Distribution of biogas technology (in which methane gas is produced by anaerobic digestion) in India led to redistribution of income to the richer members of society. Ownership of biogas plants is confined to those in the top third of income levels. This occurred because the system requires investment. Capital is in the hands of the rich, and bank loans are available only to people having at least three head of cattle ($\approx$75 percent of cattle owners in India have only two). There are currently 36,000 biogas plants in India, and the government is attempting to deal with the political problems by finding other ways to set up biogas plants. Most biogas plants require greater discipline of peasants than should be expected, and require a cadre of trained maintenance and repair people.

The plants are economic in the long term because the dung, night soil, and organic residues produce methane at low pressure for thirty to forty days. This is piped into the kitchen for heat, light, and cooking. The slurry remaining after fermentation is a fertilizer of high quality. The parasites and bacteria that often infest people are killed during fermentation. For this reason, it would improve public health to have these biogas

FUEL, FOOD, AND WORLD POLITICS

Biomass grows faster in the humid tropics than anywhere else, and countries in these regions are now among the most impoverished in the world. It is clear that the world *will* run out of liquid fuel resources during the next century. Our dependence on liquid fuels is almost irreversible, and this should give the Third World countries in the tropics a product they can sell, one that could help them break out of the cycle of poverty (31,32).

Such a shift could cause problems as well as benefits. Although these countries would benefit from cheaper fuel and would gain income, they usually have difficult debt, resource, and population problems and might be forced into a situation in which fuel production is increased at the cost of food production, causing even greater poverty and malnutrition (32,33) if their populations continue to rise at current rates of growth. Also, while current food production could provide for everyone on Earth, if politics did not interfere, it is not clear that technology could keep up with explosive population growth (see Chapter 17). This could pose a political problem for the developing countries and for the world as well. Many of these countries have deemphasized the importance of agriculture, thereby depressing farm costs, subsidizing food, and encouraging farmers to migrate to cities, where they generally form a poor lower class. The possibility that the poorer countries might default on their loans in the early 1980s gave rise to the specter of a world depression. Should all the loans default, with consequent bank failures, the entire world financial market could collapse.

Some help has been afforded these countries by the "green revolution," but much remains to be done. Research on cassava, the fourth most important world source of calories for people, could make a difference (34). While cassava is not economic as a fuel source in the United States or even in Brazil (an emerging Third World country) (9), it may be possible to use it for both food and fuel in the future. It grows even on depleted soils, takes water stress well, and tolerates high aluminum levels (prevalent in lateritic tropical soils) (34).

plants widely available. As more plants are installed, cattle dung, which is now an economic "free good" will, Tucker believes, become a commodity (36). This would widen the income gap in India and other developing countries still further.

There has been much concern about the denudation of hills in Nepal and the rest of the world. Ninety percent of people in developing countries use firewood as a source of fuel, and prices are climbing in most of Africa, Asia, and Latin America (37).

Cooking with wood accounts for roughly 60 percent of all energy consumed in the African Sahel, and an Indian study indicated that the figure there is closer to 80 percent, at least for the six villages studied (38). Where wood for fuel is part of the market economy, as much as 40 percent of household expenditures are for wood (38). It is partly for this reason that digesters have been so widely introduced in developing countries. A more direct way to influence wood use may lie in designing and marketing more fuel-efficient stoves for urban and rural cooking. Traditional open fires yield only 5 to 10 percent of the energy stored in the wood (38). Doubling or tripling this by innovative stove design would appear feasible, and would greatly reduce the

Some geographic areas have a great number of biomass options—for example, the temperate forest. Other areas, such as the polar regions, have little potential for biomass. Nevertheless, in every world region, there is some potential.

There are many plants that are not used widely but that hold out the possibility of nutritional adjuncts to the diets of the entire human race. Mankind is heavily dependent on just a few crops: wheat, rice, corn, and potatoes, for each of which over 300 million tonnes are harvested each year; barley, sweet potato, and cassava, each harvested at over 100 million tonnes yearly; grapes, soybean, oats, sorghum, and sugar cane, each harvested at over 50 million tonnes annually; millet, bananas, tomatoes, sugar beets, rye, oranges, and cottonseed (for oil), each produced at over 25 million tonnes each year; and even smaller crops such as apples, yams, peanuts, watermelons, cabbage, onions, beans, peas, mangoes, and so on (35). This list comes close to exhausting the standard items available in grocery stores.

Plants such as grain amaranth, oca (a yam-like tuber), yam beans (jicama), arracacha (a celery-like root), nunas (popping beans), and quinoa were grown by the Aztecs and the Incas for food; in modern times, their use has languished (35). The pejibaye palm of Central America produces a nutty fruit that is ideal for healthy eating (35).

The groundnut (*Apios americana*) was once an important American Indian food, and the bambara groundnut (*Voandzeia subterrania*) is even now widely grown in Africa. Both taste good and could serve as dietary supplements elsewhere (35); they are similar to peanuts. In semi-desert areas, the tepary bean, the marama bean, and Ye-eb could help feed people where no other food plants grow easily (35).

Dessert fruits abound in the tropics. Many South American fruits have been neglected in the north—naranjilla, cocona, pepino dolce, tree tomato, cape gooseberry (of the potato family); cherimoya, soursop, sweetsop, and atemoya (of the *Annonacia* family); black sapote; carambola (native to southeast Asia); and plantain (35). Some of these fruits are becoming more readily available as immigration from Central and South America increases.

pressure on forests and woodland. Research so far is inconclusive as to whether stoves are indeed more efficient than the open fire. Consensus among experts is that the focus of stove introduction programs should shift to urban areas, and that stoves should be mass produced and portable (38). The original emphasis was on the rural population, but experience has shown that the stoves are often quickly abandoned for lack of qualified repairmen or because they seem too complicated to the user. Also, much experience on "mud stoves" (heavy-weight clay stoves made of natural materials) shows that use does not diffuse well, and the stove's proper use would involve changes in traditional methods of preparation. Despite such problems, stove upgrading or replacement holds great promise for reducing wood consumption worldwide.

Deforestation is proceeding apace in Brazil, in spite of concern expressed by some experts that the loss of trees in great numbers might cause a decrease in the amount of oxygen in the atmosphere, and an increase in carbon dioxide (see Chapter 23). The Agency for International Development has been trying to persuade Third World countries that it would be useful to save tropical rain forests. Ironically, an ill-conceived operation in Hawaii is chewing up several acres of tropical forest a day to supply wood chips for electricity generation (39). This certainly provides a poor example for Third World countries. Better forest husbandry and further study of agriculture in the tropics shows promise for reducing the risks associated with the opening of new areas to settlement by farmers (40).

SUMMARY

Biomass energy is a form of solar energy already exploited by people. A century ago, it accounted for over 90 percent of U.S. energy; as of 1979, energy consumption from other sources had grown so much that biomass totaled a paltry 2 percent. Plants are as much as 2 percent efficient in changing sunlight into usable material (for sugar cane, the efficiency is 2 percent; for corn, 1 percent; and most others are less than 1 percent efficient). Production of ethanol or methanol from wood is economically and energetically more viable in countries exhibiting less reliance on fossil fuels than the United States. Whether or not such production in the United States breaks even is a tossup.

It is unlikely that our present economy could be run on liquid fuels purely from biomass. The prospects for chemicals from biomass are considerably better.

Recently discovered plants hold some promise of help for fuel and resource shortages. Their discovery underlines the concern for widespread cutting of tropical forests and the attendant loss of species.

Some forms of biomass energy are well suited to developing countries, especially biogas digesters, which are widespread in China and India. However, even the introduction of what seems "benign" appropriate technology may have deleterious effects on society.

REFERENCES

1. D. M. Gates and W. Tantraporn, *Science* 115 (1952): 613; C. G. Granqvist, *Appl. Optics* 20 (1981): 2606; and C. G. Granqvist, *Phys. Teacher* 24 (1984): 372.
2. J. R. Bolton and D. O. Hall, *Ann. Rev. Energy* 4 (1979): 353.
3. R. Radmer and B. Kok, *BioScience* 27 (1977): 599.
4. J. R. Gosz, R. T. Holmes, G. E. Likens, and F. H. Borman, *Sci. Am.* 238, no. 3 (1978): 93.
5. D. Pimentel et al. (Review Panel on Biomass Energy), *Solar Energy* 30 (1983): 1.
6. M. M. Waldrop, *Science* 211 (1981): 914.
7. S. Bronstein, *New York Times,* 23 June 1985.
8. Anonymous, *New York Times,* 16 Dec. 1984.
9. C. E. Hewett et al., *Ann. Rev. Energy* 6 (1981): 139.
10. AP dispatch, *New York Times,* 13 Oct. 1985.
11. J. W. Shupe, *Science* 216 (1982): 1193.
12. J. Goldemberg, *Science* 200 (1978): 158; and A. L. Hammond, *Science* 195 (1977): 564 and 566.
13. J. Gomes Da Silva et al., *Science* 201 (1978): 903.
14. C. C. Burwell, *Science* 199 (1978): 1041.
15. N. Wade, *Science* 204 (1979): 928.
16. S. E. Plotkin, *Environment* 22, no. 9 (1980): 6.
17. F. F. Hartline, *Science* 206 (1979): 41.
18. D. Pimentel et al., *BioScience* 28 (1978): 376; and D. Pimentel et al., *Science* 212 (1981): 1110.
19. C. C. Burwell, *Science* 199 (1978): 1041.
20. D. L. Klass, *Science* 223 (1984): 1021.

21. L. F. Diaz, G. M. Savage, and C. G. Golueke, *CRC Critical Reviews in Environmental Control* 14 (1985): 251.
22. W. S. Hillman and D. D. Culley, Jr., *Am. Sci.* 66 (1978): 442.
23. E. S. Lipinsky, *Science* 212 (1981): 1465.
24. H. R. Bungay, *Science* 218 (1982): 643.
25. C. W. Hinman, *Science* 218 (1984): 1445.
26. J. D. Johnson and C. W. Hinman, *Science* 208 (1980): 460.
27. C. W. Hinman, *Sci. Am.* 255, no. 1 (1986): 32.
28. K. Schneider, *New York Times,* 28 March 1986.
29. T. H. Maugh II, *Science* 194 (1976): 46.
30. T. H. Maugh II, *Science* 206 (1979): 436.
31. W. G. Pollard, *Am. Sci.* 64 (1976): 509.
32. H. Cleveland and A. King, *Environment* 22, no. 3 (1980): 47.
33. L. Brown, *Environment* 22, no. 4 (1980): 32.
34. J. H. Cock, *Science* 218 (1982): 755.
35. N. D. Vietmeyer, *Science* 232 (1986): 1379.
36. J. B. Tucker, *Environment* 24, no. 8 (1982): 13.
37. E. P. Eckholm, *Losing Ground* (New York: W. W. Norton & Co., 1976); and S. Postel, *Natural History* 94, no. 4 (1985): 58.
38. F. R. Manibog, *Ann. Rev. Energy* 9 (1984): 199.
39. C. Holden, *Science* 228 (1985): 1073.
40. J. J. Nicholaides III et al., *BioScience* 35 (1985): 279.

PROBLEMS AND QUESTIONS

True or False

1. Schemes for utilization of biomass are environmentally sound by their very nature.
2. It is probably not a good idea to gather all parts of the tree for use when an area is clearcut because the ground will be more susceptible to erosion.
3. Biogas digesters are suitable for use only in undeveloped countries.
4. Distillation of ethanol from corn or sugar cane is likely to be a more energy-efficient process in a developing country.
5. It is unlikely that plants will be able to use more than one-tenth of the sunlight falling on them to build new tissue.
6. Photosynthesis in bacteria does not liberate oxygen.
7. Solar photons with wavelengths much greater than 700 nm cause too much heating in leaves, and so are not absorbed.
8. Development and dispersal of low-technology, improved-efficiency stoves will probably play a major role in the solution to the firewood crises in developing countries.
9. Chemicals derived from biological sources are purer and more effective than those from industrial processes.
10. Some organic chemicals derived directly from plants can be used as fuels without further processing.

Multiple Choice

11. Which of the following materials could be supplied without extensive processing from plants already known?
    a. Transmission fluid
    b. Rubber
    c. Adhesives
    d. Diesel fuel
    e. All of the above-named materials could be supplied from plants without extensive processing.

12. What would be the consequence of introducing anaerobic digestion technology into Indian villages?
    a. A more equitable income distribution within Indian villages
    b. Involvement of the poor in the cash economy
    c. Further degradation of Indian cropland
    d. Production of gas for cook stoves and heating by Indian families
    e. Increase in infestation by parasites among village inhabitants

13. About what proportion of energy consumed in the less developed countries comes from wood?
    a. 1/10
    b. 1/4
    c. 1/3
    d. 2/3
    e. 95/100

14. About how efficient are most food crops at utilizing sunlight?
    a. 0.5%
    b. 3%
    c. 5%
    d. 30%
    e. 50%

15. Energy use in the United States is approximately 1.5 TkWh$_e$. Roughly what area of trees (in hectares) would be necessary to supply this energy?
    a. 1 million
    b. 5 million
    c. 10 million
    d. 50 million
    e. 100 million

16. Which of the following statements about the efficiency of photosynthesis in plants is *not* true?
    a. Less than half the incident sunlight is used by plants on which it falls.
    b. Some solar energy absorbed by plants is released as heat.
    c. Some solar energy is not absorbed by chloroplasts.
    d. Some energy absorbed is used in plant growth and maintenance.
    e. The overall efficiency of photosynthesis is about 1 percent.

17. Which of the following gases did *not* make up a fairly large component of Earth's primordial atmosphere?
    a. Methane
    b. Oxygen
    c. Carbon dioxide
    d. Helium
    e. Hydrogen

18. Why are most plants green?
    a. They absorb the green light in incident sunlight.
    b. They are envious of animals, because it's not easy being green.
    c. They absorb their green color from moss living symbiotically with them.
    d. They shine with an internally produced green light.
    e. They absorb the red and blue in incident sunlight.

Discussion Questions

19. The food balance in the world is slightly positive (that is, production is marginally sufficient to feed the world population, if equitably distributed). Discuss how mass famine in Ethiopia is (or is not) consistent with this assertion.
20. Why are leaves green and not black?
21. Discuss the problems of introducing "appropriate technology" into developing countries. Why are some people likely to benefit more than others?
22. Why should many noncommercial species of plants be preserved?
23. Estimate the total area necessary to supply the United States with its fossil fuel needs, if the diesel fuel tree discovered in Brazil were the sole source of energy. (In 1984, the United States used 7.1 Gbbl of oil; each barrel has a volume of about 200 liters.)

17

THE ENERGY COST OF AGRICULTURE—
A CASE STUDY

This is another chapter on solar (biomass) energy, with emphasis on differing practice between "primitive" and technological agriculture. Energy use in the agricultural system is examined, in particular in terms of a comparison of energy output to energy input. The agricultural revolution of the developed countries (mechanization, application of chemistry, and application of genetics) and its advantages and disadvantages are discussed.

KEY TERMS   *swidden agriculture*
LDC
DES
genetic engineering
green revolution
triglyceride
ammonia
productivity
stability

Our remote ancestors found it advantageous to settle down and farm rather than depend on hunting, which their ancestors had done. The concentration of people into large agricultural units allowed the small surplus per family to be gathered to support a local priesthood or gentry. This upper class of nobles was then free from the everyday labor of farming. In return, the nobility and priesthood protected the farmers from evils both natural and spiritual. (The nobles rode out to protect the farmers from marauders. The priests propitiated the gods.)

Thus the excess productive capacity allowed the beginnings of culture. Only men of leisure could be scholars, remembrancers, scribes.

TRADITIONAL AGRICULTURE

One way to estimate the agricultural efficiency of our forebears is to study contemporary peoples using traditional approaches to farming. The Tsembaga tribe of New Guinea has been the subject of such a study (1). The tribe lives in the interior New Guinea highlands and is relatively unaffected by modern farming ideas.

Human energy input in the Tsembaga farming method was calculated by closely examining the tribe's practice of swidden (slash-and-burn) agriculture. A section of forest would be cut down, and the residue burned to fertilize the soil (because tropical soils are thin and do not retain water well). People work at the burning, at erecting fence, at planting, weeding, and transporting their produce back to the village. The total energy input per acre was estimated to be about 650 kWh, while the total energy output per acre was estimated to be about 11,400 kWh. This is a ratio of seventeen energy units of output to one of input. Much of the harvest is fed to pigs; only about two-thirds (7000 kWh) is consumed by people.

Humans need between 40 and 600 kilojoules per day (1500 and 2500 kcal/day) to live. If we use the higher figure to estimate Tsembaga needs, an acre supports 6¾ people for one year. The entire Tsembaga population requires only about 30 acres at full productivity (that is, new fields; an old field is nowhere near as productive, and such fields are allowed to revert to forest). About 90 acres of old and new field are in use each year (1). Since food is invariably lost to pests, there is some excess production.

Pigs use a substantial share of the tribe's production. In turn, they provide a rich protein supply in emergencies such as sickness or bereavement. It appears necessary in any case that the people eat some meat (2).

The overlords in Europe had resources at their command because their peasants produced more than they ate. We can guess that overlords in the New Guinea highlands probably could not last long with their style of agriculture. The small tribal excess could not support "conspicuous consumption." If customs were altered and wide areas were planted, the soil would soon be exhausted. Tribal behavior seems admirably suited to its clime. We might guess that these tribes are semidemocratic of necessity.

An important aspect to remember is that this "primitive" agriculture is well suited to its environment and produces about seventeen units of energy output for each unit of human energy input. Many so-called primitive agricultural systems are quite productive. This level of productivity holds even for the pastoralists of the Sahel (sub-Sahara lands) before the advent of sophisticated European agricultural ideas (3), which were unsuited to conditions there.

AGRICULTURE IN NORTH AMERICA

Let us draw the contrast clearly by shifting our focus to modern North American agriculture. This is appropriate because many people enthusiastically advocate exporting these techniques to the poor countries of the world (in euphemistic parlance, less developed countries, or LDCs). While such technique transfer presents a tantalizing picture to those convinced of the superiority of technology, we must see if it has any basis in reality.

TABLE 17.1
Yields and energetic efficiencies of different crops.

Food	Method	Yield	Input	Efficiency
Corn	U. S., 1945	1.77	0.58	3.04
Corn	U. S., 1975	4.48	1.84	2.44
Corn	Mexico, manpower	1.61	0.16	10.02
Corn	Mexico, oxen	0.78	0.23	3.34
Rice	U. S., 1970	5.03	3.67	1.37
Rice	Philippines, 1970	1.43	0.44	3.26
Rice	Philippines, water buffalo	0.36	0.02	16.00
Yams	Tsembaga	0.34	0.02	16.50
Cassava	Africa, manpower	4.59	0.17	26.88

NOTE: Efficiency is defined as yield/input; yield and input are given in kilojoules per hectare.
SOURCE: Reference 4.

North American agriculture has seen three revolutionary changes in the last century or so. The first is the substitution of machines as a replacement for human and animal labor.

Table 17.1 (4) shows how energy efficiency varies for different methods of production. Note that primitive methods are much more energy-efficient than modern agricultural methods, although they do involve an investment of many more hours of human and perhaps animal work. Mexican corn farmers may have to work over 1100 hours per hectare as compared to 17 hours per hectare in the United States (4).

Several sources of information (5–7) will be used to place agricultural technology in its proper perspective. Corn is a good crop to study. It is an American native crop, it produces more digestible nutrients per hectare than any other food crop, and it is the most efficient in the use of technology (8).

Census Bureau data (9) are shown in Figure 17.1 for the last three decades. The most striking trend is an increase in production per acre and a decrease in the labor component of the total energy cost, which have characterized the last thirty years. The farm yield itself is substantially increasing (10), as Figure 17.2 also shows for several important crops.

The second revolution in North American agriculture occurred with the application of chemistry to farming. The last half-century has seen the introduction of pesticides such as DDT and many more modern chemical pesticides. This has had at least a short-term effect, reducing pest populations (although pests have often developed resistance to the pesticides). Another gift of chemistry is the wider use of fertilizer in crop management.

The third revolution in North American agriculture is the greatly increased pace of improvements in genetics. New varieties of standard farm crops and domestic animals and new types of crops may be cultivated in regions previously unusable for the crop. All three changes have resulted in a decrease in the number of farm workers. With even a 1900 style of agriculture, three-quarters of the population would have to work in agriculture to feed the rest. Today, fewer than one worker in twenty in North America works in farming.

FIGURE 17.1
The growth in output per hour and output per acre between 1950 and 1982 is shown compared to a 1977 value set to 100 (left axis scale). To find the number of person-hours of labor required by the agricultural sector of the economy between 1950 and 1982, use the right axis scale.

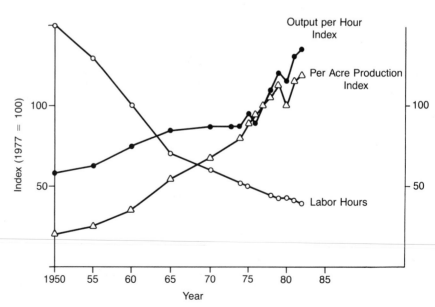

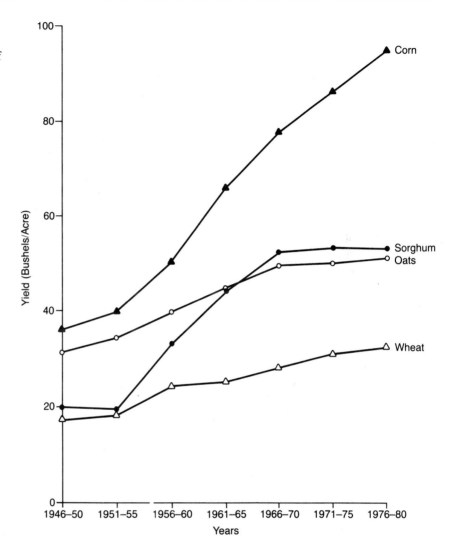

FIGURE 17.2

The five-year-averaged yield, in bushels/acre, of selected grains from 1946–50 to 1976–80.

Several examples illustrating the decrease in human labor over four decades are examined in Figure 17.3 (11). The decreases shown are due to increased energy input and to genetic engineering. For example, cotton was hand-picked until the introduction of cotton harvesters in the 1950s. Machines replaced many poor blacks who had been employed for farm labor and the replacement of people by machine was at least partly responsible for the black exodus to the northern cities. Tobacco farming has undergone

the least improvement, because of all the human care required. The leaves are placed by hand and are dried in sheds. Anyone who has passed through the tobacco-growing areas of Connecticut has seen the fields covered by burlap (to bleach the tobacco for cigar wrappers) strung at great cost in human effort.

Cattle and hog raising shows the more recent effect of feedlots, genetic factors on productivity, and the use of diethylstilbestrol (DES), which is a growth hormone used to promote faster weight

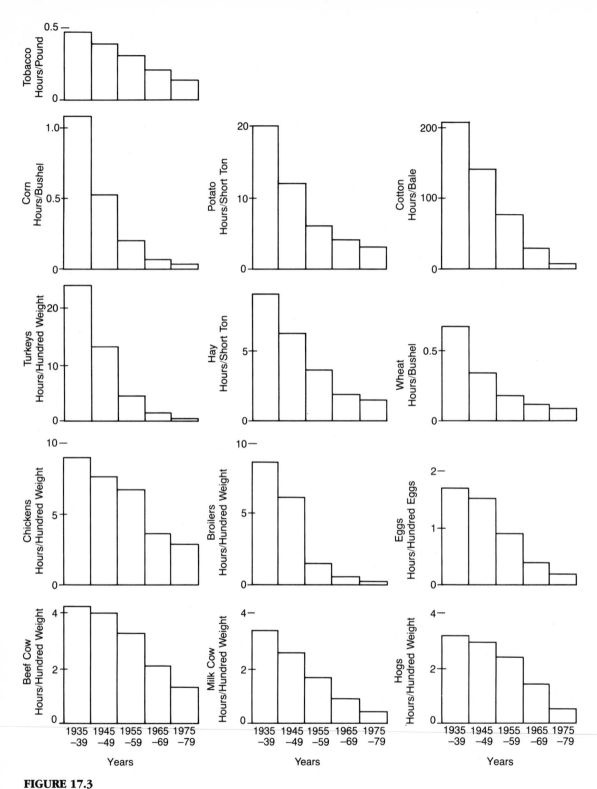

FIGURE 17.3
The decrease in labor per unit output (five-year-averages) is shown for various agricultural products by decades, from 1935–39 through 1975–79.

FEEDLOTS

It is appropriate to note here that feedlots are a source of pollution as well as productivity. Manure that was once spread on farmland now is a sewage treatment problem for the affected communities—a double loss. If feedlots were more widely scattered, manure could be used as fertilizer in surrounding farms, occasioning a great energy savings over artificial fertilizer, as much as 4 million kJ/acre (5). The solid waste problem is not the only one the feedlot presents, as anyone who has ever been downwind of one knows.

As we saw in Chapter 2, when we eat cattle, we are secondary consumers. In the feedlots, the cattle are fed grain that could feed human beings. Thus we lose 90 percent of the feed value of the grain. When cattle are raised on the range, on the other hand, they generally eat grass. Grass is indigestible to people. Since this grassland cannot support a settled agriculture, raising livestock on it is a rational way for people to reap some benefit from an otherwise unusable resource. It would even be possible to switch to a system in which all animal protein is raised using pasture and grazed range, but the production of meat and milk would be reduced by half (14). Such a change would require a change in lifestyle as meat and milk became rarer and more expensive.

In addition to all the disadvantages of feedlots enumerated above, it has recently been shown that feedlot animals are deficient in essential fatty acids and supersufficient in adipose (triglyceride) fatty tissue (2). The fatty acids are essential to human brain development. All animals make some from their diets. The cow has a brain that is only 0.05 percent of its body weight and can make a sufficient amount for its meager needs. The brain of a human being is 2 percent of body weight. A person's brain cannot properly develop unless that person eats food containing the fatty acids or has a supplement. Free-ranging animals provide us with a source of food for brain tissue, while feedlot animals do not (2). Furthermore, the triglycerides seem to have a negative effect on the cell structure of the heart and vascular system, which can lead to heart attacks for heavy meat-eaters.

With all these liabilities of the feedlot system, you may well ask why feedlots exist. Their continued existence (and prosperity) is a combination of an historically artificially low price for energy (as we saw in Chapter 13), and the "tragedy of the commons" effect, both of which allow feedlots to be economically feasible.

gain. DES was regulated and finally banned by the Federal Drug Administration after it was implicated in carcinogenesis. In the feedlots, many animals are kept close together, and their food and water requirements can be delivered mechanically, cutting down on human labor.

The spectacular decrease in human labor in raising poultry can be explained in much the same fashion. "Chickens" are raised on a farm in large buildings, but get some exercise walking around. "Broilers" (sometimes called "Arkansas chickens") are chickens raised in tiny cages with no room for exercise. The same holds for egg-laying chickens. They are kept in tiny cages and fed, watered, and cleaned up mechanically.

Both cereal crops and animals have been subject to genetic manipulation since the advent of sedentary agricultural societies. Modern wheat and corn bear scant resemblance to the native grasses they originally were. Modern cattle are

THE GREEN REVOLUTION, FERTILIZER, AND EROSION

Use of fertilizer and irrigation are fundamental to the "green revolution" we are encouraging poor countries to adopt (15). Fertilizer use increased from 13 Mtonne in 1950 to 110 Mtonne in 1984 (16). Yields would drop by a third if no fertilizer were applied to the world's fields. As we have seen, fertilizer and irrigation are the two most energy-expensive components of the American agricultural system. While Indian farmers (for example) use more energy than is generally acknowledged (17), most of this is human energy. It is difficult to imagine that most Indians can afford to import oil stocks to supply high-grade energy and fertilizer on scales required for wide adoption of the "green revolution" techniques, especially since it would vastly increase the number of unemployed. Because the revolutionary techniques are capital-intensive (i.e., the initial investment is high), only relatively wealthy farmers or landowners can implement them in any case (18). Only the Punjab province in India, which was relatively well off beforehand, seems to have reaped real benefits from the green revolution. This feat has allowed India to triple wheat production between 1964 and 1979, so that India is now self-sufficient in wheat (14). Fertilizer application does relatively more good in India than in the United States, since the smaller the energy expenditure to begin with, the greater the incremental results. Fertilizer use in India has skyrocketed; eighteen times more is used now than in 1960. The United States and India have roughly the same cropland area: The United States still uses four times the nitrogen (8.3 to 2), over eight times the phosphate (5.0 to 0.6), and over fourteen times the potash (4.4 to 0.3) that India does. This spectacular increase in fertilizer use is not limited to India; LDC fertilizer use doubled between 1972 and 1982 (13).

In many cases, since green revolutionary techniques favor the large-scale landowners, who are concerned with immediate profit rather than the welfare of the land, poor farming practices result as small-scale landholders are bought out or tenants are evicted. This is partly responsible for the worldwide problems of erosion.

Even in the United States, erosion is a problem. Erosion under normal tillage can exceed 100 tonne/hectare (and in the Palouse region of the Pacific Northwest, 225 tonne/hectare) (19). Often, application of large amounts of fertilizer can compensate to some extent for loss of soil. In the United States, the 1982 National Resources Inventory found 44 percent of the nation's cropland losing topsoil in excess of its tolerance level; perhaps a third of cropland topsoil has been lost already (14). Our rivers carry about 22 billion tonnes of topsoil into the oceans each year (an estimated 8 billion tonnes was lost from soil in pre-agricultural America) (20). If such erosion continues, U.S. agriculture could experience a 10 percent decrease in productivity over the next century (21).

It should be clear that the mission of exporting American agricultural techniques ought to be considered carefully in light of these adverse ramifications. We must consider which sort of research results are suitable for export to underdeveloped countries. Also, while it is advantageous for developed countries to use fossil fuel in the agricultural system to free human labor for other purposes, this is probably not true in developing countries. LDCs need to find new jobs for the large numbers of newly landless, untrained laborers put out of work by mechanization.

bred to gain weight or produce milk with great success. Our ancestors did much of this breeding, but modern agriculture has systematized and accelerated the changes. Genetic adaptation accounts for about half the increased yield since 1935 (12).

In use or on the horizon are other successes of modern agricultural research. The "green revolution" followed development of crops for LDCs with spectacularly higher yields, although the newly developed strains of cereals require substantial amounts of fertilizer during growth. China increased yields by partly restoring private sector rewards to productive farmers. New developments that bode well for future crop production from LDCs include genetically engineered vaccines, viral insecticides, weeds turned into forage, additives or genetic strains that raise the productivity of dairy animals, and development of new hybrids that extend the geographical growing range of a crop (13).

The trends shown in Figure 17.3 are obviously long-term and substantial. They are mostly due to the spectacular increase in energy input associated with reducing human labor: tractors, herbicides, pesticides, and fertilizers. We see in Figure 17.4 (6) that, between 1930 and 1960, the increase in energy input nearly doubled the output. Table 17.2 shows how the increasing energy input and the increasing crop energy output per acre compare over the period from 1945 through 1970. Note the steady trend of the drop in efficiency, the ratio of output per unit input.

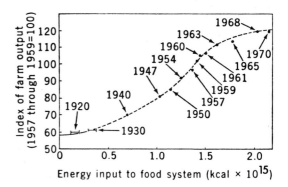

FIGURE 17.4
Farm output (1957–59 is set to 100) in terms of energy input to the U.S. agricultural sector from 1920 to 1970. (Ref. 6, copyright 1974 by AAAS)

Irrigation and Fertilizer

Between 1945 and 1970, the uses of the input energy have changed profoundly (Figure 17.5). Table 17.3 presents energy inputs for selected crops and livestock (14). The human labor component has become all but invisible. The energy cost for combines and tractors (machinery), while rising in absolute numbers from about 210 kWh to about 490 kWh, has decreased proportionally because of the spectacular growth in fertilizer use. There has been a decrease in direct energy use (gasoline, electricity, drying, transportation) in proportion to the energy used on crop care (nitrogen, phosphorus, potassium, insecticides, herbicides). In 1970 about 260 MJ (1300 kWh)

TABLE 17.2
Energy input and output per hectare (in MJ) and ratio, 1945–1970.

Year	Energy Input	Energy Output	Ratio
1945	221.1	818.7	3.70
1950	288.2	915.0	3.18
1955	369.8	987.3	2.67
1960	451.3	1300.3	2.88
1965	535.6	1637.4	3.06
1970	692.0	1950.5	2.82

SOURCE: Reference 6, copyright 1974 by AAAS.

FIGURE 17.5
Comparison of the energy inputs to the agricultural sector, 1945 and 1970.

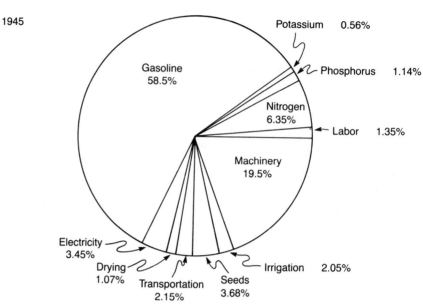

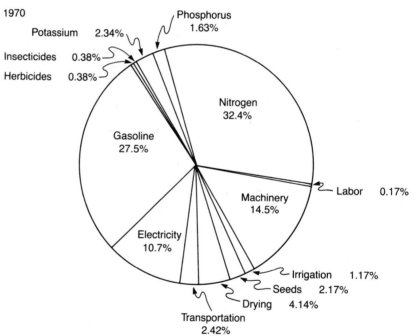

per acre went for artificial fertilizers. This fossil-fuel energy allows labor to be used for other purposes in the economy. Without it, the United States would need to have a majority of the population engaged in farming.

We may remark here that, although the *average* irrigation cost per acre is not large, only about 4 percent of all corn is irrigated. If we had chosen a crop such as watermelon, the proportion of irrigated field would have been much

TABLE 17.3
Energy input, in kilojoules per kilogram of production.

Corn	1800
Wheat	143
Average fossil energy,	
U. S. grain	1430
Broilers	9300
Milk	11,200
Eggs	17,000
Beef	20,800 (feedlot beef 24,000)
Pork	41,999
Average fossil energy	
U. S. animal	18,200

SOURCE: Reference 14, copyright 1980 by AAAS.

higher. On the basis of the 1970 energy cost for irrigating the average cornfield (40 kWh), the actual cost for an irrigated field is about 1000 kWh per year. On an irrigated field, over half the energy is used for fertilizers and irrigating. We might say that the reason the energy return on corn is so high is that only 4 percent of the fields are irrigated.

Irrigation is a further problem for lack of available water: only 30 percent of potentially irrigable land has enough water available. This arises from the maldistribution of runoff. About a third of the world's total runoff (precipitation minus evaporation) comes from tropical South America (the entire continent is only 15 percent of the world land area). Southwest Asia, North Africa, Mexico, temperate South America, the southwest United States, and Australia, with 25 percent of the world land area, have only 5 percent of the runoff (23). In Chapter 11, we discussed a few problems of irrigation.

Another consideration is that the over-application of fertilizer can decrease yield efficiency. Figure 17.6 (5) shows that the maximum in energy output/energy input (efficiency) occurs with an application of about 120 pounds of nitrogen per acre. Actual crop yields do increase (somewhat) with further fertilizer application. In consequence, if the price of fertilizer is artificially

low (because of artificially low energy costs), farmers are forced to increase energy expenditure per acre (decreasing the energy output/energy input ratio) in order to maximize yield (which is directly proportional to dollar return).

Our final remark on fertilizer is that it can cause pollution problems when it runs off the land. Barry Commoner (24) showed how the nitrate load of the Missouri River depends on farm runoff, and vividly pictured the adverse health effects of fertilizers in the ground water (25). In any case, we can save energy and help the soil by applying animal manure to the soil (5) or by using "green manure," that is, rotating crops of legumes instead of supplying nitrogen or ammonia. In addition to relieving the health hazard posed by runoff and the enhanced eutrophication that results from the high phosphate levels in lakes, these latter two methods save essentially the entire energy cost of the fertilizer.

So-called organic farms, which do use crop rotation and manure spreading, have been shown to be essentially as efficient, in terms of total dollar return, as more conventional farms (26). Although the yield is somewhat lower on the

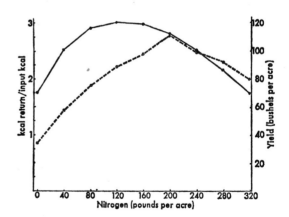

FIGURE 17.6
Ratio of food energy output to farmer energy input as amount of nitrogen fertilizer applied increases (solid line); per-acre yield as amount of nitrogen fertilizer applied increases (broken line). Note that yield actually decreases if too much fertilizer is applied. (Ref. 5, copyright 1974 by AAAS)

"organic" farms, the fertilizer costs are absent also. The two effects are in balance (economically) at the present time. "Organic" farmers do not eschew all modern methods. For example, they also use grain dryers (the newer hybrid corn is wetter than older varieties, and so takes more energy to dry).

Farms and Energy Efficiency of the Agricultural System

It has been estimated that, although farms themselves consume only 4 percent of the U.S. energy supply, 13 to 15 percent of all the energy use takes place in the food system (27). We can compare the distribution of energy to the various stages of the food system in 1940 (Figure 17.7a) and 1970 (Figure 17.7b) (6). The changes between 1940 and 1970 indicated in the figures took place smoothly, as we see in Figure 17.8 (6). Food energy produced has kept pace with population growth, while energy use has far outstripped the gain in population, as well as the food energy ultimately obtained.

Actually the estimate of energy use in the food system is probably too low because we have not

FIGURE 17.7
(a) Energy in the food system, 1940.

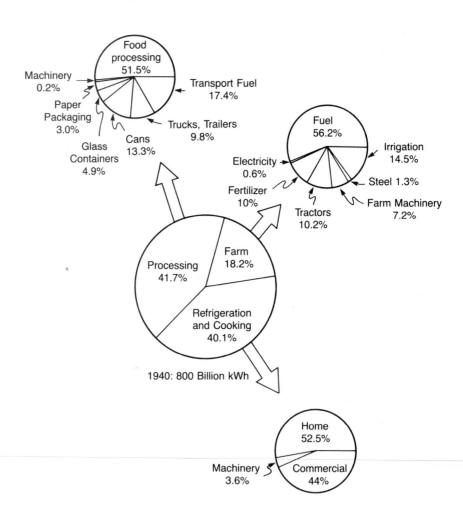

charged the energy costs of constructing buildings used for processing, storage, and distribution of the food. Also, a part of the fuel production is lost in processing, and a great deal of fuel is used in the production and distribution of food; the proportion representing food production should be charged to the system. Other factors neglected include exports and waste disposal.

Energy gained in American agriculture has been less than energy spent for quite a long time, as is seen in Figure 17.9 (6). An estimate for 1973 (28) for the United States (shown in Table 17.4) reveals that 7.5 times more energy is used in the

food system than is obtained from it. Other countries have differing experiences: The British ratio is estimated at 2.5 to 1 (29).

Much of the actual energy used in the food system is spent in packaging and handling of the food and in transport of the food from farm to distribution facility to markets and stores and from the point of sale to home or restaurant. Another large energy input to the food system is in refrigeration and cooking.

There are many ways to minimize the amount of energy put into the food system, some labor-intensive, such as applying pesticides by hand,

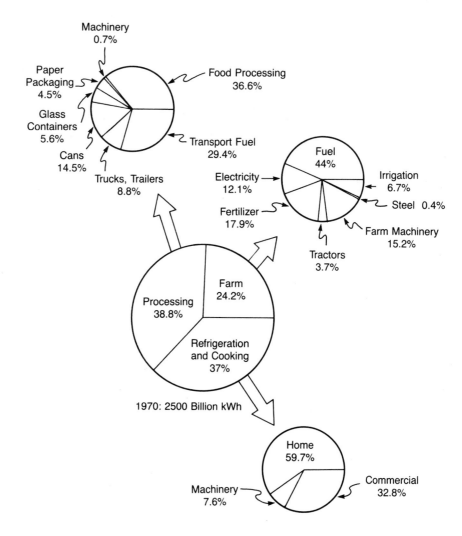

(b) Energy in the food system, 1970.

Machinery 0.7%

Paper Packaging 4.5%

Glass Containers 5.6%

Cans 14.5%

Food Processing 36.6%

Transport Fuel 29.4%

Trucks, Trailers 8.8%

Electricity 12.1%

Fertilizer 17.9%

Fuel 44%

Irrigation 6.7%

Steel 0.4%

Farm Machinery 15.2%

Tractors 3.7%

Processing 38.8%

Farm 24.2%

Refrigeration and Cooking 37%

1970: 2500 Billion kWh

Home 59.7%

Commercial 32.8%

Machinery 7.6%

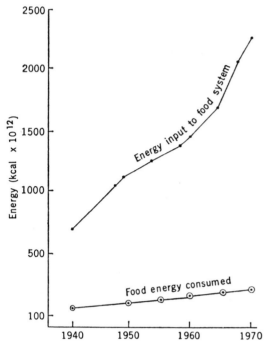

FIGURE 17.8
Energy input to food system and food energy consumed (terakilocalories) from 1940 to 1970 (Ref. 6, copyright 1974 by AAAS).

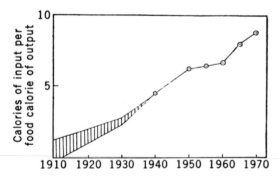

FIGURE 17.9
Ratio of energy input to food energy consumed, 1910 to 1970 (Ref. 6, copyright 1974 by AAAS).

TABLE 17.4
Energy cost of the food production and distribution system.

Component of Food System	Input Energy Cost (MJ/person)
Agriculture	0.36
Processing	0.65
Transport	0.05
Wholesale, retail costs	0.31
Home use	0.60
Total	1.96
Output energy of food	0.26

and some conservation-minded, such as using chemicals—fertilizers and pesticides—only as necessary. The use of "organic" farming methods, in which the farmer uses natural fertilizer and rotates crops and uses natural, biological pest control, decreases the energy input, again at an increased labor cost. Some changes can be made through crop research and genetic alterations; for example, by work on increasing the protein content of the crop, on increasing the length of time the crop is harvestable. A change in cosmetic standards of American consumers in buying produce (more as is the case in Europe) could save otherwise unusable fruits and vegetables, thus reducing waste and saving energy.

A vast amount of energy could be saved in marketing in the United States, by use of rail transport in place of truck transport and by use of a reduced amount of packaging or substitution of paper packaging for packaging made of petroleum-based products.

The ultimate user could help save energy by demanding standardized returnable bottles for products marketed in glass (see Chapter 19), as well as by use of more energy-efficient refrigeration (or returning to the sort of refrigerators that must be defrosted by hand) and self-cleaning ovens. Both energy-efficient refrigerators and self-cleaning ovens are much better insulated than the more conventional products, and the additional purchase cost is soon regained from the lower operating cost.

FERTILIZERS

There are two major fertilizers used to enhance plant growth: nitrogen and phosphates. Some nitrogen is fixed by legumes—plants such as alfalfa. Legumes have bacteria living in a symbiotic relationship with them that put atmospheric nitrogen into the soil. Some nitrogen fertilizer is created by lightning, and this falls on all the world's soils. Most nitrogen used as fertilizer is in the form of ammonia (NH_4), which is easily assimilated by plants. Ammonia is made from methane (CH_4) and air at high pressures. It is convenient to use natural gas, which is mostly CH_4, as the starting point for producing ammonia. The natural gas also supplies the energy needed for ammonia production. The production of ammonia takes a large amount of energy. Since much natural gas is flared anyway, especially in underdeveloped areas of the world, production of fertilizer provides a useful product while decreasing waste.

Even in this country, we must question past practices. For example, fertilizer is apt to get more, not less, expensive. Oil is necessary as both a raw material and an energy source for fertilizer manufacture. The United States uses the equivalent of 4.5×10^{11} cubic feet of natural gas yearly to slake the thirst for fertilizer. Since ten times this amount is flared—or burned off—each year in the Persian Gulf, North Africa, Nigeria, and Venezuela, there is a large potential fertilizer supply (22). The gas flared in the Middle East is not really likely to become available any time soon, however, because it requires expensive installations to produce fertilizer from gas, and there is no reason to expect the cost of fertilizer to decline if it did become available.

Phosphates are necessary to plant and animal growth. Energy is transferred in a living organism by the compound ATP, a phosphate. Most phosphate is mined in Morocco and in Florida (which was once part of Africa).

We are living on past savings of energy, the deposits of fossil fuel accumulated over half a billion years (see Chapter 13). To feed the world at a U.S. level would take 80 percent of the current world energy expenditure; looked at another way, the total known reserves of fossil fuel would be exhausted in only thirteen years if we used the energy to feed the world on the American standard (27).

American agriculture appears to have reached the stage at which further energy input does not substantially increase output (refer to Figure 17.4) unless a new technology is discovered. In the past, there have been examples of such strides: breeding of cereal strains that mature earlier; development of cereals, such as triticale, that have a better amino acid balance (6,15,30); and so on.

Stability and Productivity

One of the difficulties we face as agriculturalists is that all manmade ecosystems are fragile. Our aim is maximum *productivity;* nature attempts to maximize *stability.* By stability we mean that a change induces forces that tend to counteract the change. The most stable natural ecosystems seem to be the most complex, and to have the lowest productivity. This occurs because, in the natural state, everything feeds on or depends on everything else. From a human perspective, this is low productivity. Simplified ecosystems typified by farm fields produce easily accessible excess, which we human beings see as productive.

The Arctic, with few species, is much less stable than temperate grassland or forest (bulldozer tracks dating to World War II still scar the Arctic tundra). In a complex system of diverse species such as is found in the tropics, virtually any scar would be healed (in Brazil the forest is already trying to overgrow the recently built Amazonian road system).

To place the productivity of agriculture in perspective (even our cultivated lawns are quite productive [31]), an alfalfa field is over seven times as productive as a pine forest; tropical rain forests and the oceans have zero productivity (32). Economic forces such as economies of scale drive farmers to monoculture—forces made stronger by the three revolutions referred to early in the chapter (33). Because of the simplification (emphasized in monoculture) we impose on nature, our system is very unstable. There was a corn blight epidemic that struck the American Midwest in 1972. It was disastrous because of the overwhelming acceptance afforded one particular strain of hybrid corn. Were several types of corn planted together, such diseases would have a relatively harder time spreading. In a natural setting, similar plants are not usually close to one another, inhibiting the spread of diseases and rots. A human analogy to this might be the danger of flu. In a preschool or retirement community, the effect of flu is much stronger than in a community of mixed ages. Because children and older persons are more susceptible, the disease becomes more contagious. A field of one sort of plant runs a danger similar to these groups. This is an even worse problem when the same crop is planted on the same fields year after year (some 20 percent of corn crops are grown this way) (33). The blight has emphasized for expert and novice alike that we must keep an inventory of plants of differing genetic background so that we can use the qualities of the "wild" grain to enhance the qualities of the cultivated grain. Monoculture also leads to increased soil erosion and prevents the farm from reaching its maximum agricultural potential.

SUMMARY

"Primitive" forms of agriculture, characterized by heavy reliance on human labor, are very efficient. American agriculture uses fossil-fuel energy as a substitute for human energy, and is therefore much less efficient. The "green revolution" has exported high-technology American-style agriculture around the world. It has been most successful where farmers had sufficient capital available to take advantage of it.

In addition to problems associated with high fossil energy input, the number of plants eaten has decreased and the number of strains of plants grown has decreased markedly. With this sort of monoculture, one disease can wipe out national grain supplies (as almost happened in the United States in 1972 with corn). Wild strains of domesticated plants should be preserved so that genetic diversity is not lost.

REFERENCES

1. R. Rappaport, in *Energy and Power* (San Francisco: Freeman, 1972), 69.
2. R. Allen, *Ecologist* 5 (1975): 4.
3. R. Baker, *Ecologist* 4 (1974): 170; M. B. Coughenour et al., *Science* 230 (1985): 619.
4. D. Pimentel and E. L. Terhune, *Ann. Rev. Energy* 2 (1977): 171 and Tables 2, 3, 4, 5, and 7; and F. S. Roberts, in *Models in Applied Mathematics,* vol. 4 of *Life Sciences Models,* ed. H. Marcus-Roberts and M. Thompson (New York: Springer-Verlag, 1984), 269.
5. D. Pimentel et al., *Science* 182 (1973): 443; and *Science* 187 (1975): 561.
6. J. S. Steinhart and C. E. Steinhart, *Science* 184 (1974): 307.

7. a. Department of Commerce, *Statistical Abstract of the United States, 1972* (Washington, D.C.: GPO, 1984).
    b. Department of Commerce, *Statistical Abstract of the United States, 1984* (Washington, D.C.: GPO, 1984).

8. S. H. Wittwer, *BioScience* 24 (1974): 216.

9. Reference 7b, Table 1183, p. 670.

10. Reference 7a, Table 1010, p. 608; Reference 7b, Table 1193, p.676.

11. Reference 7a, Table 1002, p. 604; Reference 7b, Table 1185, p.672.

12. J. S. Boyer, *Science* 218 (1982): 443.

13. D. Avery, *Science* 230 (1985): 408.

14. D. Pimentel et al., *Science* 207 (1980): 843.

15. N. C. Brady, *Science* 218 (1982): 847.

16. L. R. Brown, *Natural History* 94, no. 4 (1985): 63.

17. R. Revelle, *Science* 192 (1976): 969.

18. "Prester John," *Ecologist* 4 (1974): 304; P. R. Kann, *Wall Street Journal,* 18 Nov. 1974; P. R. Jennings, *Science* 186 (1974): 1085; N. Wade, *Science* 186: (1974) 1091 and 1186; T. T. Poleman, *Science* 188 (1975): 510; and R. B. Trenbath, *Ecologist* 5 (1975): 76.

19. M. R. Gebhandt et al., *Science* 230 (1985): 625.

20. E. C. Wolf, *Natural History* 94, no. 4 (1985): 53; and H. M. Peskin, *Environment* 28 (1986): 30.

21. W. E. Larson, F. J. Pierce, and R. H. Dowdy, *Science* 219 (1983): 458.

22. V.K. McElheny, *New York Times,* 1 Sept. 1974.

23. R. Revelle, *Sci. Am.* 231, no. 3 (1974): 160.

24. B. Commoner, in *Global Effects of Environmental Pollution,* ed. F. Singer (Dodrecht, Neth.: D. Reidel Publ. Co., 1970), 70.

25. B. Commoner, *The Closing Circle* (New York: Alfred A. Knopf, 1971).

26. R. Reed, *New York Times,* 20 July 1975.

27. D. Pimentel et al., *Science* 190 (1975): 754.

28. E. Hirst, *Natural History* 82, no. 10 (1973): 21. See also E. Hirst, *Science* 184 (1974): 134.

29. K. Blaxter, *New Scientist* 65 (1975): 697.

30. L. R. Brown, in *The Biosphere* (San Francisco: Freeman, 1970), 93; and J. H. Hulse and D. Spurgen, *Sci. Am.* 231, no. 2 (1974): 72.

31. B. Webster, *New York Times,* 2 May 1976.

32. J. P. Holdren and P. R. Ehrlich, *Am. Sci.* 62 (1974): 282.

33. J. F. Power and R. F. Follett, *Sci. Am.* 256, no. 3 (1987): 78.

PROBLEMS AND QUESTIONS

True or False

1. The Tsembaga of New Guinea obtain one order of magnitude more energy from their agriculture than they put into it.

2. A *trophic pyramid* describes the hierarchies in the food chain in terms of decreasing efficiency in direct utilization of sunlight.

3. There are no circumstances in which it is better to raise cattle, since there are such energy problems involved in cattle-raising.

4. It takes about three energy units to produce one energy unit of corn at the cornfield.

5. Where land must be irrigated, the irrigation is accompanied by heavy energy expenditure.

6. The amount of fertilizer applied to give maximum energy return (output/input) is the same as the amount giving maximum yield (amount of crop/acre).
7. The amount of energy it takes to run the food production and distribution system in the United States is roughly an order of magnitude greater than the energy value of the food produced.
8. Biological systems that are very productive are also very stable (resistant to changes).
9. Ammonia for fertilizer is produced from crude oil.
10. Erosion can be caused by the application of large amounts of fertilizer.

Multiple Choice

11. Which animal listed produced the most efficiently in terms of human labor in the period 1935–39?
    a. Chickens
    b. Broilers
    c. Beef cows
    d. Milk cows
    e. Hogs

12. Which animal listed produced the most efficiently in terms of human labor in the period 1975–79?
    a. Chickens
    b. Broilers
    c. Beef cows
    d. Milk cows
    e. Hogs

13. In comparing the hours of labor required for corn and wheat in the periods 1935–39 and 1975–79, which of the following is true?
    a. Corn was more labor-efficient in both periods.
    b. Wheat was more labor-efficient in both periods.
    c. Corn was more labor-efficient in 1935–39, but wheat was more labor efficient in 1975–79.
    d. Wheat was more labor efficient in 1935–39, but corn was more labor efficient in 1975–79.
    e. It is impossible to judge from the information given in the chapter which was the more labor-efficient.

14. About what proportion of the energy used on an irrigated field is used for irrigation and fertilizer?
    a. 10%
    b. 25%
    c. 50%
    d. 80%
    e. 95%

15. In modern agriculture, gasoline, nitrogen fertilizer, and machinery represent large energy inputs. Which of the following orders these in terms of increasing proportion of energy used (1970)?
    a. Machinery, gasoline, nitrogen fertilizer
    b. Gasoline, machinery, nitrogen fertilizer
    c. Machinery, gasoline, nitrogen fertilizer
    d. Nitrogen fertilizer, gasoline, machinery
    e. Machinery, nitrogen fertilizer, gasoline

16. About how many billion tonnes of topsoil are estimated to be lost from the land area of the United States each year?
    a. ½
    b. 2
    c. 8
    d. 22
    e. 45

17. Which of the ways listed is the best feasible way to assure that cereal crops will *not* suffer from epidemics?
    a. Plant crops in restricted areas (all corn together, all wheat together, etc.).
    b. Plant crops with one strain of seed.
    c. Save germ plasm of wild grasses related to cereals.
    d. Intercrop: Plant several different crops intermingled with one another.
    e. Develop new cereal crops faster than plant diseases can strike the old ones.

18. Which of the following is *not* a way to save energy in the agricultural system?
    a. Eliminate self-cleaning ovens.
    b. Apply pesticides by hand.
    c. Use rail transport in preference to truck transport where possible.
    d. Eliminate frost-free refrigerators.
    e. Increase the protein content of the product.

Discussion Questions

19. Suppose American agricultural technology could be transported to India. Assess the impact of this on India and the Indians.
    a. What would this do to farm employment?
    b. India currently imports most oil needs. If modern technology is grafted on, what would happen to oil imports? What consequence would this have on the balance of payments?
    c. Fertilizer is made from petroleum. If India increased fertilizer use to U.S. levels, what consequences would there be?
    d. If there were a food shortage, and India had to choose to pay for food now or oil for food next year, what choice do you expect they would make?

20. The United States spends about ten energy units of fossil fuel to put one energy unit on the table. How long can this be done? About how long could it be done if the entire world did it (Americans make up a sixteenth of the world population)?

21. What are some of the consequences of crop monoculture?

18

ENERGY STORAGE AND ENERGY ALTERNATIVES

In this chapter, we examine many different technologies pertaining to the storage of energy. Most of these are innovative, unconventional technologies, and they give hints of developments to come (even if only a few of these do become viable). In addition, we consider proven alternatives for generating electricity, including geothermal energy sources. We conclude with a discussion of the "hydrogen economy," an alternative system for energy distribution.

KEY TERMS   *pumped storage*
load management
type I superconductor
type II superconductor
phase change
latent heat of fusion
PACER
thermal regeneration
magnetohydrodynamics
volume energy content

We have discussed the rise of the utilities as monopolies as an outcome of the difficulty in storing electric energy. In this chapter, we present and evaluate several feasible ideas as well as more speculative energy storage possibilities. The most exciting recent development, which has many implications for the electric distribution system as well as for energy storage, is the discovery of "high-temperature" superconductivity.

ENERGY STORAGE

In Chapter 11, we discussed the use of pumped storage as a way for utilities to even out hour-by-hour fluctuations in use of electric energy by customers. Recall that base load is only about 50 percent of peak demand. Base load is also the cheapest electricity. It is in everyone's interest to increase base load.

Load management may be the cheapest way to smooth out demand. In load management, the utility may start and stop home electric storage water heating systems by radio or by timer. This uses the customers' own use of electricity to flatten out the load curve (1); storage water heaters are large so that they can act as energy storage devices. Water has a high specific heat, so a large volume can lose substantial energy and not lower its temperature by much. Typical units include tanks for pressurized hot water, floor slab heaters, and individual room ceramic units. A comparison of conventional and solar hot water cost with and without storage is given in Figure 18.1 (1). Some American and European utilities use this method (1). When I lived in Germany, I noticed that some people who had electric water heaters would use them only at night for hot water; I understand that this is common in rural areas in the United States as well. The electricity rates were lower in Germany in the evening. This self-regulation helps smooth out demand in a smaller way than the use of home heating systems.

There are other strategies for storing energy during off-peak hours, some speculative, others proven. These strategies allow substitution of stored energy for new energy sources because they allow an overall increase in efficiency of electricity generation. Some of the devices we discuss can be used on a scale smaller than a large generating facility.

A variant on the pumped storage idea is that one could excavate huge underground reservoirs in hard rock at a depth of about a kilometer below

ground (1). As energy is needed, water can be allowed to fall into the underground reservoir. Off-peak energy can then be used to pump it back up. While this does not eliminate all environmental objections, it is superior in many respects to aboveground pumped storage facilities. It is also attractive because the excavation and tunneling methods that would have to be used already exist (1), but it is much more expensive than aboveground reservoirs.

It is also possible to store hot water to be used to generate peaking power. This was used in Berlin-Charlottenburg in the 1930s. The water was stored in steel vessels called Ruth accumulators at high pressures, and released as energy was needed (1). There have even been proposals to use natural aquifers to store hot water in the ground (2). It was estimated that such storage facilities would cost only $40 per installed kW.

Superconductors for Energy Storage and Electrical Transmission

In 1911, H. Kammerlingh Onnes discovered that certain materials at very low temperatures—several kelvin—exhibited very strange properties. Resistance to the flow of electric currents suddenly disappeared at some "critical" temperature. Onnes found that the resistance of mercury vanished at 4.2 kelvin, for example. Since these very cold materials are perfect conductors of electricity, they were called superconductors (as we mentioned in Chapter 9).

Over the years since 1911, progress has been made in understanding the mechanism by which superconductivity works. John Bardeen, Leon Cooper, and J. Richard Schrieffer explicated the mechanism in the 1950s and won the 1972 Nobel Prize for their theory of superconductivity, called the BCS theory, after their names.

In the BCS theory, the electrons in the superconductor are able to conduct electricity with no resistance because they form "Cooper pairs." The material that becomes a superconductor consists of a lattice of atoms from which the electrons come. When the temperature is high, the atoms vibrate from thermal energy about their equilib-

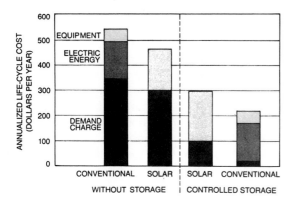

FIGURE 18.1
The cost of heating water for residential use may be decreased by the addition of hot-water storage. The "demand charge" is the amount charged by the utility to assure backup generating capacity. (From "Energy-Storage Systems" by Fritz R. Kalhammer. Copyright © 1979 by Scientific American, Inc. All rights reserved.)

rium conditions. When the temperature becomes low enough, this thermal vibration is of very small amplitude. The pairing of electrons is made possible by their interactions with the lattice. As an electron moves between two lattice atoms, it attracts them electrically and causes them to move toward the position the electron occupied between them. These lattice atoms in turn present a region of higher positive charge to the paired electron, attracting it. The result is that the pair can transfer energy to and from the lattice in such a way as to experience no resistance to the paired motion.

The temperature of the superconducting transition for the materials investigated by Onnes was very low. Later research discovered materials of steadily higher transition temperature until, by the 1960s, it was around 23 K for a niobium–tin alloy (Nb_3Sn). Indeed, the BCS theory seemed to indicate that transition temperature could not be much higher than already found. This material (Nb_3Sn) could be worked, and most superconducting devices made up until 1987 used it. This remained the highest transition temperature seen until late 1986, when G. Bednorz and A. Müller (3) reported a material—a perovskite mineral made of barium, copper, lanthanum, and oxygen—with a transition temperature of 36 K. After such a long spell of stagnation in transition temperatures, this publication set off a race to find other materials having higher transition temperature. Bednorz and Müller won the 1987 Nobel Prize for their work.

In March 1987, the report of superconductivity at 95 K in the so-called 123 compounds (materials made of yttrium or some other rare earth element, two barium atoms, three copper atoms, and about seven oxygen atoms—the 123 refers to the relative amounts of the elements present) electrified physicists working in low-temperature research (4). It also set off incredible excitement among theoretical physicists trying to learn more about how the new superconductors might work. Subsequent investigation identified another sort of material, known as a 2122 compound, which exhibits superconductivity at about 110 K: it was made of bismuth, calcium or aluminum, strontium, copper, and oxygen. Another material, known as a 2223 compound, made up of thallium, calcium, barium, copper, and oxygen, exhibits superconductivity up to 125 K (5).

The new materials had not been previously investigated for superconducting behavior because the oxygen atoms in the mineral were expected to make it act as an insulator and to prevent superconductivity. It was exactly these defects— oxygen vacancies—in the material that were later recognized to be the key to its puzzling properties. By mid-1988, there were reports of superconductivity at over 200 K (although the reports were not reproducible by other investigators in other laboratories), and even tantalizing hints of actual room-temperature superconductivity. True room-temperature superconductors would revolutionize technology, making superconducting transmission lines and magnetically levitated trains feasible, and holding the promise of incredible advances in computing power for the next generations of computers.

Other strange properties of superconductors have to do with their response to magnetic fields, and several different types may be distinguished. The superconductors known as type II superconductors would be used in storage applications in superconducting magnets, because type I superconductors lose their superconductivity in the presence of small external magnetic fields.

The *Meissner effect* (the expulsion of all external magnetic fields from the inside of the superconductor) is an unambiguous signal for real superconductivity, as opposed to a precipitous drop in resistance from some other cause. It is now the key test used to determine which materials are superconducting.

Superconducting Transmission Lines The most promising use of the new superconductors will be in superconducting transmission lines (see Chapter 9). Up to now, such transmission lines were very expensive because they had to be cooled using liquid helium. The new superconductors exhibit their marvelous behavior above the temperature of liquid nitrogen, which is much cheaper to produce because its vaporization temperature is so much higher. Cooling to this temperature will be much easier and less expensive, making transmission lines feasible. If the rumors of room-temperature superconductivity prove true, it would be possible to replace present transmission lines with superconducting lines as soon as usable materials can be fabricated. Also, room-temperature superconductors could be buried underground and would probably be more acceptable to the public than the large above-ground transmission lines. This will save the perhaps 10 to 15 percent of Joule heating losses attendant on short-distance transmission, as well as the 2 percent losses suffered in long-distance transmission. This can be a very important development because of the recent rise in the amount of "wheeling," or interutility transfers of electricity through intervening transmission lines (6). The most important consequence of the high-temperature superconductors is the fact that the current sent through each transmission line could rise spectacularly—the lines would be able to transfer much more power than current technology allows.

Superconducting Storage Magnets

If magnets are made of material that can become superconducting at low temperature, very strong magnetic fields can result. The powerful magnets used in high-energy physics experiments are often superconducting magnets. A large accelerator called the superconducting supercollider (SSC) is being planned. It will probably be a circle about 50 km in diameter ringed by superconducting magnets in order to accelerate particles to energies of 10 TeV, over a thousand times more energy than is bound in the mass-energy of a proton. A major cost in running current superconductivity magnets is cooling them to low enough temperatures.

In the energy storage mode, the magnet's current would be increased when there was an excess of energy available, and then the current would be withdrawn as needed. The energy would

be stored in the magnetic field. With enough sufficiently large magnets, a great amount of energy could be stored.

Large normal magnets could theoretically be used to store energy, but in actuality, the cost of cooling the magnets using cold water makes the venture moot. With superconducting magnets, at least short-term energy storage has become feasible. Because the materials are superconductors, the stored currents circulate with no loss. Of course, their low temperatures must be maintained at some energy cost. The new superconductors raise the promise of larger and much cheaper magnets. The technology is developing rapidly, especially since the discovery of perovskite superconductors. The magnets would need to store in excess of 10 GWh to be economically feasible (1).

Other Possibilities of the New Superconductors

One of the most important gains from new superconducting materials is their use in computers. Currently, there is a limitation on computer packing because of the problem of heat generated by the currents flowing in the devices. This should allow smaller, faster supercomputers; the speed of computation is beginning to approach the limitation decreed by the finiteness of the speed of light, $3 \times 10^{14} \mu m/s$. Since typical sizes are micrometers, typical times to transmit information must be greater than 10^{-14} seconds. Current computers have the capability of processing in times smaller than nanoseconds.

Another possibility that could be realized is that of the magnetically levitated trains. Research was virtually abandoned in the past because of the then-expected high costs. Room-temperature superconductors would reduce the cost by a very large amount.

Energy Storage in Flywheels

Some energy has been stored in flywheels for a long time. The car engine stores energy in a flywheel between cylinder strokes to keep the engine running. The car's flywheel is connected directly to the crankshaft and continues to rotate even after the cylinder has stopped driving it. This rotational energy provides work to compress air in the next cylinder in the sequence so that it can fire, and on and on.

The flywheel is old technology. As children, we played with "friction" cars; these cars had a small flywheel to store the energy. What is new about the flywheel is that extremely strong plastics and epoxies have been developed. They allow the design of flywheels specifically for storing energy, and the total mass need not be so great as present solid flywheels in order to store the same amount of energy. Large amounts of energy could be stored with no danger of the flywheel's flying apart (7). These new flywheels could be used to store huge amounts of energy from power plants to be drawn on for peak demand; in smaller versions, they could be used to run cars—home electricity could be used to start the flywheel rotating rapidly, and the family car could be driven on the energy stored on the flywheel, and so on. Instead of the corner gas station, all the traveler would need is the wall outlet to power his travels. Energy normally lost in braking could be returned to the flywheel, making driving more energy-efficient.

Energy Stored in Phase Changes

As we saw in Chapter 7, when materials change from solid phase to liquid phase (as, for example, ice becoming water) they lose some of their internal structure. In a solid, the atoms are fixed in place. They do not move around freely throughout the solid. The attractive forces between atoms prevent this. In a liquid, however, some of this structure is lost. While atoms in a liquid feel attractive forces, they are not bound in one particular place. The atoms can slide around other atoms to a certain extent.

Materials tend to occupy the state of lowest energy; it therefore takes energy to "unlock" the atoms from their relative stability in the solid and allow them the relative freedom in the liquid. For example, to change ice at 0°C to water at 0°C requires an energy input of about 333 kilojoules for each kilogram of water. This is called the latent heat of fusion of water.

The same kind of considerations hold in analyzing the transition from liquid to gas. In the gas, the interatomic attractive forces are weak indeed. Thus, a certain amount of energy per unit mass (the latent heat of vaporization) must be paid to change liquid to gas. For water at sea level, the latent heat of vaporization is 2260 kilojoules per kilogram.

In energy storage applications, it is useful to use a material that is solid at room temperature and has a melting point slightly above room temperature. Energy can be fed in at a low temperature and be used merely to change the phase of the liquid without requiring the temperature to rise. Recall from Chapter 7 that the temperature is a constant during a change of phase. These devices are called *heat-of-fusion* devices. As an example of the large amount of energy that can be stored in a change of phase, we note that it takes as much energy to melt ice at 0°C as it does afterward to heat the resulting water to 80°C.

Of course, we are looking for a material that melts at slightly above room temperature. One such material is sodium sulfate (8), which has a melting point of 31°C and a latent heat of fusion of about 213 kilojoules per kilogram. Other materials, such as Glauber's salt, are suitable as well. The material to use would vary from application to application. The point is that the latent heat is large, and the thermal energy can be stored even with small temperature changes.

The proposal for the solar energy farm discussed in Chapter 15 (9) included liquid sodium phase changes for energy storage. Liquid sodium has a rather large heat capacity (amount of heat held per unit change in the temperature). Energy generated during off-peak hours could be used to heat a large amount of liquid sodium. The heat could later be used to run a generator (8,9).

Pumped Storage Using Compressed Air

One ploy is similar to the pumped storage of water: Off-peak power is used to pump air into a sealed underground cavern to a high pressure. This high pressure then drives turbines as the air pressure is slowly heated and released; the resulting power produced may be used at peak hours. There are many geologic formations that can be used in this scheme, and it has the advantage that it does not involve huge, costly installations. The first working compressed air storage facility was built in Huntorf, West Germany, by Nordwestdeutschland Kraftwerke, a local utility (1). The air is stored in two salt caverns with a total volume of 300,000 m³ at pressures of about seventy times atmospheric pressure. It can produce about 290 MW for roughly two hours for peak demand (1). For each kilowatt-hour output, 0.8 kWh$_e$ is spent in air compression and cooling (the gas gets hot on compression and must be cooled before storage to prevent heat fracturing of the salt formation), and 1.6 kWh$_t$ is used for reheating (1).

Research in progress is aimed at reducing the energy penalty for storage to acceptable levels. Typical proposed underground pumped storage plants should have a capacity of 1 to 2 GW, with six to ten hours of storage (10). Schemes with and without intermediate storage reservoirs have been considered; the most economical choice appears to be the system without intermediate storage, but with multistage pumps (10).

Batteries for Energy Storage

The oldest electric energy storage device is the battery (see Chapter 5). The lead–acid battery used in automobiles is capable of producing 50 to 100 W/kg for total stored energy of 25 to 35 Wh/kg. It will withstand about 200 to 300 charge–

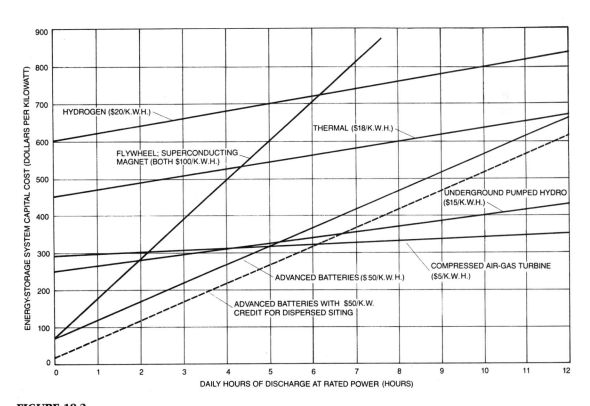

FIGURE 18.2
There is great variation in costs to utilities, as shown in this screening curve. (From "Energy-Storage Systems" by Fritz R. Kalhammer. Copyright © 1979 by Scientific American, Inc. All rights reserved.)

discharge cycles (11), although new designs could extend this to 3000 cycles (12). The batteries are very heavy, so battery-powered cars are penalized because they have so much mass to move. Also, lead is expensive, and a utility wishing to use storage batteries would find purchase of significant storage capacity beyond its means. Several different options are being developed. General Electric is working on a 100 kWh sodium sulfur electrolyte battery (1). Gulf and Western is working on a 50 Wh prototype zinc chloride battery (1). Researchers at Argonne National Laboratory are developing a lithium–aluminum sulfide battery with a molten lithium electrolyte that would operate at 425°C to 500°C (11). The prototypes produce 70 to 100 Wh/kg at powers of 50 to 150 W/kg, and can be charged and discharged 300 to 1500 times (11).

Among all these storage alternatives, there are choices to be made for particular purposes. Figure 18.2 (1) presents the analog of the screening curve for various energy storage systems. The figure indicates why compressed air storage, of all the alternatives considered, has actually advanced to the commercial phase.

Still more energy storage alternatives exist to minimize diurnal fluctuations (1): Refrigerators could run at night to chill water or make ice that could be used to cool the house during the next day; natural ponds could be used to store warm or cool water that could be used when desired. Some home ground water heat pumps store cooled and heated water in the ground near the house, increasing efficiency of the system both winter and summer as the reservoirs are reversed with the season.

NOVEL ENERGY ALTERNATIVES

Shape-Changing Devices

Even more speculative sources of energy are available. There are materials that retain an apparently (permanent) plastic deformation strain (that is, appear to retain a new shape) below some critical temperature T_c; above T_c, the ma-

terials return to their original shape. In the return to the original shape, heat is absorbed and converted into mechanical work (13). There are about twenty elements whose alloys exhibit this martensite transition (13), one of which is a nickel–titanium alloy called *nitinol*. A motor can be built with nitinol that uses the temperature difference between hot water and air to drive a shaft. Model engines and demonstration kits are available (14).

PACER

One of the more intriguing proposals for an immediate energy source, although it had a brief existence, is PACER (15,16). Scientists from Los Alamos National Laboratory claimed that reactor fuel and electricity could be produced by repeated underground thermonuclear explosions. They proposed detonating at least two devices a day in a mile-deep cavity, most likely in a salt formation. The superheated steam produced from the proposed million tons of water in the cavity would power a regular generating plant on site and breed reactor fuel. PACER devices would have yields in the range from 10 to 100 kilotons. One of the problems with the proposal is that it would have invited arguments similar to those over nuclear plants as to safety. There is also the fear that the cavity could collapse, releasing the radioactivity inside, contaminating the ground water, and so on. As a breeder, it could have ended fuel worries forever if it had not been scuttled. If we assume one 50 kiloton explosion at 10 percent efficiency per day, we get a total energy output of (50 kilotons) (4.19 × 10^{12} J/kiloton) (0.1)/(2.6 × 10^6 J/kWh) = 6 × 10^6 kWh per day for an average power output of 6 × 10^6 kWh/24 h = 240 MW (17). At a cost of $40,000 per nuclear device, this would have produced energy at a cost of about 7 cents per kWh, comparable to current prices.

Solar Power Satellite

A proposed 10 GW solar power satellite (SPS) would sit in geosynchronous orbit (Figure 18.3). It would be shadowed by Earth less than 5 per-

FIGURE 18.3
Orbiting solar energy
satellite. (U.S. Depart-
ment of Energy)

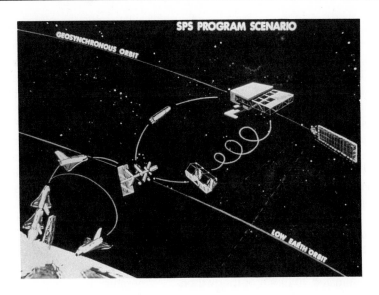

cent of the time at that distance. In the proposal, a 100 km² array would convert solar energy to electricity, then convert the electricity to microwave energy. The microwave energy would be beamed at low intensity down to a 100 km² antenna on Earth. This satellite array would take 0.13 percent of total world refined aluminum, require five hundred shuttle launches with concomitant pollution, and could cause health effects from spillover of straying microwaves (18). The analysis shows an energy ratio of energy delivered over lifetime to primary energy for construction of around two for SPS; such a ratio is always less than one for fossil-fuel plants (18). If one neglects the energy cost of running fossil-fuel plants, then they *appear* superior to the SPS. Another similar proposal is for an orbiting large reflecting sheet that could be used to illuminate cities at night, obviating the need for street lighting (and perhaps reducing crime as well). Of course, this could affect the Earth's heat balance, but so does burning coal, oil, or uranium.

Thermal Regeneration as an Energy Source

In thermally regenerated electrochemical systems, reactants for cells are regenerated by thermal energy from a heat source flowing through a device into a heat sink. There is a membrane (beta-aluminum solid electrolyte) that divides a closed vessel into two regions and a pump. Liquid sodium fills the upper region, maintained at a temperature of 900 to 1300 K. The lower region contains liquid sodium and vapor at 400 to 800 K. Electric leads are connected to the porous electrode and the high-temperature liquid sodium (19). The sodium from the condenser enters a hot zone and absorbs external thermal energy. The difference in temperature causes a pressure difference across the membrane, which in turn forces sodium ions to the low-pressure surface (19). These devices, called AMTEC for "Alkalai Metal Thermoelectric Converter," operate at 20 to 40 percent efficiency and can operate for up to 2000 hours (19).

PROVEN ENERGY ALTERNATIVES

Geothermal Energy

We now turn to geothermal energy, a source from Earth itself. The interior of the Earth is hot. It became hot originally as the Earth formed out of the primordial dust cloud, and radioactive decay of heavy elements in the Earth's interior keeps it hot. The natural heat flow at the Earth's surface is estimated at 3.5×10^4 kWh/s (20), with a total energy of 230 TkWh stored in the Earth to a depth of 3 km (20), and 350 million TkWh to a depth of 10 km (21). In most places, the hot rock is far from the surface; where the heat flow is greatest, there is the potential for exploitation of the geothermal resources. There are quite a few geothermal plants operational now: Lardarello, Italy (400 MW); Cerro Prieto, Mexico (645 MW) (22); Broadlands, New Zealand (145 MW) (23); and four fields in the Philippines, among others (24).

In the United States, Alaska, California, and Hawaii have the greatest concentration of exploitable geothermal resources. (We expect that the geothermal resources of Yellowstone National Park will remain inviolate.) The United States has an estimated 70 EkWh of thermal energy stored in the upper 10 km of the crust (21). Not even a small part of this amount is truly exploitable (that is, appears as hot rock near the surface). The most optimistic forecasts envision an eventual domestic geothermal supply of 400,000 MW (21).

Iceland has used geothermal heat since 1925 to provide heating for homes in Reykjavik and in other sites having hot magma near the surface. Boise, Idaho, and Klamath Falls, Oregon, have been using geothermal energy for space heating and hot water for almost a century.

Exploitation of Hawaii's geothermal resources is just beginning. There is a 3 MW generator at Puna, near Mauna Loa (25), currently producing electricity. More exploitation of the field appears likely, since the field has an estimated 3 GW-centuries of thermal energy available (25).

California leads the world in geothermal energy development. There are twenty-two plants at the Geysers, producing a total of 1650 MW_e (24). This is the single largest field in the world, and the resource appears here as "dry steam" at the surface (it contains little or no water). There are also 10 MW_e installations in Brawley and on the Salton Sea near Niland; there is a 45 MW_e binary cycle (i.e., heat exchanging) facility near El Centro; and a 47 MW_e double flash facility (in which the steam runs two turbines in sequence)

near the same location (26). All of these resources appear in the form of huge pockets of hot brine. The Salton Sea water is approximately one-quarter salt, which is extremely corrosive and clogs the interior of the installation's pipes. Nevertheless, since the brine is at such high temperature, 230°C, it has the capability of producing a lot of energy, and engineers are confident that they can develop the resource. If the pumped-up brine is dumped onto the ground, it could poison plants and pollute ground water, so all the brine is reinjected back into the ground (26). Such high salinity as is found in these areas is rarely found in geothermal fields.

In other developments, the city of Boise is expanding its distribution facilities for low-temperature hot water for space heating from geothermal wells; Elko, Nevada, uses low-temperature hot water from wells to heat downtown buildings and for irrigation; and San Bernardino, California, uses geothermal hot water for heating in the city's wastewater treatment plant (24). One of the larger fields, at Broadlands, New Zealand, which produces 145 MW, has been studied for effluents (23). Large amounts of arsenic, mercury, hydrogen sulfide, and carbon dioxide are expelled from the plant. The river temperature sometimes rises as much as 6°C. The Broadlands facility has a thermal efficiency of 7.5 percent—this geothermal plant attains only 25 percent of its maximum thermodynamic efficiency, while a typical fossil-fuel plant reaches 70 percent of its maximum thermodynamic efficiency (23). New Zealanders are considering scaling back their exploitation of geothermal energy near Rotorua in order to encourage increased tourism. It is clear that geothermal energy is not always the best alternative.

One could build a geothermal device in any suitable place—one with a sufficient volume of hot dry rock—by fracturing a large volume of rock, pumping cold water down into the fracture, and drawing heated water up to run a generator. Such a system was tested for nine months in New Mexico (21). This approach suffers from problems of mineral caking in the pipes in the same fashion that natural geothermal plants do. It would

FIGURE 18.4
Geothermal resources in
the United States.

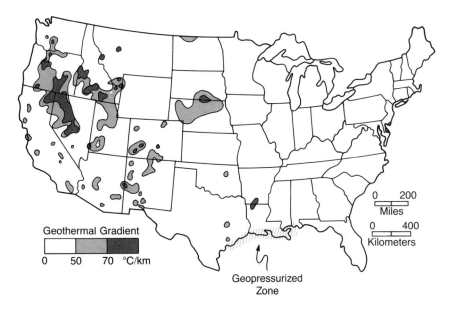

Geothermal Gradient

0   50   70   °C/km

0   200
Miles

0   400
Kilometers

Geopressurized
Zone

produce no effluent, however, since the water would be reinjected into the central well for reuse.

Most of the potential geothermal resources in the United States are west of the Rockies. These are areas of great surface heat flow, primarily in Idaho and Oregon, with smaller scattered areas in Nebraska, New Mexico, Nevada, and Colorado. Reasonable heat fluxes are available throughout most of the West and in small areas of New Jersey, Delaware, Maryland, and Virginia (see Figure 18.4).

Three to 6 kilometers under the coastal rock of Texas and Louisiana is hot water under high pressure. The hot water could run turbines to give some "geopressure" energy, and the water contains dilute natural gas. It has been estimated that the deposits could supply methane equivalent to a supply of six to fifty-five years at the current use rate (27). However, it is very dilute and will probably not be a large resource.

MHD

Magnetohydrodynamics (MHD) provides electric energy directly. Hot gas is "seeded" with a chemical agent to ionize it. The hot gas carries charge with it, producing a direct current with

no moving parts. Since MHD is not spectacularly efficient (20 to 30 percent) (28), and since efficiency increases with temperature, it is probably best used as a topping cycle (that is, as a cycle at the high-temperature end to get some energy there, with use of the hot gas from the MHD generator to run a normal power plant to generate energy). In this mode, the net efficiency can be much greater ($\approx$50 percent) for the combined cycle than for either part alone. The amount of water necessary for cooling is less for the combined cycle. The sulfur in the gas effluent combines with the seeding agent (usually potassium) so that the total amount of sulfur emitted is much less with MHD, although the smallest particles may be increased in number (28).

Burning Garbage

Another proven technology is the garbage-to-energy conversion. It is well known that garbage produces methane at a rate of 10^4 cubic feet per ton. The Garrett pyrolysis process produces about a barrel of oil, 140 pounds of ferrous metals, 120 pounds of glass, and varying amounts of gas (power gas) per ton of waste (29). This sort of list makes it clear that garbage can be more than just something to throw away. Seventy to 80 percent of municipal waste is combustible (30). It has been estimated that incineration of all U.S. municipal waste could supply 2.5 percent of the total U.S. energy use (31). The resource-poor Japanese have made their own study, which indicates that it is possible to recover 0.3 tonnes of fertilizer, 0.3 tonnes of paper pulp, 30 to 40 kilograms of iron, and over 7 cubic meters of fuel gas per tonne of waste (32). The Japanese have made the processing of most garbage compulsory.

It is known that municipal wastes have been burned to generate steam at least since 1890, and to generate steam and electricity since 1896 (30). It is appropriate that the first plant generating electricity was in Europe (Hamburg, Germany), since Europe has adopted this method of dealing with municipal wastes with much more vigor and interest than the United States. Most garbage-to-energy plants in the United States use direct burning of garbage to reduce the volume, meanwhile attempting to intercept valuable resources before or after burning. This method has been adopted in many communities in the United States and Europe because of the shortage of landfill space (and the opposition of practically everybody to new landfills). The cost of disposal in landfills rose by a factor of 6 between 1980 and 1985 to $15/ton in California and $30/ton in Minnesota (33); and Philadelphia has had to truck some of its refuse 350 miles to western Ohio (34) and must pay an average of $90/ton to dispose of garbage. Disposal cost for garbage in northern New Jersey has risen as high as $102.50/ton (35).

European countries historically have been short of space for people, let alone for landfills; in the United States, the cost of landfill disposal dropped far below incineration, especially from the early 1950s to the late 1970s (30). In addition, European municipalities generally supply electricity as well as collect garbage, while in the United States, these services are supplied by different entities. As a result, Europeans could more easily see the benefits of burning trash.

Still, it is clearly advantageous in the U.S. economy today to burn trash rather than send it to landfills. American trash even has more heating value than European trash (an average of 10.3 MJ/kg versus 8.3 MJ/kg [29]), as well as being generated at greater per capita rates. The amounts of trash involved are stupendous. New York City alone sends 20,000 tonnes of garbage into landfills every day and must somehow dispose of another 8 million tonnes a year (36).

Costs of disposal using energy generation generally prove higher than originally estimated, both here and abroad (37,38). Many cities have trash-burning power plants, among them Hempstead, Long Island; Albany, New York; Baltimore, Maryland; St. Petersburg, Florida; Saugus, Massachusetts; North Andover, Massachusetts; Minneapolis-St. Paul, Minnesota; and Columbus, Ohio. Forty-five trash-burning plants were operational as of mid-1985, with about as many more under construction (39). Current U.S. capacity is about 33 kilotonnes per day (40).

Because there are so few trash-burning facilities in the United States, there are no typical examples. As an example of publicly owned plants, consider the trash-burning facility in Columbus, Ohio. Three shredders are scattered about the city, with an additional 50 ton/h capacity shredder in the plant itself. City trash is shredded at the satellite facilities; trash brought by private haulers and other municipalities is shredded at the plant. All shredding facilities have had problems with bedspring coils and with long lengths of nylon-reinforced hose. For this reason, several employees are assigned to fish out such objects in addition to separating out large items, such as washing machines, and hazardous items, such as compressed gas cylinders and gasoline. Other ferrous metals are separated electromagnetically. Glass is held to a small proportion of trash to prevent formation of clinkers (solid hot masses of material) and hot spots in the combustors.

The Columbus trash-burning plant was designed to operate on a 10 percent mixture of coal. At present, coal is used only for startup and for maintaining one boiler on standby, in case of pressure loss in one of the trash-fired boilers. The boiler temperature is low, only 265°C. At present, the plant uses about 95 tonnes of coal and around 1200 tonnes of refuse per day. The energy content of the trash burned in Columbus averages 10.5 MJ/kg (about one-third that of the coal). The plant operates only two turbines at any one time with the current trash volume, but the improvements in progress should increase plant capacity to about 1800 tonnes/day (41). This will increase generating capability from 60 MW to 90 MW. The ability to generate its own energy saves the City of Columbus 1.25 million dollars per month, as well as decreasing the volume sent to landfills by a factor 5 to 6 (41), and providing a small income from salvaged metals. An abandoned water-filled quarry next to the site is used for cooling water. The facility meets all the requirements of the EPA.

Baltimore, Maryland, has a privately constructed trash-to-energy plant. The owner, Signal Environmental Systems, runs several other facilities in Florida, Massachusetts, and New York. It sells the electricity from its 60 MW turbine generators to Baltimore Gas and Electric Company and distributes hot water to the downtown heating loop run by Thermal Resources of Baltimore, Inc. Despite its for-profit nature, the plant was partly financed through the sale of tax-exempt government bonds.

In this plant, the trash is not presorted. The raw refuse is simply dumped onto the furnace feed hopper. The trash is burned at high temperature (almost 1400°C) to break down and burn most noxious chemical compounds. The facility can process as much as 2 kilotonnes of refuse per day, but it is currently processing at about 80 percent of its capacity. The clinkers are sorted to recover ferrous metals, which are sold; the remaining aggregate is dumped.

Not all trash-burning facilities have met with EPA approval. The Hempstead, New York, facility was opened with great hopes in 1979 (42), but had to be closed in 1980 because of problems with dioxin emissions (35,37). Environmentalists across the country have since argued against such plants because of the possibility of dioxin emission. The Hempstead plant was relatively small, processing 4000 tons of garbage a day into 2 GWh of electricity. It was built by a private company, Parsons and Whittemore, which had agreed to take trash for payment of $16 per ton, and then give back 25 percent of its recycling earnings and 40 percent of its electricity sales to the town (42).

Other plants have had problems as well. There have been several explosions in the facilities in Akron, Ohio, and in satellite shredders in Columbus, Ohio, as well as unanticipated problems in operating and feeding materials through. This idea clearly has great virtue if the plastics causing the chlorine and dioxin problems can be kept out of the garbage stream (43). The New Jersey legislature may mandate a deposit on plastics to prevent them from being thrown away at all (43). Another way to prevent problems is to burn the trash at temperatures above 1000°C, because this causes complete combustion of all components of the plastic; however, this does have the effect of increasing emissions of nitrogen oxides (see Chapter 22). This high-temperature combustion method has also been used by Signal in its plants; its Peekskill, New York, plant registered the lowest dioxin emissions on record according to the New York State Department of Environmental Conservation (34). The use of garbage to generate energy is more attractive in densely settled metropolitan areas, and in these locations, this method should probably be pursued. Separation is clearly to the advantage of all facilities, and is not extraordinarily expensive.

AN ALTERNATIVE ENERGY DISTRIBUTION SYSTEM?

The *Hindenburg,* a German zeppelin, burned spectacularly upon its landing in Lakewood, New Jersey in 1936. Hydrogen in bags was used for buoyancy of the airship. The *Hindenburg* disaster followed from the explosion of the hydrogen in the bags. The live radio report and the films of the disaster effectively ended the era of lighter-than-air ships.

The energy content per unit volume is low for hydrogen in comparison to the hydrocarbons, but its energy content per unit mass is over twice as large. (Hydrogen has only 10 percent of the mass of the lightest hydrocarbons.) The comparison of total available work for various compounds is presented in Table 18.1. The favorable energy-to-mass ratio for hydrogen has led to the proposal that hydrogen replace gasoline as a motor fuel. Because of the large volume of hydrogen in comparison to other fuels, and its extreme flammability, there are storage problems for vehicles that would carry the hydrogen as fuel. The best alternative is an iron and titanium alloy that "soaks up hydrogen like a sponge" at twenty times atmospheric pressure; at five times atmospheric pressure, hydrogen is released as fuel (44). The metal hydrides clean the hydrogen as it is stored; the major problem is the expense of the metal and the weight of the storage container (44–46).

One advantage of a hydrogen-powered vehicle is that it is nearly pollution-free. If hydrogen is burned in pure oxygen, the result is energy plus water. No other chemical is exhausted. However, it is more likely that hydrogen will be burned in air; in this case, nitrogen oxides will also be produced. The combustion temperature can be

TABLE 18.1

Comparison of available work from various fuels (MJ/kg).

Hydrogen (H_2)	114
Carbon (C)	32.9
Carbon monoxide (CO)	9.23
Methane (CH_4)	50.5
Ethane (C_2H_6)	48.4
Propane (C_3H_8)	47.5
Ethylene (C_2H_9)	47.3
Octane (C_8H_{18})	46.1
Gasoline (C_8H_{18})	2.7

SOURCE: Reprinted with permission from *Handbook of Chemistry and Physics*, 51st ed. Copyright 1970, CRC Press, Inc. Boca Raton, FL.

set low, which minimizes release of this class of pollutants. Hydrogen-powered cars run much cleaner than gasoline-powered cars, and hydrogen-powered airplanes are cleaner than kerosene-powered jets.

In view of the *Hindenburg* disaster, you may be tempted to think hydrogen very unsafe. In fact, sixty-five of the ninety-odd passengers survived! A test with armor-piercing bullets of a hydrogen metal hydride storage container resulted in a short-lived flame and a 2 percent hydrogen loss; in a similar test with gasoline, a 2-meter-long flame burned for five minutes (32). Also, because the hydrogen is less dense than air, burning hydrogen rises straight up. It was for this reason that there *were Hindenburg* survivors and that the shuttle *Challenger* crew cabin survived the hydrogen explosion to plunge into the Atlantic, where it broke apart. All the hydrocarbons in liquid form are much denser than air and spread as they burn, which can be much more dangerous.

The other storage methods for hydrogen are compressed gas in cylinder (which can present problems if the cylinder is damaged), and as a liquid (but since hydrogen liquefies at 20 K, it costs a lot of energy to liquefy) (45,46). Large volumes of hydrogen can be stored at low cost underground. If cars are run on hydrogen, the carburetors must be replaced and the compression ratio improved (45,46). This is easily done by a trained mechanic.

A further proposal has been made to replace our present electric distribution system with a system of pipeline distribution of hydrogen. For long-distance energy transmission, Gregory and Pangborn (46) have calculated that pipeline transmission of hydrogen costs somewhat more than natural gas but is much less expensive than overhead electric transmission lines. A 210 km pipeline in the Ruhr region of West Germany has operated reliably since 1938 (46).

We may compare the present cost of natural gas delivery to that of electricity. Typical transmission costs for natural gas run 1/300 cent per kWh per 100 miles of transmission. For electricity, average distribution costs are typically 0.74 cents/kWh out of a total cost of 1.63 cents/kWh (44). Table 18.2 shows a comparison of distribution costs in various energy distribution systems, including the estimated cost of a hydrogen system (47).

The most important objection to this scheme is the shortage of hydrogen and the high energy cost of obtaining it. It would take roughly 3×10^{10} m³ of hydrogen gas to replace totally the electricity used in a given year. Most proposals would use the output of present large power plants to obtain hydrogen in some way (for example, by dissociation of water). An economic analysis based on various strategies (48) found that the proposed hydrogen distribution system was the cheapest (see Table 18.3).

The hydrogen would be supplied to the home over present gas pipelines. The only problem here is that the present valves would leak hydrogen if they were used. Many home appliances would still operate electrically; fuel cells such as those developed by NASA (see Chapter 6) could burn hydrogen to produce electricity. If a hydrogen-powered auto is used, a fill-up could take place at home.

Research is progressing on ways to generate hydrogen directly from photosynthesis (see Chapter 16). As emphasized in Chapter 15, it has

TABLE 18.2

Cost of various energy transport modes, in cents/kWh/100 miles.

Method	Local	Long Distance
Methane by pipeline	0.2	0.01
Hydrogen by pipeline	0.22	0.011
Electricity by high voltage	0.85	0.07
Gasoline by tanker	0.23	0.033

SOURCE: Reference 47.

TABLE 18.3

Cost of alternatives for supplying energy for residential users.

Alternative	Cost 1973 dollars (1985 dollars)
Nuclear—electric and gasoline	822 (1297)
H_2 from nuclear and coal	815 (1290)
Nuclear to H_2	928 (1055)
Coal to H_2	486 (613)
Coal to methane	515 (642)
Nuclear—electric only	815 (1504)
Coal to electricity and gasoline	764 (1239)

SOURCE: Reference 48, copyright 1973 by AAAS.

not been possible to provide energy storage, or to break water into its components, hydrogen and oxygen, except biologically. Thus, the method adopted is the use of biological chloroplasts (thylakoid membranes), which are trapped on filter paper by platinum (45). The result is a material that is a composite: part biological, part metal. The electrons given off in the chloroplasts go into the platinum. Large amounts of both hydrogen and oxygen are produced, but eventually hydrogen evolution tapers off. If the hydrogen production can be sustained, it could provide a foundation allowing solar generation and storage of hydrogen. This would mesh very well with the proposed switch to a hydrogen energy economy.

Power plants could work more efficiently in the case of a hydrogen system simply because hydrogen can be stored, in contrast to electricity. Thus, the entire load can be taken to be base load. Time delays in the distribution system and inconstant demand are then easily handled.

SUMMARY

Energy storage can be helpful in leveling off the daily utility demand curve. Among the currently used methods are load management using home heating systems, pumped water storage, and compressed air storage in underground caverns. Storage methods under discussion include use of superconducting magnets, epoxy flywheels, and use of shape-memory devices.

Speculative energy sources include solar power satellites, geopressure energy, and AMTEC conversion. Proven energy sources are geothermal energy, MHD topping cycles, and the burning of trash. The United States leads the world in installed geothermal power stations. Trash-burning power plants are growing in number because of the increase in price of electric energy and because burning offers reduction in the volume of material that must be taken to landfills, prolonging landfill lives by a significant factor.

Hydrogen burns cleanly and is safer than hydrocarbon fuels. The use of an energy distribution system utilizing hydrogen seems feasible for several reasons: The natural gas pipelines could be used to transport hydrogen with relatively minor modifications; storage for hydrogen is cheap, so its use will automatically alleviate the storage problem electricity has; and hydrogen is cheaper to transport than electricity.

REFERENCES

1. F. R. Kalhammer, *Sci. Am.* 241, no. 6 (1979): 56; F. R. Kahammer and T. R. Scheider, *Ann. Rev. Energy* 1 (1976): 311.
2. W. J. Schaetzle, C. E. Breet, and J. M. Ansari, in *Alternative Energy Sources II,* ed. T. N. Veziroglu (Washington, D.C.: Hemisphere Publ. Corp., 1981).
3. J. G. Bednorz and K. A. Müller, *Z. Phys. B* 64 (1986): 189.
4. M. K. Wu et al., *Phys. Rev. Lett.* 58 (1987): 908; see also A.L. Robinson, *Science* 235, (1987): 1137.
5. C. W. Chu et al., *Phys. Rev. Lett.* 60 (1988): 941; Z. Z. Sheng and A. M. Hermann, *Nature* 332 (1988): 138; R. Pool, *Science* 240 (1988): 25 and 146; and A. Khurana, *Phys. Today* 41, no. 4 (1988): 21.
6. A. Pollack, *New York Times,* 11 August 1987, 12 August 1987, and 13 August 1987.
7. R. F. Post and S. F. Post, *Sci. Am.* 229, no. 6 (1973): 17.
8. A. L. Robinson, *Science* 184 (1974): 884.
9. A. B. Meinel and M. P. Meinel, *Phys. Today* 25, no. 1 (1972): 44.
10. S. W. Tam, A. A. Frigo, and C. A. Blomquist, in *Alternative Energy Sources II,* ed. T. N. Veziroglu (Washington D.C.: Hemisphere Publ. Corp., 1981).
11. A. Chilenskas and R. K. Steinenberg, *Logos* 23 (1985): 10.
12. B. Sorenson, *Ann. Rev. Energy* 9 (1984): 1.
13. A. A. Golestaneh, *Phys. Today* 37, no. 4 (1984): 62; R. D. Spence and M. J. Harrison, *Am. J. Phys.* 52 (1984): 1144; and A. D. Johnson and J. L. McNichols, Jr., *Am. J. Phys.* 54 (1986): 745.
14. Model heat engines: TINI Sales, Box 1431, Lafayette, CA 94549; demonstration kits: Nitinol Devices, 1436 View Point, Escondido, CA 92027.
15. *New York Times,* 27 July 1976.
16. W. D. Metz, *Science* 188, (1975): 136.
17. D. Hafmeister, *Am. J. Phys.* 47 (1979): 671.
18. R. A. Herendeen, T. Kary, and J. Rebitzer, *Science* 205 (1979): 451.
19. T. Cole, *Science* 221, (1983): 915.
20. P. E. White, *Geothermal Energy,* U.S. Geological Survey Circular 519, 1965.
21. P. Kreiger, *Ann. Rev. Energy* 1 (1976): 159.
22. J. Lear, *Saturday Review,* 5 Dec. 1970, p. 53.
23. R. C. Axtmann, *Science* 187 (1975): 795.

24. C. Flavin, *Futurist* 19, no. 2 (1985): 36; DOE circular CE-0117 (1985).
25. J. W. Shupe, *Science* 216 (1982): 1193.
26. I. R. Straughan, *Middle Management Perspectives Program* (Rosemead, Calif.: Southern California Edison, 1985).
27. R. A. Kerr, *Science* 207 (1980): 1455.
28. J. Melcher, *Environment* 20, no. 2 (1978): 20.
29. T. H. Maugh II, *Science* 178 (1972): 599.
30. L. F. Diaz, G. M. Savage, and C. F. Golueke, *CRC Critical Reviews in Environmental Control* 14 (1985): 251.
31. C. F. Golueke and P. H. McGauhey, *Ann. Rev. Energy* 1 (1976): 257.
32. *New York Times,* 3 Nov. 1976.
33. D. Morris, *Alt. Sources Energy* 80 (1986): 14.
34. W. K. Steven, *New York Times,* 9 March 1986.
35. A. A. Narvaez, *New York Times,* 4 August 1987.
36. J. Barron, *New York Times,* 11 Dec. 1983.
37. J. T. McQuiston, *New York Times,* 10 Jan. 1982.
38. J. Mayer-List, *die Zeit,* 18 June 1982.
39. S. Bronstein, *New York Times,* 23 June 1985.
40. D. Marier, *Alt. Sources Energy* 80 (1986): 19.
41. H. A. Bell, Plant Administrator, Div. of Electricity, City of Columbus, Ohio, 1985. Pamphlet, "Trash to Electricity—an Environmental Plus."
42. R. Smothers, *New York Times,* 14 Jan. 1979.
43. M. Bishop, *Newark Star Ledger* (Newark, N.J.), 24 Dec. 1984.
44. J. J. Reilly and G. D. Sandrock, *Sci. Am.* 242, no. 2, (1980): 118. See also W. Sullivan, *New York Times,* 29 March 1978.
45. R. S. El-Mallakh, *Environment* 23, no. 3 (1981): 30.
46. D. P. Gregory and J. B. Pangborn, *Ann. Rev. Energy* 1 (1976): 279.
47. T. H. Maugh II, *Science* 178 (1972): 849.
48. W. E. Winsche, K. C. Hoffman, and F. J. Salzano, *Science* 180 (1973): 1325.
49. E. Greenbaum, *Science* 230 (1985): 1373.

PROBLEMS AND QUESTIONS

True or False

1. Waste material is an important energy resource.
2. Geothermal energy does not involve any sort of pollution.
3. It will be some time before superconducting magnets are developed enough to be used for energy storage.
4. Devices using latent heat are useful because the energy storage takes place at one temperature.
5. The energy value of municipal waste is comparable to that of lignite (soft brown coal).
6. An energy distribution system using hydrogen as the carrier involves evolution of no pollutants, since hydrogen burns with oxygen to form water.
7. Energy storage is an important component of any system almost always as base load.
8. The use of flywheels to store energy in the spinning mass is a totally new concept of energy storage.

9. Storage of energy as compressed air in underground rock formations is more attractive than storage of water because more energy per unit volume may be stored in compressed air.

10. Most lead–acid batteries such as used in cars are not generally suitable as the "fuel train" of cars because the batteries can be recharged only a few hundred times.

Multiple Choice

11. Hydrogen is safer to store than
    a. propane.
    b. kerosene.
    c. gasoline.
    d. liquid natural gas.
    e. all of the above energy sources.

12. If the average power use in North America is 1 kW per person, about how long would our stockpile of nuclear weapons (about 20,000) last if PACER were to provide all the power?
    a. 1 day
    b. 20 days
    c. 100 days
    d. 1000 days
    e. 20,000 days

13. Which of the following methods of energy storage is unlikely to be useful right now?
    a. Air is compressed and stored in exhausted natural gas formations.
    b. Water is pumped uphill into a reservoir.
    c. Water is electrolyzed into hydrogen and oxygen, which are stored in tanks.
    d. Radiant energy is stored as chemical energy in carbohydrates.
    e. Epoxy flywheels are spun up to very high rates of rotation.

14. Which method(s) of thermal energy storage appear(s) most promising for large-scale use?
    a. Storage of hot water in underground aquifers
    b. Storage of thermal energy in materials that change phase
    c. Storage of large amounts of hot water in many residences, allowing utilities to shift load when necessary
    d. Storage of energy in very cold materials such as superfluid helium at 4 K
    e. Storage of thermal energy in the specific heat of Glauber's salt

15. For what reason(s) given below was trash-burning adopted more readily in Europe than in the United States?
    a. The heat content of European trash is greater than that of American trash.
    b. The amount of available land for landfills in Europe is much smaller than that in America.
    c. Europeans are more innovative than Americans.
    d. The amount of waste generated per person is greater in Europe.
    e. It is easier for anaerobic digestion of trash to occur in American wastes than in European wastes.

16. Which of the following is *not* useful in setting up a trash-burning generating facility?

a. Separation of large metal appliances before burning

b. Holding the proportion of glass to under 20 to 30 percent to prevent clinker formation

c. Sale of ash or residue from the power plant for use as paving material

d. Removal of compressed gas cylinders

e. Increasing the proportion of petroleum-derived plastics in the refuse stream because of their substantial heat of combustion

17. Which fuel listed is greatest in heat of combustion?

a. Propane

b. Municipal trash

c. Carbon monoxide

d. Gasoline

e. Methane

18. Which fuel listed has the least heat of combustion?

a. Propane

b. Municipal trash

c. Carbon monoxide

d. Gasoline

e. Methane

Discussion Questions

19. Discuss the environmental hazards of geothermal installations.

20. What dangers could the PACER plan pose for the environment?

21. Why should pipeline transmission of natural gas or hydrogen be any less expensive than electric transmission? What questions would you have for someone proposing a hydrogen energy economy?

22. Are all energy storage devices inefficient, in that they take more energy in to "recharge" than is put out as useful work? Discuss whether or not energy storage is cost-effective.

19

CONSERVATION: AN IMPORTANT ENERGY SOURCE

In considering the supply of energy available to heat or cool houses or produce electricity, it is important to understand that major savings have come from using less to do the same amount— a practice known as conservation. Uses of energy that make do on less reduce the need to build new generating facilities, saving everyone money. In this chapter, we consider different ways of achieving this goal.

KEY TERMS   *energy conservation*
visual acuity
incandescent
fluorescent
R-value
thermal conductivity
superinsulated house
infiltration
payback time
passive solar
direct gain
Trombe wall
sunspace
radiant heat
pulsed combustion furnace
condensing furnace

heat pump
Rolladen
demand water heater
life-cycle costing
variable-speed motor

HOW "CONSERVATION" DIFFERS FROM "CONSERVATION OF ENERGY"

In Chapter 3, we discussed the definition of energy and the principle of conservation of energy: that energy can be neither created nor destroyed, but rather that the forms of energy can be interchanged, one transformed to another. Since the first modern energy crisis in 1973, people have been speaking of energy conservation in another sense. What is meant by conservation in this chapter is the provision of goods or services with a smaller drain than at present on high-quality energy resources (ultimately fossil fuels).

The science of thermodynamics has shown us that the efficiency of any heat engine will be less than that of the Carnot engine operating between the same high and low temperatures. Even with the Carnot engine, there must be waste heat rejected. With real engines, more energy is wasted than in a Carnot engine. If we could devise a more efficient real engine, we could reduce the amount of waste heat, and thereby increase useful work, for the same amount of heat taken from the high temperature reservoir. Thus, more of the fossil-fuel energy would have done useful work. Similarly, a device that produced the same amount of work for a smaller electricity input would have saved fossil fuel, which would not have to have been burned. These measures to increase useful work for the same energy input or to decrease the amount of fuel for the same energy output are conservation measures. The 55 mile per hour speed limit is a conservation measure: It requires a smaller amount of fuel to go the same distance at a slower speed.

Most people view conservation measures as "doing without." These people would rather not have to think about such an unattractive topic. They are not interested in energy per se but only in having life made as comfortable and convenient as possible, and moreover, getting the comfort at the lowest cost. In many cases, however, rather than making people have to "do without," energy conservation provides similar or greater comfort to people at reduced cost. People's preconceptions may prevent them from being able to get what they want.

Often people do not have a unified concept of the entire energy delivery system. Consider home heating by gas as an example. Gas is brought into the home by pipeline. It is brought to the furnace. To use the furnace for heating, we need a pilot light, a thermostat, a valve in the furnace, a plenum, a forced air system (fan), and furnace ducts and vents. This delivery system must be considered as a whole in design, not as a series of separate components, if it is ultimately to function best as a unit. People who do think in terms of systems are more likely to realize the positive aspects of conservation.

In this chapter, we discuss conservation in lighting, in space heating, and in air conditioning, as well as passive solar techniques, conservation in action in West Germany, and conservation in industry.

SAVINGS IN ILLUMINATION

According to a study by General Electric (GE), about 5 percent of all the energy generated in the United States is spent on lighting (1). (This estimate is slightly larger than reported by the Stanford Research Institute report on energy conservation [2].) With lighting taking such a large share of total energy use, any savings here would be welcome.

The first suggestion we can make is for people to turn out lights as they leave a room. When a 100 W incandescent bulb is turned on, some energy is used to heat it up. In fact, it is said to take 2.13×10^{-5} kWh$_e$ to heat up the 100 W light bulb (3). Since 1 kWh is 3.6×10^6 joules, 2.13×10^{-5} kWh is about 76.7 joules. This is the amount of energy it would take to lift a 1 newton weight 76.7 meters or a 1 pound weight 56.3 feet—quite a lot. To determine how long you should be out of a room before you plan to turn out the 100 W bulb, recall that if power consumption is constant, the total amount of energy used in a time t is energy = Pt, where P is the constant power. The energy used in turning on the light bulb is 76.7 joules; at a power of 100 W, this amount of energy is used in a time of only 0.77 seconds. That is, if you were to leave the room for one second, you would still save energy by turning off the light bulb. This is not quite true because there is a surge that occurs every time the bulb is turned on. This surge fatigues the filament, shortening bulb life. A good compromise here is probably a minute: If you plan to be out of a room more than a minute or two, turn off the bulb!

Residences and business establishments should use sunlight for illumination as much as possible. A business office, for example, could use photocells to turn out the lights if the illumination from outside is bright enough. Such a device has the added advantage that it reduces the cooling load of the office.

Levels of illumination recommended to architects have been repeatedly raised over the years until they are about twice the European levels of illumination (4). New York City's recommended level of illumination (in lumens per square foot) increased from 20 in 1952 to 60 by 1971 for the schools (5). Modern office buildings typically have illumination levels of 80 to 100 lumens/ft² (1). Such higher illumination levels are not necessary. Visual acuity levels do increase somewhat with increases in light, but as Figure 19.1 makes apparent, acuity increases only a few percent in doubling lighting levels. There is probably room for energy savings here. In addition to energy conservation, we must consider the comfort of people using the lighting and the space in determining proper illumination levels. People apparently feel better about working where the light is brighter and feel best working in natural light.

A third area of energy savings involves the decision about which type of light bulb to use. Table 19.1 indicates the efficacy of the various types of light bulbs (1). Fluorescent lighting uses less than a quarter as much energy as incandescent lighting. A normal 100 W incandescent bulb

TABLE 19.1

Efficacy of illumination measured in lumens/W$_e$.

Type	Power	Efficacy
Incandescent lamp	40 W	12
	100 W	18
Fluorescent lamp	40 W + 13.5 W	59
	75 W + 11 W	73
Sodium vapor lamp	180 W + 30 W	154
Mercury vapor lamp	400 W + 26 W	53.5

NOTE: The lumen measures effective illumination light per unit time. The ballast power demand is also listed where applicable.
SOURCE: Reference 1.

puts about 95 percent of its power into heat and only 5 percent into light. A fluorescent bulb uses about 80 percent of its power in heat and about 20 percent of its power as light (6,7). For this reason, about 70 percent of our commercial lighting is fluorescent (7); however, overall, two-thirds of our lighting is incandescent and one-third fluorescent (1). New incandescent bulbs introduced by GE are three times as efficient and last four times as long as ordinary light bulbs but sell for around $10 (8,9). Such bulbs are cost-effective relative to ordinary bulbs, even if used only half an hour per day; at an hour's use per day, the return on investment is 13 percent, even at an electric cost of 6 cents/kWh (10). Also important are new designs that could increase electrical energy-to-light efficiency by a factor as high as 4 in fluorescent lamps (11).

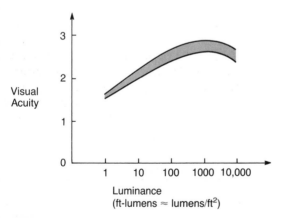

FIGURE 19.1

Visual acuity in people depends on the luminance, measured in lumens/ft². Acuity becomes saturated, following a logistic-shaped curve so that, beyond a certain level of illumination, extra light does not contribute much to acuity.

SAVINGS IN SPACE HEATING AND AIR CONDITIONING

As we can see from Table 19.2 (1), space heating and air conditioning account for about 20 percent of U.S. energy use. Space heating is the largest home consumer of energy. In fact, space heating, water heating, refrigeration, and cooking combined use 95 percent of commercial energy and 85 percent of residential energy (10). Space heating now accounts for about a third of our energy

TABLE 19.2
Perspective on efficiency
of energy use.

Use	% U.S. Fuel Consumption	Second-law Efficiency
Space heat	18	0.06
Process steam (industrial)	17	0.25
Auto transport	13	0.1
Truck transport	5	0.1
Water heat	4	0.03
Air conditioning	2.5	0.05
Refrigeration	2	0.04

SOURCE: Reference 1.

use, while in 1900 it accounted for about half, and in 1850 it comprised the huge majority (7,12,13). Clearly, energy savings even on an individual level can be extremely important. In Chapter 7, we saw that second-law efficiencies are a better measure of energy efficiency than the usual first-law efficiency. The very small second-law efficiencies in the U.S. energy economy, shown in Table 19.2, indicate vast room for improvements.

There have been several studies of the effects of government measures and energy prices on conservation using an energy model developed at Oak Ridge National Laboratory (13,14). These studies indicate that increasing efficiency and insulation measures will save a substantial amount of energy cumulatively: four years' worth of direct energy use in buildings by the year 2000. It will also create new jobs (an estimated 425,000 overall).

To illustrate the possible effect of price rises, a rather early study (15) claimed that sharp gas price rises were ineffective in causing changes in household energy use. Despite that, gas furnaces available on the market are now up to 95 percent (first-law) efficient, while pre-1980 models are only 50 to 60 percent efficient (16). There is a time lag between the cause and the response in many of these matters. By the end of the 1970s, many appliances were much more efficient than at the start of the decade: Refrigerators are 45 percent more efficient; freezers, 48 percent; dryers, 8 percent; room air condi-

tioners, 12 percent; and dishwashers, 50 percent (9,17).

Appliances, particularly refrigerators and freezers, still have great potential for energy savings (17,18). In fact, even on current models, the connection between price and energy efficiency is essentially nonexistent (18). The role of certain states, such as California, in setting energy standards has had an exemplary effect on national efficiences. By 1992, refrigerators sold in California will be required to be 150 percent more efficient than those sold in 1972, while by 1986, refrigerators overall had been improved by over 70 percent relative to 1972 (19).

Insulation

Insulation is one of the most important factors in reducing leakage of heat from the home. About 12 percent of conductive heat losses are through the ceiling, and 17 percent through walls (20). With greater recognition of the increasing expense of energy, accepted insulating standards have been increasingly tightened. The 1965 FHA standards allowed 0.6 kWh/10^3 ft³/day heat loss in the installation of home insulation. By 1970, this had been reduced to 0.45 kWh/10^3 ft³/day; and by 1972, to 0.3 kWh/10^3 ft³/day. Present buildings consume about 40 percent more than they would if the 1972 standards had been enforced earlier (21); if all buildings met these standards, there would have been a 13.5 percent national decrease in energy consumption.

R-VALUES

In heat conduction, the rate of heat flow depends on the temperature difference between sides, the thickness, and the area in contact. The greater the temperature difference, the greater the heat flow. The greater the area in contact, the greater the heat flow. The shorter the distance of conduction (the thickness), the greater the heat flow. The connection between heat flow and these quantities is called the *thermal conductivity* or the *thermal resistance,* depending on how the relationship is written. We may write:

heat flow rate $= kA \, \Delta T/t$, or alternatively,

R(heat flow rate) $= A \, \Delta T$,

where k is the thermal conductivity, R is thermal resistance, A the area in contact, t the thickness, and ΔT the temperature difference. These equations describe the heat conduction rate through a plate of material of face area A and thickness t, when the two sides of the plate differ in temperature by ΔT.

For fixed values of R, A, and t, the heat flow increases if ΔT increases. Therefore, the warmer your house is kept relative to the outside temperature, the more heat you will lose by conduction. Since making the walls of a house thicker is not usually feasible, and since the area in contact with the outside is fixed, we can see that the heat flow can be reduced (according to the equation) either by increasing R for a fixed ΔT, decreasing ΔT for fixed R, or some combination of the two. R may be increased by adding insulation to ceiling, walls, and floor. You can decrease ΔT for at least part of the day by turning down the furnace when no one is home or at night when everyone is sleeping.

As mentioned in Chapter 7, if we look at the temperature profile of a window exposed to cold outdoors with a heated inside, there is a boundary layer of air of intermediate temperature on both sides of the window. If the air outside is in motion, some of the heated outside air is moved away rapidly, increasing the rate of heat flow (or, alternatively, effectively decreasing the R-value). The same principle applies to a house. When a cold wind blows, the boundary layer is much smaller, and the effective R-value decreases.

Dry air does not conduct very well. Most insulation works by trapping air so that it cannot move. Dry, relatively still air an inch thick has an R-value of about 1. Completely motionless air an inch thick has an R-value of 3 to 4. Some selected R-values are listed in Table 19.3

The R-value is in common use, rather than the conductivity k, because R-values add. This is easily seen from our defining equation. For a given fixed heat flow rate and area A, we have, for each layer, the equation as given. If all the individual ΔTs

In Canada, which generally experiences extremely cold winter temperatures, there was an 11 percent decrease in energy per household between 1976 and 1982 (22). Had *no additional insulation* (compared to previous practice) been installed in Canada after 1978, the 1985 demand would have been the equivalent of 60 to 80 million barrels of oil higher (22). The installation of such additional insulation alone accounts for half the decrease in energy use. Canadian homes are now much better insulated than was the case in bygone years (see Table 19.3 for the economic

TABLE 19.3

R-values of common construction materials per inch of material.

Material	R-value/Inch
Air	1.44
Rock wool (batt)	3.38
Fiberglass (batt)	3.16
Rock wool (blown)	2.75
Fiberglass (blown)	2.20
Cellulose (blown)	3.67
Vermiculite	2.20
Perlite	2.75
Wood (av. pine)	1.28
Wallboard	1.0
Brick	0.11
Glass	7.2

SOURCE: Reprinted with permission from *Handbook of Chemistry and Physics,* 32nd ed., Copyright 1950, CRC Press, Inc. Boca Raton, FL, and D.O.E., *Insulation* (Washington, D.C.: GPO, 1980).

are added up, we get expressions like $(T_1 - T_2) + (T_2 - T_3) + (T_3 - T_4) + \ldots$ Clearly the sum of all these is just the overall ΔT. If we add the ΔTs on the right-hand side, we must also add the terms R (heat flow rate) on the left-hand side to get

$$(R_1 + R_2 + R_3 + \ldots) \text{ (heat flow rate)} = A\Delta T;$$

this shows that the R-values add.

To illustrate how this property of additivity might be used, consider a typical house wall. From the inside out, there is still air along the wall (R-0.68); wall board (R-0.45); air space (R-1.01); sheathing over the studs (R-1.32); wood siding (R-0.81); and the outside air that moves, thus contributing very little (22). Ignoring the contribution of outside air, which depends on the weather conditions, the R-value for the house wall is

$$0.68 + 0.45 + 1.01 + 1.32 + 0.81 = 4.27.$$

Clearly, adding R-11 or R-19 insulation will decrease heat loss substantially for this house.

The units used to measure R-values, as presented in the table, are common English units, ft² °F/(Btu/h). In the metric system, the unit would be m² °C/W.

incentives for Canadians). Some houses are constructed with R-50 in the walls, R-70 in the ceiling, and R-10 in the floors (23) (see box). Such homes are said to be *superinsulated*. The additional costs of such housing range from 5 to 10 percent of the purchase price; the mean cost of such extras in Minnesota is $44/m² of living space (typical homes have 138 m² of living space) (10).

Typical North American houses built in 1980 have an annual energy use of 122 kJ/m²/degree-day (°C). The energy use in the superinsulated Minnesota homes averaged 51 kJ/m²/degree-day

(°C) (10). There are also economies with super-insulated homes that are not readily apparent. For example, the heating system of such a home is much smaller than that of an ordinary home built in a similar climate. In Uppsala, Sweden, a 130 m² superinsulated house uses only 110 MJ/m²/yr as compared to the usual 800 MJ/m²/yr (24). Sweden is marketing superinsulated homes in the United States. They are prefabricated and shipped in crates. A Swedish-built house costs about $30 to $35/ft², comparable to the late 1980s cost of new American housing (25). However, the additional costs of superinsulation are probably not justified except in areas of sustained winter cold. Figures 19.2 a and b show the number of heating and cooling degree-days for the United States and the recommended insulation thicknesses.

In 1973, Hirst and Moyers (20) analyzed homes in Atlanta, Minneapolis, and New York in terms of minimum property standards (MPS), revised MPS (RMPS), and "optimal" economic maximization. Possible savings are substantial; the New York results presented in Table 19.4 give a representative picture of the poor state of insulation in typical houses prior to the substantial rise in prices for petroleum. Such savings from insulation remain possible today. California has some of the highest thermal resistance (insulation) standards in the country. For homes, the standards are R-30 for ceilings, R-19 for walls, and use of R-1.5 double-glazed windows (26). These are much greater than envisaged in the 1973 study

as the "optimal" values, and reflect the higher fuel prices resulting from the 1973 and 1979 crude oil price hikes.

A recording method called PRISM has been developed at the Center for Energy and Environmental Studies at Princeton. PRISM could be used to gather valid data and evaluate how well or poorly various conservation methods work (27). Several experiments showed that it was possible to save as much as 10 percent of home energy use by a single visit from a "house doctor" (28,29), and that major rebuilding efforts could save as much as 20 percent (28,30). Some utilities are even supporting insulation measures for customers as a way to avoid the cost of building new generating plants (31).

Of course, insulation does not necessarily prevent or even minimize heat flow to the outside. Air infiltrates a house through cracks and when a door is opened for entry or egress (convective losses). As a result, there is some approximately steady rate at which heat must be supplied to a house, in addition to that which would be necessary to maintain a temperature difference between inside and outside because of conduction through exterior surfaces. This heat is used to warm up the outside air to interior room temperature. This depends to some extent on weather conditions outside (25). In times of high wind, transfer of air will be greater than at other times.

It has also been found that more heating is necessary on cloudless, still nights than on cloudy, windy nights. This indicates that radiation is an

TABLE 19.4
Energy comparison of several insulation alternatives.

	MPS		RMPS		MAX	
	Gas	Electric	Gas	Electric	Gas	Electric
Wall	0	R-6	R-6	R-6	R-11	R-11
Ceiling	R-6	R-6	R-11	R-11	R-11	R-19
Floor	no	no	yes	yes	yes	yes
Storm windows	no	no	no	no	yes	yes
Energy reduction (%)	0	0	29	19	49	47

SOURCE: Reference 20, copyright 1973 by AAAS.

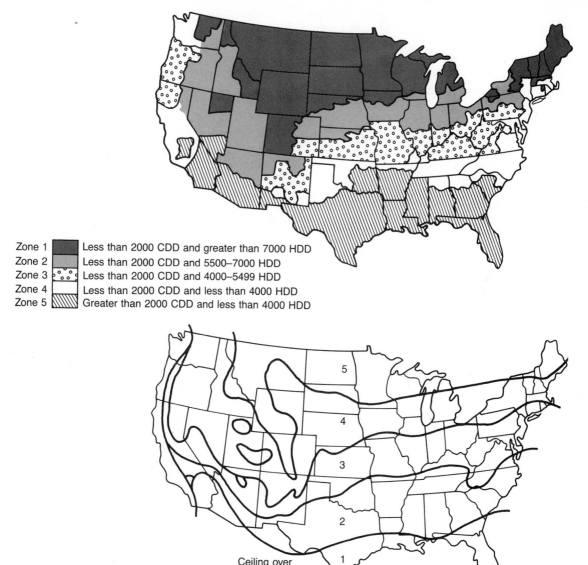

Zone 1	Less than 2000 CDD and greater than 7000 HDD
Zone 2	Less than 2000 CDD and 5500–7000 HDD
Zone 3	Less than 2000 CDD and 4000–5499 HDD
Zone 4	Less than 2000 CDD and less than 4000 HDD
Zone 5	Greater than 2000 CDD and less than 4000 HDD

Heating Zone	Attic Floors	Exterior Walls	Ceiling over Unheated Crawl Space or Basement
1	R-26	Full-wall	R-11
2	R-26	insulation,	R-13
3	R-30	which is 3 1/2″ thick,	R-19
4	R-33	is approximately	R-22
5	R-38	R-14	R-22

FIGURE 19.2

(a) Weather zone map of heating degree days (HDD) and cooling degree days (CDD). One degree day is referred to as 65°F. HDDs are given as the number of degrees the mean daily temperature falls below 65°F on each day. CDDs are given as the number of degrees the mean daily temperature rises above 65°F on each day. HDDs and CDDs are cumulative. (U.S. Census Bureau). (b) Climatic zones map for determining winter insulation needs. Zones 1 and 2, in the sunbelt, need more insulation than indicated to repel summer heat. (U.S. Department of Energy) important energy loss. Radiation loss is large through windows; for double- or triple-glazed windows, it is larger than any other source of energy loss. In fact, about 5 percent of total national energy consumption has been attributed to windows (27).

Indoor Air Quality

Infiltration of outside air is one of the important problems in space heating and air conditioning. Air exchange accounts for about 40 percent of heat loss in the home (1). (Estimates range from 25 to 50 percent [4].) One problem is control of those areas getting outside air: The kitchen and the bathroom have high ventilation requirements. If infiltration could be reduced elsewhere, and directed to high-use areas, large energy savings could result.

The American Physical Society study on efficient uses of energy (1) found that total energy supplied to heat and humidify the incoming air is almost as large as (75 percent of) the energy supplied to compensate for the conductive losses through the shell. The study targets 0.2 exchanges of air per hour as sufficient for ventilation requirements while minimizing energy waste. Existing infiltration rates in American homes are about 1 to 1.5 changes per hour (28). The superinsulated Swedish home has only one-fifth of an air change per hour (25). Because houses are so leaky, it is important to try to minimize the leaks. A technique called pressurization appears to be very useful in finding leaks (32). With the house closed up, a powerful fan over- or underpressurizes the house to a much greater extent than expected for normal weather. The leaks can be found easily from influx or efflux of air by inspection and then plugged.

It has been estimated that capital expenditures on thermal upgrading of existent structures can pay for themselves in five years (or less, if energy costs resume their upward spiral). An analysis of the payback time for simple efficiency changes in appliances, which have not yet been made, show that even with short payback times, the changes are not made (33). Consumers and producers seem to demand unrealistically short payback times in making energy-conserving choices (10,33). Time and again, savings from efficiency investments were underestimated (34). Consumers underinvest in energy efficiency.

Of course, with superinsulated houses, it is necessary to provide for sufficient new air. It is reasonably easy to put in a heat exchanger to minimize heat loss (23). But with most people at home for sixteen hours a day, any deterioration in air quality at home can be very worrisome (35–37). For example, urea formaldehyde foam insulation has been abandoned in the United States and Canada because of high formaldehyde levels in homes (37). Kerosene heaters can cause air quality degradation and emit toxic and carcinogenic gases (36,37). Gas stoves can produce carbon monoxide levels of 200 to 700 $\mu g/m^3$ (the U.S. National Ambient Air Quality Standard is <100 $\mu g/m^3$). Brickwork can emit radon, a radioactive gas with decay products that can easily become lodged in the lungs (35–38). (This is discussed in more detail in Chapter 24.) Higher levels of eleven chemicals have been found inside homes than in the outside air. These are generally from solvents used in the house and stored indoors (35,38,39). The exposure to in-home pollutants for people who live in clean-air environments is about the same as for people who live in dirtier areas (39). It has been estimated that as many as 10,000 people die each year of cancer brought on by exposure to radon in the home. Even if the inhabitants do not smoke, all these considerations imply the necessity of sufficient air exchange.

PASSIVE SOLAR ENERGY AND CONSERVATION

In Chapter 15, we discussed active solar energy in residences. We have deferred to this chapter the discussion of passive solar energy, because modern building practice works with both conservation and passive solar on the same basis. It makes sense to treat them simultaneously, so that information on both will be available to those of you who want to design a new home, or repair or rebuild your old house, as well as to architects.

The Romans, in response to deforestation of the land around Rome, were forced to bring wood from farther and farther (up to 1000 km) for heating. In response to this pressure, the Romans began to design and build solar homes using glass (40,41). Since that first use of passive solar heat, the technology was developed and then lost twice. The losses occurred in the Dark Ages, as much knowledge was extinguished, and again in the late nineteenth to the late twentieth century, in the age of cheap energy.

There are three basic types of passive heating systems: direct gain systems, storage wall systems, and sunspaces. In passive solar systems, it is essential to have a net heat gain from the windows or other openings during the heating season, and helpful to have a net loss over the course of the cooling season.

Direct solar heat gains are possible through windows, which presently account for 30 percent of all heat losses (1). On a sunny day in winter, the peak radiation energy flux from the sun can be about 1 kW/m² on the ground. Figure 19.3 shows the average energy gain possible through windows facing in various directions (1). Another factor influencing the amount of sunlight received is the overhang on the window; it is possible to design windows so that summer sunlight, when the sun is high in the sky, is not incident on the window, while winter sunlight, when the sun is low in the sky, is captured by the window. Deciduous trees can also shield a house from summer sun while in leaf, and let winter sun through bare branches.

As mentioned briefly in Chapter 15, Trombe walls—masonry walls for storing thermal energy, with the outer surface blackened and glazed— make good sense. When the sun shines, the masonry (which has a large heat capacity) can absorb solar energy. This thermal energy is re-emitted in all directions as infrared radiation and can help make the house more livable. Water walls (same idea as the Trombe wall, but using barrels of

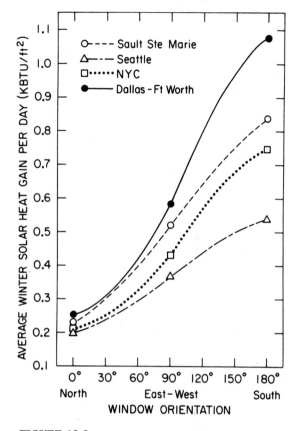

FIGURE 19.3
The daily average solar heat gain for four U.S. cities. (From A.I.P. Conference Proceedings #25, *Efficient Uses of Energy,* copyright 1975 by American Institute of Physics)

water instead of masonry to store the solar energy) and solaria (sunspaces) also help to mitigate day–night temperature swings. A 0.35 m thick wall will be warmest after the heat has been conducted all the way through twelve hours later (39). It has been found that the Trombe wall with natural circulation provides the greatest heat flux to the living space (42).

The idea of the Trombe wall is old. Middle Eastern architecture used (as did Latin American) thick-walled mud-brick houses in hot regions to provide natural cooling by day and warming by night (since the desert sky is usually clear, warm objects radiate their heat away very quickly [43]). Old architectural ideas emphasize courtyards (to help midday air circulation), narrow streets (to prevent large areas from being baked in the sun and to protect passersby), and houses built chock-a-block, sharing walls to prevent too much exposure to the incident solar energy.

The sunspace, a 1970s idea, is actually a revival of the sun porch. Sun porches, first developed by the Greeks, were brought to a high stage of art in America in the eighteenth and nineteenth centuries, only to vanish in the early twentieth century (41). Sunspaces are south-facing vertical walls with brick backing, installed to take advantage of incident solar energy. Such spaces can be very pleasant, in addition to helping to decrease home energy consumption. It is important not to have too much glazed area, or sloped glass with little or no effective shading in sunspaces, in order to prevent overheating in summer (44).

Radiant Heating

The time delay inherent in massive masonry or stone structures is well known. Cobblestone houses take two months to warm up or cool down (45). Meanwhile, they provide a comfortable environment for people because radiated heat makes people feel more comfortable. Warm walls allow the air temperature to be reduced with maintenance of comfort (46,47). The temperature used to determine comfort is the mean radiant temperature, the average of room air temperature and wall temperatures. This would seem to indicate that the folks who are insulating their cobblestone houses on the inside (45) are ignoring the better return on their investment they would get if they were to install heating onto the walls to maintain their temperature and take advantage of insulation on the stone even in winter. This is becoming more common practice in Germany (47). A technique applied both here and in Europe involves radiant heating through the floors with plastic pipe (46,47). The pipes are polypropylene or polybutane, and medium-temperature water (30°C–50°C) is circulated through the pipes,

which are embedded in the concrete floor (46). The system scheme is shown in Figure 19.4 (46).

Quartz electric heaters increase comfort levels by providing radiant heating. A more speculative proposal along these lines was the suggestion that homes use microwave radiant energy to provide warmth directly (48). It is said to be more effective than infrared, and works by providing a radiant temperature component to raise comfort levels. However, there may be hazards to such use of microwaves (49).

The number of people in the house can also influence heating requirements. People are approximately 100 W heat sources, as we saw in Chapter 3. It is for this reason that we see open windows and doors in winter at crowded parties.

Passive Solar Techniques of Earlier Times

Many natural heating and cooling techniques were developed in the Seistan region (on the Iran–Afghanistan border), where the windmill was invented in the eighth or ninth century (windmills reached Europe by the twelfth century). Some of these techniques can work elsewhere as well. Thick adobe walls act as insulators and will mitigate temperature swings (as mentioned above). Adobe wind towers act to cool in all sorts of summer weather. If there is no wind at night, the warm air rises and draws in cool air (50,51). If a wind blows, the circulation is reversed (51). In daytime, the cool tower causes air to sink, causing a flow of cooling air. If the air passes over water, it cools still more as it picks up water vapor (51) and loses latent heat of vaporization. Air vents or roofs serve a similar function, causing hot air near the ceiling to be pulled out, cooling the building (51). Wetted bushes placed outside windows can cool a room by 8°C if a breeze is blowing (50). The Seistanis also designed natural refrigerators and freezers, although they used stagnant water, which often led to health problems.

The open areas in buildings in Saudi Arabia's semidesert areas have high walls with openings reminiscent of Iranian towers. The air may be

FIGURE 19.4
Slabs may be heated by pumping warm water through embedded plastic pipes to provide radiant heat to room inhabitants.

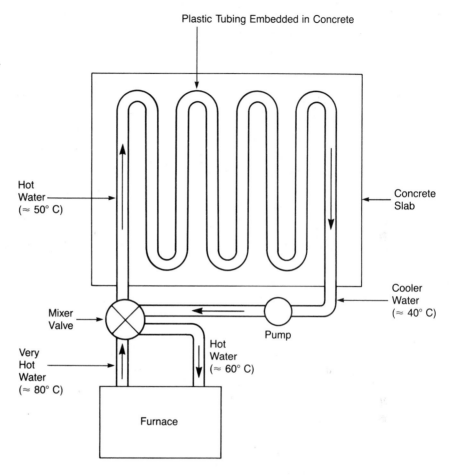

excluded from the rooms or brought inside by sliding wood doors. The streets are winding, and the houses have courtyards just as in desert areas, for the same reasons (43).

In hot and humid regions of Saudi Arabia, similar needs and solutions were found. The louvered "Rowshan," which is ariel- or bay window-like, allows air to flow freely into the room while not collecting heat the way a sunspace would, and providing protection against glare (43). The conical huts seen in hot and humid areas in Africa have roofs of dried thatch that serve much the same purpose—the hot air rises to the peak of the roof, allowing the cooler air to condition the air below naturally (43).

Before leaving the topic of passive solar energy, we might mention that much research on solar energy in less developed countries focuses on its use in agriculture. The use of passive solar energy in construction of poultry sheds (52) and for drying of grains or fruits (53) has been very popular.

In this country, the 1970s brought an explosion of developments in passive solar heating. In order to see which techniques (conservation and passive solar) performed best, the Buildings Energy Use Compilation and Analysis project was initiated. The results of the class-B monitoring program were reported in several places (40,44,54,55). The most important findings were:

Heating costs in multifamily dwellings are smaller per family than costs in single-family dwellings;

heating costs in passive solar homes, which ranged from \$320 to \$915/m^2, are comparable in cost to conventional homes in the same area, which ranged from \$430 to \$760/m^2;

median annual savings are about 500 to 300 kWh;

passive solar met about 37 percent of the total heating load;

thermal loads ranged from 0.83 to 1.53 W/m^2/degree-day (°C);

there was little difference among the types of passive solar device (Trombe wall, direct gain, sunspace) used;

good overall performance of passive solar devices and climate are not connected;

proper site selection, collecting area, and orientation of collectors are very important system characteristics;

movable insulation was not used properly, but has the capability of providing great energy savings in cloudy, colder climates; and

proper insulation is essential to performance.

A program backed by the Bonneville Power Administration demonstrated that retrofitting of houses with insulation and other conventional measures of energy saving was cost-effective (56), as other studies had shown (33,40). The study found that, as compared to an audit model, actual savings due to storm windows and heating ducts were greater than the predictions of the model. The actual savings due to storm doors, caulking, and weather-stripping are much smaller than predicted. Overall, predicted savings were 6200 kWh/yr, while actual mean savings were 4130 kWh/yr, for an average \$2100 savings. Part of the reason for this may have been that many homes that had used wood as a heating supplement—wood was not accounted for in pre-change surveys—switched back to conventional heating.

SAVING ENERGY IN HOME HEATING AND COOLING

Furnaces

Another simple method for reducing heat use is the timing of the furnace. For example, the only reason for heating a house above about 15°C (60°F) while people are sleeping is so that they are not uncomfortable when they get out of bed. By setting a timer to turn the furnace on half an hour before awakening, energy wastage can be cut. (There is no truth to the myth that it is less energy-expensive to keep the house at a constant temperature by running the furnace continuously. Energy loss is proportional to the inside–outside temperature difference, as discussed earlier in the chapter.) If no one is at home during the day, the timer could also be used to turn the furnace on shortly before anyone was expected home.

Even the location of heating ducts can be important for energy conservation. There are energy losses through convection at the window surfaces. If ducts are placed directly beneath windows, losses are increased.

We might recall that the average efficiency at a conventional power plant is about 33 percent, and the efficiency of local electric transmission is about 91 percent. Resistive house heating is 100 percent efficient with floorboard or baseboard heat, but only about 50 percent efficient for ceiling heat. Thus the overall efficiency of electric heating is 30 percent or less, except if heat pumps are used.

The use of gas- or oil-burning in-house facilities can be 40 to 95 percent efficient, with an average of about 60 percent (16,20). The upper figure would represent full-capacity use; typical usage is about half that. Soot buildup can soon lead to substantial loss of efficiency, typically to 50 percent of that expected (4). Some units are as low as 35 percent efficient (57). People should have their gas furnaces checked to get maximum heating from each cubic meter of gas.

The newest gas furnaces are much more efficient than the older ones. The technology of furnaces has improved substantially in the 1980s. The major change has been the development of pulsed furnaces and of condensing gas furnaces (separately or together) (10,17,58).

In the pulsed combustion furnace (58), a cylindrical chamber with one open end and one valved end is supplied with air and gas. A spark starts ignition. The burning of the gas causes the pressure of the air in the chamber to rise, which closes the valve. Combustion continues until the air is exhausted, when the effluent rises. This reopens the valve, and the chamber refills with air. Since it is hot, there is no need for another spark. The furnace cycles on and off in this model.

It operates at resonance, pulling in air as the valve opens, burning the gas–air mixture, and exhausting the effluent to begin again (58). First-law efficiencies of 95 percent are possible with a pulsed furnace.

The pulsed furnace is rather simple to supply, and of simple construction. However, the furnace is somewhat noisy, and the valve (which may be inaccessible) is a maintenance problem, since it opens and closes half a billion times a heating season (58).

Much energy goes up the stack as water vapor in an ordinary furnace. In some new furnaces, this additional thermal energy is extracted from the flue gases, which increases the efficiency (10,17,58). There are several ways of achieving

FIGURE 19.5
Schematic design of a pulsed combustion condensing furnace. These furnaces provide first-law efficiencies of ≈95%. (Courtesy of Lennox Industries, Inc.)

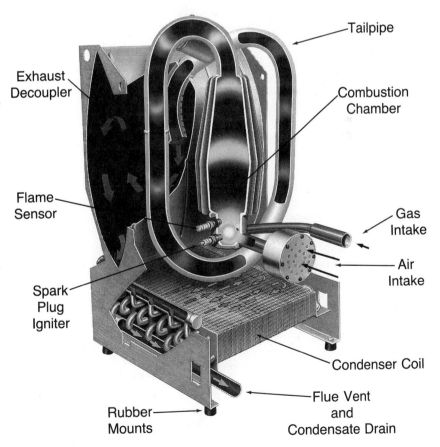

this: stack damping (temperature-activated); vent damping; use of a power burner (a blower forces combustion gases through the furnace); and heat exchanging (17). Some condensing furnaces are over 92 percent efficient. When condensing flue furnaces are used, the exhaust gas is so cool, ($\approx 40°C$) (10), that one may use a PVC pipe to eliminate the gas through a sidewall of the house (10). One of the first models combining pulsed combustion and condensation is shown in Figure 19.5 (17).

Wood Heating

With more people using wood for energy nowadays (up to a third of all households in New England use wood for some space heating) (59), it is important to burn wood more efficiently. Normal fireplaces are only about 10 percent efficient, and there is a net loss in energy from a wood fire when the outside temperature is less than $-5°C$ because of the escape of warm air up the chimney (60). However, new ideas such as use of outside air instead of inside air in the fire, and use of more efficient grates—the "Texas fireframe" is about 30 percent efficient (61)—can also save energy.

Most people who do substantial heating with wood use enclosed wood stoves modeled after the sort first developed by Benjamin Franklin. Franklin touted the stove's energy conservation potential (62):

Wood, our common fuel, which within these 100 years might be had at every man's door, must now be fetched nearly 100 miles to some towns, and make a considerable article in the expense of families.

We leave it to the political arithmetician to compute, how much money will be saved to a country, by its spending two-thirds less on fuel; how much labor saved in cutting and carriage of it; how much more land may be cleared for cultivation; how great the profit by the additional quantity of work done, . . . and to physicians to say, how much healthier thick-built towns and cities will be, now half suffocated with sulfury smoke, when so much less of that smoke shall be made, and the air breathed by the inhabitants so much purer.

Modern stoves, with air-tight enclosures lined with firebrick, can provide warmth for hours on relatively little fuel. People, especially in rural and suburban areas, have installed wood stoves for partial heating of their homes. In the richly forested Pacific Northwest, many city dwellers have wood stoves.

This success of stove technology has caused a problem for many cities and towns in the West, such as Vail and Denver, in Colorado. In Portland, Oregon, on winter days nearly 50 percent of ambient particulates come from residential wood burning (59). Woodstove emissions have essentially wiped out air quality gains from control of industrial emissions (63). Residential stoves emit much more polycyclic organic material than commercial burning (59), and these agents are known carcinogens. Oregon has tackled the problem by first comparing performance of current stove models, and then mandated a 50 percent reduction in particulate emission by July 1986 and a 75 percent reduction by July 1988 (63) in stoves sold in retail stores. This state regulation initiative will probably result in sales of much cleaner stoves nationwide after 1988.

Heat Pumps

Figure 19.6 (64) reminds us, as we learned in Chapter 7, that the freezing point of water and room temperature are very close in absolute temperature, which is a measure of their thermal energy. This means that the amount of thermal energy of an object at room temperature is only marginally greater (≈ 7 percent) than the thermal energy of that object at freezing. It would be wonderful to be able to "piggyback" on all that energy around us even when the temperature is low.

Logs in the fireplace (or even home furnaces) take fuel energy, turn it into high-temperature thermal energy (in the vicinity of the flame), and then mix it inefficiently with interior air to warm us.

The heat pump can use some low-grade ambient thermal energy to increase the efficiency

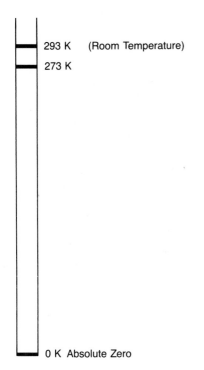

293 K   (Room Temperature)

273 K

0 K Absolute Zero

FIGURE 19.6
Room temperature is only about 7% greater than outdoor freezing temperature. The thermal energy in the environment at temperatures between 273K and 293K can contribute to heat release in a warm (293K) environment when a heat pump is used.

of providing thermal energy at room temperature. A heat pump's operation is similar to that of a refrigerator. The refrigerator cools itself by heating the room it is in. The heat pump warms the inside of a house by cooling outside air (or water), and cools inside air by heating outside air (or water).

Thus, the heat pump allows us to use the thermal energy in the outside (cold) environment, do work on it, and exhaust the thermal energy plus the work at higher temperature, warming the inside of the house. Since there is so much thermal energy in the outside environment relative to absolute zero temperature, the electric

energy input (in theory) can be only 7 percent of the net energy delivered to the interior of the house. Of course, there is a factor of 3 loss in generating the electricity in the first place, and no heat pump is even close to ideal. Nevertheless, an electric heat pump that warms the inside of a house by cooling ground water uses only 45 percent as much fuel as a furnace that is 75 percent efficient for the same amount of heating, *despite* the factor of one-third efficiency loss at the generating plant. A large-scale heat pump using ambient air and having exhaust heat recovery can be an efficient heating or cooling device for large buildings or groups of houses.

Air Conditioning

The insulation put in to reduce heating loss reduces energy loss in air conditioning as well. For the 1972 insulation standards, there are 18 percent savings for electricity and 26 percent savings for gas air conditioners as compared to the 1970 standards (19). Most popular models of window air conditioners use about 4 kWh$_e$ while running. (Commercial consumption for air cooling far exceeds residential consumption.)

The air-cooling component of energy consumption has been doubling about every five years. The saturation index had reached about 40 percent for air conditioners as of 1968 (2), and rose to 60 percent by 1986. Air conditioners have spurred the growth of the American "sunbelt" by making it possible for northerners to live in the South comfortably.

Many models of air conditioner were tested for efficiency. The models ranged from 4.7 to 12.2 Btu/kWh$_e$. Of the ninety models tested at 10,000 Btu/h capacity, the lowest energy use was 880 W; the highest, 2100 W (20). In Washington, D.C., this would entail an energy cost of 976 kWh$_e$ higher for the highest power user as compared to the lowest. At the 1972 Washington cost of 1.8 cents/kWh$_e$, this is an added yearly running expense of $17.57 (and more recent energy costs are more than four times as expensive). If we

What will happen if we need to use less and less energy? Must we lower our standard of living? Will we suffer?

It is difficult to deal with these questions because Americans have no experience in contemplating such possibilities. I spent over a year living in Karlsruhe, West Germany, and I experienced the quality of life in a society that has had to live with expensive energy for quite some time.

One of the most striking things about Europe, even to the casual traveler, is the railroad net knitting the continent together. In the United States, we have systematically dismantled much passenger rail service and erected interstate highways and airports. Europe has its superhighways and airports as well, but it has not allowed the energy-conserving railroad network to decay (Figure 19.7). I was impressed by my first encounter with the Deutsche Bundesbahn (DB), as the railroad in Germany is called. The service offered is good, air fares between European cities are much higher than train fares, and the DB even makes an effort to advertise that using the train saves energy and protects the environment. The system is better patronized than ours because of better reliability and more attention to comfort.

The second thing to strike me as I came to Karlsruhe was the streetcar and bus network (Figure 19.8). Service is extraordinarily good by American standards, beginning early in the morning and continuing late into the night. The system is quite heavily traveled despite the charge of DM 1.50 (90 cents) per ride. There is some grumbling that the service ends too early (it runs only until about 1 A.M.),

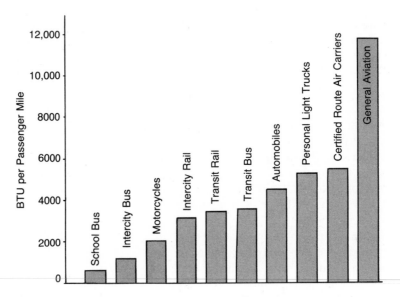

FIGURE 19.7
Efficiency of passenger transport by various modes in 1984. (U.S. Department of Energy)

FIGURE 19.8
Strassenbahn Haltestelle (streetcar stop) on Karl-Wilhelm-Platz, the stop nearest the author's apartment.

forcing someone staying out late to take the taxis, which are very expensive indeed for the people involved, but much cheaper than running empty trolleys.

Part of the reason for the success of mass transit in European cities such as Karlsruhe may well be the lack of parking facilities. Street scenes such as shown in Figure 19.9 are common to the German cities I have visited, and I can attest to the frustration of driving about searching for a parking place. Cars are often illegally parked for lack of an alternative.

It is at this point that I should perhaps say a bit more about the cost of energy and its relationship to modes of transportation. Europe, and West Germany in particular, has had to import virtually all its oil for many years. Germany pays for its crude oil in dollars. As the mark becomes stronger, then, relative to the dollar, the real cost of oil for Germans should fall. It has not. In addition, at the time I was there, food and housing costs in Germany were about twice as high as in America. A half-liter of beer in a local bar cost DM 2.40 ($1.40), a half-kilo of butter about DM 5 ($3.00). The perceived cost of gasoline (relative to normal living costs) should have been low. Yet, as was painfully obvious when I filled the tank of my VW Beetle, gasoline costs in Germany are over twice those in America. Regular gas cost DM 1.16 per liter (about $2.60 per gallon). The whopping European gas price, which is (and has been) responsible for the smallness of the European cars, is set artificially high by a massive tax on gasoline. That is, the government, to save foreign exchange, has taxed gasoline to encourage gasoline (energy) saving, creating the market for smaller, more efficient, cars.

My VW Beetle was called a gas-guzzler by my German friends (I regarded Beetles as economy cars when I lived in America). In Germany, the norm for mileage

(a)

(b)

FIGURE 19.9

(a) A view of Akademiestrasse looking west. The tilted cars are parked on the sidewalk. (b) A view of Akademiestrasse looking east. The sign with the cross and double arrow underneath indicates that parking is forbidden. The author's VW was among the illegally parked cars.

is set by the DIN (Deutsche Industrie Norm) in a manner similar to that of the U.S. EPA: Autos are run in simulation at 90 and 120 km/h and in city traffic. The results are given in terms of the numbers of liters used per hundred kilometers. Perhaps this "reciprocal" relationship to the American standard has something to do with the psychology of the two peoples, in that in America we think that bigger is better (more miles per gallon), while in Germany they think that smaller is better (fewer liters used per hundred kilometers). People want to buy cars that use only 5 to 6 liters/100 km (40 to 45 miles/gal) instead of the 10 to 11 liters/100 km (20 to 24 miles/gal) typical of the Beetle. The German auto industry was certainly not able to ignore the public's demand for energy-efficient cars in the face of gasoline prices.

A surprising number of people bicycle to work in Karlsruhe. The terrain is flat in the Rhine valley, and the weather is usually mild, which may encourage biking. Certainly many more students bike in Karlsruhe than at Ohio State University (Figure 19.10). I encountered many ordinary people, old as well as young, riding bicycles on the street, carrying groceries and doing other things matter-of-factly, in addition to those riding for pleasure. The bicycle seems integrated into the city's life in Karlsruhe.

It is not only in the field of transportation that Germans save energy. Natural gas and electricity are also expensive, and because a lot of energy gets used in water heating, it is common to heat water at the point of use. German washing machines and dishwashers have heaters and use only cold water input. In many homes there is a reservoir above the sink in which water is heated electrically or by gas as it is needed. It is turned on only when in use, which does mean an annoying five to ten minute wait for hot water if it is an electric heater. In the bathroom, hot water is supplied by a gas burner (Figure 19.11) so that the flow of hot water can be continuous on demand. These are now available in the United States. I recently bought a demand water heater to replace the storage water heater to supply my

FIGURE 19.10
Glimpse of the bicycle parking place by a student apartment house, Ha-Di-Ko, in Karlsruhe. The entire open area is filled with bicycles.

FIGURE 19.11
A European bathroom water heater. A shower curtain is visible to the right in the photo.

family with hot water. The unit, shown in Figure 19.12, has worked very well since hookup was completed. Many companies are now beginning to offer demand heaters (66), even though they cost about $600 (three times as much as a conventional water heater).

Washing machines can be set for cool-water wash for delicate fabrics (30°C), warm-water wash for normal clothes (60°C), or so-called cook wash for bedding and underwear (90°C). Heating at the point of use is much more efficient than central water heating because there is no energy loss in transmission. The temperature can be closely controlled and is set only as high as necessary. For example, there are different chemical reactions taking place in the detergents at the different temperatures, as anyone who has used a cold-water soap in a hot wash can attest if he can escape from the mountains of foam. The temperature thresholds are standard for German detergents, which makes the standard washer temperature settings convenient. Most washers have a special heavy-duty spin cycle because it is much more energy-efficient to dry by spinning than by heating. Although most washers are also dryers, the drying cycle seems not to be used very often (at least by my acquaintances, who said that a sun-dried wash smelled better).

The refrigerators in German homes are much smaller than American ones. The one in my apartment was 60 cm broad, about 60 cm deep, and 85 cm high. It had a storage capacity of only about 1/5 m³, including the freezer, merely half the volume of my American refrigerator alone. Most Germans shop every day or every

FIGURE 19.12
The author's demand water heater in his house in Delaware, Ohio. These water heaters are now widely available in the United States.

other day for their food and thus do not need as large a volume as we do with our weekly shopping habit. Grocery stores and bakeries are dispersed throughout every urban area. Farm-fresh produce is available from the farmers themselves.

Not many people own freezers in Germany. Germans prefer to buy their produce and meat fresh. The refrigerators, and what freezers there are, are not usually the frost-free type. By eschewing frost-free refrigerators, the Germans save energy. By buying in local stores to which they bike or walk, they minimize the volume they have to pay to cool, also saving energy. And of course, with fewer things in the refrigerator, the door is opened less often, again reducing energy loss as well as reducing entropy gain.

Most homes in West Germany are heated by hot-water radiators, and the individual units are separately adjustable. In my apartment house, each radiator was fitted with an energy use gauge, and the tenant was charged for exact use (this feature is handy when a dwelling is shared, as many are). Many people have a thermostat on each of the heaters in a room. The thermostat can be set to some temperature (say 18°C); if the room is warmer than 18°C because there are a lot of people there, or a lot of light bulbs are burning, or sunlight is streaming in, the valve connecting the unit to the hot water from the furnace is closed. If the room is colder than 18°C, hot water from the furnace flows until the room temperature is 18°C. One manufacturer of such a device claims an average of 20 percent saving on fuel over a winter.

Another peculiarity of German houses, which contributes to energy conservation, especially in older homes, is the *Rolladen*. In older houses, the Rolladen are made from strips of wood that are fastened together and set into a channel on the outside of the window. The Rolladen functions something like a roll-top desk. With the Rolladen rolled up on a shaft inside a housing over the window, the window is uncovered. As the Rolladen are unrolled, they travel down the channel outside the window, making a cover for the window (see Figures 19.13 and 19.14). The Rolladen are generally let down at night, and are up during the day so that light and/or air can get in. In newer homes and apartment buildings, the Rolladen are made of thin plastic, and the housing containing the shaft is not insulated; these are not as effective as the older ones in saving energy. The Rolladen may also be lowered during daylight to block sun and keep the house cooler in summer. Rolladen are a great idea, and I would like to see them sold more widely and cheaply in the United States. They serve several purposes at once: providing shade and sunlight as desired; reducing heat loss by putting an additional air space between inside and outside and by blocking the escape of radiant energy; providing extra protection

FIGURE 19.13
A view of a Rolladen in an apartment house.

FIGURE 19.14
A close-up view of a partially rolled-up Rolladen, showing the shaft and metal channel.

against break-in; and providing protection from vandalism. High-cost metal versions of the Rolladen are now being sold in the United States.

Another idea used extensively in restaurants and cafes during the winter is the hanging separating the door from the rest of the inside. A thick cloth or plastic curtain is hung from a frame so that it makes a semicircle about the door, or when there is a vestibule, it closes it off from the establishment proper. This cuts down on drafts (making the patrons more comfortable) and is substantially cheaper than double doors, while effective in preventing heat loss to the outside (Figure 19.15).

With this, as with some of the other ideas I've discussed, the Germans have some simple solutions to problems of comfort and energy conservation worth considering. At the very least, they give us a basis upon which to draw to answer questions on the realistic possibilities for energy saving.

FIGURE 19.15
The woman in the photo is pushing aside a plastic hanging as she leaves Konditorei Dahmen in Bonn. Such hangings can save energy and improve comfort by preventing large-scale air transfers to or from outside.

were to assume a ten year life and constant energy cost (the actual energy cost has risen substantially between 1972 and the present), the lowest power user would have cost $175.70 less to run. We can thus justify spending a large proportion of the savings attributable to cheaper running on buying the more efficient model. An added factor is that the lower priced models have no thermostat. Thus the air conditioner tends to operate continuously, cooling a room to a temperature lower than desired. This decreases the air conditioner lifetime. As with furnaces, average efficiencies of air conditioners have been increasing in recent years (≈3 percent per year) (17).

COMMERCIAL AND INDUSTRIAL ENERGY CONSERVATION

Buildings

Many office buildings are energy-guzzlers. However, in the wake of the energy crises of the 1970s and 1980s, businesses began to consider life-cycle costing (that is, using continuing as well as acquisition costs to determine which alternative was cheaper overall). One of the first such buildings was Citicorp Tower. Most buildings in Manhattan require an energy input of 2800 MJ/m²/yr;

the Citicorp building uses less than 40 percent of that amount (67). Glass panes are double thick, with a reflective coating on the glass.

As with all buildings with a floor area over about 90,000 m², enough waste heat is available that the building must be cooled all year round (67,68). The phenomenon of large buildings needing to be cooled year-round is not hard to understand.

In the large building, energy inputs are determined by the sun, by the people inside the building, and by the amount of lighting (and, of course, heating) done. Recall that each person is radiating heat equivalent to a 100 watt bulb, and every electric apparatus dissipates heat. Since these buildings are well lighted, and much energy is emitted as heat, lighting produces much indoor heating.

In large buildings, people on a floor provide a heat flow of about 10 W/m², and lighting, a heat flow of over 30 W/m² (69). The heat loss is due to conduction through the walls, ceiling, and floor, radiation through the windows, and infiltration and exfiltration of air (which must be warmed as it invades the building and carries heat away convectively as it leaves the building).

If losses are less than gains, the building cools to some new equilibrium temperature if not heated. If losses equal gains, nothing occurs. If gains exceed losses, the building warms up to some new equilibrium temperature if not cooled. The former situation is analogous to the ease of evaporation from a material of large surface area to volume ratio discussed in Chapter 7: Evaporation occurs rapidly. The latter situation is analogous to the case of evaporation from a material of small surface area to volume ratio: The volume is so large that evaporation proceeds quite slowly. The larger the building, the smaller its surface area to volume ratio, and the less important the surface is to any process. Thus, above some volume (90,000 m² × 3 m, the interfloor height), a building will always produce more waste heat than it is able to dissipate, even in winter. It must then be cooled at all times. Another problem is

the maldistribution of thermal energy in such a building: It is very hot at the center, and cold at the periphery.

The low cost of energy in the early 1970s caused construction of many "dinosaurs." Numark and Bartlett documented some of the diseconomies of a building constructed in 1973 for student recreation at the University of Colorado (70). They demonstrated that use of the heat rejected in freezing water in the ice rink could be used to heat shower water in the same building at an estimated annual savings of $40,000.

Computers in the Citicorp Towers reallocate thermal energy from people, lights, and computers to the periphery of the building where it is needed in winter, saving energy. The Citicorp building is still worse than a typical Swedish office building (≈900 MJ/m²/yr), and pales in comparison to the Folksam building in Stockholm, which uses only 150 MJ/m²/yr (24).

The Swedes use a system called "Thermodeck," in which warm or cool air is supplied through hollowed-out regions in concrete floor slabs. Even though every Swedish office must have a window, the triple-glazing used in the windows assures a solar gain without corresponding radiation loss, and the building can be occupied virtually without heating in winter (26). While it must be cooled in summer, the cooling can occur in the evening, when the rates for electricity are lower due to the large heat capacity of the Thermodeck slabs (26).

Conservation in Industry

Many industrial processes are still very inefficient (23), but energy used in industry in 1982 was at the lowest level since 1967 (71), about a third less than had been predicted. The net reduced growth came from slower growth in the economy (1.4 percent/yr decrease); efficiency improvements (1.2 percent/yr decrease); and changes in the composition of the output. Overall, productivity increased about 18 percent for all the forms, and more than 30 percent for fossil and wood-derived fuels. Between 1973 and 1982, about one year's oil consumption was saved (71). A projection for the Canadian economy is shown in Figure 19.16 (22).

Of course, the 1981–1983 recession played a part in this drop in energy use by cutting demand (and, hence, production). Little money has been put into increased energy efficiency (15). Also, the mix of energy use in industry is changing. In 1980, manufacturing accounted for 85 percent of U.S. industrial energy use (72); in 1984, it was 78 percent (73). In 1980, mining accounted for 7 percent of U.S. industrial energy use (72); in 1984, it was 10 percent (73). In 1980, construction accounted for 4 percent of U.S. industrial energy use (72); in 1984, it was 6 percent (73). The remainder, 4 percent in 1980 (72) and 6 percent in 1984 (73), is used in agriculture. Most of this energy use is for process heat. Redesign using heat recovery and other forms of conservation will probably decrease the amount of low temperature process heat generated directly (72,73).

Industry in the United States is becoming less dependent on materials processing, which should indicate a shift to more value added per unit material processed (72). However, comparisons with Japan and Europe imply that the United States still has a long way to go (a factor of roughly 2) to equal foreign efficiencies (72). Streamlining

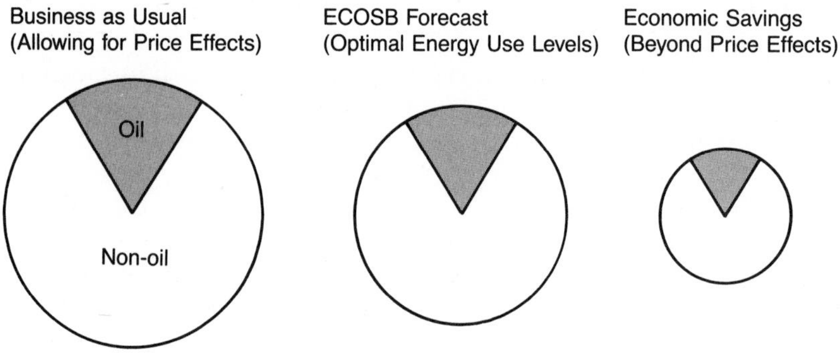
Business as Usual
(Allowing for Price Effects)
ECOSB Forecast
(Optimal Energy Use Levels)
Economic Savings
(Beyond Price Effects)
Oil
Non-oil

	Business as Usual	ECOSB Forecast	Economic Savings
Total Consumption			
Petajoules (Joules × 10^{15})	11,075	9000	2075
Barrel Oil Equivalent per Day	4.9 Million	4.0 Million	0.9 Million
Oil Sand Plant Equivalent (125,000 Bbl/Day)	40	32	8
Electric Generating Stations Equivalent (2000 MW)	175	141	34
Oil Consumption			
Petajoules	1700	1335	365
Barrels per Day	770,000	600,000	170,000
Oil Sand Plant Equivalent (125,000 Bbl/Day)	6.2	4.7	1.5

FIGURE 19.16
Canadian forecasts for energy consumption for the year 2000.

of operations is still possible. New, more efficient, equipment may be ordered.

There is always room for improvement, and one promising place is in AC motor drives (74). Almost all AC motors are currently constant speed (3600 revolutions per minute), and since about a third of all electricity consumption is for motors, any small improvement in efficiency will pay large dividends. Constant speed motors are not well suited to a majority of applications. For example, in a pump with a constant-speed motor, a valve is partly shut to cut down the flow rate by increasing the resistance, dissipating energy. The variable-speed AC motor is easily controlled, and can be made to match the demand much more closely. Although it is more costly to buy initially, it is an energy-saver in the long run. More such innovations are needed.

Potentially, the greatest savings follow if one can design new, more efficient processes to replace older ones. The chemical industry does this routinely. Large savings in fabrication of steel can be made by forming it, while hot, to near its final desired shape, rather than reheating many times (75). Large-scale savings in petroleum refining are also possible in principle (73).

Energy-intensive industries in the United States do not face a bright future. One expects to see relocation of these industries to the energy sources (just as iron ore was once brought to coal-rich Pittsburgh). If U.S. industry researches and adopts new technology, the future is relatively bright. If not, the United States will become even more a debtor nation, with sorry consequences for us all.

SUMMARY

By energy conservation, we mean measures for getting more for the same energy price or getting the same at a lower energy cost. Reduction of illumination levels and the purchase of longer lasting, energy-efficient incandescent or fluorescent bulbs seems advisable.

Passive solar techniques show great promise for residential energy savings. Buildings should be properly insulated. Recommended R-values increase generally as we go from south to north. Increased insulation generally saves energy and money. Heating of interior masonry walls or concrete floors can increase comfort at modest cost, as can construction of Trombe walls in homes, strategic placement of windows with appropriate overhangs, and the use of furnace timers. In some areas, natural ventilation can save energy.

The experience of other countries that have higher energy costs than the United States indicates that it is possible to live reasonably in residential communities with a lower per capita energy expenditure.

Business and industry learned about life-cycle costing in the 1973 and 1979 energy crises. Thus, the newer generation of buildings is much more energy-efficient than the previous generations.

Innovations in energy use devices can also help conserve valuable energy. Industry will become less materials-dependent than now, with more emphasis on value added. The long-term health of the economy demands rationalization, replacement of inefficiencies, and innovation of technology.

REFERENCES

1. R. H. Socolow et al., eds., *Efficient Use of Energy, The APS Studies on the Technical Aspects of the More Efficient Use of Energy* (New York: American Institute of Physics, 1975).
2. Stanford Research Institute, *Patterns of Energy Consumption in the United States* (Washington, D.C.: GPO, 1972).
3. This number is attributed to personnel at Consolidated Edison of New York, *New York Times,* 3 Feb. 1974.
4. C. A. Berg, *Science* 181 (1973): 128.

5. New York City Board of Education Manual of School Planning, as quoted in Reference 1.
6. P. Hammond, *Science* 178 (1972): 1079.
7. C. M. Summers, in *Energy and Power* (San Francisco: Freeman, 1971).
8. M. deC. Hinds, *New York Times,* 27 Jan. 1980.
9. J. Hlusha, *New York Times,* 23 Nov. 1980.
10. R. H. Williams, G. S. Dutt, and H. S. Geller, *Ann. Rev. Energy* 8 (1983): 269.
11. S. Berman, in *Energy Sources: Conservation and Renewables,* ed. D. Hafemeister, W. Kelly, and B. Levi (New York: American Institute of Physics, 1985), 247.
12. C. Starr, in *Energy and Power* (San Francisco: Freeman, 1971).
13. E. Hirst and B. Hannon, *Science* 205 (1979): 656; and E. Hirst and J. Carney, *Science* 199, (1978): 845.
14. E. Hirst et al., *Ann. Rev. Energy* 8 (1983): 193.
15. R. L. Lehman and H. E. Warren, *Science* 199 (1978): 879.
16. L. Hodges, *Phys. Teacher* 22 (1984): 576.
17. H. S. Geller, in *Energy Sources: Conservation and Renewables,* ed. D. Hafemeister, W. Kelly, and B. Levi (New York: American Institute of Physics, 1985), 270.
18. A. H. Rosenfeld, in *Energy Sources: Conservation and Renewables,* ed. D. Hafemeister, W. Kelly, and B. Levi (New York: American Institute of Physics, 1985), 92.
19. S. Prokesch, *New York Times,* 12 Feb. 1986.
20. E. Hirst and J. C. Moyers, *Science* 179 (1973): 1299.
21. G. A. Lincoln, *Science* 180 (1973): 155.
22. B. L. Cohen, *Am. J. Phys.* 52 (1984): 614. G. T. Armstrong, *Bull. Am. Phys. Soc.,* 30 (1985): 32.
23. J. Ames, *Fine Homebuilding* 24 (1985): 65.
24. T. R. Johansson et al., *Science* 219 (1983): 355.
25. M. Cunningham, *Insight,* 29 Sept. 1986.
26. D. Hafemeister, *Am. J. Phys.* 55 (1987): 307.
27. M. P. Fels, *Energy and Buildings* 9 (1986): 5. See also S. Selkowitz, in *Energy Sources: Conservation and Renewables,* ed. D. Hafemeister, W. Kelly, and B. Levi (New York: American Institute of Physics, 1985), 258.
28. G. S. Dutt et al., *Energy and Buildings* 9 (1986): 21.
29. M. L. Goldberg, *Energy and Buildings* 9 (1986): 37.
30. E. Hirst, *Energy and Buildings* 9 (1986): 45. Also, from the same issue of *Energy and Buildings,* see L. S. Rodberg, p. 55; M. J. Hewett et al., p. 65; and C. A. Goldman and R. L. Ritschord, p. 89.
31. AP dispatch, *New York Times,* 16 Sept. 1984; R. Smiley, *New York Times,* 11 Nov. 1984; and D. Roe, *New York Times,* 11 Nov. 1984.
32. G. S. Dutt, in *Energy Sources: Conservation and Renewables,* ed. D. Hafemeister, W. Kelly, and B. Levi (New York: American Institute of Physics, 1985), 122; and M. Sherman, also in *Energy Resources: Conservation and Renewables,* p. 655.
33. M. D. Levine et al., in *Energy Sources: Conservation and Renewables,* ed. D. Hafemeister, W. Kelly, and B. Levi (New York: American Institute of Physics, 1985), 247.
34. M. L. Savitz, *Energy and Buildings* 8 (1985): 93.
35. P. E. McNall, Jr., *ASHRAE J.* 28, no. 6 (1986): 39, and 28, no. 7 (1986): 37. See also J. E. Snell, P. R. Achenbach, and S. R. Petersen, *Science* 192 (1976): 1305.

36. W. J. Fisk et al., Berkeley report LBL-16493 (1984). See also R. G. Sextro, A. V. Nero, and D. T. Grimsrud, in *Energy Sources: Conservation and Renewables,* ed. D. Hafemeister, W. Kelly, and B. Levi (New York: American Institute of Physics, 1985), 229.

37. J. D. Spengler and K. Sexton, *Science* 221 (1983): 9. The authors discuss the Canadian ban; the Consumer Product Safety Commission banned urea formaldehyde insulation in September 1982.

38. D. T. Harrje and K. J. Gadsby, *ASHRAE J.* 28, no. 7 (1986): 32. See also A. V. Nero, Jr., *Sci. Am.* 258, no. 5 (1988): 42.

39. P. Shabecoff, *New York Times,* 14 July 1985.

40. D. E. Claridge and R. J. Mowris, in *Energy Sources: Conservation and Renewables,* ed. D. Hafemeister, W. Kelly, and B. Levi (New York: American Institute of Physics, 1985), 184.

41. S. Craviott, in *Alternative Energy Sources III,* vol. 2, ed. T. N. Veziroglu (Washington, D.C.: Hemisphere Publ. Corp., 1984), 47.

42. J. K. Nayak, N. K. Bansol and M. S. Sodha, *Solar Energy* 30 (1983): 51.

43. K. Talib, in *Alternative Energy Sources III,* vol. 2, ed. T. N. Veziroglu (Washington, D.C.: Hemisphere Publ. Corp., 1984), 57.

44. M. Holtz et al., *Solar Age* 10, no. 10 (1985): 49.

45. S. W. Jacobson, *New York Times,* 23 Dec. 1982.

46. D. Adelman, *Fine Homebuilding* 22 (1984): 68.

47. H. H. Wiechmann and Z. Varsek, *Ziegelindustrie International* 5 (1982): 307.

48. R. V. Pound, *Science,* 208 (1980): 494.

49. K. R. Foster and A. W. Gray, *Sci. Am.* 255, no. 3 (1986): 32.

50. J.-L. Bourgeois, *Nat. Hist.* 89, no. 11 (1980): 70.

51. M. N. Bahadori, *Sci. Am.* 238, no. 2 (1978): 144.

52. E. A. MacDougall, in *Alternative Energy Sources II,* ed. T. N. Veziroglu (Washington, D.C.: Hemisphere Publ. Corp., 1982), 931. In the same volume, see also N. E. Collins and W. E. Handy, p. 937, and C. Benard and D. Gobin, p. 949. In *Alternative Energy Sources III* (1984), see E. A. MacDougall, p. 44, and E. M. Abu El-salam, p. 447.

53. M. N. Ozisik, B. K. Huang, and M. Toksoy, in *Alternative Energy Sources II,* ed. T. N. Veziroglu (Washington, D.C.: Hemisphere Publ. Corp., 1982), 891.

54. C. A. Goldman, *Energy and Buildings* 8 (1985): 137.

55. J. D. Balcomb, *Energy and Buildings* 7 (1984): 281.

56. E. Hirst et al., *Energy and Buildings* 8 (1985): 83.

57. H. Hottel and J. Howard, in *New Energy Technology—Some Facts and Assessments* (Boston: MIT Press, 1970).

58. R. A. Macriss, in *Energy Sources: Conservation and Renewables,* ed. D. Hafemeister, W. Kelly, and B. Levi (New York: American Institute of Physics, 1985), 247.

59. C. E. Hewett et al., *Ann. Rev. Energy* 6 (1981): 139.

60. B. Gladstone, *New York Times,* 22 Dec. 1977.

61. L. Cranberg, *Am. J. Phys.* 49 (1981): 596.

62. B. Franklin, *An Account of the New Invented Pennsylvania Fireplaces, Philadelphia, 1744* (as quoted in E. R. Berndt, *Ann. Rev. Energy* 3 (1978): 225). Quotations are from pages 1 and 32 of the original.

63. J. F. Kowalczyk and B. J. Tombleson, *J.A.P.C.A.* 35 (1985): 619.

64. R. H. Socolow, *Ann. Rev. Energy* 2 (1977): 239; and R. H. Socolow, in *Energy Sources: Conservation and Renewables,* ed. D. Hafemeister, W. Kelly, and B. Levi (New York: American Institute of Physics, 1985), 15.

65. G. J. Aubrecht II, *Phys. Teacher* 21 (1981): 30.

66. P. duPont, *Energy Auditor and Retrofitter* 3, no. 1 (1986): 32.

67. A. J. Parisi, *New York Times,* 29 Oct. 1977.

68. T. Kelly, *Canada Today,* March 1982.

69. A. H. Rosenfeld and D. Hafemeister, in *Energy Sources: Conservation and Renewables,* ed. D. Hafemeister, W. Kelly, and B. Levi (New York: American Institute of Physics, 1985), 148.

70. N. J. Numack and A. A. Bartlett, *Am. J. Phys.* 50 (1982): 329.

71. R. C. Marlay, *Science* 226 (1984): 1277.

72. M. Ross, *Ann. Rev. Energy* 6 (1981): 379.

73. M. Ross, in *Energy Sources: Conservation and Renewables,* ed. D. Hafemeister, W. Kelly, and B. Levi (New York: American Institute of Physics, 1985), 347.

74. D. J. BenDaniel and E. E. David, Jr., *Science* 206 (1979): 773.

75. J. R. Miller, *Sci. Am.* 250, no. 5 (1984): 32.

PROBLEMS AND QUESTIONS

True or False

1. Space heating at the present time in the United States requires more per capita energy than it did in 1850.

2. It will probably take a homeowner about ten years to realize an overall economic return from the installation of insulation in his home.

3. A fluorescent lamp is about four times as efficient in converting electric energy to light as a normal incandescent lamp (for the same power use).

4. Based on the evidence from abroad, office buildings in the United States could be made much more energy-efficient than they are at present.

5. An increase of 1 percent in efficiency of a turbine would be a great boon to energy efficiency.

6. The higher the inside temperature of a house in winter, the more heat the house loses to the outdoors in any given time.

7. Passive solar energy homes are much more expensive than conventional homes to construct.

8. Adobe buildings are well suited to the northern Mexican climate.

9. Home design in a desert area should probably involve few windows and include a central courtyard.

10. Since the newest furnaces are 95 percent (first-law) efficient, there is little room left for improvement in furnace design.

Multiple Choice

11. Suppose that an oven lasts ten years. For a given heating effect, the least efficient oven draws 1000 watts. The most efficient uses 450 watts. Assuming that the oven is used 700 hours annually (at a mean rate as given above), and that energy cost 6 cents/kWh, how much *extra* could you afford to pay for the most efficient model?
    a. $ 420
    b. $ 231
    c. $ 189
    d. $ 42
    e. None of the above

12. Which is most effective in diminishing heat losses from a house for R-30 insulation?
    a. Fiberglass
    b. Rock wool
    c. Cellulose fiber
    d. Urea-formaldehyde foam
    e. All of the above are equally effective in diminishing heat losses.

13. You are a homeowner considering an energy conservation improvement to your house. You plan to be there permanently. If you are a "typical" homeowner, which of the payback periods listed below (in years) is the maximum you would consider in deciding whether or not to make the improvement?
    a. ½
    b. 1
    c. 5
    d. 10
    e. 20

14. Small buildings, such as one-family homes, are almost certainly
    a. net consumers of ambient solar energy.
    b. well insulated, if constructed in the 1950s.
    c. having to radiate their extra heat burden to the environment.
    d. generating a heat flow from lighting comparable to that from the furnace.
    e. in need of thermal energy from the sun or an in-house furnace.

15. Large buildings in winter, such as office buildings, are
    a. net consumers of ambient solar energy.
    b. likely to cool to near outside temperature if no heating is supplied.
    c. likely to heat up because of the large surface area exposed to the air.
    d. likely to heat up due to large internal energy inputs from lighting and people.
    e. likely to cool due to increased rates of conductive, convective, and radiative heat transfer.

16. Why does "energy conservation" have a bad reputation among the public?
    a. All conservation measures are costly.
    b. No retrofits have been able to save money for consumers.
    c. The public does not appreciate the systems nature of the energy distribution network.
    d. People associate conservation with the hated 55 mile per hour speed limit.
    e. People think of conservation as involving suffering and lack of comfort.

17. Which of the following probably has the greatest efficiency?
    a. A coal furnace
    b. A wood fire in a Franklin stove
    c. A wood fire in a Texas fireframe
    d. Resistive ceiling heat
    e. A pulsed gas furnace

18. You are going to use a material to fill the volume between the outer wall and the walls of your rooms to insulate your house. Which of the following materials would you probably want to use?
    a. Fiberglass
    b. Blown cellulose
    c. Vermiculite
    d. Rock wool
    e. Perlite

19. You wish to heat a child's room in winter. Which method of space heating is probably the safest and most effective?
    a. Quartz electric heater
    b. A fan blown electric heater
    c. A kerosene heater
    d. A propane heater
    e. A natural gas heater

20. Which of the following is *not* passive solar in nature?
    a. Trombe wall
    b. Water wall
    c. Heat pump
    d. Sun porch
    e. Greenhouse

Discussion Questions

21. Is it possible that experience gleaned in other countries, in which energy prices have historically been high, could be applied in America? Give some examples if possible from your own knowledge of other countries.

22. Why are people so blasé about exposure to toxic chemicals in their own homes? Do you ever think about the turpentine or the can of gasoline in the basement?

23. How does a demand water heater save energy?

24. Can you think of other ways architects can make buildings more livable and more cheaply? If voluntary effort is not forthcoming, what strategies could be pursued to achieve this end?

25. Which of the lessons on energy conservation I learned in Germany was the biggest surprise to you? Why?

20

RECYCLING AND REUSE

We extend the discussion of the last chapter, of the effect of conservation on energy supplies, to the discussion of the economy of recycling and reuse. Urban wastes of today will probably be mines for resources in the future. Reuse is clearly preferable even to recycling, if the political will to institute reuse can be summoned. This chapter forms the final part of the segment of this book devoted to examination of resources (Chapters 11–20).

KEY TERMS   *recycling*
bauxite
beneficiation
taconite
hematite
cullet
throwaways
bottle bills

The amount of waste material thrown away in the United States is staggering. About half a megatonne per day is generated (over 2 kg per person per day). In this country alone, then, a volume equivalent to piling about 75 m of garbage onto an area 100 m by 100 m (about the same volume occupied by five hundred middle-class American homes) is added to landfills each day. This means that a *cubic kilometer* of garbage must be disposed of every three years or so. America is truly the throwaway society (see Tables 20.1 and 20.2).

Once you think of the huge amounts of trash being generated, it must occur to you that some of the stuff being thrown away could be valuable. Indeed it is. Previous generations of Americans threw away far less. My mother and father can remember the scrap-metal collector with his horse-drawn wagon. He *bought* anything metal and then recycled it. The rag-picker would come around periodically buying old clothing and paper. Such people made a good living (1), and performed a useful service. These sorts of service industries contracted after the end of World War II. Resources and energy were so cheap that it seemed to make sense to people to squander them.

Not everyone was content with this new American way of doing things. The counterculture youth of the late 1960s questioned their parents' ways

TABLE 20.1

Composition of a typical refuse pile by percentage by weight.

Material	Percentage
Paper	56.01
Food wastes	9.24
Plant wastes	7.56
Glass, ceramics	8.50
Plastics	3.50
Wood	2.52
Metals	7.53
Dirt, vacuum cleaner catch	2.52
Plants, oils, removers	0.84
Rags	0.84
Leather	0.42
Rubber	0.42
Misc.	0.10

SOURCE: Paul Sarnoff, *The New York Times Encyclopedic Dictionary of the Environment* (New York: Times Books, a Division of Random House, Inc., 1973), Table 24, p. 261.

with peace and war; they saw in their parents' attitudes the effect of grasping materialism, which permeated the entire society. As a result, many people turned to a simpler life and to the values of their grandparents. The *Whole Earth Catalog* (2) gave advice on returning to the land and on reuse of materials. These attitudes in turn affected ordinary citizens, making them more aware of the price they were paying for their material comforts: air pollution, water pollution, destruction of the environment. In 1970, groups of people gathered across the country to celebrate the first "Earth Day" and helped raise consciousness even

more. The time was ripe for the return of recycling, and in small towns across America, recycling groups were formed. (One such group of recyclers works in Delaware, Ohio, where I live. I joined the board of Waste Watchers/Delaware Recyclers when I moved to Delaware in 1976 and have been involved ever since.)

INDUSTRIAL WASTES AND ENERGY IN INDUSTRY

While residential wastes are large in amount, industrial wastes loom large as well. The U.S. EPA estimated the 1970 total waste from residential, commercial, and institutional establishments as 228 million tonnes; industrial wastes totaled 100 million tonnes; mineral processing wastes totaled 1545 million tonnes; and agricultural wastes totaled 2073 million tonnes (3).

Industry also consumes tremendous amounts of energy. Figure 20.1 shows the distribution of energy among the economic sectors of the economy (4). The electrical and chemical industries are among the largest users of energy and have traditionally been most conscious of its costs. For the industrial sector (not including the electrical utilities), energy use grew at an annual rate of 3.8 percent, slower than any other sector, in the 1960s. Energy use increased from 5.6 GkWh$_t$ in 1960 to 7.6 GkWh$_t$ in 1968.

The industries offering the most energy-saving possibilities are those with the largest energy-to-labor ratios: the chemical, automobile, paper, and construction industries (5). As Table 20.3 shows, these are among the largest energy consumers.

TABLE 20.2

Percentage of gross residential and commercial discards by weight and total gross waste, 1960–1980.

Material	1960	1965	1970	1975	1980
Paper	18.8	15.8	16.0	15.5	20.6
Glass	1.4	1.2	1.3	2.7	4.9
Metal	0.5	1.0	1.2	4.4	4.1
Aluminum	—	—	1.3	8.7	14.8
Rubber	5.7	12.5	8.2	6.9	3.9
Total amount, Mtonne	39.5	47.6	59.6	61.6	70.7

SOURCE: Department of Commerce, *Statistical Abstract of the United States, 1984* (Washington, D.C.: GPO, 1984), Table 361.

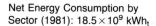

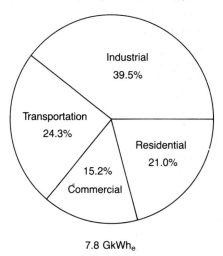

Net Energy Consumption by Sector (1981): 18.5×10^9 kWh$_t$

Industrial 39.5%

Transportation 24.3%

Residential 21.0%

15.2% Commercial

7.8 GkWh$_e$

FIGURE 20.1

The breakdown of 1981 energy use among the sectors of the economy.

The labor-intensive industries, such as the leather and leather products industry, are the places in industry for which the energy growth rate is smaller than the growth rate of value added. According to Table 20.4 (6), industry in general is becoming more economically efficient in energy use. The sectors becoming less energy-efficient comprise less than 2 percent of the total energy consumption (6).

We know that it is possible to save substantial amounts of energy. Japanese steel firms used an average of 30 percent less fuel energy per tonne of shipped steel than the U.S. average in 1978 (7). The chemical industry's new organic chemical plants used 40 percent less fuel per tonne of output in 1981 compared to 1971 (7). The paper industry sees an overall 20 to 40 percent reduction in energy use for new plants coming on line (7).

Much industrial waste of energy occurs because industries often find it cheaper to leak energy than to modify or replace outmoded equipment. Industry is expending some capital funds on pollution control at present; Table 20.5 gives the amounts spent for pollution control (4), and Table 20.6 indicates total investment compared to pollution abatement by sector. The highest expenditures are occurring in industrial applications where retrofit is the only option. Where *new* machinery is being installed, it often incorporates pollution control at modest additional expense. In some cases, it can even save

TABLE 20.3

Industrial fuel consumption by major users (trillion Btu).

	Coal	Natural Gas	Petroleum Products	Electricity	Total Industry Group Energy
Primary metal industries	2838	836	306	1291	5298
Chemicals and allied products	666	1219	1426	1626	4937
Petroleum refining and related industries		1012	1589	225	2826
Food and kindred products	263	593	134	338	1328
Paper and allied products	467	341	211	280	1299
Stone, clay, glass, and concrete products	406	449	87	280	1222
Subtotal	4640	4477	3753	4040	16,910
All other industries	976	4781	721	1572	8050
Total	5615	9258	4474	5612	24,960

TABLE 20.4

Comparison of value added in manufacturing with power and fuel consumption in the industrial sector (feedstocks not included).

Standard Industrial Classification	Ratio of Value Added in 1967 to Value Added in 1962*	Ratio of Energy Used in 1987 to Energy Used in 1962	Ratio of Increase in Energy Used to Increase in Value Added	Rate of Growth in Value Added: 1962 to 1967 (Percent per Year)	Rate of Growth of Energy Used: 1962 to 1967 (Percent per Year)	Ratio of Rate of Growth in Energy Used to Rate of Growth in Value Added
Major group 20—food and kindred products	1.20	1.13	0.942	3.7%	2.5%	0.676
Major group 21—tobacco manufactures	1.19	2.00	1.681	3.5	14.9	4.257
Major group 22—textile mill products	1.26	1.28	1.016	4.7	5.1	1.085
Major group 23—apparel and other finished products made from fabrics and similar materials	1.32	1.67	1.265	5.7	10.8	1.895
Major group 24—lumber and wood products, except furniture	1.31	1.58	1.206	5.5	9.6	1.716
Major group 25—furniture and fixtures	1.39	1.25	0.899	6.8	4.6	0.677
Major group 26—paper and allied products	1.30	1.25	0.962	5.4	4.6	0.852
Major group 27—printing, publishing, and allied industries	1.35	1.40	1.037	6.2	7.0	1.129
Major group 28—chemicals and allied products	1.39	1.31	0.942	6.8	5.5	0.809
Major group 29—petroleum refining and related industries	1.50	1.22	0.813	8.5	4.1	0.432
Major group 30—rubber and miscellaneous plastics products	1.49	1.29	0.866	8.3	5.2	0.627
Major group 31—leather and leather products	1.19	1.00	0.840	3.5	0.1	0.029
Major group 32—stone, clay, glass, and concrete products	1.21	1.16	0.959	3.9	3.0	0.767
Major group 33—primary metal industries	1.37	1.20	0.876	6.5	3.7	0.569
Major group 34—fabricated metal products, except ordnance machinery, and transportation equipment	1.52	1.43	0.941	8.7	7.4	0.851
Major group 35—machinery, except electrical	1.63	1.29	0.791	10.3	5.2	0.505
Major group 36—electrical machinery, equipment, and supplies	1.50	1.41	0.940	8.5	7.1	0.835
Major group 37—transportation equipment	1.27	1.23	0.969	4.9	4.2	0.857
Major group 38—professional, scientific, and controlling instruments; photographic and optical goods; watches and clocks	1.37	1.25	0.912	6.5	4.6	0.708
Major group 39—miscellaneous manufacturing industries	1.70	1.29	0.759	11.2	5.2	0.464
Sector Total	1.38	1.24	0.899	6.7	4.4	0.657

*Constant dollars.

SOURCES: 1963 Census of Manufactures.
1967 Census of Manufactures.
Stanford Research Institute.

TABLE 20.5

Total pollution control expenditures, 1973–1981, billions of constant (1984–85) dollars.

1973	52.9	1978	67.5
1974	54.6	1979	69.0
1975	59.0	1980	68.5
1976	62.3	1981	67.7
1977	63.6		

SOURCE: Department of Commerce, *Statistical Abstract of the United States, 1984* (Washington, D.C.: GPO, 1984), Table 365.

money as well as energy. For example, gas-fired vacuum furnaces use about 25 percent of the fuel used in past models as a result of new research. With introduction of the vacuum furnaces during the 1960s, the steel industry-wide energy use per ton of coal decreased one-fifth from 1959 to 1969, from 9.5 MWh_t to 7.7 MWh_t (6). Substantial increases in energy costs here have caused installation of new equipment for small steel mills while forcing closing of many obsolescent mills (8).

TABLE 20.6

1982 investment for pollution control, billions of 1982 dollars.

	Total Expenditures	Pollution Abatement Expenditure
Total nonfarm business	316.43	8.49
Manufacturing	119.68	4.72
Durable goods	56.44	1.76
Primary metals	7.46	0.76
Blast furnaces, steelworks	3.47	0.41
Nonferrous metals	2.71	0.30
Fabricated metals	2.59	0.04
Electrical machinery	10.62	0.15
Machinery, except elec.	12.89	0.18
Transportation equipment	15.16	0.40
Motor vehicles	7.92	0.32
Aircraft	6.04	0.03
Stone, clay, and glass	2.61	0.08
Other durables	5.13	0.15
Nondurable goods	63.23	2.96
Food, including beverage	7.74	0.38
Textiles	1.33	0.03
Paper	5.97	0.30
Chemicals	13.27	0.67
Petroleum	26.69	1.50
Rubber	1.71	0.04
Other nondurables	6.52	0.04
Nonmanufacturing	196.75	3.77
Mining	15.45	0.52
Transportation	11.95	0.14
Railroad	4.38	0.08
Air	3.93	0.01
Other	3.64	0.05
Public utilities	41.95	3.00
Electric	33.40	2.89
Gas and other	8.55	0.11
Trade and services	86.95	0.09
Communications and other	40.46	0.02

SOURCE: Department of Commerce, Bureau of Economic Analysis.

The cost of transportation has the potential for great energy savings. More than 20 percent of the energy cost of industry is in bringing energy to the materials processer (9). The shipment of raw ore, and its associated mining costs, are not included in estimates of the energy cost of ore refining. Since ore includes non-useful material, shipping of ore can be very expensive; the lower the grade of ore, the more substantial the expense.

Why Recycle?

There is money to be saved by recycling of scrap metal. We are not, however, recycling in great amounts, as Table 20.7 shows. Recycling did increase in the 1970s. By 1977, the recycling of lead increased to 49 percent, that of copper, to 40 percent; and that of aluminum, to 25 percent (10). These numbers are on the rise. Worldwide, as of 1984 (11) 40 percent of lead, 33 percent of aluminum, 33 percent of copper, 10 percent of chrome, and 6 percent of zinc are recycled. Of the materials listed in Table 20.7, lead and zinc do not save much energy per se in recycling. However, since lead and zinc are difficult to find in appreciable concentrations, a reasonable fraction of production represents recycled material.

Aluminum   Consider the aluminum beer or soda can. At one time, the aluminum was in com-

pound in the ground, in bauxite deposits. It had to be dug up, transported to a smelter, refined, sent to a mill, formed in sheets, and sent to the can manufacturer—all this before the brewer buys the cans to put beer into them. If aluminum is recycled, it has to be transported from scrap dealer to foundry, melted, sent to a mill, and so on as indicated above.

From this description, it does not seem that recycling of aluminum should be a great saving. Why am I such an avid recycler of aluminum cans? The reason becomes clear when the first few steps in aluminum production are examined. Much of our bauxite (aluminum ore, Al_2O_3) comes to us from Jamaica, Suriname, Canada, and Australia (3). In bauxite deposits, the ore is about 20 percent aluminum. This ore is usually transported to the coast and loaded onto ships. It is shipped to a U.S. port and brought by rail to the smelter. Thus, each tonne of aluminum transported has an additional four tonnes of waste material transported along with it. Even though the transportation is not expensive, one must pay five times more than necessary just to get the aluminum to the smelter.

The refining of aluminum is very energy-expensive, because it uses electricity (that is why aluminum is refined in regions with cheap electricity, such as the Pacific Northwest). Since 1 kWh of electric energy would cost about 3 kWh of

TABLE 20.7
Per ton energy requirements to refine selected metals, and scrap use.

Material	Energy Use (MWh$_t$/ton) (3)	Scrap as % of U.S. Consumption (12)
Mg (magnesium)	27	3%
Al (aluminum)	17.8	4
Zn (zinc)	13.4	5
Mn (manganese)	13.4	0
Sn (tin)	5.9	19
Fe (iron)	7.6	28*
Cu (copper)	8.2	24
Glass	5.1	8.5†
Pb (lead)	3.3	35

*Includes exports.
†1986 figure, M. Franklin, Franklin Associates.

thermal energy (at typical generating plant efficiencies, about 33 percent), we have to correct the energy budget to arrive at a true energy cost. We can compare energy costs of various procedures only if we choose a consistent basis for comparison. I will quote thermal energies here. This means that where thermal energies are used—for example, if fuel oil is used—the energy quoted is used as given in the table entry (or converted into joules). Where electric energy is used, the energy given in the source must be multiplied by a factor ≈3 before entry under its appropriate heading. For aluminum, the cost of refining is about 300 MJ/kg, while the cost from scrap is about 7 MJ/kg (13,14) (see Table 20.8). In fact, when one takes the energy costs of transportation into account, it will be clear that the production of aluminum from scrap must cost only about 2 percent of that from ore, as well as saving the disposal of the four-fifths of the ore material that is useless slag (and I have not counted such industrial wastes in the volume of waste to be disposed of as trash every day). While a new process developed by Alcoa should reduce energy input in smelting by one-third (7), it is still very advantageous to recycle as opposed to refining virgin ore.

Much of the cost of transport and refining of the ore arises from the dispersal of those useful aluminum atoms among all those other sorts of atoms. This configuration has greater entropy than one with the aluminum concentrated. In scrap aluminum, the atoms of aluminum are already concentrated. As we all remember from basic thermodynamics, it is easy to decrease entropy locally by paying the appropriate energy price. If aluminum is just thrown away, entropy is again increased. The energy cost of processing contaminated scrap is also much greater: ≈30 MJ/kg at ≈30 percent contamination. It would be incredibly expensive to retrieve cans from the landfills, where they would be mixed in among many other sorts of materials. The least energy-expensive way to gather aluminum is for it to be separately saved at the point of use. This means that the individual household, at the point of use, should be saving its metal separately. About 53 percent of aluminum cans produced in the United States, 600

TABLE 20.8		
Energy costs in making aluminum (MJ/kg finished aluminum).	*Ore Extracting (Mining, Drying, Shipping)*	
	Caribbean bauxite	6
	South American bauxite	9
	Average	7
	Production of Alumina from Ore	
	Caribbean bauxite	42–57
	South American bauxite	40–56
	Average	48
	Production of Aluminum from Alumina	
	Electrode { Prebaked	208–272
	Soderberg	246–277
	Cost of flourine compounds and calcining	4–10
	Average	260
	Overall Energy Cost from Ore	314
	Overall Energy Cost from Scrap	
	Pure	6
	30–40% contaminated	25

FIGURE 20.2
(a) The old method of
crushing cans at Dela-
ware Recyclers in Dela-
ware, Ohio. (b) The
new method of crushing
cans at Delaware Recy-
clers in Delaware, Ohio.

(a)

(b)

kilotonnes worth, are recycled each year (15). The same logic applies to metals other than aluminum as well as to glass and to waste paper. Figures 20.2 a and b show recycling techniques for aluminum and steel, respectively.

Aluminum consumes 71 percent of the energy used by all nonferrous metals (14). The total cost of aluminum from ore is about 43,700 kWh$_t$/ton, whereas the cost of aluminum from scrap is about 2440 kWh$_t$/ton (14). Therefore, recycling aluminum saves 94.4 percent of the total energy used in aluminum smelting. The steep rise in cost of electricity in the 1970s was responsible for the increase in aluminum recycling from 17 to 32 percent in the United States (16).

Steel While I have discussed only aluminum so far, the story for scrap steel is similar. The energy cost of steel from ore in 1979 (see Table 20.9) was 28.5 MJ/kg, while that from scrap was only 1.26 MJ/kg (less than 5 percent that from ore). However, this does not include the energy costs of mining, beneficiation, or transport. In contrast to the case for aluminum, much of the ore used for steel is produced domestically, mostly in Wisconsin and Minnesota. Taconite is low-grade iron ore mixed with minerals (it does not have a high concentration of iron), so about 3 kilograms of crude ore are mined for each kilogram of usable ore transported (12). The ore is in deep formations, and large quantities of overburden must be stripped away—an average 1 kilogram

of waste to each kilogram of crude ore (the range for the United States is 1/3 to 3) (12). In addition to the production of some 660 megatonnes of waste in mining, a gigantic amount of water is used in the pelletization (beneficiation) of taconite ore: over 3 liters per kilogram of pellet ore. This water is then dumped, mostly into Lake Superior (you may recall concern about the pollution of the lake and lawsuits over the presence of asbestos in the tailings several years ago). Energy consumed in mining ranges from 36 kJ/kg for very high quality ore ($\approx$50 percent iron) to 1.4 MJ/kg for taconite made into magnetite or hematite concentrate Fe_3O_4 (pellets) (16).

Clearly, scrap steel is very good for conserving energy; for this reason, nearly 60 percent of the total weight of steel produced comes from scrap (about half from in-plant scrap, and half from recycled steel). Scrap is very important in the newer, smaller specialty mills (8).

Waste Watchers collects a great deal of steel from old cans, motors, and so on, but it is much less valuable than the aluminum we recycle: We get less than 1 cent per pound as opposed to roughly 35 cents per pound for aluminum. This reflects the tenfold differential in the energy costs of aluminum and steel from ore.

Recycling of scrap steel costs only 4.4 percent of the energy cost of refining the steel from ore. This rerefined steel is used in place of the foundry pig iron previously used.

Copper There are two stages involved in the smelting of copper: production of matte from ore, and refining the matte to the required purity. The first process requires about 32,000 ft^3 of gas and 375 kWh$_e$ per ton; since 1 cubic foot of gas provides 1.035 Btu, or about 0.32 kWh$_t$, this means that 10,825 kWh$_t$/ton is required to produce matte. Refining the matte requires 615 kWh$_e$ and 4700 ft^3 of gas; thus the total energy cost of refining matte is 3265 kWh$_t$/ton (14). Recycling scrap eliminated the first process, for a 77 percent saving over the energy cost of smelting from ore.

The two processes discussed for copper generally occur at two separate locations. The continuous smelting of the ore can conserve thermal

TABLE 20.9
Energy costs in making steel from iron ore (MJ/kg).

Coke (0.4 kg)	11.5
Coal (0.02 kg)	0.7
Electricity (0.46 kWh)	5.6
Fuel oil (4.0 liter)	2.5
Tar, pitch (0.3 liter)	0.2
Liquid petroleum gas (0.5 liter)	0.02
Natural gas (0.13 m^3)	5.0
Coke gas (0.16 m^3)	3.0
Total	28.5
From scrap	1.26

SOURCE: U.S. Bureau of Mines.

energy here for an overall energy savings. This also reduces the emission of noxious sulfur dioxide in the process (17).

Glass In contrast to both steel and aluminum, there is not much energy savings in recycling glass. It takes almost as much energy to melt glass as it takes to melt the silicon dioxide–soda ash mixture at about 1400°C. Glass manufacturers have gradually been able to raise the amount of cullet (recycled glass) in their mixes. By 1984, a 40 percent cullet mix was not uncommon in the United States. In Switzerland, a cullet mix of 90 percent is apparently the highest in the world (18). For each 10 percent of the batch that is cullet, an energy savings of 2.5 percent results. There are advantages to using recycled glass for the manufacturer; like butter in a pan, glass melts in the furnaces before the raw materials (1200°C) and allows them to melt at a somewhat lower temperature. Also soda ash is getting more expensive (about 25 percent of the raw material is soda ash), and glass cullet allows less to be used, saving a bit of money.

The major reason for pushing glass recycling, rather than reuse, is practical. Many people simply throw glass away wherever it is convenient for them to do so. Delaware, Ohio, has a population of about 25,000; it is a city with a good school system, a university, and a large middle-class population. Even so, I pick up one to two bottles or cans a week that people have simply dropped on my lawn. It is even worse in mostly rural Delaware County. The roads are badly littered. Recycling stations can provide an alternative to rural dumping. Bottle bills (see the discussion later in this chapter) can address part of such problems.

Delaware County has a problem with rural dumping, especially since our landfills were closed and all trash is transferred to a landfill in Knox County, some distance away. Rates for trash taken to the transfer station have increased greatly over rates at our former landfill, leading some residents to scatter their garbage along the county roads. As mentioned in Chapter 18, this has happened elsewhere. The county or municipality must then pay for the cleanup, and cleanup along roads costs about ten times what it would cost if the material were concentrated in one place (the entropy problem again). Broken glass can constitute a transportation hazard.

Another problem with glass, as opposed to steel and aluminum, is its long degradation time: A steel can along a highway will rust away within five to fifty years; an aluminum can may disappear after a century or so; a glass bottle may not have disappeared even after a millenium. Recycling of glass may prevent some of this, making walks or drives more pleasant for all concerned, and saving energy as well.

FIGURE 20.3
The glass collection dumpster at the Delaware Recyclers facility.

The rationale of local recycling organizations (such as Waste Watchers/Delaware Recyclers) is that people will recognize that it is possible to save these materials at home and bring them in all at once to a recycling station that takes in all items. The glass collection dumpster is shown in Figure 20.3. At one point a few years ago, Waste Watchers was paying over twice what it received from selling glass in order to get the glass to the point at which it could be reused. We wished to remain the "universal acceptor" for recyclables, and the economics of the situation did not deter us from doing what we thought right.

Paper As Table 20.1 showed, paper and paper products may make up as much as half the volume of waste deposited in landfills. Paper ends up elsewhere as well. Recycling paper allows this volume to be used more effectively and saves some trees (and a bit of energy).

Here, too, entropy considerations are important. Mixed paper commands a much lower price than pure de-ink newsprint (used newspaper), pure computer printout, or pure rag papers. There is always a stable market for unmixed paper, while the price paid for mixed paper varies by factors of 3 to 5. This is reflective of both time and energy necessary to segregate the mixture. The paper recycling technique is shown in Figure 20.4. The situation overseas is better. In the Netherlands, about 40–65 percent of paper is recycled (19); in Germany, about 42 percent is recycled (20); while here, only 27 percent is recycled (15,21). Some paper is sold for burning, but its value for combustion is only about $20 per tonne; its value as recyclable paper runs $35 to $55 per tonne (16).

The physical quantity entropy is intimately connected to these questions of ore concentration and the throwing away of recyclable materials. Ores are used because the materials are found in concentrations significantly above the clarke (the average concentration in the crust). Ore bodies thus have a lower entropy than "average rock." The process of refining the ore introduces a further drop in its entropy. The material is then used for some purpose in products. The products are widely distributed, causing entropy to be increased. After use, the material is thrown away. This increases its entropy still more. If, instead, the material were separated, the entropy would not be increased any further.

At this point, the material can be recycled at a local recycling station, or be collected by the municipality in its separated form so that it can be recycled. Profits from municipal recycling operations could be used to prevent rises in the cost of refuse collection. Some municipalities already do this to a certain extent—for example, Urbana, Illinois; Kitchener, Ontario; and Mississauga, Ontario. Years ago, when I lived in Princeton, New Jersey, we had to separate metal trash from the rest. Municipal recycling programs seem to present a sensible approach to the problems of landfills, to enhanced recycling of materials, and to reducing cost increases for refuse collection. If this method can be implemented, it will be an appropriate solution to the problem. If this is not done, more recycling facilities should be opened, and recycling pursued as vigorously as possible.

Figure 20.4 shows how the collections of our local recycling group have increased over the last several years. Note especially the glass figure; in 1983, we recycled 27 tons of glass. A town of 25,000 will produce on average 1525 tons of glass waste per year (Table 20.10). On this basis, we are reaching only 1.8 percent of the population.

In Delaware, Ohio, alone, the volume of unbroken glass produced each year as trash is about 3900 m^3. Even if the glass were crushed, it would occupy over 2400 m^3. The space problem in landfills is even worse for large cities. A city of 500,000 would generate between 47,000 m^3 and 78,000 m^3 of glass waste volume each year. No wonder our landfills are running out of space!

Landfills have been closed in many regions of the country because of inadequate pollution control or simply because they were full. New Jersey, for example, had 350 landfills in 1977 and took out-of-state trash. By 1987, New Jersey had only 80 landfills, of which fewer than 10 took municipal garbage; as a result of landfill closing, costs

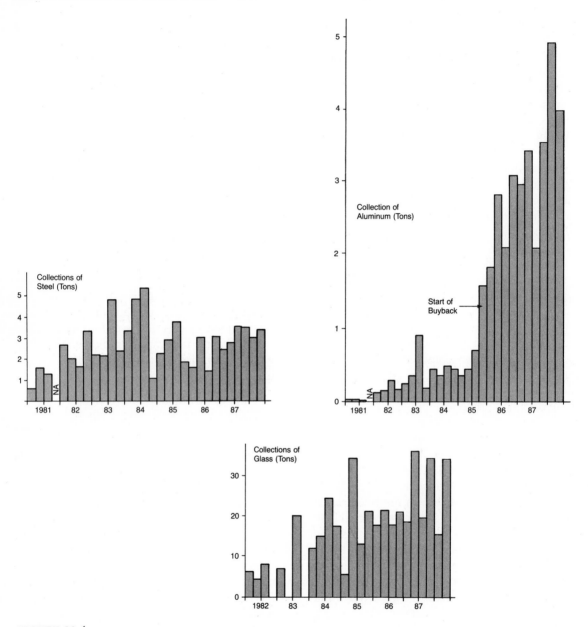

FIGURE 20.4
(a) Steel collections by Delaware Recyclers. (b) Aluminum collections by Delaware
Recyclers. (c) Glass collections by Delaware Recyclers.

of disposal jumped by factors of 2 to 5 (22).
Hempstead, Long Island, ships trash to Youngstown, Ohio, for disposal because their landfill
closed (23). The amount of trash is growing ex-

ponentially, and people are becoming aware too
late to make anything less than a complete change.

Not many of us in Delaware want a landfill
next door. The Delaware city council met solid

TABLE 20.10
Amount of glass recovered (tons).

Percent Recovery	Population				
	10,000	25,000	50,000	100,000	500,000
5	31	76	153	305	1525
10	61	153	305	610	3050
20	122	305	610	1220	6100
50	305	763	1525	3050	15,250
70	427	1068	2135	4270	21,350
100	610	1525	3050	6100	30,500

opposition from the neighbors-to-be of a landfill. This is happening everywhere. No matter how obvious it is that facilities are needed for landfills, treatment plants, and the like, people want them somewhere else: "NIMBY"! (not in my back yard).

One of the clearest illustrations of the NIMBY attitude became news in early 1987. A barge, the *Mobro,* was filled with garbage from Islip, Long Island, to send to North Carolina. This began a three-month odyssey for the unwanted garbage barge, which was refused entry everywhere it tried to dock. The *Mobro* had to return to its point of origin. The garbage itself was finally offloaded and disposed of five months after its peregrination began (24).

Capital Costs vs. Operating Expenses

Though there is great potential in our present-day midden heaps, even the most optimistic estimates show only a quarter of all discarded materials recovered by 1990 (14). The difficulty with recycling now is that "at the point of utilization, economic justification of energy consuming equipment tends to be governed by initial costs" (7). This means that operating costs are not considered in buying equipment, only the capital costs are. The operating expenses are carried on company books as overhead, and savings in operating costs possible with different equipment is thus not considered (recall the air conditioner discussed in the preceding chapter).

This phenomenon leads to unrealistic choices for accession of equipment. Perhaps the most interesting example of this sort involves the aircraft carrier *John F. Kennedy*. Presidents Kennedy and Johnson appointed Robert McNamara as Sec-

retary of Defense. A former auto executive, he introduced cost-effectiveness analysis to the Pentagon during his tenure. Such an analysis was undertaken for the decision whether the aircraft carrier should be powered by a conventional oil plant or by a nuclear plant. The ship was built using a conventional oil-fired power plant because no account was taken of the necessity of construction of oil tenders (fuel tankers), nor for the cost of their crews, nor of the continuing cost of oil (they did remember to put in the fuel cost for the nuclear option, however). This decision will continue to cost the taxpayer a considerable amount of money, especially since the dramatic rise in fuel oil costs. The ship also pays the price in maneuverability, which it would not had the Navy chosen the nuclear option.

The Diseconomy of Trash Recycling

A similar situation for trash recycling occurs because of inertia and lack of understanding of the entire process of trash collection. In terms of generation, we Americans over-package things. Approximately 10 percent of total food costs are due to packaging (21). A New Yorker generates over twice as much trash per day as someone from Hamburg, Germany, and 3.6 times as much as someone from Manila, Philippines (21).

Consider the "disposal" of trash. In other countries, such as Egypt, Mexico, and Thailand, many people make livings from trash recycling. In the United States, we see it simply as a disposal problem. However, the increasing costs of landfilling are causing a rethinking of the situation. New Jersey, which already had a significant number of municipalities begin recycling, passed a

We can look up the heat of combustion of alcohol (C_2H_5OH) in a table in a book such as the *Handbook of Chemistry and Physics* and find that it is about 320 kcal/gram molecular mass (the gram molecular mass is known as the mol). We can also find that the mol occupies a volume of about 46 milliliters—about 1.5 oz. Therefore, for pure alcohol, we can convert kcal to kWh_t to obtain a heat of combustion of 0.28 kWh_t/oz.

A 12 oz bottle of high beer (about 4 percent alcohol by volume) will have an energy content of 0.14 kWh_t. Similarly, a shot of whiskey—about 1.6 oz—which is 86 proof (43 percent alcohol) on ice has an energy content of $(0.43)(1.6 \text{ oz})(0.38 \text{ }kWh_t/oz) = 0.17 \text{ }kWh_t$. That is, a shot of whiskey has about the same energy content as a bottle of high beer.

To compare to soft drinks, we must look up the calorie content in something like the U.S.D.A. tables: cola, 145 kcal; fruit flavor, 170 kcal; ginger ale, 115 kcal; and root beer, 150 kcal. This translates into energy contents of 0.17 kWh_t, 0.20 kWh_t, 0.13 kWh_t, and 0.17 kWh_t, respectively.

Typically, then, the energy content of a beverage is about 0.2 kWh_t.

---

mandatory recycling law in 1987 (25). Without recycling, New Jersey would have been totally without landfill space by 1991 (26). Oregon has made recycling mandatory for all communities of more than 4000 people (15).

The new perception comes from a "holistic" approach to trash disposal. It is being recognized that even though municipal recycling loses money, it loses *less* money than conventional disposal of trash. According to an expert (15), per-tonne burning costs are $65 to $110, and landfill per-tonne disposal costs range from $40 up (costs for Long Island trash are $140 or more); it costs only $20 to $30 per tonne to recycle from a weekly trash collection, depending on whether or not materials are presorted (15). There is a large initial investment (capital cost) required to get the equipment necessary to operate effectively. As a result of this investment, however, operating expenses are lower in perpetuity.

REUSABLE BOTTLES AND LITTER

Some states (led by Oregon in 1972) have enacted "bottle bills"—laws specifying deposits on all bottles and cans containing beverages. This is a consequence of a recognition that collectively consumers can accomplish conservation, even if they cannot individually, and a belated realization of the huge amount of energy invested in beverage containers. In fact, as we shall see, the energy cost of the various throwaways is six to nine times the energy value of the beverage (see box).

Incredible numbers of beverage containers are made each year. In 1975, about 40 billion glass containers (7 Mtons of glass), about 10 billion steel containers (2 Mtons of steel), and about 10 billion aluminum containers (0.5 Mtons of aluminum) were made (27). About half the dollar cost of a soft drink is container cost (27). In the foregoing, we have considered only the fabrication cost of throwaways. The proportion of beverage containers in waste is growing at a rate over twice that for all refuse. There are other costs, such as the esthetic costs of litter, and the actual dollar cost of cleaning up the litter. Litter on one mile of Kansas highway contained 590 beer cans, 130 soda bottles, 120 beer bottles, and 110 whiskey bottles as well as ten tires and two sets of bedsprings (28). Bottle and can litter has been reduced by 35–40 percent by volume (≈80 percent in numbers) in states with mandatory deposits (16,29).

The cost of disposal of highway litter was four times as expensive as residential refuse disposal in 1969 (3). I was a resident of Oregon when the bottle bill went into effect, and I observed the drop in litter. Oregon highways became obviously cleaner than those in the surrounding states, at a reduced cost to the taxpayer.

Bottle Bills

In the remainder of this section, we shall examine the energy cost of the various alternative containers for beverages. In the case that we consider returnable bottles, it is desirable for the returnable bottles to go through as many trips as possible. It appears that as the use of returnables declines, a concomitant decline in the number

of trips made by returnables is observed. In the 1950s, the average number of trips per bottle was about 40; by 1972, it had dropped to 15, and by 1978, it had dropped to 8 (30). Conversely, when recycling is enforced, the average number of trips increases (30,31). In states with the bottle bill, there appears to be overwhelming support for the laws. States with bottle bills typically recycle over 90 percent of beverage containers sold (16).

Another pleasing aspect of bottle bills is the fact that enactment leads to an increase in jobs (16,21,29–33): A nationwide bottle bill would produce 80,000 jobs by one estimate. It would also save our resources. Nationwide, an equivalent of 45 million barrels of oil would be saved each year, as well as the aluminum, steel, and about 70 percent of the glass presently used to manufacture beverage containers (32). Throwaways produce more dust and sulfur dioxide, as well as nitrogen oxides two to thirty times as

FIGURE 20.5
The life cycle of a bottle. (Adapted from B. M. Hannon, *Environment* 14 No. 2, p. 11 ff, Feb. 1972. Reprinted with permission of the Helen Dwight Reed Educational Foundation. Published by Heldref Publications, 4000 Albemarle St., N.W., Washington, DC 20016. Copyright 1972.)

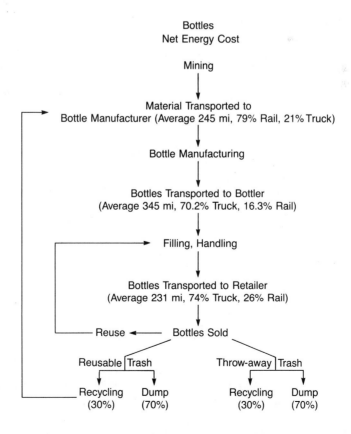

high as for returnables, and fifty times as much salty water effluent (33).

Below, we follow the analysis of Bruce M. Hannon (30) to consider the costs of the alternatives: throwaway bottles, reusable bottles, and throwaway steel and aluminum cans. In Figure 20.5, we show a chart indicating the life cycle of a bottle, and in Table 20.11, we analyze the energetics of 16 oz returnable and throwaway bottles. Note that the net cost of a throwaway is 0.63 kWh$_e$ (1.89 kWh$_t$), over nine times as much as for the beverage itself. For the returnable bottle, one use requires 0.92 kWh$_e$, eight uses cost 0.25 kWh$_e$ per use, and forty uses cost 0.18 kWh$_e$ per use (only about twice the energy content of the beverage). The analysis for throwaway steel and aluminum cans shows that they rate somewhat better than

throwaway bottles in energy use, but still far worse than returnable bottles. In Germany, it takes 22.5 times as much energy per use to supply aluminum beer cans as it does deposit bottles (33). In Tables 20.12 and 20.13, we follow Hannon's analysis for aluminum and steel throwaways.

The volume of waste clearly increases as the proportion of throwaways increases. In Germany, the beer market and the noncarbonated fruit drink market are roughly the same size. Returnable bottles make up 90 percent of sales for beer, while no deposit–no return bottles make up 77 percent of the refreshment drink market (33). The beer industry produces about 12,000 box cars of solid waste per year, while the fruit drink market produces 20,000 box cars of solid waste (33).

TABLE 20.11
Energetics of bottle beverage containers (watthours).

Activity	⅔ lb Throwaway	1 lb Returnable	
		First Use	Each Subsequent Use
Mine raw materials	56	85	
Transport to glass plant	7	11	
Glass bottle manufacture	437	666	
Bottle top	21	21	21
Transport to bottler	20	31	31
Bottling, handling	66	66	71
Transport to retailer	13	20	20
Retailer, consumer	?	?	?
Waste collection	5	1	1
Sorting, etc.	62 (30% recovery)	15	15
Total	687	915	159

Throwaway: 0.7 kWh/use

Returnable:	
1 use	0.92 kWh/use
7 uses	0.25 kWh/use
15 uses	0.21 kWh/use
40 uses	0.18 kWh/use

SOURCE: Adapted from B. M. Hannon, *Environment* 14, no. 2 (1972): 11. Reprinted with permission of the Helen Dwight Reed Educational Foundation. Published by Heldref Publications, 4000 Albemarle St., N.W. Washington, DC 20016. Copyright 1972.

TABLE 20.12
Energetics of aluminum
beverage containers*
(watthours).

Mining	30
Transport to mill	10
Production of aluminum	480
Transport to can maker (average 392 mi)	1
Can manufacture	10
Transport to bottler (average 300 mi)	1
Bottling, transport to retailer	21
Retailer, consumer	?
Waste collection	1
Total	553

0.55 kWh/use

*can weight 0.041 lb.

SOURCE: Adapted from B. M. Hannon, *Environment* 14, no. 2 (1972): 11. Reprinted with permission of the Helen Dwight Reed Educational Foundation. Published by Heldref Publications, 4000 Albemarle St., N.W. Washington, DC 20016. Copyright 1972.

TABLE 20.13
Energetics of steel
beverage containers*
(watthours).

Mining	13
Transport to mill (barge, 1000 mi)	5
Production of finished steel	223
Aluminum lid	97
Transport to can maker (average 392 mi)	2
Can manufacture (4% waste)	25
Transport to bottler (average 200 mi)	2
Bottling, transport to retailer	52
Retailer, consumer	?
Waste collection	1
Total	417

0.42 kWh/use

*can weight 0.101 lb.

SOURCE: Adapted from B. M. Hannon, *Environment* 14, no. 2 (1972): 11. Reprinted with permission of the Helen Dwight Reed Educational Foundation. Published by Heldref Publications, 4000 Albemarle St., N.W. Washington, DC 20016. Copyright 1972.

The New York bottle bill has caused the creation of "redeemers," people who are self-employed and collect bottles and cans. Their work and the effect of the bottle bill in general saves an estimated 500 tonnes of trash per day in New York City alone (out of 20 kilotonnes production). Most average citizens seem to feel this way in states with bottle bills in operation. The bottle bill seems like motherhood and apple pie to them.

The counter-argument by opponents generally focuses on four issues: inconvenience, health, job quality, and energy costs of recycling. Consumers are inconvenienced because they must save bottles; local stores must provide space and staff to handle the bottles. Some merchants have problems with vermin in their bottle storage areas. Some are concerned that the bottles could constitute a health hazard. The number of jobs overall increases with a bottle bill, but employment

FIGURE 20.6
Litter along a California highway (actual unstaged photo). The cost of cleaning up such highway litter is enormous; it has been found to be substantially less in states with "bottle bills."

of skilled workers decreases; the jobs gained are not as good as the jobs lost. It is also claimed that the energy cost of trucking the bottles back to the bottling plant and cleaning them is prohibitive. This last argument was analyzed in Table 20.11, and is favorable to the case for returnables.

The health argument seems not to be a serious problem in "bottle bill" states. The inconvenience and job quality arguments have merit. In the final analysis, one must weigh the value of esthetics and reduced road hazard against inconvenience and job redistribution.

My conclusion is that we should enact bottle bills to save energy, to create jobs, and to preserve scarce resources for more important uses. Incidentally, the return to returnable bottles favors local bottlers over national bottlers, so that this may be one way to prevent the concentration of all beer brewing into the hands of a few national breweries (20).

SUMMARY

The throwaway society creates immense mounds of trash. Not only is trash created by the throwing away of things, but it is also created in the process of extracting the raw materials themselves, and at every step in the transformation of ore into finished product.

The cost of aluminum and steel from recycled metal is less than 5 percent of the cost for ore. It is not so efficient to recycle glass, but of course reuse is very much more efficient than recycling. Universal adoption of recycling would greatly reduce the volume of waste sent to local landfills.

REFERENCES

1. A glimpse into the lives of a rag-picker's family may be had in the all-of-a-kind family books by Sidney Taylor, originally published by Follette Publishing Co., Chicago, and in paperback by the Dell Yearling Series.
2. The latest in the series is *The Last Whole Earth Catalog; Access to Tools,* ed. Stewart Brand (New York: Random House, 1980).
3. Environmental Protection Agency, *Environmental Quality–1970,* (Washington, D.C.: GPO, 1970).
4. Department of Commerce, *Statistical Abstracts of the United States, 1984* (Washington, D.C.: GPO, 1984).
5. A. P. Carter, *Science* 184 (1974): 325.
6. Office of Emergency Preparedness, *The Potential for Energy Conservation* (Washington, D.C.: GPO, 1972).
7. M. Ross, *Ann. Rev. Energy* 6 (1981): 379.
8. J. R. Miller, *Sci. Am.* 250, no. 5 (1984): 32.
9. E. T. Hayes, *Science* 191, (1976): 661.

10. Y. M. Ibrahim, *New York Times,* 14 Dec. 1972.

11. I. Meyer-List, *die Zeit,* 13 Aug. 1984.

12. E. C. Peterson, in *Mineral Facts and Problems* Bulletin 617, U.S. Bureau of Mines (Washington, D.C.: GPO, 1980).

13. H. F. Kurtz and L. H. Baumgardner, in *Mineral Facts and Problems*

14. Stanford Research Institute, *Patterns of Energy Consumption in the United States* (Washington, D.C.: GPO, 1972). Energy use is corrected for thermal and electric energy use.

15. S. Bronstein, *New York Times,* 14 Sept. 1986.

16. W. U. Chandler, *Conservation and Recycling* 9 (1986): 87.

17. J. B. Rosenbaum, *Science* 191 (1976): 720.

18. I. D. Good, *Endeavour* 10 (1986): 150.

19. P. K. deJoie, *Environment* 19, no. 7 (1977): 22.

20. I. Meyer-List, *die Zeit,* 18 June 1982.

21. C. Pollock, in *State of the World 1987,* ed. L. Starke (Washington, D.C.: Worldwatch Institute, 1987).

22. I. Peterson, *New York Times,* 16 April 1987.

23. I. Peterson, *New York Times,* 1 May 1987.

24. See, for example, P. S. Gutis, *New York Times,* 24 April 1987, 25 April 1987, 4 May 1987.

25. Anonymous, *New York Times,* 21 April 1987; J. F. Sullivan, *New York Times,* 11 Jan. 1987.

26. S. Bronstein, *New York Times,* 12 Feb. 1987.

27. G. Hill, *New York Times,* 14 Oct. 1976; R. R. Grinstead, *Environment,* 14, no. 3 (1972): 2.

28. P. Sarnoff, ed., *New York Times Encyclopedic Dictionary of the Environment,* (New York: Arno Press, 1975), Table 13.

29. J. McCaull, *Environment* 16, no. 1, (1974): 6. A similar effect was seen in Vermont following passage of the bottle bill there. See J. M. Jeffords and D. M. Webster, *Vermont 5 cent Deposit,* U.S. House of Representatives. For Oregon, see State of Oregon, Dept. of Environmental Quality, *Oregon's Bottle Bill,* the 1977 report. For Iowa, AP dispatch, *New York Times,* 26 Sept. 1984. For certain New Jersey communities, J. F. Sullivan, *New York Times,* 4 Dec. 1983. See also M. Sullivan, *The ABC's About Beverage Containers,* National Wildlife Federation, 1977.

30. B. M. Hannon, *Environment* 14, no. 2 (1972): 11.

31. L. D. Orr, *Environment,* 18, no. 10 (1976): 33.

32. R. J. Smith, *Science* 202 (1978): 34.

33. K.-H. Joepen, *die Zeit,* 24 Feb. 1984.

PROBLEMS AND QUESTIONS

True or False

1. Mixed paper is less valuable than pure rag paper or used newspapers because of its greater entropy.

2. When a country such as Jamaica builds a processing plant to make alumina from bauxite in order to create jobs, it also reduces the energy required to produce aluminum from ore.

3. Recycling steel saves less energy per kilogram than recycling aluminum.

4. In the process of throwing things away, extra jobs are created.

5. The use of life-cycle accounting is as valuable an instrument in making decisions about buying home heating equipment as it is in buying an aircraft carrier.

6. Of all litter, glass bottles are the worst because it takes so much longer for bottles to degrade than other litter.

7. Recycling of glass bottles is preferable to reuse from an energy standpoint.

8. Most of the energy used in making aluminum is electric energy used for producing aluminum from alumina.

9. The old industries such as steel have little opportunity to develop new energy-saving strategies.

10. The chemical industry is very good at saving energy because of the intimate connection between chemical research and industrial processes.

Multiple Choice

11. We learned that a deposit bottle costs 0.92 kWh$_e$ to make and 0.16 kWh$_e$ to clean after each reuse. After about how many uses will the energy content of the bottle per use be equal to the energy content of the beverage?

    a. 10   d. 1000

    b. 50   e. Never

    c. 100

12. When considering pollution control expenditures,

    a. expenditures are greatest and constitute the greatest proportion of total expenditures among the manufacturing sector.

    b. expenditures are greatest and constitute the greatest proportion of total expenditures among the nondurable goods sector.

    c. expenditures are greatest and constitute the greatest proportion of total expenditures among the nonmanufacturing sector.

    d. expenditures are greatest in the manufacturing sector, and constitute the greatest proportion of total expenditures for the nondurable goods sector.

    e. expenditures are greatest in the nonmanufacturing sector and constitute the greatest proportion of total expenditures in the nondurable goods sector.

13. Among the metals, the one with the greatest proportion recycled is

    a. lead.   d. iron.

    b. copper.   e. zinc.

    c. aluminum.

14. In comparing populations and percent of glass recovered in various cities,

    a. a city of 100,000 that recycles 10 percent of its glass recovers less glass than a city of 10,000 that recycles 70 percent of its glass.

    b. a city of 500,000 that recycles 15 percent of its glass recovers less glass than a city of 100,000 that recycles 50 percent of its glass.

    c. a city of 50,000 that recycles 50 percent of its glass recovers less glass than a city of 25,000 that recycles 90 percent of its glass.

    d. a city of 100,000 that recycles 20 percent of its glass recovers less glass than a city of 50,000 that recycles 50 percent of its glass.

    e. a city of 50,000 that recycles 25 percent of its glass recovers less glass than a city of 10,000 that recycles 70 percent of its glass.

15. It costs

    a. half as much energy to use a returnable seven times as a throwaway once.

    b. twice as much energy to use a returnable twice as a throwaway once.

    c. one-third as much energy to use a returnable fifteen times as a throwaway once.

    d. more energy to use a throwaway once than to use a returnable twice.

    e. less energy to use a returnable once than to use a throwaway once.

16. Which of the following issues are raised by opponents of recycling?

    a. Inconvenience of recyclables as opposed to throwaways

    b. Health problems associated with the recycling of dirty bottles

    c. The poorer job quality in the recycling industry as compared to the bottling industry

    d. The substantial energy costs of recycling

    e. All of the above issues are raised by opponents of recycling.

17. What is the reason for the excessive use of energy in industrial processes prior to the 1973 energy crisis?

    a. Energy was not an important component of production.

    b. Capital costs of energy-inefficient equipment were lower than those of energy-efficient equipment.

    c. Operating expenses of all energy equipment are the same.

    d. There was no innovation in industry before 1973.

    e. None of the above reasons explain this phenomenon.

18. Pollution control expenditures
    a. have risen consistently in real terms since the early 1970s.
    b. have fallen in real terms in the 1980s as compared to the 1970s.
    c. are a bit over 5 percent of all expenditures of nonfarm business.
    d. were greater in 1980 than in 1975.
    e. were at a maximum in 1981.

Discussion Questions

19. Should the energy savings involved be the only determinant of whether or not a material should be recycled?
20. In the book *Limits to Growth,* the scenario predicted a collapse of the world population, standard of living, etc., even when recycling was introduced. Is this a good reason to just drop the whole idea?
21. Explain why recycling at the point of use is the most energy-efficient form of recycling.
22. If "bottle bills" are as benign as indicated in the text, why doesn't every state have one? What disadvantages accompany reuse of bottles?
23. Give arguments for and against standardization of other types of bottles (e.g., whiskey bottles, mayonnaise jars, jelly jars), so that those bottles can be reused as well.

21

MOVING DOWN THE ROAD

Transportation is the lifeblood of economic and social exchange. The transportation system is run mainly on fossil fuel, almost 20 EJ worth, about 20 percent of the total energy use in the United States.

The present chapter considers the context of the transportation system—use of fossil energy and pollution. The main subject of the chapter is the automobile; however, mass transit, load factors, and other modes of transportation are considered as well.

KEY TERMS   *drag*
catalytic converter
stratified charge engine
photochemical smog
tetraethyl lead
Waldsterben
load factor
drag coefficient
circuitry

The American public has carried on a love affair with the automobile from its earliest days in spite of the weather, economic depression, war or peace. It has had an almost unbelievable effect on the world's way of life: changing the basis for the family; providing a new method of courtship; and giving every Walter Mitty his own chariot. One of America's great cities, Los Angeles, has developed around dependence on the private car as the mode of transportation. The flight to the suburbs gained momentum as the car came of age, so that today the American population is 73.5 percent urban or suburban (1). An astounding 18 percent of the population moves each year (2)!

THE COST OF OWNING A CAR

This "romance of the road" has occurred despite the tremendous economic burden of car ownership. If we assume (generously) that the standard car lasts twelve years and travels 10,000 miles per year, the average (1982) cost per mile is 26.7 cents (3). A subcompact would cost only 18.8 cents per mile. Over the twelve years, the total investment for transportation by private car is between $22,600 (for a subcompact) and $40,000 (for a van) (4).

As we have long known, American cars are gas guzzlers. Between 1950 and 1973, the average American car mileage declined from 18.27 miles per gallon to a low of 11.85 miles per gallon (5,6), basically because cars got heavier. By 1984, it had risen to almost 17 miles per gallon and has remained at that level (7). There is a relation between weight and mileage that seems to hold for most cars (5):

Fuel consumption (measured in gallons per thousand miles)
= 0.02227 (weight in pounds) − 1.431
= 0.01012 (mass in kilograms) − 1.431

or, roughly,

fuel consumption (liters per 100 kilometers)
= 0.0024 (mass in kilograms) − 0.34.

In those years, the average speed of cars on rural highways rose from 48.7 miles per hour (1950) to 60.6 miles per hour (1970) (7,8). Overall, the average speed rose through 1973 and then dropped as a result of the 55 mile per hour speed limit enacted by Congress in 1974, then edged upward again (Figure 21.1a) (7). Figure 21.1b shows a somewhat different perspective: the proportion of post-1974 speeders (7). This upward drift is probably partly responsible for small increases in highway fatalities in the mid-1980s after

years of decrease. Congress officially allowed the speed limit to be raised to 65 miles per hour for rural interstate highways in 1987. Later statistics will surely show the effect of speed on the death rate.

The increasing weight and increasing speed of cars during the two decades shown led to the huge petroleum distillate consumption illustrated in Figure 21.2a (9). In the ground transportation sector, buses use 1 percent of the gasoline sold, trucks use 36 percent, and automobiles use the remaining 63 percent (6). In the total transportation sector, 2×10^{19} J (5.5×10^{12} kWh$_e$) was used in 1981 (9). Over 79 percent of this energy was used by trucks and cars; 9.0 percent by aviation; 8.7 percent by inland water transport; and 2.9 percent by railroads. Only one-third of a percent of the energy is used in mass transit (buses, subways, and trolleys). The fact that gasoline is preserving its relative position in times of burgeoning jet air travel is another signal of energy profligacy (see Figure 21.2b [7,10]).

The problem with increasing speed is illustrated in Figure 21.3. As the speed of a car increases, the force of the air holding it back increases. This frictional force is called *drag force*.

FIGURE 21.1

(a) The average speed of highway vehicles, which dropped precipitously following passage of the 55 mile-per-hour speed limit, is shown here from 1960 to 1981. (b) The percentage of cars traveling at speeds greater than that indicated is shown from 1950 to 1981. This curve also shows a sharp drop after the national 55 mile-per-hour speed limit was passed in 1973.

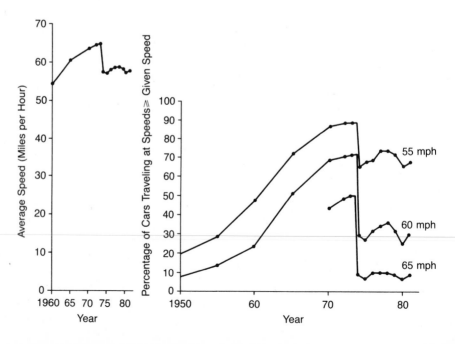

FIGURE 21.2
The relative proportions of gasoline, jet fuel, and residual and distillate fuel oils have changed over the three decades since 1955. In addition, the total value of liquid fuel used each year has grown from 235 million liters in 1955 to 575 million liters in 1985. (U.S. Department of Energy)

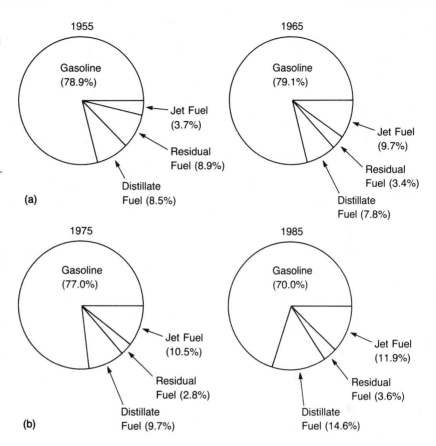

(a)

1955
Gasoline (78.9%)
Jet Fuel (3.7%)
Residual Fuel (8.9%)
Distillate Fuel (8.5%)

1965
Gasoline (79.1%)
Jet Fuel (9.7%)
Residual Fuel (3.4%)
Distillate Fuel (7.8%)

(b)

1975
Gasoline (77.0%)
Jet Fuel (10.5%)
Residual Fuel (2.8%)
Distillate Fuel (9.7%)

1985
Gasoline (70.0%)
Jet Fuel (11.9%)
Residual Fuel (3.6%)
Distillate Fuel (14.6%)

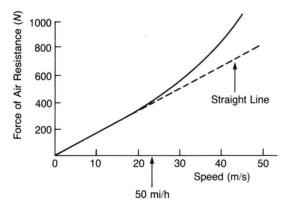

FIGURE 21.3
The drag force increases faster than speed increases as the speed increases above about 20 m/s. For higher speeds, drag force increases as the square of the speed.

The retarding force of the air grows faster for a given speed increase as the speed increases. Since the amount of work done by the car in overcoming the retarding force is equal to the retarding force times distance driven ($W_R = f_R \times d$), this work also must increase faster as the speed increases. This energy to overcome the retarding force is supplied by gasoline—the greater the retarding force, the more gasoline must be burned just to overcome that retarding force—and thus gasoline mileage must decrease.

POLLUTION FROM CARS

The advent of pollution control devices is one reason for the poorer fuel economy of recent-model cars. Engines perform less efficiently when a catalytic converter is added to the exhaust sys-

TABLE 21.1
Air pollutant levels in
the South Coast Air
Basin (tonnes/day).

| | 1985 Levels | | | | |
	Reactive Organic Gases	NO$_x$	Fine Particulates	SO$_x$	CO
Transportation	590	585	645	75	4070
Oil products, marketing	120	90	10	30	50
Commercial	310	185	30	20	150
Solvents, home heating	80	35	15	5	270
Total	1100	895	700	130	4540

| | 2007 "Baseline" Levels | | | | |
	Reactive Organic Gases	NO$_x$	Fine Particulates	SO$_x$	CO
Transportation	310	580	975	80	2375
Oil products, marketing	80	45	5	40	20
Commercial	325	155	45	45	160
Solvents, home heating	210	50	5	5	430
Total	920	830	1030	170	2980

FIGURE 21.4
The 1970 distribution of
air pollution emissions
by source (Ref. 11).
Emissions totaled 237
Mtonne.

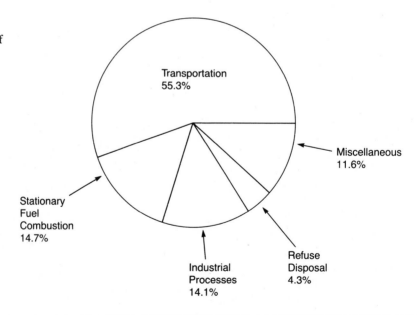

tem. In catalytic converters, the carbon monoxide (due to incomplete combustion), unburned hydrocarbons, and nitrogen oxides react at high temperature with platinum catalyst pellets. The high temperatures arise because a series of baffles slows down the exhaust gases to allow sufficient time for the catalyst to cause the pollutants to react and become carbon dioxide, normal nitrogen, and normal oxygen. This slowing causes a back pressure, increasing the pressure at the tappet through which the combustion gases exit the cylinder. Catalytic converters therefore increase gasoline consumption.

It should be said that not all strategies for pollution control cause increased gasoline consumption. The common PCV (positive crankcase ventilation) valve merely allows gas leaking from the cylinders to be reused. The stratified charge engine has a modified cylinder, which has a small chamber with fuel-rich air fired by the spark plug connected to the main cylinder. The explosion causes a large chamber with fuel-poor air to ignite (as in the Honda Civic). This design provides reduced emissions without a mileage penalty. Also see the box on organic fuels.

Cars are the leading cause of photochemical smog, and are important contributors to total emissions, as shown by the Los Angeles data in Table 21.1 (11) and the national data shown in Figure 21.4 (12). The amount of pollutant emitted increased year by year during the 1970s and early 1980s. In Table 21.1, we also see the "baseline predictions" for the South Coast Air Basin for 2007 if no new controls on emissions are instituted. Such controls will have to be instituted because, otherwise, ozone levels will exceed federal standards by a factor of 2.5, and levels of carbon monoxide and fine particulates will exceed them by a factor of almost 2. There are many strategies currently available and straightforward advances in current technology that should help reduce levels below the standards.

As an illustration of the overwhelming presence of the auto in overall air pollution, Figure 21.5 (13) illustrates the effect of the imposition of stringent pollution controls in the South Coast Air Basin of Los Angeles in 1965. If catalytic converters had been installed by 1965, the reduction in pollution would have been spectacular indeed. Note that there is a "hidden saving" made obvious in the diagram. This occurs because pollution levels would have grown without pollution control about as they had from 1955 on; the actual gain is the difference between what *would* have

happened and the drop that occurred. Figure 21.6 (14) from Los Angeles County gives additional evidence of how emissions from autos pollute urban areas.

An estimated 3 kilotonnes/day of hydrocarbons is emitted nationwide as vapor from filling station pumps; this is an indirect cause of car-related pollution and a waste of natural resource as well as an economic loss. Many California filling stations have a vapor recycling system mandated by law. The gas pump has a rubber collar that fits on the gas tank intake of the car. As gas is pumped in, the displaced vapor is pushed into the collar and collected. The EPA is planning to control such emission nationwide, although the method may be to use canisters in each vehicle rather than follow the California model (15).

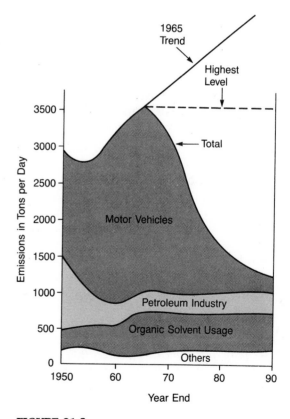

FIGURE 21.5

Actual (to 1965) and estimated emissions in the South Coast (Los Angeles) Basin (Ref. 12).

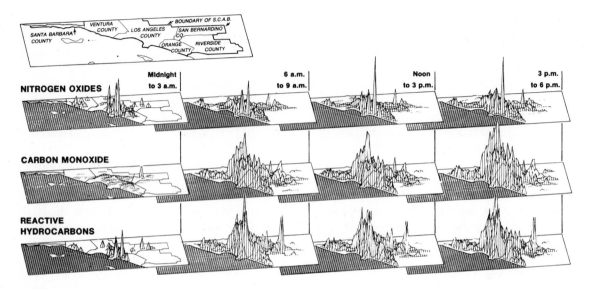

FIGURE 21.6
Profile of emissions of nitrogen oxides, carbon monoxide, and reactive hydrocarbons in the Los Angeles metropolitan area, showing the effect of commuting. (Data from G. J. McRae, cited in an article by S. Blakeslee. Copyright 1979, *Los Angeles Times*. Reprinted by permission.)

Carbon Monoxide and Autos

One way to gauge the effect of transportation on emissions (transportation-related emissions are shown in Figure 21.7) is to consider the proportion of various emissions as shown in Table 21.2 (16–18). The fear of auto pollution is clearly well justified, despite the relative harmlessness of most auto particulates as compared to those from power plants and the low levels of sulfur dioxide in the exhaust. Carbon monoxide (CO) levels are particularly high. A schematic representation of a round trip from Cal Tech to downtown Los Angeles is shown in Figure 21.8. Note how high the concentration of CO gets. In downtown Los Angeles, the concentration exceeded *100 parts per million* (ppm) on this trip. The maximum "safe" concentration allowed by national standards, 9 ppm, is discussed below. Los Angeles is not the only city with a carbon monoxide problem. Most city dwellers suffer from exposure to elevated levels of CO. One day out

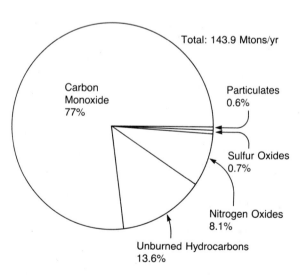

FIGURE 21.7
Proportions of pollutants emitted in transportation. Total emissions in 1970 were 131 Mtonne. (U.S. Department of Commerce)

TABLE 21.2

Emission from automobiles compared with total emissions (megatonnes/year).

	South Coast Basin[a] 1985	EPA[b] 1975	United States Plumley[c] 1973	Morgan et al.[d] 1970
CO	1.32/1.80	100.8/135.1	57.1/91.0	52.6/68.6
Unburned hydrocarbons		17.7/22.6	15.1/29.1	13.7/23.0
Reactive organic gases	0.17/0.41			
NO_x (nitrogen oxides)	0.14/0.35	10.6/20.2	7.4/18.7	6.6/15.5
Particulates	0.015/0.55	0.7/23.2	1.1/25.7	1.0/15.4
SO_2	0.007/0.04	0.9/26.5	0.7/30.2	0.6/26.8

[a]Reference 11.
[b]Reference 16.
[c]Reference 17.
[d]Reference 18.

of six, Denverites live through some of the highest levels of CO in the country: 24 ppm (19).

CO is a deadly pollutant because it can replace oxygen in the hemoglobin in the blood, and once it has done so, it is very difficult to pry out. In 1970, the government set a standard specifying an area to be unhealthy if it were subjected to levels of over 9 ppm during an eight hour period at least twice a year, or if the level exceeded 35 ppm for two or more one hour periods per year (20). Workers in New York City's garment district apparently experience a concentration of CO greater than 100 ppm every day (20). Much of

the 200 million tonnes of CO emitted each year is from cars (21,22).

Since people sitting in nearby cars are affected, it is particularly hazardous to sit in traffic jams, when about 10 percent of the gas emitted from the exhaust is CO (21). This problem could be reduced by placing car exhausts at roof level, as a diesel truck exhaust is. This placement improves dispersal of CO by a factor of 10 (21). One major problem with CO is the burning sensation in the eyes associated with its presence; another is that it interferes with judgment by hampering brain function. Since we are less able

FIGURE 21.8

Carbon monoxide levels in a trip from Pasadena to downtown Los Angeles. (A. J. Haagen-Smit, *Arch. Env. Health* 12, 548. Reprinted with permission of the Helen Dwight Reed Educational Foundation. Published by Heldref Publications, 4000 Albemarle St. N.W., Washington, DC 20016. Copyright 1966.)

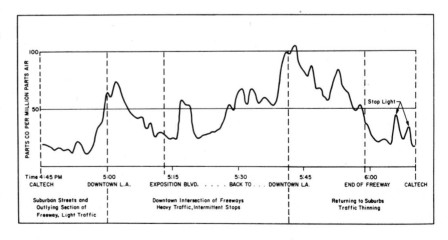

to deal with difficult situations when affected by CO, being caught in a traffic jam can also be dangerous to health because of the increased risk of accidents.

One puzzle about CO is its disappearance from the environment. Estimated worldwide CO emissions should be raising the atmospheric level by about 40 parts per billion each year. (It takes about a month for half the CO to oxidize in the atmosphere.) The CO concentration does not appear to be increasing at all (20). No one knows where it is all going (also see Chapter 23).

Carbon monoxide is a major pollutant in all cities, but other pollutants are present as well. The photochemical smog composition for Los Angeles is shown in Table 21.3 (23). We might note that it is still difficult to determine pollution levels accurately: Los Angeles readings on smog levels were found to disagree by 30 percent with the state-mandated method used by nearby counties (24). Relevant standards are given in Table 21.4. The 1975 standards represent a reduction of a factor of 10 in unburned hydrocarbons and of 6 in carbon monoxide relative to actual levels in 1960.

Lead

Another noxious pollutant emitted by cars is lead, a toxic heavy metal. Lead levels on the Greenland ice cap have increased rapidly since the advent of the industrial revolution, and particularly so since 1923, when tetraethyl lead was introduced as an antiknock fuel additive. Greenland is far from sources of pollution; pollutants trapped in the ice reflect the atmospheric levels at the time the snow was deposited. Addition of lead is the cheapest way known to enhance "octane" levels, i.e., to allow complete burning at lower temper-

TABLE 21.3

Typical concentrations in Los Angeles smog.

Material	Concentration
Water vapor	2×10^{-2}
CO_2	4×10^{-4}
NO_x	2.5×10^{-7}
CO	2×10^{-5}
O_3	2×10^{-7}
Reactive organic gases*	5×10^{-6}
Nonreactive organic gases†	2.5×10^{-6}
SO_4	2×10^{-9}
NO_3	5×10^{-9}

* Reactive organic gases excludes carbon monoxide, carbon dioxide, carbonate, and nonreactive organic gases.

† Nonreactive organic gases are CH_4, CH_3Cl, CH_2Cl_2, and the chlorofluorocarbons.

SOURCE: Ref. 23

atures and prevent knocking. In the world at large, the increase in consumption, in vehicle miles per year, has outpaced the introduction of technology as measured by decreased lead per vehicle mile. In the United States, the advent of the catalytic converter has gradually decreased the volume of leaded gasoline sold, decreasing ambient lead levels. Lead "poisons" the platinum catalyst, so it must not be used in conjunction with the converter. In Western Europe, lead levels of 0.4 grams per liter of gasoline are common (24). The highest level allowed in Germany is 0.15 grams per liter, which is the lowest in Europe (25,26). By law within the Common Market, no country may set a standard stricter than 0.15 g/l nor one looser than 0.4 g/l (27).

The German government prodded the Common Market to accept American emissions standards (shown in Table 21.4) on new cars from

TABLE 21.4

Emissions standards (grams/mile).

	Hydrocarbons	CO	NO_x	Particulates
1975 standards	1.5	15	3.1	
California standards	0.9	9	2	
Statutory 1980 standards	0.41	7.0	2.0	
Statutory 1987 standards	0.41	3.4	1.0	0.2

URBAN POLLUTION

The nitrogen dioxide in polluted areas causes more trouble than any other compound. The ozone in these areas can cause similar, though less severe, difficulties. Both compounds are very effective at intercepting sunlight in the ultraviolet region (very low wavelength). This captured energy is emitted in the infrared (long wavelength) as heat, much of which is then trapped in that region. They therefore can affect the local atmosphere heating rate. The presence of 1 ppm ozone in air can cause a 0.1°C temperature rise per day; the presence of a similar concentration of nitrogen dioxide can cause a temperature rise of 1°C per hour (12,30).

The Camp Century, Greenland, data (Figure 21.9) (31) show how the lead concentration in the atmosphere has doubled from 8 parts per hundred billion to about 16 parts per hundred billion over the last century. The only respite in growth of lead consumption after 1935 until the advent of the catalytic converter was in the period 1958 to 1962, which was probably due to the introduction of smaller American cars with lower compression ratios, as the auto makers fought the importation of smaller cars such as the Volkswagen Beetle. People who bought small cars were induced by salespeople to buy larger cars when they sought to replace their cars (32).

The U.S. Environmental Protection Agency (EPA) has acted over the years to reduce levels of lead in gasoline. In the early years of the Reagan administration, lead manufacturers attempted to get the EPA to lift allowable lead levels, citing lack of evidence of harm. The resultant storm of protest from scientists studying the effects of lead caused the retreat of the pro-lead forces. In 1984, the EPA cut allowable levels of lead in gas by 91 percent, and by 1995, lead will be banned totally as a gasoline additive (27,33). There is evidence that U.S. lead levels have been dropping since 1974 (34).

Lead levels in the U.S. decreased less than expected by regulators because car owners have been known to tamper with their air pollution control devices in order to allow use of leaded gasoline (presumably because of its lower price). About 13 percent of vehicles nationwide that were designed for use of catalytic converters have been altered so that leaded gas could be used (31). In 1979, in Southern California, it was estimated that 19 percent of air pollution control devices had been tampered with, leading to increased lead levels and an extra 100 tons of unburned hydrocarbons released into the air (35).

FIGURE 21.9

Measurement of lead levels (parts per billion) from Camp Century, Greenland, ice cores. (R. A. Bryson, *Science,* 753 [1974], copyright 1974 by AAAS)

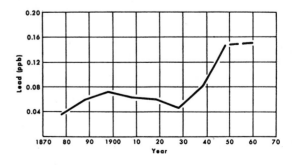

1989 on, with retrofitting of catalytic converters on older models mandatory from 1992 on; their proposal was defeated by the other countries (28). The Common Market countries did agree in 1985 to require new models with engines of displacement larger than 2 liters, such as Mercedes, to meet European standards (which are not quite American standards) by October 1988, and to require older models of such cars to meet the new standards by October 1989. Models with displacements of 1.4 to 2 liters, such as the Ford Escort or the VW Golf, have until October 1993 to meet the standards (29). The other countries eschewed the early timetable proposed by the Germans in order to allow more time to find an alternative to catalytic converters as a solution to the emissions problem.

WALDSTERBEN

The major reason for German enthusiasm for unleaded gas is not the 4.5 megatonnes of lead emitted each year into European air (26), but rather the impending destruction of the German (and Swiss) forests (*Waldsterben*) from the effects of acid rain (36–40). The connection between unleaded gas and the nitrogen oxides is, of course, the catalytic converter. Catalytic converters reduce nitrogen oxides in the exhaust: The catalyst provides a surface on which exothermic reactions can take place to reconstitute ordinary nitrogen and oxygen molecules. It also supplies the activation energy to allow the re-

TABLE 21.5

Sickened German forest, 1982–1984, in percent of area.

	1982	1983	1984
Sickly	6	25	33
Sick	1.5	9	16
Very sick, dead	0.5	1	15

SOURCE: German Ministry of the Environment, quoted in *die Zeit*.

actions to begin. Lead poisons the platinum catalyst used in the converters.

Nitrogen oxides in the air can be transformed into nitric acid by interaction with water. Over 55 percent of the 3.1 Mtonne (million tonnes) of nitrogen oxides emitted in Germany comes from cars (in contrast, 2 of every 3 Mtonne of sulfur dioxide [SO_2] presently emitted comes from power plants) (36). As discussed in the next chapter, these oxides combine with water vapor in the air to make acid precipitation, which falls over forests and weakens trees. Roughly 20 percent of the acidity in rainfall in Germany is nitric acid (25 percent in the United States) (39).

In Germany, only 8 percent of the trees seemed damaged in 1982; this rose to 34 percent in 1983 (36,37) and to 50 percent in 1984 (36). In the two southernmost states of Germany, the proportion of sick trees is higher—57 percent in Bavaria and 66 percent in Baden-Württemberg (home of the Black Forest) (35). Tables 21.5 and 21.6 (33) show how the proportions of sickened

TABLE 21.6

Sickening of trees in German forests by type.

Type	Percent of Total of German Trees	1982	1983	1984	1985
Spruce	39.2	9	41	51	55
Pine	19.9	5	44	59	57
Fir	1.4	60	75	87	88
Beech	17.0	4	26	50	55
Oak	8.4	4	15	43	52
Remainder	13.1	4	17	31	31

SOURCE: German Ministry of the Environment, quoted in *die Zeit*.

trees grew from 1982 to 1984. In Switzerland, over half of the trees in the forests are damaged, and many are expected to die (38). In the German and Swiss Alps, the problem is not only esthetic. Trees protect residents and tourists from landslides and avalanches, and the trees should, ideally, be a mix of species and ages (41).

Forests in the United States have been suffering, too. Trees have been dying in New England in areas remote from people, such as Camel's Hump in Vermont's Green Mountains (42). The haze in the Great Smoky Mountains seems to be due to the burning of coal in states distant from North Carolina (43).

While the damage to German forests can be seen in hindsight to have begun as long as twenty years ago (the diameter increase in trees in forests fell from about 20 mm per year in the mid-1960s to a few mm per year in the mid-1980s), it was not obvious to the Germans until 1981. Similar changes have occurred in the United States. An abrupt change in growth rates simultaneously observed in New England and the Southern Appalachians seems to bode the same fate for American forests (44).

Pollutants and Politics

The European countries, with the exception of West Germany, have speed limits between 100 and 130 km/h. Cars traveling 130 km/h produce emissions at twice the rate of those traveling 100 km/h (45), but because the faster car does not travel for as great a time, the total emissions are only 54 percent higher than for the car traveling 100 km/h. Part of the reason that other Common Market countries are not much inclined to rush installation of converters, as discussed above, is the lack of speed limits on the German autobahns. Adoption of a German speed limit would cut pollution levels, as these other countries point out. German politicians would rather impose catalytic converters than cut the speed limit directly. On the other hand, if cars are fitted with converters, they must be limited to 130 km/h anyway

in order to prevent overheating (26). Hot converters on cars parked near fallen leaves have caused fires, for example (46). The Germans who buy cars with converters will have to pay a performance price in any case. BMW engineers say that a converter-equipped car loses 15 percent in horsepower (47). One way or another, German speeds on the highway have to be limited if the NO_2 emissions are to be reduced.

Since converters for German cars cost 600 marks (DM600 = about $300) (26), and since the Germans intend to have no leaded gas after July 1989 (48), manufacturers and drivers had expected substantial subsidies from the government to hasten installation of converters. Both groups were disappointed. The tax that car owners now pay, DM144 per liter of engine displacement per year, is eliminated for new cars with converters. Retrofitting is to be necessary for older cars; those whose emissions drop 30 percent or more will pay DM132 per liter per year; only new cars produced without converters and older cars still without converters will pay more (DM216 and DM188, respectively) (49). The payback period for the improvement in emissions seems too long to most observers. The normal wisdom is that buyers will pay only for fuel economy savings that can realize a net return within two to three years (50). This is probably also the attitude most people have toward payments for emission control. Remember that this air is similar to a commons. Individuals will have to be coerced or cajoled into doing something that will cost them money personally to benefit everyone by preserving the forests.

The pollutants we have discussed (carbon monoxide, nitrogen oxides, sulfur oxides, and lead) were not perceived as dangerous until recent times. While this is understandable for NO_x and SO_2, it is less so for lead. Lead wine containers have long been implicated as the cause of the failure of many upper class ancient Romans to have children. More recently, lead levels in people have risen because of lead in paints, lead solder on cans, and, of course, emission of lead in car exhausts. Studies have put lead intake for prehistoric humans at 0.21 micrograms per day; Americans' level of lead intake in 1980 was 29 micrograms per day (51). Delicate measurement procedures have been used to analyze lead in tuna fish. Freshly caught tuna have concentrations of 0.3 ppb. When tuna is packed in soldered cans, levels measured in the tuna average about 1 ppm, 4000 times higher (51). While that level is considered safe by the EPA, many would argue that it is too high and advise avoidance of canned foods. Leaded paints are mostly unavailable now because the dangers of lead are widely recognized. Children often put paint flakes in their mouths, and many poor urban children suffer from lead poisoning because the buildings they live in have not been repainted since the effects of lead paint have become known or lead paint was painted over and the layers are peeling. It

has always been possible to make gasoline without lead, even in the days before the introduction of catalytic converters. Tetraethyl lead was introduced in the 1920s, and by the 1940s, research had indicated its environmental hazards. It is puzzling that the public has allowed itself to suffer exposure to lead for so long.

EFFICIENCIES OF VARIOUS MODES OF TRANSPORTATION

Autos presently account for 21 percent of *all* U.S. energy consumption. About 55 percent of travel involves trips of less than 16 km (10 miles) (52,53); 56 percent of all commuting trips involve cars with only one occupant (52), 26 percent of all commuting trips were on public transportation, and 4 percent were by other means (53). See Figure 21.10.

FIGURE 21.10
Percentage of trips of given distance and percentage of vehicle miles, according to the Personal Transportation Survey (1969–70) of the Department of Transportation.

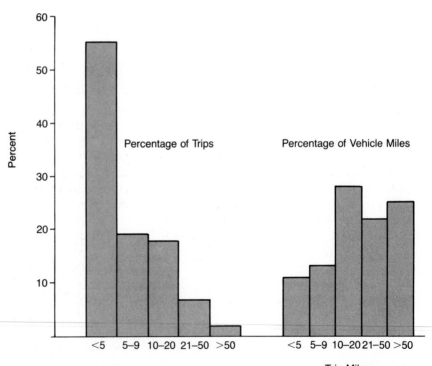

TABLE 21.7
Comparison of energy costs of various transportation modes.

Mode	Energy Cost (kJ/passenger-km)		
	Lincoln[a] 1972	DOT[b] 1987	ORNL[c] 1987
Bicycle	130		
Walking	200		
Motorcycle		2276	2080
Auto		1957	2845
Urban auto	5300		
Personal truck			3520
School bus		10,837	426
Local bus		25,416	2393
Intercity bus	1050	771	819
Mass transit railroad	2500		2301
Intercity railroad		1679	2078
General aviation		7256	7808
Certified air carriers		3175	3671

[a]G. A. Lincoln, *Science* 180 (1973): 155. Copyright 1973 by the AAAS.

[b]Department of Transportation, *National Transportation Statistics* (Washington, D.C.: GPO, 1987).

[c]M. C. Holcomb, S. D. Floyd and S. L. Cagle, *Transportation Energy Data Book: Edition 9* (Oak Ridge, Tenn.: Oak Ridge National Laboratory, 1987).

The steady increase in speed and power have, as we have seen, caused a decline in efficiency, and, as discussed above, emission control by catalytic converter causes even more. Artificially low energy costs, sustained over decades, and urban sprawl have also contributed to the inefficiencies of the system.

In Table 21.7, we see a comparison of the energy efficiencies of various transportation modes. One study (52) showed that, in the 1960s, the number of railroad passengers declined by half, the number of airline passengers tripled, and the number of auto passengers increased by a factor of 1.5. The report of the Office of Emergency Preparedness (54) estimated that improved communication facilities, urban clustering, and construction of walkways and bike paths could lead to a 15 to 20 percent energy saving for transportation.

The breakdown of fuel economy of various transportation modes is shown in Figure 21.11. We could also switch to ethanol for fuel to save energy (see box). In Figure 21.12 (60) is a comparison of the efficiencies of selected transport modes. In order to really examine the various transportation options from an informed perspective, we must know not only the energy consumption per passenger-mile, but also the load factor, or average number of passengers carried. What makes mass transit efficient is its large load factor. The cheapest way to decrease transport system cost is to increase the utilization of already extant transport.

The most important measure of economy is the number of passenger-miles per gallon. Suppose we compare a Honda Civic with an EPA rating of 41 miles per gallon with one passenger, to a Chevrolet station wagon with an EPA rating of 13 miles per gallon carrying four passengers. The Honda in this example gives 41 passenger-miles per gallon on a trip; the Chevy delivers 52 passenger-miles per gallon. The people travel more efficiently in the larger car because there are more of them—the load factor is higher.

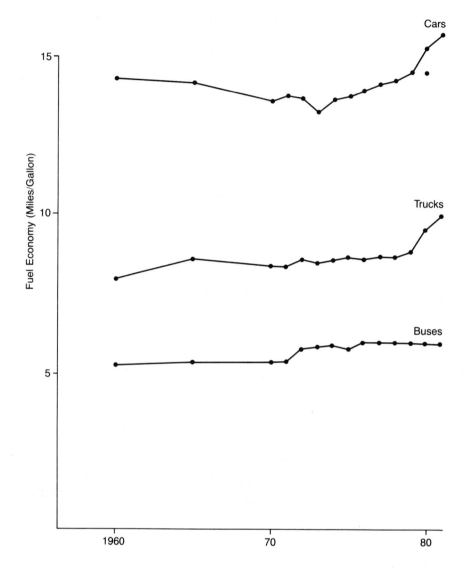

FIGURE 21.11
Fuel economy of cars, trucks, and buses in miles per gallon. (U.S. Department of Commerce)

Urban Mass Transit

The best way to transport people in urban areas is by mass transit with high load factor. Unfortunately, the load factors for urban mass transit are typically 20 to 25 percent. This happens because transit systems generally have to run during hours when there is little passenger demand. This is done to encourage usage, build ridership, and eventually raise the load factor.

Critics often argue that off-hour service should be eliminated and that buses should run only to transport people to and from work. They appear to feel that low load factors encountered in off-hours service is *prima facie* evidence that auto transport is more efficient. I argue with this view. If off-hour service were not available, some present riders would be forced to buy cars. There is a huge energy cost hidden in the auto itself (it costs about as much energy to build the car as

FIGURE 21.12
How the distribution of passenger-miles among various forms of transportation has changed, 1960–82. (U.S. Department of Commerce)

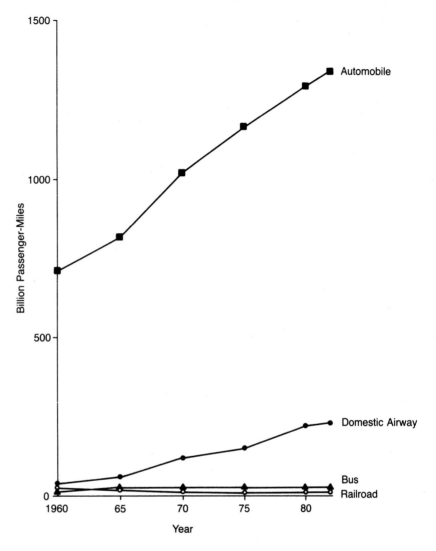

it will ultimately use in its lifetime). Hence, it is often energy-efficient to run off-hour mass transit despite the low load factor, even though it may not at first glance appear to be so. Also, critics assume that large buses will be used for service during off-hours, while a rational system apportions the size of the transportation to the expected clientele. Another alternative is call-up door-to-door delivery during off hours, but this has a high labor cost.

My argument is based to some extent on intangibles—I cannot quantify the number of people who would buy cars if service were curtailed. It is well known, however, that decreased mass transit services go hand-in-hand with increases in auto ownership. Other aspects that should be considered here include the necessity of increased street work as auto density increases; the increased pollution concomitant with running car engines cold; the crowding of cars in streets, causing increased pollution and frustration levels; and elimination of important transport services to the poor and the elderly, who often use mass transit in off-peak hours.

METHANOL AND ETHANOL

Replacement of all or part of the gasoline presently being used by methanol or ethanol is easily possible, as we discussed in Chapter 16. Mixtures of up to 15 percent ethanol (methanol) can be used as fuel in current engines with no modification (55). Methanol (ethanol) delivers, liter for liter, as much energy as gasoline (56). Both ethanol and methanol can be easily stored and conventionally transported.

Compared to gasoline, the 10 to 15 percent methanol (ethanol) mix results in improved economy and performance, lower exhaust temperature, and lower emissions (since the engine can be tuned much leaner, there can be a 50 percent reduction of CO and NO_x accompanied by about a 9 percent increase in mileage) (38,45). It is a better antiknock compound than lead. Brazil presently mixes 10 percent ethanol in its gasoline.

Cars can have their engines converted to pure alcohol use for about $100. If this is done, the converted car running on alcohol has about a twentieth the unburned fuel, a tenth the CO, and similar NO_x as compared with running on gasoline (57). Chrysler Brazil claims that their engines, run on alcohol with no pollution controls, could pass California pollution specifications (56). Emissions of a 1972 Gremlin burning gas and methanol and gas are compared in Table 21.8 (55). There is some evidence that methanol (but not ethanol) damages engine parts.

Ethanol can be made by fermentation of almost any crop containing sugar. Brazil is considering growing huge quantities of sugar cane and cassava (manioc) to supply ethanol. Assuming production of 15 tonnes of cassava per acre (which is realistic), a 30 km delivery radius could supply 2 Mtonne of feed per day to a plant that would then produce 15,000 bbl of ethanol per day (58). It is estimated that the Brazilian ethanol program could supply 250,000 to 1,000,000 jobs, mainly in agriculture (56).

The processes as used currently in the United States to produce ethanol (and methanol) have energy balances very close to (or less than) zero (59). In Brazil, the processes are energy-gainers (60). The experience of Brazil may lead to the adoption of Brazilian agriculture production methods in other Third World countries, to the betterment of everyone concerned.

TABLE 21.8
Pollutants, measured in grams per kilometer.

	Gasoline	Methanol Blend
CO	20.3	2.4
NO_x	2.0	0.22
Unburned hydrocarbons	1.38	0.20

SOURCE: T. B. Reed and R. M. Lerner. Copyright 1973 by the AAAS.

We may use load factors to find the average energy cost in the hypothetical American city, Midtown. Urban transit in Midtown is the local bus line, which we will suppose has a load factor of 25 percent, and the car, which we will suppose has a load factor of 30 percent. Buses account for only 10 percent of Midtown's in-city passenger miles; cars account for the remainder. We suppose that a fully loaded bus costs 0.25 kWh,/passenger-mile and a fully loaded car, about 0.70 kWh,/passenger-mile to run. The average energy cost of transport in Midtown for the bus is

$$(0.25 \text{ kWh}_t/\text{passenger-mi})/(0.25)$$
$$= 1.00 \text{ kWh}_t/\text{passenger-mi}.$$

and for the car is

$$(0.70 \text{ kWh}_t/\text{passenger-mi})/(0.30)$$
$$= 2.33 \text{ kWh}_t/\text{passenger-mi}.$$

Since buses account for 10 percent of the passenger miles in Midtown, the average energy cost for transport for all modes is

$$(0.1)(1 \text{ kWh}_t/\text{passenger-mi})$$
$$+ (0.9)(2.33 \text{ kWh}_t/\text{passenger-mi})$$
$$= 2.2 \text{ kWh}_t/\text{passenger-mi}.$$

If Midtown could persuade people to use buses, we could imagine a doubling of the load factor to 50 percent, in which case buses would account for 20 percent of the passenger miles. The bus transport average energy cost is only 0.5 kWh,/passenger-mile, and the average cost of transport for all modes is

$$(0.2)(0.5 \text{ kWh}_t/\text{passenger-mi})$$
$$+ (0.8)(2.33 \text{ kWh}_t/\text{passenger-mi})$$
$$= 2.0 \text{ kWh}_t/\text{passenger-mi}.$$

If Midtown could use buses at 100 percent load factor, buses would account for 40 percent of the passenger miles, and the average energy cost for Midtown would be reduced to 1.5 kWh,/passenger-mile.

Thus, the city would get a reduction of about a third in its energy use for transportation simply by *filling* its buses. If more buses are added for increased ridership, the savings are not so great, but any savings are welcome. The public trans-

portation must compete with the car's ability to take people where they want to go when they want to go. This makes it difficult to keep buses filled if they do not run often enough, and this cannot happen where the number of riders is too small to justify the number of buses. The major successful mass transit systems in North America run in the largest cities—New York, Toronto, Chicago, Washington, Atlanta, San Francisco, Mexico City.

Subways in New York have successfully carried hundreds of thousands of people per day for almost a century (although, it must be admitted, in conditions often approaching squalor during the last quarter-century). Despite the expense of subway systems (averaging $5.38 per vehicle revenue mile in 1984 [61]), the most successful modern American mass transit systems of the present are Washington's Metro and San Francisco's BART system. Both these systems are moving large numbers of people, causing economic growth near stations, and helping to relieve traffic congestion. In smaller cities such as Columbus, Ohio, or Eugene, Oregon, and in extended suburban areas, bus systems are more rational than subways. Bus systems in this sort of area have met with mixed success, and costs have grown to an average of $3.86 per revenue mile by 1984 (61). Suburban New Jersey has lost its bus service over the last quarter-century. Columbus' system is holding its own but is not a thriving concern. Eugene's system was a clear success, at least before the recession that hit Eugene (and all of Oregon) in the early 1980s.

RAISING FUEL ECONOMY

The American economy, as we indicated, depends very heavily on the automobile sector. A quarter of retail sales are automobile-related; over 10 percent of all personal consumption—and 8.5 percent of the entire gross national product—is connected to automotive expenditures (62). In 1984, the United States spent 60 billion dollars for imported oil, which produced just about the amount of gas used by cars (49). If we do not want to scrap the current system, the coming exhaustion of fossil fuels (Chapter 13) should impel us to increase fuel economy.

Americans are slowly switching to smaller cars, but the switch has profound implications for the American auto industry. Massive layoffs by the car makers in the 1974–1975 recession (in the aftershock of the first energy crisis) contributed to that recession's prolongation, as did the layoffs of the 1981–1983 recession.

As gasoline prices fell in the early 1980s, demand for bigger cars remained stronger than the auto makers anticipated. This situation led auto makers to attempt to water down Title V of the Energy Policy and Conservation Act of 1975, the Automotive Fuel Economy Program. The bill mandated that by 1985, the fleet average of the large auto makers be at least 27.5 miles per gallon (mpg). Of the big three American auto makers, only Chrysler met the standards by 1985. Ford had a fleet average of 25.9 mpg, and GM, 25.1 mpg. The credit provisions from exceeding the standard in previous years means that the penalty due would not have been as large as it could have been, but it still would have cost Ford and GM several hundred million dollars (63). The Reagan administration succeeded in its efforts to allow the required fleet average to be lowered to 26 miles per gallon.

Any possible dollar savings we make must be balanced against losses of time, comfort, and flexibility that would be encountered in the shift to a more energy-efficient mode. The old saying "time is money" has some truth to it.

Drag is a major source of energy loss, as seen above. The drag coefficient measures the ratio of drag force to area. For a flat plate, the drag coefficient is 1.17 (62). By the early 1980s, drag coefficients of cars ranged from 0.4 (for the VW Golf) to 0.55 (62,64); by 1985, U.S. models had drag coefficients between 0.3 and 0.4 (65). It is likely that a level of 0.25 will be achieved soon.

Turbocharging, in which the exhaust gases run a compressor to increase air pressure in the cylinder and allow more gasoline to be mixed in, increases efficiency by 10 to 15 percent (55).

Since kerosene contains more energy per volume than gasoline, use of diesels is good for fuel economy, although probably not so good for health because of increased emissions of NO_x and small particulate matter. Continuous torque transmissions can also contribute to energy conservation (65). A major help for the U.S. fleet fuel economy has been the shift back to manual transmissions from automatics during the 1980s (65).

The major role in increasing automobile efficiency is still the reduction of weight. A reduction of 1 percent in weight decreases lifetime energy consumption by 0.7 percent (62). This is an important connection, since passenger cars consume about ten times their weight in fuel over a lifetime (62). The reduction in weight affects the car's ride quality, but new electronically controlled air-suspension systems should help restore the smooth feeling to a car's ride (65).

All in all, available technology could raise automobile fuel economy for modified versions of cars such as the VW Golf to 70 to 80 miles per gallon (65). However, each mile per gallon improvement costs 2 to 4 billion dollars (if the retooling cost is attributed solely to this factor), so one must consider the tradeoff involved in achieving such fuel economy levels.

Economy vs. Safety

Fuel economy should not be pushed without regard to the other aspects affected by increased fuel economy. For example, for a car tuned for best fuel economy, the engine burns very hot so that NO_x emissions are a thousandfold larger than the best attainable levels, and hydrocarbon emissions are 19 percent higher than achievable (46). Similarly, for the minimum level of NO_x emission, the temperature of combustion is so low that fuel economy is 30 percent lower than the best value, and hydrocarbon emissions, 29 percent higher. Finally, to minimize hydrocarbon emission, the engine must run hot, so fuel economy is 10 percent lower than the best possible, and NO_x emissions 433 percent higher than the lowest levels (46). Clearly, a tradeoff must be

MASS TRANSIT OR HIGHWAY CONSTRUCTION? (70)

Mass transit facilities, of course, require huge capital expenditures. A study of the relative effects of spending federal money for road building, or for various other alternatives, showed that road building was just about the least effective way to spend money. In constructing the table below, the employment demands used were:

auto   4.48 man-years/passenger or ton mile
rail   5.58 man-years/passenger-mile
       2.00 man-years/ton mile
bus   3.70 man-years/passenger-mile
truck 2.79 man-years/ton mile

TABLE 21.9
Alternative results if 5 billion dollars were taken from highway construction and allocated to alternatives.

	Decrease in Energy Use (%)	Direct Increase in Employment (%)
Rail and mass transit construction	61.6	3.2
Water and waste treatment facilities construction	41.7	1.3
Educational facilities construction	37.1	4.7
National health insurance	64.0	65.2
Criminal justice and civilian safety	−3.4	53.6
Tax rebate	23.4	7.4

SOURCE: R. Bedzek et al. Copyright 1974 by the AAAS.

made in which all these costs are balanced, and an acceptable tuning point found.

Another consideration is safety. The alternatives to cars are much safer. Dying in an automobile is the most common accidental death: 20 to 25 percent of the people who die each year in the United States die because of car accidents. The next most common cause of accidental death, accidental falls, is responsible for only about 10 percent of all deaths, fire and explosion about 4 percent, and drowning about 3 percent. Railways and planes each claim a life per 100,000 people carried and cause much less than 1 percent of all deaths (66).

There is an additional safety factor to be considered when the weight of a vehicle is reduced. Bigger cars are safer. Reducing the car mass from 2000 kg to 1500 kg results in an increase of 4 to 6 percent in the chance of death or injury (46). In 1977, the car fleet on American highways had a 42 percent small car share. Had the fleet been all large cars, the 27,400 fatalities would have been reduced to 22,600; if it had been all small cars, fatalities would have risen to 34,000 (46).

There is a tradeoff at work here, too. Most car buyers ignore safety; as Lave (46) says:

> When given a choice, the vast majority of new car buyers choose an automobile with style, power, and comfort, and these preferences have been persistent in the United States and the rest of the world.

The most cost-effective method of preventing death and injury is an ordinary seat belt. Air bags cost three to four times as much per death or injury averted as installation of seat belts (46). Improvement of highway construction and maintenance practices is probably the next most cost-effective.

Economy and Pollution

It *is* possible to reduce emissions and clean the air as well. The conventional fuel economy of a car is measured according to the EPA by the emissions in grams per mile of carbon monoxide (CO), carbon dioxide (CO_2), and unburned hydrocarbons (hc) (5):

$$\text{fuel economy} = 2423 \text{ miles per gallon}/[0.866 \, (\text{hc}) + 0.429(\text{CO}) + 0.273(CO_2)].$$

Typical emissions and the effect of various proposed and adopted standards were shown previously, in Table 21.4 (67).

One sort of pollution not considered in the standards is the emission of particles of rubber, carbon black, and other fillers in tires. Some particles come from road wear also. An estimated 1.6 million tonnes of tire particles went into American air in 1968 (21). The rest came from burning old tires. Wilson (21) estimates the rate of these emissions at 0.43 tonnes of particles per million miles from road wear and 0.39 tonnes per million miles from burning of old tires.

OTHER MODES OF TRANSPORTATION

Long-distance Mass Transit

We have already mentioned mass transit as a good way to save energy and money. Long-distance buses cost about 6.6 cents per passenger-mile to run, while local buses require $1.69 per passenger-mile in 1980 (68). The fact that mass transit is so much more popular in Europe and Japan than it is in America may be due to the greater population density and the approximately two-fold greater cost of gasoline in other countries as compared to prices in the United States (69), as well as more attention paid to creature comfort in the States.

The discovery of the high-temperature superconductors (Chapter 18) may have an effect on mass transit even in the United States. Magnetically levitated trains could be run more efficiently in terms of energy use than current trains, especially since the roadbed will not need much maintenance (since the cars do not come into contact with the rails except in station areas).

Movement of Goods

Many alternatives exist for the transportation of goods to market, or for shipping of components to assembly factories, or to transport fuels. Energy

TABLE 21.10
Energy efficiencies of various freight modes (kJ/tonne-km).

	DOT[a]	ORNL[b]
Truck	10,216	1161
Class I freight railroad	1081	354
Water commerce		252
Air freight	26,008[c]	
Pipeline	790	206
Coal slurry pipeline		1990

[a]Department of Transportation, *National Transportation Statistics* (Washington, D.C.: GPO, 1987).

[b]M. C. Holcomb, S. D. Floyd, and S. L. Cagie. *Transportation Energy Data Book: Edition 9* (Oak Ridge, Tenn.: Oak Ridge National Laboratory, 1987).

[c]I have used Table 47 of *a* and Table 3.2 of *b* to construct this number.

cost (shown in Table 21.10) is one indication of the attractiveness of various commercial transportation modes, but not the only one. Time and convenience are often important considerations.

Trucks vs. Railroads The *circuitry,* or deviation from the shortest path, for rail transport is comparable to that for truck transport (1.24 and 1.21) (70). This cannot account for the switch from rail to truck transport seen in Figure 21.12 (67). Part of the reason may be that the railroad must pay its full share of facility construction and maintenance costs. The truck pays only 76 percent of its incurred costs (semis and full trailers pay only 56 percent of the actual cost for their use of interstate highways [67]). While railroads are more efficient than trucks, trucks carry a lot of traffic because they are more flexible than the railroads, in that they provide door-to-door service.

Air Traffic The air carriers constitute the fastest-growing segment in passenger-carrying in America (see Figure 21.13 [71]). The cost per person for air transport is very large, although the advent of jumbo jets has reduced this to almost that for cars (71). The lower crude oil prices of the mid-1980s did not help the airlines, which were reeling economically from the competition attendant on deregulation. The airlines know that the prices must eventually rise again. For this reason, any new development that increases fuel efficiency is of great interest. The latest development, which may increase fuel efficiency by 30 percent, is the prop-fan engine, which looks something like the old turboprops (72) but allows the plane to fly at jet speeds. The new engines should be available in 1995 and could make air travel comparable in energy use to private cars.

Transport of Fuel

Pipelines In the movement of freight, not much change has occurred since 1960 (Figure 21.13 [71]). The deregulation of freight hauling in the early 1980s caused effects that are still being sorted out, but railroads seem to be doing better relative to trucking.

Pipelines have carried more gas and oil than trains or other carriers for redistribution within a country, because they are extremely cheap in spite of a substantial capital, or initial, cost. Most natural gas burned in the U.S. Northeast and Midwest is piped from fields in Texas and Louisiana.

Pipelines also have been built between countries. Western Europe now gets a substantial fraction of its natural gas through a pipeline from the USSR. The Dutch sell natural gas to the West Germans. The Trans-Alaska Pipeline System has operated for over a decade with minor environmental impact, but a joint American–Canadian pipeline project lies in limbo because the parties cannot seem to reach agreement on the route.

A different sort of pipeline can be built to carry the coal. The coal to be transported is first pulverized and mixed with water to form a slurry. The first feasibility studies for coal slurry pipelines were done in the 1950s, and a long pipeline (over 160 kilometers) was operated successfully for seven years before unit-train coal transport was able to undercut the delivery costs (73). These pipelines proved to be extremely reliable. The Black Mesa slurry pipeline, which runs 473 kilometers (273 miles), carries 5 million tonnes of coal per year with 99 percent reliability (73). It does use a large amount of water (see Chapters 11 and 13), and many slurry pipelines carry coal from arid regions, which can cause problems. Other minerals, such as in iron and copper slurries, are currently being carried by pipelines (73).

Oil at Sea In order to run the world transportation system, large quantities of oil must be transported by pipeline, tanker truck, and oceanic tankers. The transport of petroleum fuels is extremely important in the context of the modern world. Its political importance is illustrated by the 1987 decision of the United States, Britain, and France to help patrol the sea lanes in the Persian Gulf (the source of about 90 percent of the world's oil exports) and protect tankers from the depredations of either side in the Iran–Iraq war. One could argue that the United States, which

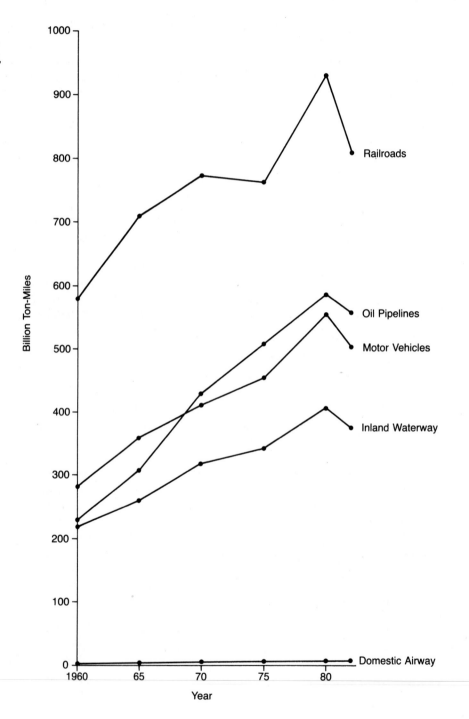

FIGURE 21.13
Transportation volume in billions of ton-miles, 1960–82. (U.S. Department of Commerce)

got involved first, was least affected by the loss of crude oil since Europe and Japan take the lion's share of Persian Gulf oil.

The "oil shocks" of 1973 and 1979 showed just how dependent on the uninterrupted flow of oil the industrialized world is. The U.S. government recognized that the interconnectedness of the world economy would cause the entire developed world to fall into another recession if substantial petroleum shortfalls were to occur.

Over 5,000 oil tankers now ply the sea lanes (74). The tanker fleet is a floating pipeline between the Persian Gulf area and Europe and Japan. Most tanker transport of petroleum occurs without incident, but accidents involving large supertankers make headlines and problems. The *Torrey Canyon* and the *Amoco Cadiz,* large supertankers lost along with the hundreds of thousands of barrels of crude oil they carried, spectacularly fouled coastlines and killed wildlife.

In addition to tanker spills, well blowouts such as in the Santa Barbara channel or the Gulf of Mexico can also foul marine environments. Sailors have spied large numbers of oil globs in areas such as the Sargasso Sea. A recent report by the National Research Council of the National Academy of Sciences (75) addresses concerns of oil in the oceans, and found, as had a previous report, that "chronic dribbling of petroleum from sloppy use by modern society is responsible for a large fraction of the input to the world's oceans."

For at least the last 100,000 years (and probably a hundred times longer), oil has been seeping into the oceans. The total from such natural sources is about 0.23 Mtonne per year. Tanker accidents, on average, account for 0.35 Mtonne per year release. Offshore production releases about 0.05 Mtonne per year. These releases are small compared to those from normal tanker operations alone, which account for 0.6 Mtonne per year, or bilge oils release of 0.3 Mtonne per year. The total release from marine transportation totals 1.3 Mtonne per year. Surprisingly, municipal and industrial wastes and urban runoff release over 1 Mtonne per year into the oceans (75).

Tanker accidents produce only slightly more pollution than natural sources, but can be studied in some detail to learn what happens to oil in the oceans. The *Amoco Cadiz* spilled about 200,000 tonnes of oil into the ocean in 1978; of that amount, 30 percent evaporated, 28 percent fouled beaches, fully 20 percent was unaccounted for, 13.5 percent remained in or on the water, 8 percent was absorbed by subtidal sediments, and 4.5 percent was biodegraded by bacteria (75).

The research now indicates that the effects of oil spills often last as long as a decade. The world's oceans seem in no danger of dying of oil pollution. However, it is still necessary to reduce total emissions from a value of over ten times the natural emission rate.

SUMMARY

Car mileage depends most strongly on car weight, but also on the profile it presents to the air (drag). In addition to emitting pollutants as they drive, cars also emit pollutants as they are refueled. The worst pollutants are carbon monoxide, which replaces oxygen in the blood, and lead, which can cause mental disorders. Nitrogen oxides and ozone can cause smog, which exacerbates lung ailments.

Load factor is important to use in assessing the efficiency of various transportation modes. A full station wagon may transport seven people for a lower energy cost per person than a subcompact carrying one person. Mass transit holds forth the promise of efficient transportation of large numbers of people.

Most forms of transport of goods and people are more efficient than cars and trucks. However, they are not as flexible or as convenient.

REFERENCES

1. *The World Almanac,* 1975 ed. (Newspaper Enterprises Assn., Inc., 1975).
2. *The New York Times Encyclopedic Almanac,* 1971 ed. (New York: *New York Times,* 1971), 191. Most moves are short distances. The 18 percent figure eliminates very short moves (e.g., within the same block).
3. Department of Commerce, *Statistical Abstract of the United States, 1984* (Washington, D.C.: GPO, 1984).
4. Table 1073, Reference 3.
5. L. Hodges, *Am. J. Phys.* 42 (1974): 456. The exact figures are for 1974 models, but the moral of the story is correct, whatever the year.
6. Table 1074, Reference 3.
7. Department of Energy, *Monthly Energy Review* (Washington, D.C.: GPO, April 1986); Department of Energy, *Annual Energy Review 1986* (Washington, D.C.: GPO, 1987).
8. Department of Commerce, *Statistical Abstract of the United States, 1972* (Washington, D.C.: GPO, 1972), Table 899.
9. G. Szegö, *The U.S. Energy Problem,* RANN report, Table N-2.
10. Table 1049, Reference 3.
11. South Coast Air Quality Management District, *The Path to Clean Air: Attainment Strategies* (El Monte, Calif., 1987); and *Air Quality Management Plan (1988 Revision),* App. III-A (El Monte, Calif., 1988).
12. Department of Commerce, *Statistical Abstract of the United States, 1972* (Washington, D.C.: GPO, 1972), Table 288.
13. A. J. Haagen-Schmidt, in *Energy Needs and the Environment,* ed. R. L. Seale and R. A. Sierka (Tucson: Univ. of Arizona Press, 1973), Figure 3.13.
14. G. J. McRae, data published in *Los Angeles Times,* 22 July 1979.
15. P. Shabecoff, *New York Times,* 17 Dec. 1985.
16. Environmental Protection Agency, *Environmental Quality 1975* (Washington, D.C.: GPO, 1975).
17. A. L. Plumley, in *Energy Needs and the Environment,* ed. R. L. Seale and R. A. Sierka (Tucson: Univ. of Arizona Press, 1973), 165ff.
18. G. B. Morgan, G. Ozolins, and E. C. Tabor, *Science* 170 (1970): 289.
19. R. Haitch, *New York Times,* 2 June 1985.
20. R. Severo, *New York Times,* 20 April 1975.
21. W. M. Thring, *Intern. J. Environmental Studies* 5 (1974): 251; and D. G. Wilson, *Intern. J. Environmental Studies* 6 (1974): 35.
22. L. S. Jaffe, in *Global Effects of Environmental Pollution,* ed. F. Singer (Dodrecht, Neth.: D. Reidel Publ. Co., 1970).
23. D. Lawson, California Air Quality Board, private communication, June 1988; M. Hoggan, A. Davidson, and M. Hsu, *1985 Summary of Air Quality in California's South Coast Air Basin* (El Monte, Calif.: South Coast Air Quality Management District, 1986); Air Quality Data, 1982, 1986 (South Coast Air Quality Management District); *Air Quality Management Plan, 1988 Revision,* App. III-A (South Coast Air Quality Management District, 1988); for earlier data, see R. D. Cadle and E. R. Allen, *Science* 167 (1970): 243.
24. Anonymous, *New York Times,* 1 Sept. 1974.
25. D. Dickson, *Science* 228 (1985): 37.
26. I. Meyer-List, *Scala,* August 1984.
27. E. Caplun, D. Petit, and E. Picciotto, *Endeavour* 8 (1984): 135.
28. P. Christ and R. Gaul, *die Zeit,* 27 April 1984.
29. D. Dickson, *Science* 228 (1985): 159.
30. P. V. Hobbs, H. Harrison, E. Robinson, *Science* 183 (1974): 909.

31. R. A. Bryson, *Science* 184 (1974): 753.

32. B. Commoner, *J. Am. Inst. Planners* 39 (1973): 147.

33. K. B. Noble, *New York Times,* 5 August 1984.

34. J. H. Trefry et al., *Science* 230 (1985): 439.

35. G. Hill, *New York Times,* 13 Feb. 1979.

36. G. Haaf et al., *die Zeit,* 26 Oct. 1984; and R. Klingholz, *die Zeit,* 8 Nov. 1985.

37. H. Hatzfeldt, *die Zeit,* 9 March 1984.

38. F. Haemmerli, *die Zeit,* August 1984.

39. S. B. McLaughlin, *J.A.P.C.A.* 35 (1985): 512, 923.

40. B. Prinz, *J.A.P.C.A.* 35 (1985): 913.

41. R. Klingholz, *die Zeit,* 21 June 1985; and Anonymous, *New York Times,* 9 Dec. 1985.

42. H. W. Vogelman, *Nat. Hist.* 91, no. 11 (1982): 8.

43. R. W. Shaw, *Sci. Am.* 257, no. 2 (1987): 96.

44. P. Shabecoff, *New York Times,* 26 Feb. 1984.

45. R. W. Leonhardt, *die Zeit,* 9 March 1984.

46. L. B. Lave, *Science* 212 (1981): 893.

47. H. Bluthmann, *die Zeit,* 7 Dec. 1984.

48. R. Gaul, *die Zeit,* 14 Sept. 1984.

49. H. Bluthmann, *die Zeit,* 1 Feb. 1985.

50. F. von Hippel, *Science* 228 (1985): 263.

51. D. M. Settle and C. C. Patterson, *Science* 207 (1980): 1167.

52. E. Hirst, as quoted in A. L. Hammond, *Science* 178 (1972): 1079.

53. E. Hirst and J. C. Moyers, *Science* 179 (1972): 1299.

54. Office of Emergency Preparedness, *The Potential for Energy Conservation* (Washington, D.C.: GPO, 1972), Appendix C.

55. T. B. Reed and R. M. Lerner, *Science* 182 (1973): 1299; E. E. Wigg, *Science* 186 (1974): 785; E. R. Holles, *New York Times,* 1 Dec. 1974; and R. K. Mullen, *Science* 188 (1975): 209.

56. A. L. Hammond, *Science* 195 (1977): 564.

57. R. K. Pelley et al., *Proc. 6th Int. Conf. on Energy Conversion Engineering,* paper 719008, AEC report TID-26136.

58. W. G. Pollard, *Am. Sci.* 64 (1976): 509.

59. R. S. Chambers et al., *Science* 206 (1979): 789.

60. J. Goldemberg, *Science* 223 (1984): 1357.

61. Department of Transportation, *Compendium of National Urban Mass Transportation Statistics for the 1984 Report Year* (Washington, D.C.: GPO, 1987).

62. C. L. Gray, Jr. and F. von Hippel, *Sci. Am.* 224, no. 5 (1981): 48.

63. M. Crawford, *Science* 228 (1985): 307.

64. J. Grey, G. W. Sutton, and M. Zlotnick, *Science* 200 (1978): 135.

65. R. K. Whitfor, *Ann. Rev. Energy* 9 (1984): 375.

66. Department of Commerce, *Statistical Abstract of the United States, 1972* (Washington, D.C.: GPO, 1972), Table 83.

67. R. W. Irwin, *New York Times,* 13 Oct. 1974; and C. Holden, *Science* 187 (1975): 818.

68. Tables 1086 and 1087, Reference 3.

69. Table 1076, Reference 3.

70. R. Bezdek and B. Hannon, *Science* 185 (1974): 669.

71. Table 1048, Reference 3.

72. R. Witkin, *New York Times,* 20 July 1986.

73. E. J. Wasp, *Sci. Am.* 249, no. 5 (1983): 48.

74. Table 1133, Reference 3.
75. J. W. Farrington, *Oceans,* Sept. 1985, summarizes the National Research Council report *Oil in the Sea: Inputs, Fates, and Effects.*
76. D. Isaacs et al., *Nature* 253 (1975): 255.

PROBLEMS AND QUESTIONS

True or False

1. The automobile contributes to pollution only through its exhaust emissions.
2. About half of all commuting cars carry more than one passenger.
3. Buses are more efficient than cars under any circumstances.
4. Railroads are less expensive for freight than ships along routes of comparable length.
5. It is technologically possible at present to utilize alcohol as a replacement for gasoline.
6. The amount of gasoline burned resisting air friction (drag) increases as speed increases.
7. Carbon monoxide is a harmful pollutant because it can cause temperatures to rise in urban areas.
8. Oil from ships' bilges is the greatest source of pollutants in the world's oceans.
9. Mass transit runs best when load factors are low.
10. Carbon dioxide is one of the least dangerous automotive pollutants for human health.

Multiple Choice

11. Figure 21.3 shows that drag force increases faster than speed; in fact, it increases as the square of speed. Given that fact, by what factor must the drag force at 70 mph be greater than that at 50 mph?
    a. 25/49
    b. 5/7
    c. 7/5
    d. 49/25
    e. 350

12. Which of the following, according to the text, is the least energy-expensive method of moving people?
    a. Seven people in a VW micro-bus
    b. A Boeing 747, half loaded
    c. An intercity bus, full
    d. A passenger train, full
    e. A Porsche racing car

13. Suppose the load factor (proportion of total number of passengers carried to those who *could* be carried) doubles. The per capita energy cost of transportation by this method as a consequence
    a. is doubled.
    b. is halved.
    c. remains the same.
    d. None of the above

14. About how much per mile does it cost to run a standard-sized car?
    a. 3 cents/mile
    b. 12 cents/mile
    c. 18 cents/mile
    d. 27 cents/mile
    e. None of the above

15. Automobiles use about what proportion of all the energy produced in the United States?
    a. 1/100
    b. 1/20
    c. 1/10
    d. 1/5
    e. 1/3

16. Urban mass transit operates with a 20 percent load factor (i.e., four-fifths empty) and represents about 3 percent of in-city passenger-miles. Urban cars operate with a load factor of 28 percent and represent 97 percent of the passenger-miles. If the load factors were 100 percent, the auto would cost 2300 Btu/passenger-mile and the mass transit, 760 Btu/passenger-mile. What is the averaged energy cost of urban transit?
    a. 630.4 Btu/passenger-mile
    b. 1530.0 Btu/passenger-mile
    c. 2253.8 Btu/passenger-mile
    d. 8081.9 Btu/passenger-mile
    e. None of the above

17. Which species of trees seems to be most susceptible to effects of air pollution?
    a. Spruce
    b. Fir
    c. Pine
    d. Beech
    e. Oak

18. By how great a factor does lead intake of present-day people differ from prehistoric times?
    a. Two orders of magnitude lower
    b. One order of magnitude lower
    c. They are of the same order of magnitude.
    d. One order of magnitude greater
    e. Two orders of magnitude greater

19. What strategy would be most effective in decreasing lead levels in the environment?
    a. Installing turbochargers on all cars
    b. Converting all cars to manual transmissions
    c. Decreasing speed limits
    d. Installing lake pipes
    e. Installing catalytic converters

20. Which of the following would help increase mileage in automobiles?
    a. Replacing manual transmissions by automatic transmissions
    b. Increasing the drag coefficient
    c. Installing catalytic converters
    d. Replacing steel by epoxy and aluminum
    e. All of the above would help increase mileage.

Discussion Questions

21. Tornados seem more prevalent recently in America (76) because American drivers drive on the right—two passing cars can set up a vortex that would grow naturally in the Northern Hemisphere. How might you lessen this unexpected consequence of driving?

22. Identify issues in the arguments between supporters and opponents of catalytic converters.

23. In a stratified charge engine, a small amount of air richly mixed with fuel is fed to an area near the spark plug. This is easily ignited by the spark. The rich mixture ignites a leaner mixture. Since the temperature of combustion is lower, NO_x emissions are low. Also, virtually all hydrocarbons burn fairly thoroughly. A Honda stratified charge engine exceeded the most stringent pollutant standards. Identify the reasons that American manufacturers have not adopted the engine.

24. What elements could contribute to the eventual growth of American mass transit systems? You might want to consider cultural change, the eventual loss of fossil liquid fuels with consequent price rises, changes in settlement patterns, etc.

22

POLLUTION FROM FOSSIL FUELS

In earlier chapters of this book, we have discussed how energy gets generated for electricity and other purposes. In Chapter 21, we discussed briefly some of the effects that automobiles might have on forests in the industrialized countries.

In this chapter, we attempt to extend the discussion of the consequences of burning fossil fuels. We consider how pollution is generated, what the health consequences can be, how it might be possible to ameliorate the effects of pollutants in the fuel, how acid rain is generated, and what its consequences can be.

KEY TERMS *thermal inversion*
adiabatic lapse rate
electrostatic precipitation
wet scrubbers
dry scrubbers
PAN smog
heavy metal tracers
acid rain
pH
wet deposition
dry deposition
acid shock

THE ATMOSPHERE

There is an ocean surrounding us—an ocean of air. The volume of the atmosphere is in fact over twice that of the world's water oceans (see box). It is useful to think of the air around us as an ocean because it emphasizes the interdependencies of the parts of the system. Experience with Lake Erie in the early 1970s, when it was written off as dead, shows that any *enclosed* body that is overused may suffer a loss of utility for all. Lake Erie has recovered somewhat, but only through great effort on the parts of the bordering states and Ontario.

People have been polluting the atmosphere for quite some time. Such pollutants come in two classes (1):

> those which take part in natural cycles and can be assimilated if their levels are trivial compared with the background levels of the natural cycles, and
> those which do not take part in natural cycles and so accumulate and cause injury.

Pollution is an unavoidable consequence of energy and materials production. Our response to pollution must take this fact into account. The prudent response is to determine the level of a given pollutant that will not accumulate injuriously and aim to produce no more than this amount. Care must be taken to recognize that materials which are not injurious pollutants in a *global* sense may still be injurious pollutants in a *local* sense.

In this chapter we shall consider in greater detail the problems of urban environments, air pollution, and acid rain (see Figure 22.1). In the following chapter we consider effects on gross properties of the entire atmosphere—the climate—of the uses to which we are putting our atmospheric ocean through the burning of fossil fuels and wood.

Air in Urban Environments

A smaller body of water is affected more severely than a large body of water when pollution is introduced into the water system. It is easy to see how small bodies of water can exist in isolation from other bodies of water. How, though, can we be allowed to make an analogy between small bodies of water and local environments of air?

The most important reason for the analogy is the knowledge that weather may be local. Los Angeles' smog does not affect San Francisco— San Francisco has its own. Each city has its own conditions, but we can generalize somewhat and note that cities are typically 2 or 3°C higher in temperature than rural areas, especially in winter, because the stone, brick, and asphalt absorb heat better than Earth, and because there is no transpiration (see Table 22.1 [2])(2–5). A study of the

FIGURE 22.1
"We don't have to worry about air pollution as long as the wind's in the right direction.," (*Die Zeit,* 26 October 1984. Used by permission, Zeitverlag Gerd Bucerius KG.)

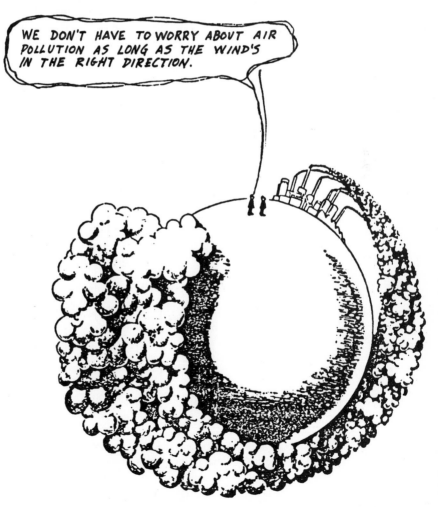

VOLUME OF THE ATMOSPHERE

The density (amount of mass in a unit volume) of air is 1.29 kg/m₃. We may estimate the mass of the atmosphere atop us in the following way. Consider an imaginary cube of air in still atmosphere. A cube of side l has area l^2 and volume l^3. The cube does not move since the air is still, despite the fact that the cube of air has weight. There must be an upward force on the bottom face of the cube by an amount just equal to the weight of the cube plus the net downward force on the top face of the cube (see Figure 22.2). The weight of the cube is just (density) $(l^3)(g)$, since the factor (density) (l^3) is just the cube mass. The force on any cube face is the product (pressure)(area), so,

$$P_{top} \times (l^2) + (density) \times (l^3) \times (g) = P_{bottom} \times (l^2)$$

or

$$P_{top} + (density) \times (l) \times (g) = P_{bottom}.$$

The pressure at the bottom of the atmosphere, then, is due to the weight of all the atmosphere pressing down from on top. Hence, imagining the atmosphere to be uniform in density, we get (for $P_{top} = 0$) an estimate of the thickness of a uniform atmosphere

$$thickness = P_{bottom}/(air\ density)\ (g)$$
$$= (1.013 \times 10^5\ N/m^2)/(1.29\ kg/m^3)(9.8m/s^2) = 8\ km.$$

Hence the volume of such an atmosphere is (surface area of Earth $= 4\pi R_E$) × (thickness), so our atmosphere has volume $6.44 \times 10^{11}\ m^3$. The total atmospheric mass is (air density) × (volume):

$$(air\ density) \times (4\pi R_E) \times (P_{bottom})/(air\ density) \times (g) = (4\pi R_E) \times (P_{bottom})$$
$$= 8.3 \times 10^{11}\ kg.$$

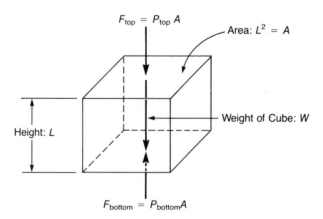

FIGURE 22.2

The vertical forces on a cube of air are due to the air pressure at the top and bottom of the cube and the cube's weight. The net upward force, $F_{bottom} - F_{top}$, is numerically equal to the weight of the cube. It is called the buoyant force on the cube.

TABLE 22.1
Comparison of urban to rural areas.

Temperature	Annual mean 0.5–1°C higher
	Winter minimum 1–2°C higher
Relative humidity	Annual mean 6% lower
	Winter 2% lower
	Summer 8% lower
Dust particles	Ten times more
Clouds	5 to 10% more
Winter fog	100% more
Summer fog	30% more
Total horizontal surface	15–20% less
Ultraviolet radiation	5–30% less
Wind speed	Annual mean 20–30% lower
	Extreme gusts 10–20% lower
	Calms 5–20% more
Rainfall	5–10% more

SOURCE: J. M. Mitchell, Reference 2.

St. Louis area's precipitation patterns showed that, while nearby farms lost income because of air pollution, they gained back more income than they had lost from the increased rainfall (5). Such increased precipitation has also been found at sites as varied as Long Island, Australia, and Washington, D.C. (6).

Local pollution comes from cars, trucks, and other moving sources; from evaporation of paint and solvents; from industrial processing; and from power plants burning coal, oil, or natural gas. Both oil and coal have some sulfur. Eastern coal contains large amounts of sulfur; this sulfur goes up the stack unless devices known as scrubbers are used; even then sulfur emissions can be substantial. Some mideastern oils are quite high in sulfur as well. Libya, for example, has been able to do very well selling crude oil in world markets because of its very low sulfur content, even though the Libyan leader Ghaddafi has made many enemies among the customer countries by his support of international terrorism.

Local pollution episodes can be bad indeed; it is more dangerous to health to live in cities than in rural areas even at "normal" times. These local differences have been apparent for some time in the case of cities surrounded by heights. Pittsburgh, Pennsylvania, for example, was notorious for its pollution as early as the turn of the century; it is in a valley surrounded by the Alleghenies. Los Angeles is in a bowl-shaped plain surrounded to landward by mountains. What has occurred more recently is noticeable local pollution even in areas not hemmed in by such topography. This results from thermally induced density differences in the air.

Thermal Inversion

In a valley, the air may become stagnant as the wind blows across the top of the hills or mountains surrounding the valley. For areas that are relatively flat, this cannot be the mechanism of stagnation. In the case of flat areas, it has to do rather with the dynamics of the atmosphere itself.

If the air is in an unstable condition, air from near the surface will rise and be replaced by cooler air sinking down from above. This is what happens in home heating; you will see that all cold air returns are at floor level. Unstable air is good for an area in that the local pollution gets carried away and diluted by the rising air. Stable air conditions in flat areas are responsible for the buildup of pollutants in the air. Pollution alerts are called when these conditions prevail.

A related condition is the temperature inversion, in which the air temperature increases with height. In this case, the air is of course stable,

HOW AIR RISES AND SINKS

In Chapter 10, we referred to the *adiabatic lapse rate*: the rate at which temperature decreases as altitude increases (for dry air)–about 10°C/km or 0.01°C/m. In normal conditions, the temperature decreases with height for a few kilometers.

Suppose that the actual lapse rate is greater than the adiabatic lapse rate for an imaginary cube of dry air. If we move the cube upwards, it would cool at the adiabatic lapse rate and become warmer than the surrounding air. Air cools as it rises because the atmospheric pressure decreases with height. As air rises, it expands; the work for the expansion comes from the air's thermal energy, causing its temperature to decrease (recall from Chapter 7 that temperature is the measure of thermal energy). Since the imaginary cube of air is warmer than the surrounding air, it is less dense and would keep on rising. This means that air under such conditions is not stable. Conversely, if the actual lapse rate is smaller than the adiabatic lapse rate, upward movement of the imaginary cube will result in the cube becoming cooler, and thus more dense, than the surrounding air. Dense air will tend to sink; it will return to its original position if it is moved from it. Air is therefore stable when the actual lapse rate is less than the adiabatic lapse rate.

The lapse rate changes if the air is wet. Cold air holds less water than warm air (which is why we use humidifiers on forced-air furnaces, as discussed in Chapter 19), so cooling air raises the relative humidity of the air. After the air has cooled (at the adiabatic lapse rate) to the point at which the air holds as much water as it possibly can at that temperature, so that the relative humidity is 100 percent, any additional cooling will cause the water to condense as drops (as we can observe from the condensation of drops on cold glasses in the summertime). As the drops form, thermal energy is released, so that the wet air will not cool as fast as dry air would.

and this stability is so strong that very high pollution levels are likely. A pollution episode, episode 104 (extreme cases are given numbers), which affected much of the eastern part of the United States for an extended period in 1969, was of this sort (4). This episode occurred in late summer, a time during which inversions are common. In hot weather, the ground (and buildings) heat more quickly than the air during the day, and cool more quickly than the air at night. Thus, at night the cool ground cools the lower layers of air, setting up a condition in which the higher the air is, the warmer it is—a temperature inversion. During the day, the hotter ground warms the air near it and causes the inversion to break up.

Episode 104 was so severe because a large high-pressure system existed over the entire eastern half of the country. Air in high-pressure regions is forced downward, thus becoming warmer, and is denser as the air molecules are forced together. This sinking takes place on top of the usual air layers, so that temperature decreases as the air rises from the ground and meets the warmer air from the high-pressure region at an altitude of a few hundred meters. Since the temperature is not decreasing with altitude, local air is trapped underneath the layer of warm air.

This isolates the air in the locality. If there are sources of pollution there, the air becomes more and more polluted the longer the high-pressure system prevents mixing with clean air.

HEALTH EFFECTS OF POLLUTION

Air Pollution Catastrophes

Conditions such as those described led to the severe pollution events of recent times. These events have served to shatter complacency, to awaken the public to the dangers of air pollution. We recognize the health problems in a severe episode, and realize that there may be a health problem even when the concentration of pollutants is not so great as in a severe episode.

The first of these extreme episodes to be documented occurred in the Meuse Valley of Belgium in 1930, and caused about sixty-three deaths; another took place in Donora, Pennsylvania (a Pittsburgh suburb) in 1948 and caused about twenty deaths. An infamous Thanksgiving weekend in 1966 in New York was responsible for an estimated 270 extra deaths (Episode 273) (4). In the pollution incident in Clairton, Pennsylvania (near Pittsburgh) in November 1975, fourteen people may have died as a result of high particulate levels (6): 929 $\mu g/m^3$, compared to national standards of 75 $\mu g/m^3$ average, and the maximum

allowed concentration of 260 $\mu g/m^3$ in one twenty-four-hour period per year (see Table 22.2). The level of pollutants in inversions can get even higher. In January 1985, the Ruhr region in West Germany suffered a five day pollution episode. In Dortmund, measured sulfur dioxide levels were reported as 1118 $\mu g/m^3$ for a long period (7). In the Western Ruhr, all traffic was banned and all factories closed. In the rest of the Ruhr, traffic was banned until 10:30 in the morning. (As the sun heats the ground during the day, the inversion disappears, only to be reestablished in the evening.)

Undoubtedly the most acute air pollution disaster (and the outstanding example of the power of such incidents to stimulate governmental action) was the December 1952 London smog that resulted in an estimated extra 4000 deaths (8,9). In this case, mortality was concentrated among the very young and the very old, the two most susceptible groups. The large number of deaths caused appointment of a Royal Commission and led to the close monitoring of mortality and air pollution data (see Figure 22.3). The less severe episodes of 1956 (about 1000 excess deaths), 1957 (about 900), and 1962 (about 700), as well as the close monitoring of death rates, made a compelling case that sudden increases in the death rate followed each abrupt change in pollution levels. The subsequent studies of both mortality

TABLE 22.2
Federal air quality standards.

	Primary	Secondary
Sulfur dioxide		
Annual	80 $\mu g/m^3$ (0.03 ppm)	
24-hour period*	365 $\mu g/m^3$ (0.14 ppm)	
3-hour period		1300 $\mu g/m^3$ (0.5 ppm)
Particulates		
Annual	75 $\mu g/m^3$	60 $\mu g/m^3$
24-hour period	260 $\mu g/m^3$	150 $\mu g/m^3$
Nitrogen dioxide		
Annual	100 $\mu g/m^3$ (0.05 ppm)	100 $\mu g/m^3$ (0.05 ppm)
Oxidants		
1-hour period	160 $\mu g/m^3$ (0.08 ppm)	160 $\mu g/m^3$ (0.08 ppm)

Period means that the air quality to be acceptable under the standards must not exceed the value given during the length of the period more than once a year.

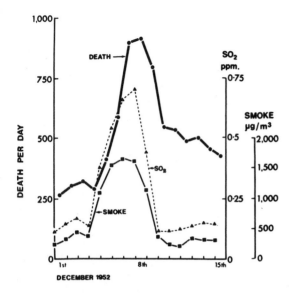

FIGURE 22.3
Air pollutant levels and excess mortality in London, December, 1952 (D. Bates, Ref. 9, by permission McGills-Queen's University Press). Over 4000 excess deaths occurred.

(death) and morbidity (sickness) seem to have been able, to some extent, to quantify the deleterious effects of urban (polluted) life on health. The 1952 disaster prompted changes in British laws and practice that have been effective in preventing a recurrence.

Quantifying Health Effects

It is known that bronchitis incidence is much higher for people living in regions of high particulate and sulfur dioxide levels (9,10). Lung cancer death rates drop (when smokers are compared to smokers or nonsmokers to nonsmokers) from 500 per 100,000 people in urban areas to 39 per 100,000 people in rural areas (10); stomach cancer death rates in New York for high-pollution regions are double those for low-pollution regions (10); and nitric acid in polluted areas has been implicated in increasing incidence of nephritis (kidney disease) as well as increasing

the level of fats in the blood (11). It has been suggested that if pollution were to decrease to minimal levels, the death rate in the United States would be reduced by 38 percent (12,13), and medical expenditures for chronic illness would be reduced as well.

Health damage is not restricted to people. Plants are affected by very low pollutant concentrations (especially for ozone). About 60 percent of the plants on land areas near freeways in San Bernardino have been severely or moderately damaged (17), and damage has been reported as far east as 120 km from the Los Angeles urban area (18). Ozone concentrations of 60 parts per billion (ppb) damage conifers. In the Black Forest in West Germany, ozone concentrations have averaged 90 ppb, with some readings as high as 270 ppb (18,19). The sick condition of German and American forests was discussed in the last chapter. As indicated in Chapter 1, recent studies predict that ozone levels near the Earth's surface will increase substantially. This comes about from a tradeoff between the decrease in ozone (especially in the stratosphere) caused by chlorofluorocarbons, methyl chloride, and carbon tetrachloride, and the increase in ozone concentrations due to nitrogen oxides, carbon dioxide, and methane (20). It appears that we will lose ozone where we need it and gain it where we do not wish it to be!

PAN smog (see box discussion) also causes damage to romaine lettuce, alfalfa, and spinach, among farm crops, as well as damage to flowers (21). Such local pollution sources as a smelter may destroy plant life in the vicinity (22).

Human and plant health damage cost money in medical bills and crop losses. Atmospheric particulates encourage corrosion, soiling of textiles, and weakening of paint and other protective coverings. By as early as 1957, Los Angeles growers' spinach crop losses totaled half a million (1957) dollars; New Jersey pollution damage in 1966 was estimated at $5 billion. A 1913 Pittsburgh study indicated a pollution cost of $20 per person per year (excluding health costs); the 1954 Beaver report estimated a cost of $35 per person

POLLUTION CONTROL DEVICES

While utilities cause a great deal of pollution, they would cause much more if it were not for application of pollution control technology. The simplest device is an electrostatic precipitator. In the electrostatic precipitator, electrodes or thin wires are held at high potential. Near the wire's surface, gas molecules can be ionized, and electrons hit other gas molecules after being accelerated in the potential, ionizing them. This releases further electrons, ionizing further gas molecules, and so on. The electrons move toward the positive wire (or electrode); the positive ions move to the negative wire (or electrode). By this process, particles in the smoke precipitate onto the negative electrode with a high efficiency as smoke rises through a chimney. However, this technique does not help control gases, which later cause acids to form in the atmosphere.

The most common large-scale technology to deal with stack effluent involves wet scrubbers. The smoke is bubbled through a limestone slurry, and a large proportion (≈ 90 percent) of the sulfur oxide is trapped as calcium sulfate. The resulting sludge, which has a volume greater than the coal ash, is then disposed of in huge holding ponds. Scrubbers are expensive to operate (14), and sludge disposal is a problem, too. Some wet scrubbers produce a corrosive liquid (14).

Of potential importance is dry scrubbing. In this technique, chemicals are added to the flue gases in solid or liquid (atomized) form to make dry, stable salts. Some settle out, bringing heavy metals with them as well, and the rest may be caught in baghouses (14), huge cloths through which effluent must pass before it can escape up the stack.

Other ways to reduce emissions include fluidized-bed combustion (discussed in Chapter 13), switching to fuels with a lower sulfur content, or cleaning coal before use by crushing and centrifuging the coal (15,16). In fluidized-bed integrated coal combustion systems, the gas from the coal is cleaned before further combustion. If one switches to coal or oil with a smaller amount of sulfur, the amount of sulfur escaping the flue is reduced. The physical cleaning of coal cannot yet provide sulfur reductions of the size possible through desulfurization of flue gases; in combination with effluent control, however, it shows promise.

The situation in regard to emission of nitrogen oxides is fairly well understood. The amounts of NO_x increase as the combustion temperature increases because formation of these pollutants is an endothermic process. It is possible to decrease NO_x emissions through several methods (14): *Staged combustion control* adjusts the amount of oxygen available at places within the combustion zone; this is standard practice for coal-fired systems, and reduces NO_x by 25 to 50 percent. *Flue gas recirculation* involves mixing of flue gas into the air supply to the boiler; this causes lowered combustion temperatures, and reduces NO_x formation by 12 to 25 percent. *Thermal denoxification* is still in the development stages. Chemicals such as ammonia or urea are put into the combustion zone or downstream effluent, causing chemical reactions in which the NO_x becomes harmless N_2.

TABLE 22.3

Comparison of per capita pollution costs (excluding health) in various studies given in constant (1984) dollars.

Study	Per Capita Yearly Cost (December 1984 dollars)
Pittsburgh, 1913	235
Beaver Report, U.K., 1954	148
Steubenville/Uniontown, 1966	279
Canada, 1965	
Average	176
Ontario	243
Toronto	319
New Jersey, 1966	240

SOURCE: W. Bach, Reference 25.

TABLE 22.4

National ambient air quality standards (NAAQS), tolerance levels, and relative toxicity.

Pollutant	Tolerance Level ($\mu g/m^3$)	Relative Toxicity	NAAQS ($\mu g/m^3$) I	II
Carbon monoxide	40,000	1	160	
Hydrocarbons	19,300	2.1	160	
Sulfur oxides	1430	28	2	15
24 hour max.			5	100
3 hour max.			25	700
Nitrogen oxides	514	78	100	
Particulates	375	107	5	10
24 hour max.			10	30

SOURCE: Selected from References 21, 22, and 26.

per year in urban areas; and studies in Steubenville, Ohio, and Uniontown, Pennsylvania, indicated an increased average cost of $84 per family in increased house maintenance, laundry, and personal grooming care (25). A similar Canadian study in 1965 indicated an average cost of over $52 per person per year (rising to $94 per person per year in Toronto) due to air pollution (12). These costs (adjusted for inflation) are presented for comparison in Table 22.3.

It is clear that pollution costs people in both money and health. In response to such evidence, the U.S. Congress authorized the Environmental Protection Agency (EPA) to propose federal air quality standards. The standards for air quality are designed to protect the health of the most sensitive groups in the population affected. In Table 22.4 the standards (26), tolerance, and tox-

icity levels are given (27). "Tolerance level" implies that below such a level, the pollutant is tolerable for some time for members of sensitive goups. Toxic substances can cause illness or death; the table indicates that particulates are a hundred times as toxic as carbon monoxide.

Damage to the Lungs

In order for atmospheric pollutants to affect us, they must enter our bodies. This usually takes place through the approximately 70 m² surface of the lungs. Particles of certain sizes are much more likely to reach and remain in our lungs. Smaller particles, those of size 1/100 to ½ μm, will be preferentially deposited in the lungs (9,28). Larger particles are trapped in the nose and upper respiratory tract. Many natural particles can

PAN SMOG

PAN—peroxyacetylnitrate—is produced in the atmosphere by the action of sunlight on the chemicals present in urban areas. These chemicals include hydrocarbons, carbon monoxide, carbon dioxide, nitrogen oxides, ozone, and sulfur oxides. Smog is a combination of ozone, carbon monoxide, PAN, and organic molecules. PAN smog affects cities mainly in the western United States, such as Denver and Los Angeles.

PAN is formed in the presence of many different chemicals, and the rates of formation of the combinations are much higher than they could be without the presence of these other chemicals. The smog also forms fine particulates (about 5 percent of organic smog vapor is particulate [23]). These "secondary" particulates are the most effective size (0.1–1 μm) to scatter light and reduce visibility by causing haziness.

PAN is formed because of the presence of hydrocarbon radicals (radicals are extremely reactive chemicals) in urban atmospheres, which combine with oxygen and nitrates produced by sunlight. Let us see how peroxyacetylnitrate may be formed in an urban area.

Methane (CH_4) is formed naturally in small concentrations in the atmosphere and in higher concentrations near cities. Methane can react with free oxygen to form hydroxyl radicals (OH) and methyl radicals (CH_3), both of which are very reactive. The free oxygen probably came from the breakup of nitrogen dioxide (NO_2) by light into nitrogen monoxide (NO) plus oxygen. The methyl radical can react with molecular oxygen (O_2) to form CH_3OO, which can again react with an oxygen molecule to form ozone (O_3) and yet another radical, CH_3O; or with nitrogen dioxide (NO_2) to form CH_3OONO_2, peroxyacetylnitrate.

Concentrations as low as a few parts per billion can cause eye irritation, and concentrations as low as a few parts per million are lethal to mice exposed to PAN for only two hours (24). PAN also has a very unpleasant odor.

---

irritate the lungs: Dust from dust storms and salt from spray evaporation are larger than 0.3 μm; particles from sun-induced gas reactions and smoke are smaller than 0.2 μm (29).

The particles penetrating the lung's defenses can cause great trouble, especially if toxic gases or liquids are present on the surface of the inhaled particle. About 30 percent of particles smaller than 1 μm remain in the lung once they are there (23). Once these particles get into the lung, the soluble particles can be rapidly transferred across the air sacs in the lungs and circulated in the blood. Small amounts of gases that dissolve in water (i.e., are soluble) will be trapped in the nose unless they are carried through on small particles. The bronchial system is illustrated in Figure 22.4.

The smaller particles put directly into the atmosphere are produced almost exclusively in engines and furnaces (23,30,31), and it is just these that are most dangerous, as we have seen (28,31,32). Small particles are also produced in the daytime by sunlight and last for about half a day before they coagulate into particles large enough to fall under gravity. (Particles larger than about 10 μm fall to the ground rapidly when the air is still.) The atmospheric distribution of particles is shown in Figure 22.5 (24,28,33). It has

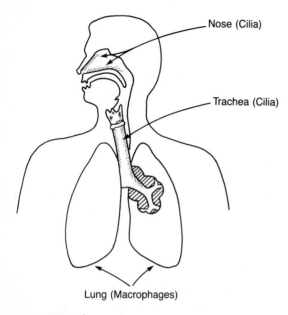

FIGURE 22.4
The human bronchial system and its defense mechanisms.

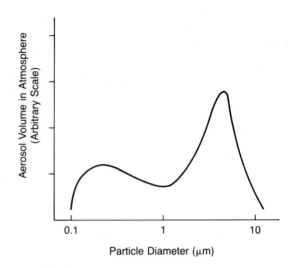

FIGURE 22.5
The distribution of total numbers of particles in the atmosphere in terms of their sizes.

TABLE 22.5
Comparison of health effects of selected pollutants.

Chemical	Concentration for Health Hazard	Infant Mortality	Cardiovascular Disease	Viral Diseases, Respiratory Tract	Chronic Bronchitis, Asthma	Lower Respiratory, Emphysema
Sulfur oxides, sulfuric acid	25 μg/m^3	X	X	X	X	X
Nitrogen oxides	8–11 μg/m^3		X	X	X	X
Lead						
Cadmium			X		X	X
Nickel						X
Beryllium				X	X	X
Mercury				X		X
Arsenic					X	
Vanadium						X
Chromium						
Asbestos				X	X	X
Polycyclic Organic matter						

two peaks—one for particles smaller than 1 μm peaking at about 0.1 μm, and the other for large particles, peaking at about 10 μm (24,28,33). There is increasing evidence that poisonous elements tend to concentrate in the smaller particles, but even dust particles can cause silicosis (23) and impair lung function. An autopsy on an Egyptian mummy showed scarred lungs from the omnipresent desert sand.

These smaller particles we are discussing, which most easily evade body defenses to penetrate deep into the lungs and are the most toxic, are most heavily concentrated in urban areas. This is so for two reasons: because of the high concentration of cars and trucks in the urban areas (60 to 90 percent of particulates from autos are fine particulates [31]) and the other sources of coal and oil combustion there; and because of condensation of particles formed by chemical reactions stimulated by the sun (smog—see box).

Research on coal combustion has found that there are two distinct peaks in the amount of mass released as fly ash. One peak occurs for large particle diameter, as expected. The surprise was the very sharp peak for particle diameter

near 0.1 μm (34). This peak occurred for all types of coal burned. Although the peak represents only a small mass release, it is responsible for most of the number of particles (and surface area) in the fly ash. The study further suggested that reducing the combustion temperature to control the release of nitrogen oxide (NO) has the desirable side-effect of reducing fine particle generation as well (34). A compilation of some environmental chemicals and their effects on health is presented in Table 22.5.

Is Pollution Local?

Urban pollution may travel long distances. One strategy for reducing *local* pollution is to construct high chimneys. This has the effect of more thoroughly mixing sulfur dioxide in the atmosphere. Tall chimneys in Britain cause sulfur dioxide pollution in Norway (10), since the pollutant is released at such a high altitude. Sulfur dioxide is a pernicious health hazard, as is seen in Table 22.6 (32). Closer to home, haze over the Great Smoky Mountains in Tennessee appears to be due to sulfate from coal burned in the Ohio Val-

TABLE 22.5 continued

Diseases of the Central Nervous System	Kidney Damage	Anemia, Fatigue	Bone Changes	Cancer	Hypertension	Skin Ulcers, Dermatitis	Visual Disorders	Gastrointestinal Disorders
X	X	X						
	X	X	X	Prostate, renal system	X		X	
				Lung		X		
X						X		
X						X	X	X
				Lung, Skin		X		
						X		
X		X		Mesothelomia		X		X
				Scrotum			X	

TABLE 22.6
Health effects associated with sulfur oxide emissions, based on postulated conversion rates of sulfur dioxide and sulfur tetroxide and EPA epidemiological data for representative power plants in the Northeast.

	Remote Location	Urban Location
Cases of chronic respiratory disease	25,600	75,000
Person-days of aggravated heart-lung disease symptoms	265,000	755,000
Asthma attacks	53,000	156,000
Cases of children's respiratory disease	6200	18,400
Premature deaths	14	42

SOURCE: C. L. Comar and L. A. Sagan. Reproduced, with permission, from the *Annual Review of Energy*, Volume 1, © 1976 by Annual Reviews Inc.

ley, and American urban and rural populations are exposed to roughly the same sulfate concentrations (33). The surface deposition of sulfur in the states in the upper Midwest and Northeast is over 8 tonnes per square kilometer, while the U.S. average is only 1.4 tonnes per square kilometer (35). One report estimated that some 20,000 people east of the Mississippi die each year as a consequence of sulfur dioxide emission (36).

Ozone levels of air entering the New York metropolitan area are often already above federal standards. New York adds its own pollutants, which travel up to 300 km through northeast Massachusetts (37). Plumes from single power

FIGURE 22.6
The boundary of a country is no defense against sulfur deposits. The countries in Europe are shown with 1978 deposits of sulfur in kilotonnes. The circles show what percentage of this total came from outside (black) and inside (white) the country. (Adapted by permission of the Spiegel-Verlag from an illustration in *Der Spiegel.*)

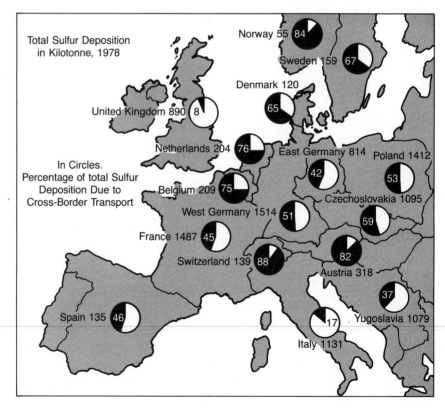

plants can cause excess ozone concentrations to distances greater than 56 km (38). A study in the New York metropolitan area indicated that ozone concentrations are regional in character, as are concentrations of PAN smog, acrolein (CH_2CHCHO), nitric acid, and other such pollutions. Control of metropolitan nitrogen oxide (NO) appeared the most effective strategy for overall reduction in air pollution in the New York metropolitan area (39).

New techniques using heavy metals as tracers (arsenic, selenium, zinc, antimony, indium, and noncrustal vanadium and manganese) are able to distinguish the area emitting the pollution by its distinctive chemical signature (40). Most coal has high selenium levels (33,40). However, manganese and vanadium may be removed from the atmosphere faster than sulfur (15).

Canada has pointed its figurative finger at the U.S. Midwest as the origin of much Canadian air pollution, and New York State has long implicated the same source as the major contributor to Adirondack lake acidification. These tracer techniques will be able to point back to the sources of pollution. The method was tested in two New England cities, one in Rhode Island, one in Vermont, and worked well. The conclusion was that local sources in the Northeast contribute about as much pollution as the Midwest in New England (even though the Midwest emits ten times as much sulfur dioxide) (40), although the Midwest contribution did dominate in Vermont. The scientists were even able to determine that a five-day pollution episode was actually a three-day local episode followed immediately by a two-day Midwest episode (40).

TABLE 22.7

Significant health risks associated with electricity generation.

Stage of Fuel Cycle	Coal	Oil	Gas	Uranium
Exploration Extraction	Mine drainage Subsidence Scarred land Pneumoconiosis of miners Accident hazards	Spills Subsidence	Risk of explosion	Subsidence Scarred land Radioactive tailings Silicosis of miners
Transport Processing	Accident risk Pneumoconiosis of workers	Spills Residuals in air (SO_2, NO_x hydrocarbons) Residuals in water (degradable and non-degradable)	Risk of explosion	Accident risk Silicosis of workers
Generation	Air pollution (SO, NO, hydrocarbons, trace elements) Water pollution (heat, non-degradable residuals) Sulfur Nitrogen			Accident risk Waste heat Radioactive waste Radionuclides in air Reprocessing or disposal of fuel
Waste Disposal	Sludge			
Ranking*	4	3	1	2

*Based on health and environmental effects per kilowatthour.

SOURCE: Table 4, ref. 13, Lave and Silverman. Reproduced, with permission, from the *Annual Review of Energy*, Volume 1, © 1976 by Annual Reviews, Inc.

Thus, what we call "local" pollution may be anything but local. Figure 22.6 (41) shows this for the countries of Western Europe. Worldwide, people contribute about 20 percent of the total atmospheric burden of fine particulates (42). As we have noted, urban areas suffer heavier precipitation than rural ones, partly because of the amounts of particles available as the nuclei for formation of raindrops (3,5,6,43).

With this health burden, which must be borne, it is not surprising that a study (44) indicates that a coal-fired power plant is up to 18,000 times more costly to health than a nuclear pressurized water reactor, as indicated in Table 22.7 (31,45). In fact, coal-fired plants emit more radioactivity than nuclear plants are allowed to emit (45), as we shall discuss further in Chapter 24.

There have been many proposals to try to clean up the air. It has been pointed out that healthy forests and grasslands can remove ambient pollutants with minimal harm as long as concentrations are sufficiently low (22). A different sort of proposal for reducing pollution has been more controversial (46): J. Heicklin of Pennsylvania State University has suggested that addition of tiny amounts of a compound diethylhydroxylamine—DEHA: $(C_2H_5)_2NOH$—to polluted air on days having a high probability of smog formation will hinder the process. Tests in laboratories have shown that reactions causing smog proceed much slower in the presence of DEHA. However, the idea of adding an additional pollutant to air in order to cut down smog has little romance. In addition, there is some evidence that DEHA may cause cell mutations (47).

ACID RAIN

It has been established that pollution can travel great distances through the atmosphere. Midwestern pollution affects New England (40); European air pollution has penetrated the Arctic (48), which at its peak is 2 percent of New York's peak pollution values; and Canada continually argues with the United States over the cross-bor-

FIGURE 22.7
This 1982 cartoon pokes fun at the Reagan Administration's cynical *laissez-faire* attitude toward the environment. (Wiley, copyright 1982, Copley News Service, *San Francisco Examiner.*)

der pollution. Figure 22.6 showed how pollution crosses European borders. Much of this phenomenon is due to the tall stacks installed to minimize the amount of local pollution; instead of falling near the stack, the pollutants are carried long distances through the air, giving sufficient time for interaction with water vapor to form acids (49,50). A droll comment on transborder acid rain is shown in Figure 22.7.

The pH Scale

The term *acid rain* refers to the deposition of sulfur and nitric oxides by rain or otherwise on plants, buildings, and the ground. Acidity is measured by measuring the concentration of hydrogen ions in the liquid. In distilled water, there are 10^{-7} ions per milliliter, which means one molecule of each 10 billion is ionized. As the concentration increases from this value, the water becomes acidic; as it decreases, the water becomes alkaline. The concentration is not very useful to characterize acidity or alkalinity. The quantity minus the logarithm to the base 10 of the concentration is taken to be the pH value (see Appendix 3). We use the pH value to measure acidity or alkalinity. Distilled water therefore has a pH $= -(\log_{10} 10^{-7}) = 7$. Vinegar has 10^{-3} ions per milliliter, and so has a pH of 3. Because the negative logarithm is used, *more* acidity means a *lower* pH. Figure 22.8 shows the pH scale and indicates that, because of atmospheric carbon dioxide, unpolluted rain is a weak acid, with a pH of 5.6.

Any precipitation with a pH value of less than 5.6 is called acid rain. The average pH of rain in the eastern United States is 4.5 (35). Acid rain comes from the gases NO_x and SO_2, as we mentioned above; sulfur causes about two-thirds of the total acidity and nitrogen oxides, one-third (35). These pollutants react with water vapor in the air to produce nitric and sulfuric acid, respectively:

$$NO_x + H_2O \rightarrow HNO_3 \,(+ \text{oxygen}),$$
$$SO_2 + H_2O \rightarrow H_2SO_4.$$

There are many possible ways for these reactions to occur in the air (they are intermediate steps in chemical reactions that take place normally in the air).

The pollution from Europe in the Arctic causes acid rain there. Even areas farther away from industrial development than the Arctic have acid rain: Hawaii (51) and Samoa (52) have significant excess sulfur deposition. Some of this sulfur comes from biological processes in the upper ocean; some may have a volcanic source.

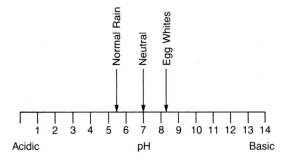

FIGURE 22.8
The pH scale.

Causes of Acid Rain

Releases of pollutants in millions of tonnes are shown in Figure 22.9 (53), and average concentrations are shown in Figure 22.10 (53). Seventy to 90 percent of acid rain appears to be manmade, and the acid in acid rain varies with the amount of sulfur in the air (54). Furthermore, increases in acid deposition parallel local increases in SO_2, and the highest concentrations of acid rain are found where SO_2 emissions are highest (15). A National Research Council Study claims that a 50 percent reduction in sulfur and nitrogen gases leads to a 50 percent reduction in acid rain (55). The only long-term study, from Hubbard Brook, New Hampshire, indicates that sulfur in the Northeast is at two to sixteen times greater concentration than in remote areas and that the water acidity is due mostly to sulfur (about 85 percent) and to nitrogen oxides (about 15 percent) (46,52,56).

There now appears to be a danger that acid rain can be a problem in the American West as well as in the East. Figure 22.11 (57) shows how sulfur concentration depends on sulfur emission. While models may disagree in details, all agree that decreases in emission lead to decreases in deposition (58). Sulfur oxide emission fell by approximately 90 percent in Arizona and New Mexico during a copper smelter strike in 1980, and in consequence, measured sulfate levels fell by over half from the previous summer (15). The acid rain in the West comes from nitrogen oxides from cars (implicated in the German forest disaster as well) and from sulfur dioxide from copper smelters and utilities. There is great concern for thousands of lakes in the Rockies, especially because of the opening of the Nocozari copper smelter in Mexico just south of the border. The plant emits about a half-million tonnes of SO_2 each year (50 percent more than any U.S. source) (53,59). Mexico and the United States agreed in July, 1985 to control pollution from the three giant smelters located in the border area, but the results of the agreement remain to be determined (60).

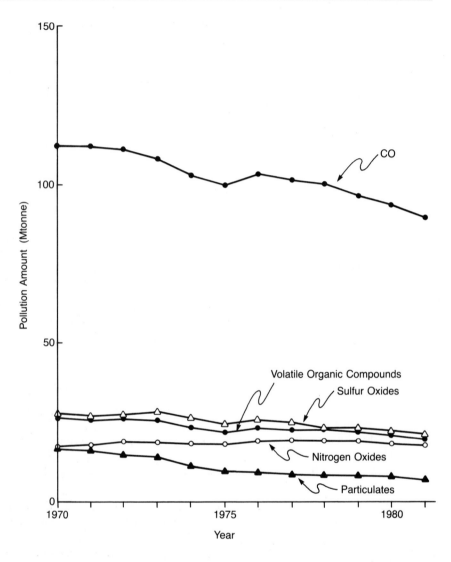

FIGURE 22.9
Release of pollutants (Mtonne) has been declining slowly since 1970. (U.S. Environmental Protection Agency)

Sulfur Release

Canada and the United States each have reason to complain of and to one another. The Sudbury smelter of INCO, Ltd. in Canada produces about 20 percent of Canadian SO_2—about 1 percent of total sulfur emissions from all human activity (49). The United States sends Canada 3.4 times as much sulfur as Canada sends the United States (61).

Much of the increase in the Northeast is due to sulfur from the Midwest. Utilities dump some 16 million tonnes (Mtonne) of SO_2 into the atmosphere east of the Mississippi (62); Ohio alone released 5.1 Mtonne of SO_2 in 1978 (63). By 1980, Ohio's emissions had fallen to 3.8 Mtonne, but Ohio emissions were still higher than those of any other state (15), about 12 percent of entire national emissions. The other Ohio River Valley states contributed large amounts as well: Illinois, 2.1 Mtonne; Indiana, 2.8 Mtonne; Pennsylvania, 3.0 Mtonne; and West Virginia, 1.5 Mtonne (15). The emissions from these six states are *half* the total emissions of the entire United States.

Figure 22.12 shows the best estimate of pH in North America developed under a study prepared prior to the joint American–Canadian

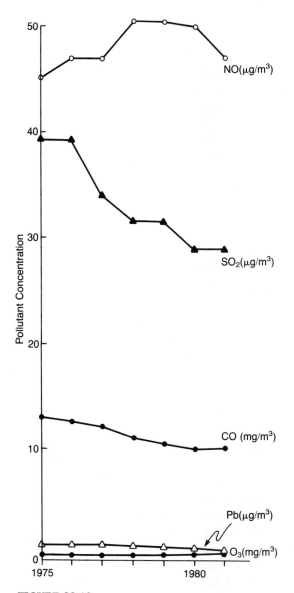

FIGURE 22.10
Average pollutant concentrations in air, 1975–81.
(U.S. Environmental Protection Agency)

Memorandum of Intent on Transboundary Air Pollution (64), and Figure 22.13 shows the actual distribution of emissions (65). Figure 22.14 (64) shows the agreed-to limitations for 1980 in kilotonne/yr. Regions of pH greater than 4.2 are centered around the area of highest SO_2 emission, as shown also in Figure 22.15 (53).

The amount of sulfur released by Ohio is very large. West Germany, containing five times as many people as Ohio, has only 3.5 Mtonne of sulfur deposited each year (65). Ohio is working on sulfur removal technology, but West Germany is commissioning a coal facility at Buschhaus (which caused German cabinet crises) that will emit 12 grams of SO_2 per cubic meter of smoke— thirty times the official standard (66). In a few years, Germany will produce more SO_2 than Ohio.

Sulfur pollution has been around since the dawn of the industrial revolution, as recorded in the Greenland ice cap, and acid rain was described as early as 1852 (67). The sulfuric and nitric acid load from combustion has health effects. The pollution episodes in the Meuse Valley, in London, and in Donora were so deadly because of acidic pollutants. Inadvertent experiments in Yokkaichi, Japan, and in Ontario have established that lung disease and hospital admissions are correlated with acidic concentra-

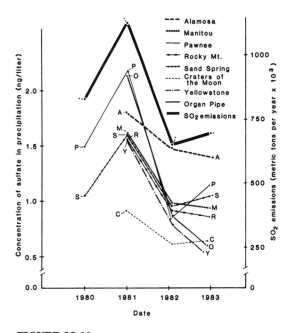

FIGURE 22.11
Correlation between sulfur emissions from smelters in the American west and sulfur in rain, 1980–83.
(M. Oppenheimer et al., copyright 1985 by AAAS)

FIGURE 22.12
Best estimate of rain-weighted pH for 1980, according to a Joint Canadian-American study. (B. Neumann, copyright 1984 by Butterworths Publishers)

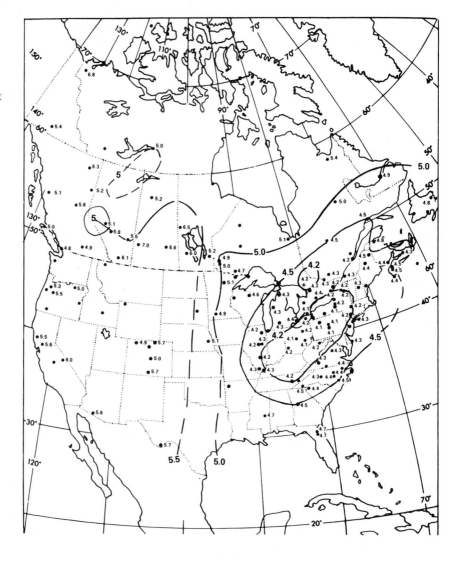

tions and decrease as distance from the source increases (68). Trees are affected as well: 60 μg of SO_2/m^3 leads to leaf function harm, and at 80 $μg/m^3$ the tree becomes deeply sick (64). The West German forests are suffering profoundly; as many as 70 percent of the trees are affected (69). With European sulfur emissions increasing, the problem is likely to get worse. What happened to the German forests, which was long brewing, but which was not really noticed until 1981, was a very rapid deterioration. There are indications that eastern forests in the United States are in decline. Softwoods at high elevation are dying or failing to reproduce in the Southeast as well as the Northeast (19,70). About 10 percent of annual forest growth in the United States takes place in areas of high sulfur deposition (69). The Forest Service has documented a large-scale, rapid, simultaneous 20 to 30 percent drop in growth rate for at least six species of coniferous trees. This pattern was observed in West Germany twenty years ago (19,70).

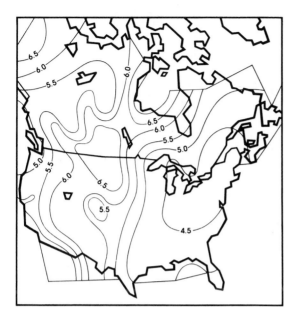

FIGURE 22.13
Actual distribution of acid rain. (Courtesy Ontario
Ministry of the Environment)

A parallel problem is that increased damage
to trees seems accompanied by increased concentrations of heavy metal in the soil
(19,64,67,70,71). Europe is leading the United
States in damage to trees because there has been
a chronic problem in Europe for years (pH of
approximately 3.4 has been stable for fifty years
[72]). The heavy metal problem, mixed with the
acidity of the precipitation, can cause problems
for soils, especially poor soils. There is some
indication, however, that soils can produce acidity independent of acid rain (73).

EFFECTS OF ACID DEPOSITION

Wet and Dry Deposition

There is disagreement about how much of the
problem is wet deposition (that is, in actual rainfall or snowfall) and how much is due to dry
deposition. The problem is that measurement of
dry deposition is not possible because physical

collectors cannot duplicate natural surfaces, such
as leaves (59,74). Conventional collectors in Ontario measure only about 60 percent of the sulfate
since some particles are too fine to precipitate
out (74). At Hubbard Brook, about a third of the
deposition is in dry form, and in many lakes,
more sulfur flows out of them than flows into
them in feeder streams. This indicates that sulfur
is being absorbed from the air on the surfaces.
Dry deposition may even be half the total sulfur
deposition. The acidification of the Elbe River
basin in central Europe appears to be caused by
dry deposition of sulfur dioxide and runoff from
chemical fertilizers (75). Further research in
progress on this question should resolve it.

Of course, other damage occurs to metal and
to stone monuments (67,71,76). Damage was
shown to occur to German monuments and old
buildings built of stone, even at concentrations
below 25 $\mu g/m^3$. It is not only old stone that
suffers; a tablet erected in Alexandria, Virginia,
in the 1920s shows extreme damage where it has
been exposed to rain (Figure 22.16). Carbon-
containing particles cause acceleration of the destructive work of acid rain (77).

Costs of repairing the damages add up. A study
by the Organization for Economic Cooperation
and Development in 1981 estimated these costs
as $805 million for Belgium and Luxembourg,
$462 million for Denmark, $316 million for
France, $873 million for the Netherlands, $4.331
billion for the United Kingdom, and $7.042 bil-
lion for West Germany. Such sums inspire awe.

Acid Lakes

Acid rain can lead to acid lakes. The average
acidity of Adirondack lakes (from all sources) has
increased over the last fifty years by a factor of
40 (41). Only 4 percent of these lakes were acid
during the period 1920–1937, but by the 1970s,
over half were acid (78,79). Some 50,000 lakes
in Canada may be destroyed by acid rain (80)
(see Figure 22.17[78]), and many other lakes in
the Northeast have shown increases in acidity
(50). Nine percent of lakes in the Northeast have

FIGURE 22.14
Limitations on SO$_2$ emission for 1980 in kilotonnes/year, according to the Memorandum of Intent. (B. Neumann, copyright 1984 by Butterworths Publishers)

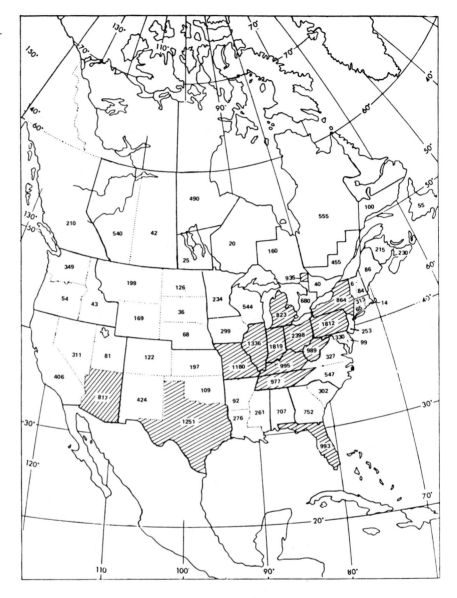

pH less than 5.5, while 20 percent of the lakes in Florida have such low pH values (54). Lakes do have the capability of neutralizing some acid. However, about one-fifth of lakes in the Northeast and one-third of Florida lakes were extremely vulnerable to acid attack (54).

If acid snow falls on a region during winter, the spring snowmelt can bring great increases in the amount of acid water flowing into lakes, with corresponding decreases in pH. The results of a study of a lake in Ontario are shown in Figure 22.18 (78). Even if the lake itself is not acidic, the shock can cause problems for lake fauna, especially fish (81).

While the health effects of airborne sulfur and nitric oxides are not totally certain, the effect of

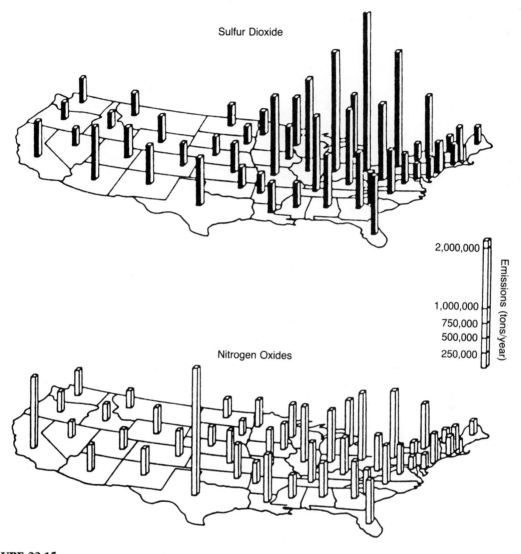

FIGURE 22.15
The emission of sulfur and nitrogen oxides for 1980 on a state-by-state basis. (U.S. Office of Technology Assessment)

the acids on bodies of water and the bodies of its inhabitants is much better known. As the concentration of acid increases, fish experience reproductive difficulties. Any lake with a pH below 7.0 is somewhat acidic. Below a pH of 5.5, most species are endangered, and the lake is classified as acid. Fish growth is retarded, the salt balance

in the blood is set awry, calcium is depleted from the skeleton (leading to malformation), mercury may be absorbed (15), and aluminum clogs fish gills (61,67,68,82).

Aluminum is often found in water at small concentration. Its concentration increases as pH decreases (79). Why should it be implicated so

FIGURE 22.16
The stone tablet has been damaged where it's exposed to acid rain.

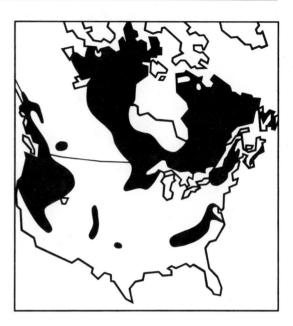

FIGURE 22.17
The approximate areas of North America with lakes having little ability to neutralize acid. (Courtesy Ontario Ministry of the Environment)

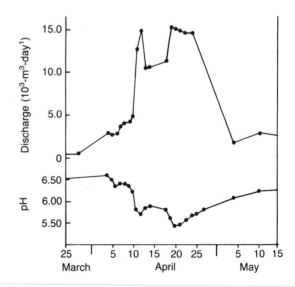

FIGURE 22.18
Spring pH depression in one stream flowing into Harp Lake, Muskoka, Canada. The spring runoff produces a severe acid shock to stream biota. (Courtesy Ontario Ministry of the Environment)

strongly in toxicity to fish in acid lakes? We know that aluminum is toxic in humans. It is directly associated with syndromes that mimic senility, which occur in patients with chronic kidney failure, dialysis encephalopathy, and osteomalocia (82), and it is implicated in certain brain disorders, such as senile dementia, Alzheimer's disease, and Parkinson's disease (68). Patients with such brain malfunction have high aluminum content in their brain tissue (called neurofibrillary tangles). No aluminum deficiency has ever been identified in living organisms (82). Aluminum appears to be harmless to most animals as long as calcium levels are maintained. When calcium is in short supply, aluminum may replace it, causing the problems already mentioned (68). Aluminum seems to be more the problem than the acid in the lakes itself (68). Perhaps for this reason, liming (adding calcium to) lakes to ameliorate damage from acid rain does not restore the lake (and, of course, it is only a stopgap measure) (67). Concentrations of aluminum that are lethal to half the animals, LC_{50}, are presented in Table 22.8 (82) for several compounds.

A study was conducted in which a Canadian lake was artifically acidified over an eight-year period. Fish became more obviously misshapen as the acidity increased. It appears that in this lake, irreversible changes in the ecosystem occured at a pH of 5.8 (earlier in acidification than had been believed before); no fish reproduced at pH values below 5.4 (83). Below a pH of 4.5, most life is gone except for algae, moss, and fungi. The lake can remain clear, or it may be matted with algae blooms and mosses.

The final word on acid rain and acid lakes has not yet been spoken. There is enough circumstantial evidence to warrant joint American–Canadian action to reduce sulfur and nitric oxides and to warrant the European Community's taking action to try to save European forests before it is too late. There is evidence that reduction in sulfur dioxide levels gives at least as great a decrease in the acidity of rain (84).

Economic Impact of Sulfur Emissions Abatement

Evidence mounts that control of acid rain (and sulfur emission) will provide *gains* as well as losses, even for the state of Ohio, which is the major sulfur emitter. A cost/benefit study done for the U.S. Congress shows opportunities for benefits to Ohio far in excess of the costs necessary to control sulfur emissions. The study pondered many scenarios, including a national switch to mandatory scrubbers and to fuel switching as an alternative to installation of some scrubbers (85).

In the scrubber-only scenario, Ohio would see a net employment increase of over 6000 due to creation of jobs in high-technology scrubber suppliers, and would realize cumulative profits after ten years of over $600 million; the entire nation would gain more than 75,000 jobs and would gain cumulative profits of almost $12 billion (85). In the scrubbers-plus-fuel-switching option, Ohio would lose almost 4000 jobs (most in coal-mining areas) but still gain cumulative ten year profits of $440 million; nationwide, 72,000 jobs would be created, and cumulative profits would total just under $10 billion (85).

In Ohio, 95 percent of all electricity comes from coal. This net profit to Ohio includes the effects of the increased electric bills necessary to cover the cost of scrubber installation ($4.2 billion for scrubbers only, or $3.0 billion for scrubbers plus fuel switching). By 1985, Ohio utilities had already spent $2 billion for pollution control; even the larger figure does not greatly exceed that (85). The beneficial effect to the state could be even greater if government officials and private industry seize the advantage to be gained in the opportunity.

While it would be nice to be absolutely certain about the causes of acid rain before we take action, as Abelson (18) has suggested, action is often

TABLE 22.8 Concentration of aluminum compounds fatal to half the experimental animals.

Compound	pH	Animal	LC_{50} (ppm)
$AlCl.6H_2O$	7–8	Goldfish	1.0
$AlCl_3$	7.4 ± 0.1	Trout	0.56
		Goldfish	0.15
$AlCl_3$ aqueous	4.9–5.2	Trout fry	0.5–1.0

SOURCE: T. K. Morris and G. L. Krueger, Reference 82, copyright 1984 by Butterworths.

necessary in life before absolute certainty is writ large. John Snow used a novel graphic technique to implicate one public pump on Broad Street, London, as the source of a cholera epidemic (Figure 22.19) (85). He acted on his knowledge by removing the pump handle. The end of the cholera epidemic was his proof (67).

Clearly, there is evidence of effect here. The amount of sulfuric acid in the environment appears to be proportional to SO_2 concentrations in the air. Regional levels of sulfuric acid follow from increases or decreases in emissions (decreases in sulfur emissions from western U.S. smelters lead to increases in rain pH), with the highest sulfuric acid deposition coming in regions of highest sulfur oxide emission. The downturns in pollutants, illustrated in Figures 22.9 and 22.10, seem to be correlated with the economic recession of the early 1980s. We know

there is a connection. It is probably time to act on this knowledge, to disarm the pollutant pump.

SUMMARY

Urban areas suffer pollution from cars, trucks, industrial processing, and electric utilities. Urban areas are generally warmer as well as more polluted than rural areas. Many great conurbations are in valleys or are surrounded to one side by hills. There, air may become trapped by temperature inversion. In the normal case, air temperature decreases as one moves upward in the first few kilometers of air. Occasionally, conditions arise in which air temperature increases with height, the inverse of the usual condition: There is a temperature inversion. In this condition, the air is so stable that it is quiescent, trap-

FIGURE 22.19
John Snow's graph of cholera cases implicated the pump on Broad Street (marked *X*) as the source.

ping all the pollutants within the urban area. Such extreme events may cause sickness or even death.

Health damage has been documented for people living in urban areas. Plants are also affected adversely by pollution—especially ozone and acid rain. In addition, pollution causes property damage in the billions of dollars.

Large particles cause exhaust gases to *look* awful. These particles are trapped effectively by the lungs' defenses. Small particles may not appear to be in the exhaust gases, but they can cause more severe injury than the larger particles and set the stage for serious lung disease.

While local air pollution is bad, we find that pollution respects no boundaries. Midwestern acid rain falls in Canada and New England. Britain's air pollution invades Norway, also as acid rain. The pH scale is used to measure acidity. A pH of 7 is neutral; when pH is less than 7, the liquid is acidic; when pH is greater than 7, the liquid is alkaline. Acid rain has damaged German forests, and U.S. forests are showing symptoms similar to those observed in Germany one or two decades ago.

Lakes with a pH less than 5.4 have no reproducing fish. Many lakes in the eastern United States have such pH values, but lately, acidity has been observed in the West as well. With the pollution problems identified, and the health consequences of pollution known reasonably well, the time has come to address the issue. The innocent bystander should not have to live with the adverse consequences. Those who benefit from emission of pollutants should pay the cost of cleaning the air.

REFERENCES

1. National Academy of Sciences, *Mineral Resources and the Environment* (Washington, D.C.: N.A.S., 1975).
2. J. M. Mitchell, in *Man's Impact on the Climate,* ed. W. H. Matthews, W. W. Kellogg, and G. D. Robinson (Cambridge, Mass.: MIT Press, 1971), Table 8.1, pg. 167. W. P. Lowry, *Sci. Am.* 217, no. 2 (1967): 15.
3. P. V. Hobbs, H. Harrison, and E. Robinson, *Science* 189 (1974): 909.
4. V. Brodine, *Environment* 13, no. 1 (1972): 3.
5. W. Sullivan, *New York Times,* 28 Feb. 1977; and B. A. Franklin, *New York Times,* 13 July 1986.
6. T. B. Smith, in *Man's Impact on the Climate,* ed. W. H. Matthews, W. W. Kellogg, and G. D. Robinson (Cambridge, Mass.: MIT Press, 1971).
7. R. Kirbach, *die Zeit,* 1 Feb. 1985.
8. E. W. Kenworthy, *New York Times,* 2 May 1976; and Department of the Interior, *Kaiparowits, Environmental Impact Statement,* Sec. III (Washington, D.C.: GPO, 1975).
9. D. V. Bates, *A Citizen's Guide to Air Pollution* (Montreal: McGill-Queens U.P., 1972).
10. M. W. Thring, *Intern. J. Environmental Studies* 5 (1974): 251; J. W. Sawyer, *Environment* 20, no. 2 (1978): 25; and R. Frank, in *Perspectives on Energy,* ed. L. C. Ruedisili and M. W. Firebaugh (New York: Oxford Univ. Press, 1975), 166.
11. C. Holden, *Science* 187 (1975): 818.
12. E. Goldsmith, *Ecologist* 7 (1977): 160.
13. L. B. Lave and E. P. Seskin, *Science* 169 (1970): 723; and L. B. Lave and L. P. Silverman, *Ann. Rev. of Energy* 1 (1976): 601.
14. L. F. Diaz, G. M. Savage, and C. G. Golueke, *CRC Critical Reviews in Environmental Control* 14 (1985): 251.
15. R. R. Gould, *Ann. Rev. Energy* 9 (1984): 529.
16. R. D. Doctor et al., *J.A.P.C.A.* 35 (1985): 331.

17. J. L. Marx, *Science* 187 (1975): 731.
18. P. H. Abelson, *Science* 226 (1984): 1263.
19. S. B. McLaughlin, *J.A.P.C.A.* 35 (1985): 512.
20. P. M. Solomon et al., *Science* 224 (1984): 1210; and T. H. Maugh II, *Science* 223 (1984): 1051.
21. J. N. Pitts, Jr., in *Advances in Environmental Sciences,* ed. J. N. Pitts, Jr., and R. L. Metcalf (New York: John Wiley & Sons, 1969).
22. W. H. Smith, *Environmental Pollution* 6 (1974): 111.
23. P. F. Fennelly, *Am. Sci.* 64 (1976): 46.
24. E. R. Stephens, in *Advances in Environmental Science,* ed. J. N. Pitts, Jr., and R. L. Metcalf (New York: John Wiley & Sons, 1969). See also M. W. McElroy et al., *Science* 215 (1982): 13.
25. W. Bach, *Atmospheric Pollution* (New York: McGraw-Hill Book Co., 1972).
26. *Federal Register* 38 (1974): 235.
27. R. L. Babcock, Jr., *J.A.P.C.A.* 20 (1970): 653.
28. D. F. S. Nautsch and J. R. Wallace, *Science* 186 (1974): 695.
29. A. J. Haagen-Smit, in *Energy Needs and the Environment,* ed. R. J. Seale and R. A. Sierka (Tucson: Univ. of Arizona Press, 1973).
30. SCEP (Study of Critical Environmental Problems) report, in *Man's Impact on the Climate,* ed. W. H. Matthews, W. W. Kellogg, and G. D. Robinson (Cambridge, Mass.: MIT Press, 1971).
31. F. P. Prera and A. K. Ahmed, *Respirable Particles* (New York: Natural Resources Defense Council, 1978).
32. C. L. Comar and L. A. Sagan, *Ann. Rev. Energy* 1 (1976): 581.
33. R. W. Shaw, *Sci. Am.* 257, no. 2 (1987): 96.
34. M. W. McElroy et al. *Science* 215 (1982): 13.
35. D. Hafemeister, *Am. J. Phys.* 50 (1982): 713.
36. *Air Conservation Newsletter* 92 (1977).
37. J. L. Warner, *Science* 191 (1976): 179.
38. D. D. Davis, G. Smith, and G. Klauser, *Science* 186 (1974): 733.
39. W. S. Cleveland and T. E. Graedel, *Science* 204 (1979): 1273.
40. K. A. Rahn and D. H. Lowenthal, *Science* 228 (1985): 275; *Natural History* 95, no. 7 (1986): 62.
41. *Der Spiegel* 47 (1981): 96; 48 (1981): 188; and 49 (1981): 174.
42. J. T. Peterson and C. E. Junge, in *Man's Impact on the Climate,* ed. W. H. Matthews, W. W. Kellogg, and G. D. Robinson (Cambridge, Mass.: MIT Press, 1971).
43. J. E. Jiusto, *The Physics of Weather Modification* (College Park, Md.: AAPT, 1986).
44. L. B. Lave and L. C. Freeburg, *Nuc. Safety* 14 (1973): 409.
45. J. P. McBride et al., *Science* 202 (1978): 1045.
46. T. H. Maugh II, *Science* 193 (1976): 871.
47. B. J. Finlayson and J. N. Pitts, *Science* 192 (1976): 111.
48. R. A. Kerr, *Science* 212 (1987): 1013; and K. A. Rahn, *Natural History* 93, no. 5 (1984): 30.
49. G. E. Likens et al., *Sci. Am.* 241, no. 4 (1979): 43.
50. R. Partrick, V. P. Binetti, and S. G. Halterman, *Science* 211 (1981): 446.
51. J. M. Miller and A. Yoshinaga, *Geophys. Res. Lett.* 18 (1981): 779; and R. A. Kerr, *Science* 212 (1981): 1014.

52. S. J. Nagourney and D. C. Bogen, in *Meteorological Aspects of Acid Rain,* ed. C. M. Bhumralkar (Boston: Butterworths, 1984).

53. Department of Commerce, *Statistical Abstract of the United States* (Washington D.C.: GPO, 1984).

54. E. Marshall, *Science* 229 (1985): 1070. See also R. E. Newell, *Sci. Am.* 224, no. 1 (1971): 32.

55. R. A. Kerr, *Science* 221 (1983): 254.

56. J. V. Galloway, G. E. Likens, and M. E. Hawley, *Science* 226 (1984): 829.

57. M. Oppenheimer, C.B. Epstein, and R.E. Yuhnke, *Science* 229 (1985): 859.

58. M. Sun, *Science* 228 (1985): 34.

59. R. W. Shaw, in *Meteorological Aspects of Acid Rain,* ed. C. M. Bhumralkar (Boston: Butterworths, 1984).

60. M. Sun, *Science* 229 (1985): 949.

61. R. Howard and M. Perley, *Acid Rain* (Toronto: Anansi, 1980).

62. E. Marshall, *Science* 221 (1983): 241.

63. *OMRA Newsletter,* 1978.

64. B. L. Niemann, in *Meteorological Aspects of Acid Rain,* ed. C. M. Bhumralkar (Boston: Butterworths, 1984).

65. J. S., *Scala* 1 (1983): 7.

66. J. Nawrocki, *die Zeit,* 30 Nov. 1984.

67. E. Gorham, in *Meteorological Aspects of Acid Rain,* ed. C. M. Bhumralkar (Boston: Butterworths, 1984).

68. T. H. Maugh II, *Science* 262 (1984): 1408.

69. S. Postel, *Futurist* 18, no. 4 (1984): 39.

70. P. Shabecoff, *New York Times,* 26 Feb. 1984; and H. W. Vogelman, *Nat. Hist.* 91, no. 11 (1982): 8.

71. Environmental Resources Ltd., *Acid Rain* (New York: Unipub, 1983).

72. P. Winkler, in *Chemistry of Particles, Fogs, and Rain,* ed. Jack L. Durham (Boston: Butterworths, 1984).

73. E. C. Krug and C. R. Frink, *Science* 221 (1983): 520.

74. R. A. Kerr, *Science* 211 (1981): 692.

75. T. Paces, *Nature* 315 (1985): 31.

76. P. Lewis, *New York Times,* 3 August 1984.

77. M. del Monte and O. Vittori, *Endeavour* 9 (1985): 117.

78. Environment Canada, *Downwind: The Acid Rain Story* (Ottawa, 1981).

79. M. Uman, in *Acid Rain: How Serious and What to Do,* ed. D. Hafemeister (College Park, Md.: AAPT, 1986).

80. D. Martin, *New York Times,* 26 Feb. 1984; and E. Marshall, *Science* 217 (1982): 1118.

81. J. N. Galloway and G. E. Likens, *Atmos. Env.* 15 (1981): 1081.

82. T. K. Morris and G. L. Krueger, in *Meteorological Aspects of Acid Rain,* ed. C. M. Bhumralkar (Boston: Butterworths, 1984).

83. D. W. Schindler et al., *Science* 228 (1985): 1396.

84. M. Oppenheimer, in D. Hafemeister, ed., *Acid Rain: How Serious and What to Do* (College Park, Md.: AAPT, 1986).

85. A. A. Cook and J. D. Rosenberg, *Amicus J.* 8, no. 2 (1986): 5.

86. E. R. Tutte, *The Visual Display of Quantitative Information* (Cheshire, Conn.: Graphics Press, 1983), 24.

PROBLEMS AND
QUESTIONS

True or False

1. Pollution is strictly a local phenomenon because pollutants are so quickly washed out by the rain.
2. Air pollution is inherently a problem of perception rather than reality, because not all people are affected by it.
3. Nitrogen oxides are implicated in the formation of both smog and acid rain.
4. Under normal circumstances, aluminum is not a health hazard for people or animals.
5. Tall smokestacks reduce local pollution.
6. Acid rain is a problem because it can cause stone to weaken and ultimately disintegrate.
7. Acid rain can react with the soil and cause pollution of lakes and streams.
8. Acid rain is found in regions far from the places the pollution originated.
9. Rain is acidified when ozone encounters clouds.
10. Tall chimneys on major pollution sources have played a part in the increased incidence of acid rain.

Multiple Choice

11. Which of the following power plants probably emits the most radioactivity?
    a. Gas-fired
    b. Hydroelectric
    c. Coal-fired
    d. Nuclear reactor
12. Acid rain has poisoned lakes and destroyed monuments. Which chemical(s) below is(are) implicated in acid rain?
    a. Carbon monoxide
    b. Carbon dioxide
    c. Hydrocarbons
    d. Nitrogen oxides
    e. All of the above chemicals are implicated.
13. Smog is a persistent problem in some cities. Which of the following is *not* involved in the formation of smog?
    a. A temperature inversion
    b. Sulfur oxides
    c. Nitrogen oxides
    d. Unburned hydrocarbons
    e. Sunlight
14. Which of the following is *not* a reason to believe that reduction of emissions will decrease acid rain?
    a. Ambient sulfuric acid concentrations are related to ambient SO_2 levels.
    b. Sulfate levels decrease when emissions from metals smelters decrease.
    c. Dry deposition rates may be greater than wet deposition in areas of low sulfur concentration.
    d. Regional trends in sulfur dioxide emissions follow trends in sulfuric acid levels.
    e. Rain with the lowest pH is found in regions with highest emissions of sulfur dioxide.
15. Which body organ(s) are most sensitive to small particulates in the air?
    a. Eyes
    b. Mouth
    c. Nose
    d. Lungs
    e. Brain

16. In a temperature inversion,
    a. temperatures are below zero.
    b. cold air rests above warm air.
    c. air moves upward very rapidly.
    d. warm air rests above cold air.
    e. rain often occurs.

17. About what density of particulate matter would probably prove dangerous over a long period of time (several months)?
    a. 25 $\mu g/m^3$
    b. 75 $\mu g/m^3$
    c. 100 $\mu g/m^3$
    d. 250 $\mu g/m^3$
    e. 1000 $\mu g/m^3$

18. About what density of particulate matter would probably prove dangerous over even a short period of time (days)?
    a. 25 $\mu g/m^3$
    b. 75 $\mu g/m^3$
    c. 100 $\mu g/m^3$
    d. 250 $\mu g/m^3$
    e. 1000 $\mu g/m^3$

19. Which of the following pollutants are most toxic to people's health?
    a. Carbon monoxide
    b. Carbon dioxide
    c. Sulfur oxide
    d. Nitrogen oxide
    e. Particulates

20. Which of the following states emits the most sulfur?
    a. Pennsylvania
    b. Ohio
    c. Texas
    d. Illinois
    e. Florida

Discussion Questions

21. What consequences might result from the cleanup of particulates in the air?
22. Why are smaller particles from combustion more dangerous to health?
23. Why has the acid rain problem become more apparent in the last several years?
24. Because cities are heat islands, pollution in the city is reduced by drawing air from the surrounding countryside inward, and using the cleaner air to replace air that has risen because it is warm. Is this sufficient reason to be complacent about the worldwide redistribution of heat caused by human activities?
25. What other automotive pollutants are there in addition to those discussed here? What effect might they have?

23
CLIMATE

In the last chapter, we considered local and regional atmospheric pollution and its interactions with the atmosphere. Here we turn our attention to the effects of the energy circulation in the atmosphere. This involves weather and climate. Weather behaves unpredictably from day to day, but predictably from season to season. Climate, the predictable seasonal pattern of the weather, changes slowly.

What causes climate to change? How is energy transferred in the global system? We examine the interconnection of air and ocean, and look at possible cosmological and geological mechanisms of climate change.

KEY TERMS   weather
                 teleconnection
                 stratosphere
                 troposphere
                 climate
                 Hadley cell
                 Coriolis effect
                 albedo
                 Southern Oscillation
                 El Niño
                 interglacial
                 Little Ice Age
                 isotope ratios

CLIMAP
cryosphere
cosmological
anthropogenic
solar constant
sunspot cycles
eccentricity
obliquity
precession
Milankovitch cycle
isostatic uplift
nuclear winter
aerosol

WEATHER

Weather is a *local* phenomenon. It is the combination of events such as the ambient temperature and its changes, precipitation or the lack of it, or cloudiness, occurring over a small time span in a circumscribed area. The world's weather is nothing more than the sum total of all these brief occurrences at each place, each one of which causes some other effect later somewhere else. Weather *is* "predictable" in a sense, because one of the best ways to find out what weather is likely to be on any day is to look at the weather experienced on that day in the past. Figure 23.1 (1) illustrates this principle. The temperature for 1980 and 1986 in New York City fluctuated mostly within the range of "normal highs" and "normal lows" (these are long-term means of the high and low temperatures for each of the days). The seasons show clearly in Figure 23.1.

Weather forecasts have, historically speaking, a not-very-good track record. The reason is that everything in the atmosphere is connected to everything else. If I want to predict the weather in New York for tomorrow, and I wish to model it using the physics of atmospheric motion, I would consider as input weather conditions in the Atlantic through the midline of the continent (including the Gulf of Mexico). To predict one day's worth of weather in western Europe involves conditions in Greenland, West Africa, and the Urals; a week's weather forecast involves the weather conditions at the North Pole, in Honolulu and Rio de Janeiro, in Jakarta and Teheran (2). Without modern supercomputer facilities, there would be no hope of success: The weather is predicted by putting in "initial" weather conditions in a grid pattern on the globe, and at several height levels. Patterns among the grid points are traced in time. The more grid points, the better the weather is traced, but the more calculations are required. The new North American model, called RAFS (Regional Analysis Forecast Systems) uses points 80 km apart on land and 320 km apart over the oceans, and it has sixteen vertical layers. Running the weather on the old model, LFM (Limited-area Fine Mesh), with grid spacing 190 km and seven layers, was eight times faster but somewhat cruder (3).

Events occurring in the Pacific can affect weather in the U.S. Midwest. Long-distance connections between other distant spots abound: These are called *teleconnections*. In addition to the North American teleconnection to the Indian monsoons, Greenland and Europe are teleconnected. There is a high correlation between Eurasian snow cover in October and November and winter air temperature the following winter in Siberia, as well as the intensity of the summer monsoon (4). Even stranger correlations seem

to exist: Weather forecasts for the British Atlantic coast err by a significantly greater amount when the number of thermal (slow) neutrons in the lower stratosphere and troposphere decrease substantially (5).

The *stratosphere* is the region in which temperature rises with height; this is above 8 km at the poles and 18 km at the equator. The *troposphere,* in which the temperature decreases with height, is the part of the atmosphere below the stratosphere. The troposphere contains air at breathable density, water, and clouds. Above the stratosphere are the mesosphere and the thermosphere. The composition of the atmosphere is uniform in the troposphere and stratosphere.

While weather forecasts of longer than about a week cannot yet be taken seriously (3), we have argued that weather history is a good guide to what will happen. Weather patterns may exist in two different, but equally stable flows—a sort of equilibrium, which may bring similar weather day after day, then suddenly switch to the other (6). While weather fluctuates from day to day, the pattern persists and determines mean values, which do not change greatly.

Weather history emphasizes year-to-year comparisons rather than day-to-day ones. As Figure 23.1 showed, the temperature varies by the day as well as by the season. If we look over a year, though, we would expect the mean temperature to be stable from year to year. Indeed, the inset in the upper left hand corner of Figure 23.1 shows that this is so—the mean annual temperature is about 11°C. It is clear that deviations from this mean occur every day in New York. If we look at the weather over other places and average the data over a reasonable number of years (ten, twenty, thirty, or fifty, say), we would get a general idea of the weather over that place. Similar deviations or variations will occur in other localities, and the patterns of these variations are important to consider. For example, while the mean annual temperatures of San Francisco and Cincinnati are about the same (7)—13.8°C (57°F) versus 13.6°C (55°F)—the deviation of the temperature from the mean is much greater for Cincinnati (9.3°C)

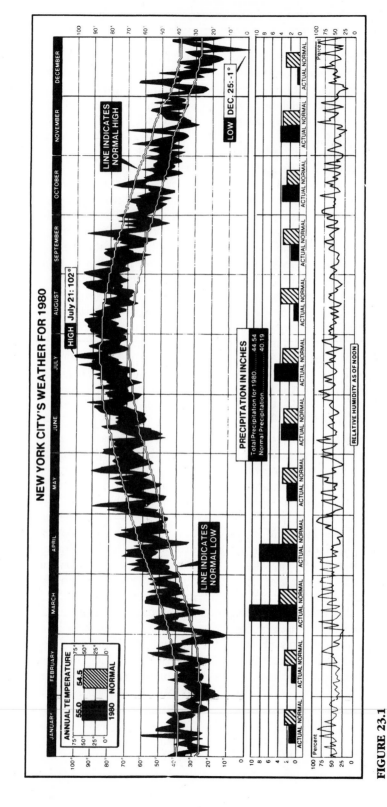

FIGURE 23.1
New York City's weather for 1980 and 1986 (copyright 1981, 1987 by the New York Times Company. Reprinted by permission.).

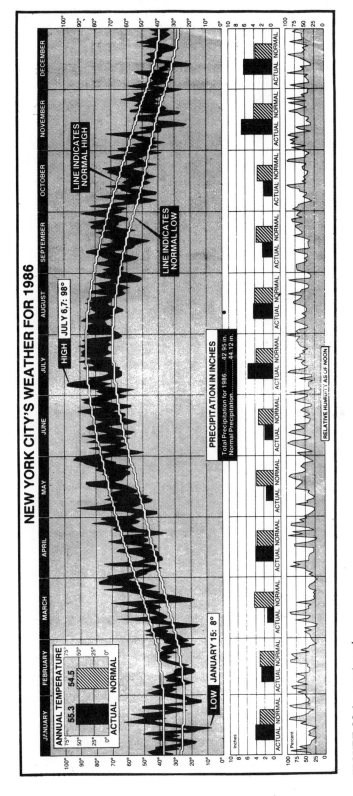

FIGURE 23.1 continued

FIGURE 23.2
The disequilibrium be-
tween the amount of so-
lar radiation intercepted
per m² at the poles and
the equator drives the
world's weather.

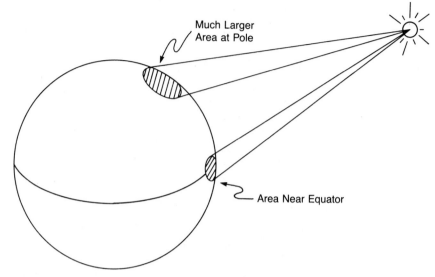

Much Larger
Area at Pole

Area Near Equator

than for San Francisco (3.1°C). This generalized weather we call *climate*. We will use the average temperature, the extremes of temperature, the amount of precipitation, and other such indicators to describe the climate.

Of course, one of the most important things about climate is that it is always changing. The numbers that describe the climate do not have to change much to be important (8–11). A study of the mean annual temperature in Iceland over the past millenium shows that a 1°C decrease reduces the length of the growing season by about two weeks and the number of growing-degree-days by about 25 percent (8).

What is ultimately responsible for the weather? It is, as indicated in Chapter 15, the sun. The Earth absorbs solar radiation from the sun, which serves to heat it and drive the winds and the weather. Wind patterns arise because the poles intercept very small amounts of light per m² while the equatorial regions capture large amounts per m² (Figure 23.2). The net result is an influx of ≈60 W/m² in the equatorial regions and an efflux of ≈100 W/m² at the poles (12). The heated air at the equator rises and cools as it moves to higher latitudes, slowly sinking back to ground level, where it recirculates back to the equator.

There it is reheated and recycled. The air that sank north or south of the equator dragged more northerly or southerly air with it, setting up another circulation pattern. This air rises at still higher latitudes, dragging local air along to set up still another circulation pattern. These seg-

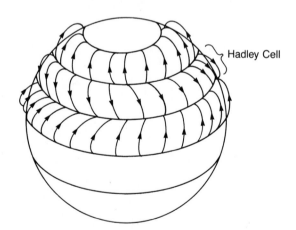

Hadley Cell

FIGURE 23.3
So-called *Hadley cells* partition the atmosphere between the equator and the poles. (Technically, only the cell adjacent to the equator is a Hadley cell.) The boundaries of the cells are known as *convergence zones*.

RADIATION BALANCE AND EARTH TEMPERATURE

The Earth appears from the sun to be a disk—it looks flat, just as the full moon does from Earth. The area of the disk is πR_E^2, where R_E is the Earth's radius. The total energy absorbed for a solar constant of 1.35 kW/m² is, assuming an albedo of 30 percent,

$$(1 - 0.3)(1.35 \text{ kW/m}^2)(\pi R_E^2).$$

A body at absolute temperature T radiates according to the Stefan-Boltzmann law as

$$(\text{total surface area})(5.67 \times 10^{-11} \text{ kW/m}^2 \text{ K}^4)(T^4).$$

The Earth's total spherical surface area is $4\pi R_E^2$. Thus, since πR_E^2 is a common factor,

$$(0.7)(1.35 \text{ kW/m}^2) = 4(5.67 \times 10^{-11} \text{ kW/m}^2 \text{ K}^4)(T^4).$$

Therefore,

$$T^4 = \frac{(0.7)(1.35 \text{ kW/m}^2)}{4 (5.67 \times 10^{-11} \text{ kW/m}^2 \text{ K}^4)} = 4.17 \times 10^9 \text{ K}^4,$$

or

$$T = 254 \text{ K}.$$

We may roughly estimate the entropy increase as solar energy runs into, and Earth's radiation out of, the Earth system. By definition, the increase or decrease in entropy consequent in transferring a certain amount of heat is that amount of heat divided by the absolute temperature of the fluid. We make the approximation that Earth's temperature is constant: Since the sun's temperature is roughly 6000 K and the Earth's is roughly 250 K, an amount of heat as it is radiated by the sun will cause an entropy increase about twenty-five times as great as the entropy decrease consequent on the reception of that radiation by the Earth.

regated regions of air are called *Hadley cells*. This circulation, which serves to redistribute the sun's heat, combined with the effects of Earth's rotation (the *Coriolis effect*), sets·up the familiar wind patterns such as trade winds and the furious forties (Figure 23.3). Of course, this is a much simplified description of the heat circulation in the atmosphere.

Radiation Balance

Since the Earth absorbs solar radiation with a spectrum of radiation corresponding to that from a body of about 5800 K (see Figure 15.4), which is the sun's surface temperature (as discussed in Chapter 15), the sun's radiation causes the Earth to rise in temperature. The temperature does not rise indefinitely, as is obvious to all of us residing on Earth. The Earth's average temperature has been constant to within 15°C over millions of years. The Earth must therefore emit radiation to space as the sun does in order to maintain this long-term balance. Because the temperature of the Earth is so much lower than that of the sun, it emits radiation with a spectrum appropriate to a body of temperature about 300 K (infrared radiation). A quick calculation, just a little more sophisticated than that in Chapter 15 (box), shows that the radiation balance alone would predict the Earth's temperature about 45 K too low (below the freezing point of water).

As has been emphasized in Chapters 1, 15, and 16, not all the sun's energy reaches the surface of the Earth. Much of the infrared and ultraviolet radiation is absorbed (see Figure 23.4): the ultraviolet by atmospheric ozone, the infrared principally by carbon dioxide and water vapor in the air.

Let us consider solar radiation of wavelengths in the range 300 nm (0.3 μm) to 3.0 μm (the gap in Figure 23.4). The intensity of solar radiation at the Earth's distance from the sun is about 1360 W/m^2. As discussed in Chapter 15, the Earth's effective area is just a quarter of its total surface area. Not all solar radiation that reaches the Earth makes it through the atmosphere. About 30 percent of this radiation is reflected (we call the proportion of light reflected the *albedo*). Suppose we call the available intensity 100 units; then 30 units are reflected. This takes place from clouds (20 units), from the air itself (6 units), and from the surface (4 units), as illustrated in Figure 23.5 (12,13). Since the Earth's temperature is not zero, it emits infrared radiation in the wavelength range 6 to 60 μm (about 5100 TW) (12–14). As the figure shows, some of this radiation is absorbed by the carbon dioxide, water, and other trace

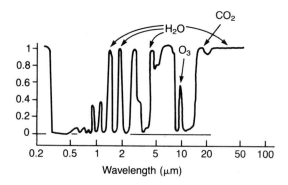

FIGURE 23.4
Atmospheric absorptance at sea level during clear weather.

gases in the air, and some of this is reradiated to the Earth. Altogether, 38 units are radiated to space from the gases, and 6 units go directly to space from the surface. Since most of the surface is ocean, the ocean absorbs most of the incoming radiation. Some of this is transferred to the atmosphere (7 units) in the form of turbulent diffusion from land and ocean (13,14). Finally, water is evaporated, absorbing latent heat of vaporization and transferring this latent heat to the clouds as it recondenses to water (23 units, or ≈4400 TW) (12–14). The clouds then radiate 26 units into space. Note that the temperature of the upper atmosphere is 255 K, just about what we calculated when we took albedo into account.

FIGURE 23.5
The mean annual radiation and heat balance of the atmosphere, based on satellite measurements and conventional observations. The total incoming radiation is given a value of 100, and all fluctuations are measured relative to this.

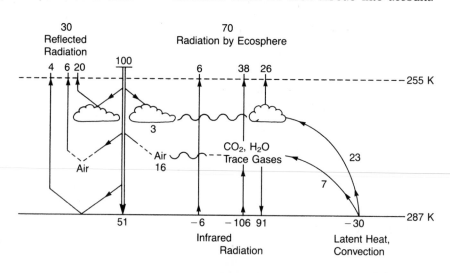

There are two sorts of El Niño events that occur with no periodicity (19). Their advent seems difficult to predict, although there are indications that the 1987 El Niño might have been predicted. In both sorts of ENSO events, warm water is brought to the west coast of South America and can cover the cool, fertile water for extended periods (six to eighteen months).

Late in the year before El Niño (for the more common variety), the Southern Oscillation index begins to fall as the south Pacific high in Tahiti weakens, causing the southern trade winds to weaken (15). The sea surface temperature gets higher than normal off Peru, and the anomaly spreads to the west. Easterly winds exert a frictional drag on the surface of the ocean, giving rise to the sea level difference of ½ m. The wind circulation cell is driven by the high sea surface temperature in the western Pacific (14): The winds pick up water vapor as they flow westward; they drop it in the western Pacific as rain, and the air rises, completing the circulation.

As a result of the high sea surface temperature peaking in late spring off South America, about 2.5°C above normal, fewer nutrients are transported to the surface, and so productivity may decline to 20 percent of what it was along longitude 95° W and to 5 percent along longitude 92° W. The anchovy catch crashes, and sometimes, as in 1982–1983, fur seal pups die, and the Christmas Island seabird reproduction fails (18). As the anomalies drift westward, the easterlies die at midyear, and the ENSO decays over a period of months (15).

The ENSO of 1982–1983 was of the rarer type (the last previous one was in 1940–1941). It began in the Pacific and spread east to the coast; it started in May 1982 (fall in the Southern Hemisphere), not in the late spring around Christmas. In the western Pacific, the Indonesian low had moved north to the equator to meet the intertropical convergence zone as it moved south (18). The intertropical convergence zone (ITCZ) is the boundary between the northeast and southwest trade winds (see Figure 23.2). As the convergence zone moved across the warm water, they intensified each other. Warm wet air rose, supplying latent heat to the air, causing winds that brought in warm water from far away (15,18).

The wind change allowed the warm water that had been piled up in the western Pacific to head back along the equator toward Peru, where it covered the cool layer and pushed it down, causing the rise in sea surface temperature (15). In the 1982–1983 episode, the sea surface temperature was 7°C higher along the coast than normal, and in the central Pacific it was 5°C higher than normal (19).

In this episode, there was a drought in Indonesia, eastern Australia, and Melanesia; rainfall in Peru was the heaviest in 450 years, and the southwestern United States had one of the wettest winters on record (15). Many fish perished or failed to reproduce as primary production fell to 5 percent of pre-ENSO values (20).

The surface temperature is higher than that by 32°C because of reradiation from the carbon dioxide, water vapor, methane, and other atmospheric trace gases (13).

The Oceans' Weather Connection

We mentioned that there is a teleconnection between the South Pacific's weather and that in midwestern North America. This occurs through a weather pattern called the *Southern Oscillation,* which is a seesawing of the coupled South Pacific ocean–atmosphere system first described by Sir G. T. Walker in the 1920s. Tahiti, in the central Pacific, and Darwin, Australia, are at opposite ends of the Southern Oscillation (15). As put by Walker (16):

> When pressure is high in the Pacific Ocean it tends to be low in the Indian Ocean from Africa to Australia; these conditions are associated with low temperatures in both these areas and rainfall varies in the opposite direction.

Actually Walker found three oscillations, one in the North Atlantic, one in the North Pacific, and the Southern Oscillation.

Modern studies of the oceans involve determination of sea surface temperature and atmospheric pressure. The sea surface temperature is highest in the western Pacific of any of the large ocean basins, and the central equatorial Pacific is the key region in ocean–atmosphere coupling (15).

Under normal conditions, the ocean off Peru exhibits cool temperatures (due to upwelling from the Humboldt current, a cold current in the upper ocean that flows toward the equator along South America) and has a lower sea level than in the western Pacific because the trade winds blow water westward, a shallow surface mixed layer, and a slow temperature decrease with increasing depth (15,17). The friction of the current causes an upwelling of cold water, which brings nutrients up from below and leads to high biological productivity of various fish species (17).

An interesting ocean temperature phenomenon known as El Niño occurs every three to seven years, usually around Christmas time, off the Peruvian coast (*El Niño* means "the child" in Spanish and is a reference to the birth of Christ). Fishermen notice it because of the failure of their catches during these occurrences. The appearances of El Niño are connected to anomalies in the Southern Oscillation and the seasonal migration of the intertropical convergence zone (15,17–19), so the events are now called ENSO for El Niño–Southern Oscillation.

The ENSO is the most obviously dramatic weather event involving the ocean. Hurricanes and typhoons are driven by similar mechanisms of uplift and deposition of latent heat energy. The circulation of heat through the ocean is important for mitigating large temperature variations in the mid-latitudes. The oceans act to damp rapid changes (12,21) to the benefit of humanity.

OUR CLIMATE

The climate is described by the mean and the variability of quantities measuring variables of the atmosphere, oceans, and land masses, such as temperature, rainfall or snowfall, or extent of snow cover. Climate depends on the region considered and the time interval chosen. We think of our current climate as normal. But climate is not permanent, and our "normal" everyday climate is not normal for the Earth (22). We are in a time of rather benign climate, one which has existed for only about 10 percent of the last 2 million years (23,24). We live in what is called the Holocene interglacial, between two glacial ages.

To illustrate how precarious our position is, how variable the climate is, consider the cases of the "Middle Ages optimum" and the "Little Ice Age." In the Middle Ages optimum (900–1050 A.D.) (13), the Vikings settled on an Iceland that was about 60 percent forested, ice free, and could grow its own grain. They even sent a colony from the Iceland colony to Greenland. The Caspian Sea was 32 m lower than it is today (13). At this time, the world's mean temperature was about 1°C greater than now (13). In the "Little Ice Age" (1550–1850), the world's mean temperature was about 1°C less than now (13,25). The Icelandic colony in Greenland disappeared (9,13,22,24), and Iceland itself was locked in pack ice for six to nine months a year (13). There were general famines in Iceland, Norway, and Finland (13,24–27) and crop failures and political unrest in much of Europe. Breughel and other Dutch artists painted icy winter landscapes in Holland (Figure 23.6). Glaciers in the Alps, Scandinavia, and Alaska were at their greatest extents in thousands of years (13).

As part of the effect of a previous cold period, Great Plains Indians abandoned their villages about 1200 A.D., and Mesa Verde Pueblos left their homes in about 1300 A.D. due to a long-term drought in the Great Plains (28). The tree line in Canada moved from north of today's to south of today's (28). As late as the 1850s, Taos, New Mexico, had the rainfall that Madison, Wisconsin, has nowadays (28).

Furthermore, climate is slowly changing elsewhere in the world. There is evidence for a steady progressive desiccation in the sub-Saharan zone of Africa from at least 1800 (29). The droughts of the 1970s and 1980s may be part of the continuation of a long trend that will presumably get worse.

How can we know the temperature record without time travel? We have several methods: some are chemical; some involve the study of plankton and pollen; and some are models that can be checked against data generated in the first two methods. As an example of a chemical method, consider the fact that oxygen has two stable isotopes, ^{16}O and ^{18}O. The ratio of ^{18}O to ^{16}O in a sample is an indicator of the global total of glacial ice. Because ^{18}O is heavier, ^{16}O is preferentially evaporated and so is deposited preferentially on ice sheets (30). Figure 23.7 (30) shows two different determinations of the isotope composition of ocean floor sediments. Note that different analyses in different parts of the world agree on the main features. While there is some disagreement on exactly how sensitive a measure it is (since a shift in precipitation could affect the results), the general features have been accepted by everyone (30). Ice cores from Greenland and Antarctica have been analyzed for the ^{16}O to ^{18}O ratio (see Figure 23.8c [31]) in a similar way.

Data from analysis of plankton and pollen are shown in Figures 23.8 a,b,d, and 23.9 a,c (31). Plankton speciation changes as temperature changes. By studying the changing numbers of fossils of various groups deposited in sediments,

FIGURE 23.6
Breughel painted scenes of Holland winters which indicate that the climate was more severe in the sixteenth century. (P. Breugel, *The Hunters in the Snow;* Kunsthistorisches Museum, Vienna)

FIGURE 23.7
Different determinations of composition of ocean-floor sediments versus age of samples. The two independent sets of data are remarkably similar. (C. Covey, reproduced by courtesy of the International Glaciological Society, from *Annals of Glaciology,* 5, 43 [1984].)

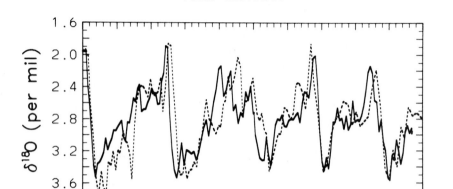

one can infer the water temperature. In a similar way, plant pollen reflects temperature conditions at some particular location.

The data developed by the CLIMAP (Climate: Long-range Investigation, Mapping, and Prediction) Project can then be compared to the models of glacial climate, to see how realistic the data are. Figure 23.10 (32) shows the results of the

calculations for 18,000 years ago at the end of the last glacial period. The resemblance between calculated and reconstructed sea surface temperature is striking. It means that the models of the atmosphere used to do the calculations are reasonable. The temperature in the Greenland ice sheet was $-32 \pm 2°C$ during the last ice age, about 12°C colder than now (33).

FIGURE 23.8
Estimated climate records for the last 135,000 years. (National Academy of Sciences). (a) Estimate of the summer sea-surface temperature. (b) Percentage of tree pollen among all pollen in a Yugoslavian lake. High pollen counts occur under warmer, drier conditions. (c) Air temperature above the Greenland ice cap is reflected in oxygen-isotope ratios. (d) Negative values accompany ice cap melting, indicating warmer climate. (e) Generalized estimates of sea level. High levels indicate warmer periods.

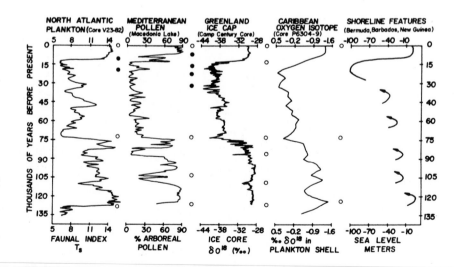

FIGURE 23.9

Estimated climate records for the last 1,000,000 years. (National Academy of Sciences). (a) Oxygenisotope curves reflect global ice volume. (b) Percentage of calcium carbonate ($CaCO_3$) in Pacific cores. Low values are associated with dissolution of bottom waters and warmer climate. (c) Glacial ages are associated with measures in salinity levels and numbers of plankton preferring high-salinity water. (d) Soil types of Brno, Czechoslovakia: examples range from loess soils, which indicate extreme cold (type 1), to soils of temperate savannas (type 5).

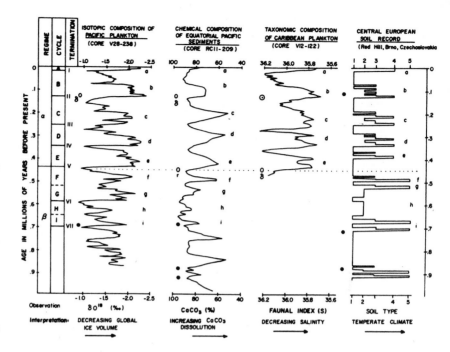

FIGURE 23.10

February monthly mean sea surface temperature differences in kelvin. Top: observed difference. Bottom: reconstruction by CLIMAP. (Manabe and Broccoli, reproduced by courtesy of the International Glaciological Society from *Annals of Glaciology,* 5, 100 [1984].)

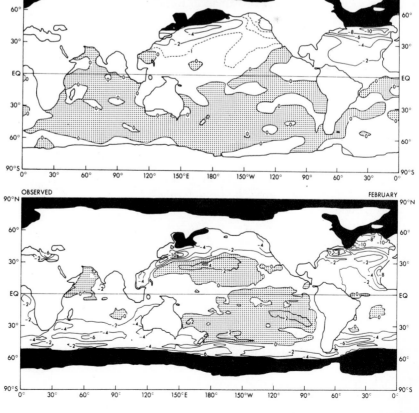

The Physics of the Atmosphere

The actual global climate system is made up of the interaction of the atmosphere, the hydrosphere (mainly the oceans), the cryosphere (the polar regions), the surface of the Earth, and the biosphere. The only outside influences are the sun's radiation and the action of gravity. In the old days of climate modeling, only the atmosphere was considered. Later models incorporate the essential effect of the oceans and the important effects of the cryosphere (12).

The physics is inserted by first looking at the important variables (for example, the atmospheric density about a particular point in space within the atmosphere, the velocity of this packet, the pressure at that point, the temperature at that point, and so on). The physical laws are then applied to a little packet of air and the equations developed to express mass conservation, Newton's second law, the first law of thermodynamics, the water vapor balance equation, and the description of the air as an approximation to an ideal gas. For an ideal gas, the pressure divided by the product of the density and temperature at any point is a constant number.

In addition, the forces in Newton's second law vary with latitude because of the Earth's rotation. The trade winds arise because air traveling south from the north (in the Northern Hemisphere) is deflected east as the Earth appears to try to rotate out from underneath the air.

The resulting equations (except for the ideal gas equation) involve changes in the properties of air at each point (they are called differential equations), and every point in the atmosphere is connected to every other point. Furthermore, the equations are nonlinear, so there are no simple closed-form solutions. Solving these equations (even crudely) demands large amounts of computer time. The initial values of the variables are set, for instance, by using remote sensing stations scattered around the globe. The solution of the equations is the global weather.

As previously discussed, the calculations for the world's weather are based on a coarse grid. Phenomena smaller than the grid size but that influence the weather must be represented in the calculations if the state of the atmosphere is to be described (3,12,34). The weather is difficult to predict because of the extreme sensitivity of the system to initial conditions; these conditions cannot be exactly known because of missing data, and small errors can grow to be very large ones because of the nonlinear character of the equations.

The atmospheric heat engine is driven by the mismatch in energy received at the equator and the poles. Most of the energy taken to the polar regions is carried by water's latent heat of vaporization. Because the specific heat of water is much greater than that of air, the oceans store roughly twenty times as much heat as the air and minimize extreme changes (3,12).

The winds blow harder in the winter. As a result, the Earth is, in a manner of speaking, pushed backward, and its angular momentum varies (that of the system Earth plus atmosphere is constant). For example, the air can push on mountains, forcing a change in rotation rate. The meterological conditions cause the day to be longer in January than in July (by 0.7 ms) (12).

Causes of Climate Change

Now that the record shows obvious climate changes, we want to find out what the causes for these changes might be. The causes may be classified generally as cosmological, geological, and anthropogenic (or human-caused). By cosmological change, I mean change in which the Earth is influenced by extraterrestrial agencies. By geological change, I mean changes in the Earth, as in, for example, the effects of continental drift on land distribution. Finally, in Chapter 24, we shall examine the evidence for climate change caused by our activities and that of our ancestors. Despite this classification, it is not entirely possible to segregate the causes; there is always cross-category feedback.

Cosmological Change
Changes in the Solar Constant:
Possible cosmological causes are many. The solar system could be passing through stationary dust clouds (35,36), or the solar constant (see Chapter

2) may be *in*constant (37). The number of sunspots (magnetic storms on the solar surface) changes regularly in eleven-year cycles. Figure 23.11 (36) shows the sunspot number as it has changed during the course of over two hundred years. Droughts in the midwestern United States seem to be correlated with the number of sunspots (9,26). London annual mean temperature may be determined by sunspot number; and mean rainfall seems to be correlated as well (35,38). While there is no acceptable physical mechanism for short-term changes in the sun to influence weather, especially since the changes on this scale are small relative to total output, the influence

appears to have some observational bases (36). The number of sunspots, however, seems to be connected to secular increases and decreases in the solar constant of as much as 1 to 2 percent (9,36,39). It is probable that a 1 percent change in the solar constant would cause the polar caps to melt; a 2 percent change in the other direction would cause glaciation (12).

There is a period, called the Maunder minimum (1645–1715), during which sunspots seem to have virtually ceased to be observed. There is some evidence that Chinese astronomers continued to observe sunspots during this period, but no evidence has been found [for phenomena

FIGURE 23.11
Variations in annual mean Zürich sunspot number. (Herman and Goldberg, Ref. 36)

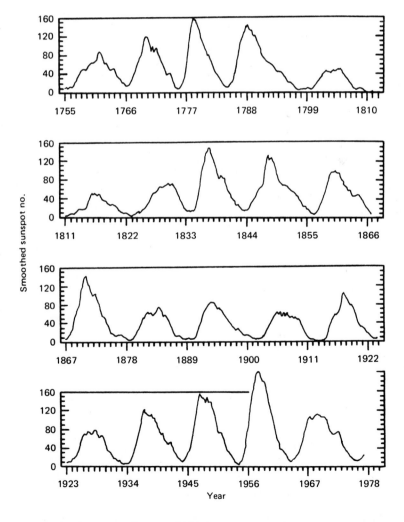

such as the Maunder minimum are reputed to be from records of the deposition of sandstone in Australia 700 million years ago]. The eleven year cycle is seen [39]). If the Maunder minimum exists, it may be an important contributor to climatic change. (Note that the Maunder minimum occurred during the 1550–1850 coolest part of the Little Ice Age.) The predicted temperature variations are too small to explain the whole of the surface temperature anomaly (39,40). The record of ^{14}C production as seen in tree rings is reflective of the number of sunspots. The 860 year record shows that there may be at least two other minima in sunspot activity—the Spörer minimum (1416–1534) and the Wolf minimum (1282–1342) (41).

Cosmological Change
Milankovitch's Cycles:

It was an old idea, resuscitated by Miltun Milankovitch in 1930 (42), that variations in the Earth's orbit may cause changes in insolation. The changes are too small (there is a variation due to eccentricity, for example, of only 0.3 percent over the last million years [43]) to believe that they can actually cause ice ages. Milankovitch met these objections by suggesting that the insolation at high latitudes, which changes by 20 percent, could cause great changes in climate by affecting the amount of ice and snow cover in the Northern Hemisphere.

The Milankovitch theory (41–45) holds that changes in the orientation of Earth's rotation axis and its orbital shape fix the timing of the glaciations by changing the distribution of sunlight on the Earth. It is now generally agreed that the effect is the primary cause of glacial–interglacial changes in an ice age climate because of development of recent evidence supporting the theory. The Milankovitch theory predicts a glacial episode to occur in four to five thousand years (46,47).

The Earth's rotation axis precesses (changes direction) with respect to the fixed stars with a period of 21,000 years. (This is known as the precesssion of the equinoxes, and is the reason for the "Ages" of the Zodiac—we are now entering the Age of Aquarius.) This change is im-

portant because it fixes the points on the orbit at which summer and winter occur. The Northern Hemisphere's winter now occurs near the Earth's closest approach to the sun, moderating its severity. At midlatitudes, this effect is equivalent to a change of 10 percent in intercepted sunlight (48). A 23,000 year cycle was found in the deposit of carbonaceous sediment (the remains of plankton) in the Pacific (45). Plankton should be more numerous when the sun shines more and less numerous when the sun shines less. Hence, the density of plankton skeletons is greater when more sunlight is available, and this is detected in sediment.

The axial tilt or obliquity, now 23.5°, changes from 22.1° to 24.5° with a period of 41,000 years. There are indications of a 42,000 year cycle in fossil pollen from old lakebeds (44). The rise and fall of very large lakes leaves sedimentary evidence of past climates. An analysis covering 40 million years (49) found a 44,000 year cycle corresponding to tilt (as well as 25,000, 100,000, 133,000, and 400,000 year cycles). Increasing the tilt causes colder winters and hotter summers. Current winters are slightly milder than mean conditions (43).

The *eccentricity,* which indicates how elliptical the Earth's orbit is, changes in a 105,000 year cycle. The Earth's eccentricity is now 0.017, and it varies between 0.005 and 0.06 (43). The greater the eccentricity, the more pronounced the effects of the other changes. Studies of coral terraces indicate high sea levels 80,000, 105,000, and 125,000 years ago (44).

In 1976, a study based on use of two deep sea cores to measure ^{18}O (which together represent 468,000 years of history), showed the existence of periodic signals of 105,000 years, 41,000 years, 23,000 years, and 19,000 years (44). The 100,000 year cycle, which was supposed just to modulate the others, was the *major* feature of the climate record; it explained 50 percent of the observed variance. The 40,000 year cycle explained another 25 percent of the observed variance, with 10 percent explained by the 23,000 year cycle. The work of Hays, Imbrie, and Shakleton established the existence of the 100,000 cycle in sea sediment cores (44,50), and later workers have seen this cycle as well (48,49,51).

The 100,000 year cycle should not cause the variations observed if the couplings are linear, since the eccentricity should merely in that case change the size of the precession effects. The changes in these three quantities (obliquity, precession index, and eccentricity) are shown in Figure 23.12 (36).

A breakthrough in understanding came in 1980, when the Imbries (50) showed that a very simple model caused sufficient nonlinearity to give a good simulation to the data. They chose to look at polar ice sheets, because that is the only obvious part of the climate system for which response times are anywhere near those from the

orbital driving forces. Ice in glaciers takes more time to accrete than to melt, which introduces delays into the interactions. As the Imbries put it:

Consider the radiation–climate system as a pot of water over fire, with the heat input varying over time. If the input were constant, the water would approach an equilibrium with time constant, T (50).

There is a lag of 9 ± 3000 years between the orbital input and the 41,000 year component of the response. In their fine-tuned simple model (with only four parameters describing the eccentricity, obliquity, precession, and time delays), the Imbries found a warming time delay of 10,600

FIGURE 23.12
Changes in the precession of the equinoxes over the last 500,000 years (Herman and Goldberg, Ref. 36). (a) Percentage deviation from the mean June 21 sun-Earth distance. (b) Obliquity of the ecliptic, or axial tilt (degrees). (c) Orbital eccentricity.

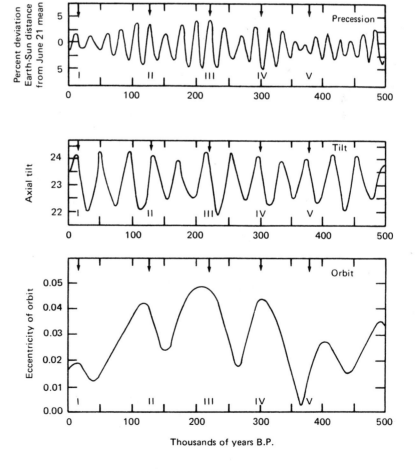

Thousands of years B.P.

years and a cooling time delay of 42,500 years. Thus, computer models seem to be verifying Milankovitch's guess.

There are still conceptual problems. Why is there a sudden onset of ice ages 65 million years ago? Why does the Imbrie model give too low a power to the 100,000 year cycle and too much to a 400,000 year cycle? Some guesses have been advanced in addition to that of the ice sheet extent. There may be a resonance due to the fact that bedrock subsides under the weight of ice sheets and rises when the sheets vanish (isostatic uplift). If there is a 10,000 year response time, the 100,000 year cycle will be enhanced (43).

Recent indications are that carbon dioxide levels are intimately involved in the Milankovitch approach to climate (47,51,52,53). Changes in carbon dioxide levels lag behind changes in orbital geometry by about 5000 years and precede changes in ice volume (48,52). This seems to be in accord with models of the atmosphere, which give 3000 year delay times (53). Adding the effects of carbon dioxide to the "pure" Milankovitch predictions improves the agreement between prediction and actual data tremendously (48).

The minor questions remain, but the idea has been a great success. In the next chapter, we will discuss yet another effect enhancing the 100,000 year cycle. We must now turn to ways ice ages can begin, which can be the initial condition feeding into the Milankovitch theory.

Geological Changes   Geological changes include isostatic uplift (discussed just above), mountain-building and continental uplift, changes in the latitudes of continents, changes in the characteristics of the ocean, volcanism, and alterations in the circulation of material in the core. Some 200 million years ago, there was only one continent on Earth. It fractured, and the pieces are still moving about on the plates making up the Earth's outer surface (see Chapter 12). It is clear that continental masses at high latitudes are heavily glaciated (Antarctica, Greenland). The continents that drifted poleward became more susceptible to glaciation.

The Earth reflects on average about 30 percent of the incident sunlight back into space; we say it has an albedo of 30 percent. Pack ice and snow has an albedo of 80 percent, while the ground the ice covers has an albedo of only about 20 percent. The mean annual coverage of the poles by ice has been about 30 Mkm^2 in recent times (54). In the last full glacial, mean annual coverage was 60 to 70 Mkm^2 (54,55). Satellite monitoring of the ice cover has revealed that the mean annual coverage increased from 32.9 Mkm^2 in 1970 to 36.9 Mkm^2 in 1971. This area remained enlarged for several years. Only seven years more of such an increase might have reestablished the glacial surface albedo and could have initiated or accelerated the change to a glacial era (25,54) in our "marginally interglacial" climate. The snow and ice not only reflect away incoming infrared radiation, but also consume substantial latent heat as they melt, and reduce the energy and moisture exchanged between the atmosphere and the surface insulated from it by the overlying ice (4,56). The energy deficit from snow is in the lowest 2 km of the atmosphere (4). Figure 23.13 (57) shows that the increase was not sustained; nevertheless, it is possible that, without the human-caused effects discussed in the next chapter, we might have been entering another glacial age. The data shown were obtained from satellite monitoring using microwaves (because those wavelengths are not affected by the omnipresent cloud cover). Clouds are also very important in cooling, since they contribute over half the Earth's albedo (58). We have already mentioned how important water vapor is in warming, so the tradeoffs must be examined closely.

It has been shown (8,59) that, over the past 100,000 years, the temperature over the Greenland and Antarctic ice sheets decreases as the amount of volcanic dust falling on them increases. The measured strength of sunlight shows the effects of great volcanic eruptions (59,60). The most stunning historical example is of the year without a summer: 1816, which was known to those who lived through it (61) as "eighteen hundred and froze to death." Prices of grain rose

FIGURE 23.13
Seasonal cycle of the extent of Antarctic sea ice. (H. J. Zwally, reproduced by courtesy of the International Glaciological Society from *Annals of Glaciology*, 5, 191 [1984].)

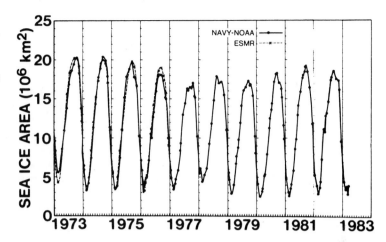

substantially in both America and Europe in the wake of crop failures (made worse in Europe because of the just-ended Napoleonic Wars). There is also some evidence (61) that the world's first cholera pandemic was partly due to what happened when the volcano Tambora on Sumbawa put around 100 cubic kilometers of dust into the atmosphere and caused a worldwide temperature drop of 1°C (24).

Does dust always cause a temperature drop? Atmospheric scientists continue to make models of atmospheric behavior, which they test by comparing model predictions with actual measurements. Model calculations show that small particles tend to cool the Earth's surface, and big particles, to warm the surface for normal atmospheric aerosols. Calculations based on the model of the 1963 Mt. Agung eruption agree very well with the observations (62), which gives modelers some confidence that they are on the way to understanding these phenomena.

The big particles in the air tend to settle out quickly; while they are present, they cause warming, but they soon disappear. Small particles in the stratosphere have a mean time of stay of 0.5–1 year (62). Since big particles of dust quickly leave the atmosphere, and small particles remain for long times, then dust particles injected by volcanic action or air circulation (such as a hurricane or nuclear explosion in the atmosphere)

cause the Earth's surface to cool and the stratosphere to become warmer (62,64).

These models were often tested in simulations of Mars' atmosphere, since the weather on Mars is less complex than that on Earth because of the absence of clouds and exposed bodies of water. Comparisons of model calculations and the results of lengthy Martian dust storms, in which surface temperatures plunge, led to the development of the idea of nuclear winter (see box).

The lesson here is clear: The more dust and particulate matter in the atmosphere, the greater the drop in temperature. Thus, a sustained siege of volcanic activity could trigger a glacial episode. Evidence that this is the case comes from the Deep Sea Drilling Project (67), which found that the rapidly oscillating climatic conditions of the last 2 million years is synchronized with eras of greatly increased volcanic activity.

Other evidence abounds from the Roman era to today. Greenland ice cores show that there were three years of acidic fallout about 50 B.C., at a time when histories of Rome report a dimming of the sun after Caesar's assassination (68). The sulfur released by volcanoes can cause increases in levels of sulfuric acid in the atmosphere. The sulfur and other volcanic aerosols injected into the stratosphere during a volcanic eruption can cause global cooling (62). Ice cores from 1601 and 1602, during the depths of the

EFFECTS OF DUST

In order to study the effect of dust (or aerosols, as such particles are called), it is necessary to examine three properties of the atmosphere (62): the *optical depth,* which is a measure of the size and number of particles present in a column of air; the *single scattering albedo,* which is a measure of the fraction of light intercepted and scattered by a single particle; and the *asymmetry parameter,* which indicates the portion of the light scattered forward and the portion scattered backward. The intensity of light traversing an aerosol-laden air column is attenuated (reduced) exponentially as the ratio of distance traveled to optical depth. For most atmospheric particles, the single scattering albedo for visible light is 0.9 to 1.0; in the infrared it is about 0.1 (62). For high values of the single scattering albedo, aerosols cause cooling. As the single scattering albedo is lowered, progressively smaller cooling effects are seen. If it is lowered still further, aerosols begin to cause heating. In laboratory studies on air, the single scattering albedo is large enough that aerosols cool; direct studies in the troposphere, however, show that the value of the single scattering albedo is so low that aerosols warm (56). It is at its lowest in urban areas. Particles of soot in the troposphere (of anthropogenic origin) absorb four times as much light as expected from theoretical calculations (63).

The asymmetry parameter is +1 if all light is scattered forward, and −1 if all light is scattered backward. In the visible region, the asymmetry parameter is about 0.7, while in the infrared it is less than about 0.5 (62).

Scientists measure these parameters in the laboratory and in the atmosphere to use as input to calculations of the effects of dust in the atmosphere. The feedback between experiment and theory helps meteorologists make improved analyses of atmospheric effects.

Little Ice Age, show sharp peaks of sulfuric acid, too (68), indicating increased volcanic activity.

In our century, there were a much greater number of eruptions from 1900 to 1925 and 1955 to 1980 than from 1925 to 1955 (69). Between 1945 and 1970, the average number of eruptions per year doubled (69). If the optical depth of the atmosphere increased enough to reduce the direct beam at the Earth's surface by 5 percent, the resulting increase in diffuse radiation would cause a cooling of −0.85°C. The increasing carbon dioxide levels (see Chapter 24) should have caused an increase of ≈0.35°C; the combined result of −0.5°C compares well with the measured change of −0.3°C observed between 1945 and 1975 as the measured value of the solar constant decreased 5 percent because of particulate emission by volcanoes and human activity (67).

Radiocarbon dating results suggest increased volcanic activity 600 years ago (Little Ice Age), 1200 years ago, 3700 years ago (coincident with a long-term failure of the Indian monsoon, which is tied to the Southern Oscillation), and 8000 years ago (a time of ice stillstand) (69).

More recent evidence comes from the eruption of El Chichón in Mexico. The cloud was produced in spring of 1982 and, in contrast to that of Mt. St. Helens, spewed up into the stratosphere beginning between 22 and 26 km high and spreading to an altitude of 32 km (70). It

MARTIAN WEATHER AND NUCLEAR WINTER

The Martian dust storms observed by *Viking* landers and the fly-bys led to development of better models for the reaction of the atmosphere to airborne aerosols (62,63,65), which are important for studying particulate pollution as well. A paper authored by scientists who studied the Martian atmosphere, Professors Turco, Toon, Ackerman, Pollack, and Sagan (65), best known as TTAPS, asserted that a nuclear exchange could put aerosols into the stratosphere, from both the explosion itself and the smoke from the firestorms expected to follow a detonation.

The TTAPS study indicated that even small (or "limited") nuclear exchanges could produce enough cooling to decrease the temperature on the surface substantially. The threshold for this effect may be at yields as low as 100 megatons.

Their baseline for comparison was a 5000 megaton exchange. Such a large-scale nuclear war could be followed in their model by plunging surface temperatures—a 25°C drop in summer surface temperature. This could lead to widespread freezing and starvation of people and animals. The cloud cover would remain for months, and temperatures would stay below normal for over a year. Combined direct effects and subsequent effects lead the TTAPS group to question the survival of humanity itself in the case of nuclear war, since the aerosols would be transported worldwide.

Studies done since have tended to agree that mankind faces severe problems if nuclear weapons are used. More realistic three-dimensional models (34) appear to give results that are not so severe as the original TTAPS estimate—only a 10°C to 15°C drop in temperature is predicted for a war in summer. Critics had found problems with TTAPS updraft transport of particles, but use of tested thunderstorm models indicate that about 40 percent of particles produced could be transported into the lower stratosphere, where they have long residence times (64). Updrafts would be twice those found in normal thunderstorms, and downdrafts, almost ten times as strong; both are at heights far exceeding those involving thunderstorms: 8–10 km instead of 2 km. Other workers have found uncertainty in the amount of material that would actually be burnt in a nuclear exchange; the severity of the "winter" depends on this sensitivity (66).

even spread south of the equator, warming the stratosphere by 3°C (71). Unfortunately, the El Chichón eruption coincided with the appearance of the 1982–1983 El Niño, so that climatic effects were not so clear cut as they would have been had the two events not happened at once. It might be noted that the sea surface was 1°C cooler than usual. Since the El Chichón eruption was relatively small, the cooling effect would also be expected to be relatively small.

SUMMARY

Weather describes the local atmospheric state over short periods of time. Climate describes the state of the annual and seasonal average weather; these averages are stable for long times. Weather is driven by the sun's energy input and the difference between insolation per unit area of the poles and the equator. The energy flux of the Earth is in long-term balance—as much is radiated away by the Earth as is absorbed, or the mean temperature would have to increase or decrease steadily (and, of course, this is not observed).

Some solar energy is absorbed directly by the oceans and some is transferred by the winds to the ocean. The interplay between atmosphere and ocean over long distances is evidenced by ENSO events. The large heat capacity of the oceans helps to damp rapid fluctuations in weather and climate.

Climate has changed over periods of hundreds of years. The mean temperature in the Little Ice Age was only 1°C cooler, but there were large effects felt, especially toward the poles.

The movement of continents (which is driven by the motion of the plates) from equator to pole or vice versa can cause climate change. The solar constant, if it changes, can cause climate change. The Milankovitch cycles have clearly caused climate changes, even though the exact mechanism enhancing the 100,000 year cycles is still a subject of scientific debate.

Volcanic activity injects aerosols into the upper atmosphere. These aerosols lead to a lower surface temperature. Hence, climate is affected by agencies outside the atmosphere itself. In the next chapter, we consider the consequences of human intervention in the atmosphere–climate system.

REFERENCES

1. Anonymous, *New York Times,* 11 Jan. 1981; *New York Times,* 5 Jan. 1986.
2. T. H. Maugh II, *Science* 220 (1983): 39.
3. J. J. Tribbia and R. A. Anthes, *Science* 237 (1987): 493; R. A. Kerr, *Science* 228 (1985): 40.
4. J. E. Walsh, *Am. Sci* 72 (1984): 50; R. A. Kerr, *Science* 216 (1982): 608.
5. R. Reiter and R. Sladkovic, *Arch. Met. Geoph. Biocl.* A33 (1985): 297.
6. B. Reinhold, *Science* 235 (1987): 437.
7. Department of Commerce, *Statistical Abstract of the United States, 1984* (Washington, D.C.: GPO, 1984).
8. R. A. Bryson, *Science* 184 (1974): 753.
9. S. H. Schneider, *The Genesis Strategy* (New York: Plenum, 1976).
10. R. A. Bryson, *Ecologist* 6 (1976): 205.
11. A. L. Hammond, *Science* 191 (1976): 455; and J. W. Sawyer, *Environment* 20, no. 2 (1978): 25.
12. J. P. Peixoto and A. H. Oort, *Rev. Mod. Phys.* 56 (1984): 365; W. L. Gates, in *Projecting the Climatic Effects of Increasing Carbon Dioxide,* ed. M. C. MacCracken and F. M. Luther (Department of Energy, DOE/ER-0237, 1986), 57ff; and M. E. Schlesinger, also in *Projecting the Climatic Effects,* p. 281ff.
13. U. von Zahn, *Alexander von Humboldt-Stiftung Mitteilungen* 39, no. 9 (1981): 15.
14. R. E. Newell, *Am. Sci.* 67 (1979): 405.
15. E. M. Rasmussen, *Am. Sci.* 73 (1985): 168.
16. G. T. Walker and E. W. Bliss, *Mem. Royal Meteorol. Soc.* 4 (1932): 53.
17. R. T. Barber and F. P. Chavez, *Science* 222 (1983): 1203.
18. C. S. Ramage, *Sci. Am.* 254, no. 6 (1986): 77; and R. A. Kerr, *Science* 216 (1982): 608, and *Science* 221 (1983): 940.
19. M. A. Cane and S. E. Zebiak, *Science* 228 (1985): 1085.
20. R. T. Barber and F. P. Chávez, *Nature* 319 (1986): 279.
21. T. P. Barnett, in *Climatic Change,* ed. J. Gribben (New York: Cambridge Univ. Press, 1978).

22. J. Gribben and H. H. Lamb, in *Climatic Change,* ed. J. Gribben (New York: Cambridge Univ. Press, 1978).

23. J. J. Walsh, *BioScience* 34 (1984): 499.

24. S. W. Matthews, *National Geographic* 150 (1976): 576.

25. W. W. Kellogg and S. H. Schneider, *Science* 186 (1974): 1163.

26. J. Norwine, *Environment* 19, no. 8 (1977): 7.

27. C. B. Beaty, *Am. Sci.* 66 (1978): 452.

28. R. A. Bryson, *Nat. Hist.* 89, no. 6 (1980): 65.

29. V. A. Todorov, *J. Clim. Appl. Met.* 24 (1985): 97; D. Winstanley, *Weatherwise* 38 (1985): 74.

30. C. Covey, *Ann. Glaciology* 5 (1984): 43.

31. Panel on climatic variations, U.S. Committee for the Global Atmospheric Research Program, National Research Council, *Understanding Climatic Change* (Detroit: Grand River Books, 1980).

32. S. Manabe and A. J. Broccoli, *Ann. Glaciology* 5 (1984): 100.

33. D. Dahl-Jensen and S. J. Johnson, *Nature* 320 (1986): 250.

34. S. H. Schneider, *Sci. Am.* 256, no. 5 (1987): 72.

35. D. H. Tarling, in *Climatic Change,* ed. J. Gribben (New York: Cambridge Univ. Press, 1978).

36. J. R. Herman and R. A. Goldberg, *Sun, Weather, and Climate* (Detroit: Grand River Books, 1980).

37. S. Sofia, P. Demarque, and A. Endal, *Am. Sci.* 73 (1985): 326.

38. J. A. Eddy, *Science* 192 (1976): 1189, and *Sci. Am.* 236, no. 5 (1977): 80; and R. A. Kerr, *Science* 231 (1986): 339.

39. S. H. Schneider and C. Maas, *Science* 190 (1975): 741; W. W. Kellogg, in *Man's Impact on the Climate* ed. W. H. Matthews, W. W. Kellogg, and G. D. Robinson (Cambridge, Mass.: MIT Press, 1971).

40. G. E. Williams, *Sci. Am.* 255, no. 2 (1986): 88.

41. M. Stuiver and P. D. Quay, *Science* 207 (1980): 11.

42. C. Covey, *Sci. Am.* 250, no. 2 (1984): 38.

43. M. Milankovitch, in *Handbuch der Klimatologie,* ed. W. Koppen and R. Geiger (Berlin: Borntraeger, 1930).

44. J. D. Hays, J. Imbrie, and N. J. Shackleton, *Science* 194 (1976): 1121.

45. R. A. Kerr, *Science* 201 (1978): 144.

46. T. D. Herbert and A. G. Fisher, *Nature* 321 (1986): 739.

47. W. L. Gates and M. C. MacCracken, in *Detecting the Climatic Effects of Increasing Carbon Dioxide,* ed. M. C. MacCracken and F. M. Luther (Department of Energy, DOE/ER-0235, 1986), 1ff.

48. N. G. Pisias and J. Imbrie, *Oceanus* 29, 4 (1986): 43.

49. P. E. Olsen, *Science* 234 (1986): 842.

50. J. Imbrie and J. Z. Imbrie, *Science* 207 (1980): 943.

51. R. A. Kerr, *Science* 234 (1986): 842.

52. R. A. Kerr, *Science* 234 (1986): 283.

53. D. A. Short and J. C. Mengel, *Nature* 323 (1986): 48.

54. G. J. Kukla and H. J. Kukla, *Science* 183 (1974): 709; W. Sullivan, *New York Times,* 22 Feb. 1976.

55. W. L. Gates, *Science* 191 (1976): 1138.

56. G. J. Kukla, in *Climate Change,* ed. J. Gribben (New York: Cambridge Univ. Press, 1978).

57. H. J. Zwally, *Ann. Glaciology* 5 (1984): 191.

58. E. Raschke and J. Schmetz, *Reports of the Deutsche Forschunggemeinshaft* 1 (1984): 10.

59. J. M. Mitchell, in *Global Effects of Environmental Pollution,* ed. F. Singer (Dodrecht, Neth.: D. Reidel Publ. Co., 1970); and W. H. Matthews, *Intern. J. Environmental Studies* 4 (1973): 283.

60. H. H. Lamb, *Ecologist* 4 (1974): 10.

61. H. Stommel and E. Stommel, *Sci. Am.* 240, no. 6 (1979): 176.

62. O. B. Toon and J. B. Pollack, *Am. Sci.* 68 (1980): 268. See also F. M. Luther and R. G. Ellingson, in *Projecting the Climatic Effects of Increasing Carbon Dioxide,* ed. M. C. MacCracken and F. M. Luther (Department of Energy, DOE/ER-0237, 1986), 25ff.

63. A. D. Clarke and R. J. Charlson, *Science* 229 (1985): 263.

64. W. R. Cotton, *Am. Sci.* 73 (1985): 275.

65. R. P. Turco, O. B. Toon, T. Ackerman, J. B. Pollack and C. Sagan, *Science* 222 (1983): 1283 and *Sci. Am.* 251, no. 2 (1984): 33. This study is known as TTAPS after the first initials of the authors' last names.

66. J. E. Penner, *Nature* 324 (1986): 222.

67. J. P. Kennett and R. C. Thunell, *Science* 187 (1975): 434.

68. W. Sullivan, *New York Times,* 9 August 1981.

69. R. A. Bryson and B. M. Goodman, *Science* 207 (1980): 1041.

70. R. A. Kerr, *Science* 217 (1982): 1023.

71. R. A. Kerr, *Science* 219 (1983): 157.

PROBLEMS AND QUESTIONS

True or False

1. It is unlikely that continental drift has anything to do with episodes of glaciation.

2. Times of great volcanic activity have a colder climate in general than times in which volcanoes are generally quiescent.

3. The albedo of ice-covered ground is higher than that of bare ground.

4. Increasing average temperature could cause cooling locally because clouds (which reflect light) will form.

5. As one goes higher in the troposphere, the air temperature rises.

6. Weather in one place can cause effects in places very far removed from the first place.

7. It is a well-established scientific law that sunspots can affect the weather on Earth.

8. The atmosphere–hydrosphere–cryosphere system responds essentially immediately to any changes (i.e., on a time scale not more than an order of magnitude larger than the event itself).

9. Snow-covered areas often cannot be measured because of extensive cloud cover.

10. There is no modern evidence that geological changes can cause weather changes.

11. Studies of weather patterns on other planets could provide useful data to scientists studying Earth's weather.

Multiple Choice

12. What orbital factors can affect the climate in the Milankovitch theory?
    a. The eccentricity of Earth's orbit
    b. The length of the year
    c. The number of hours in the day
    d. The resonance between the moon's orbit and the Earth
    e. The Coriolis effect

13. In normal El Niño–Southern Oscillation events,
    a. the Peruvian anchovy catch increases.
    b. the storms normally occurring in the ITCZ are much weaker than usual.
    c. the cold water current along the South American coast sinks.
    d. sea surface pressure varies rhythmically.
    e. the duration is approximately one month.

14. If the Earth's absolute temperature were to double, by what factor would the radiated power go up?
    a. 2
    b. 4
    c. 8
    d. 16
    e. 20

15. The main force driving the weather is
    a. the ocean.
    b. the rotation of the Earth.
    c. the ellipticity of Earth's orbit.
    d. the sun.
    e. Zephyrus.

16. The radiation from the sun is mostly in the visible region. The radiation from the Earth is in which region?
    a. Gamma ray
    b. X-ray
    c. Ultraviolet
    d. Visible
    e. Infrared

Discussion Question

17. It has been observed that the twenty-two-year sunspot cycles are correlated with periodic droughts in Nebraska (9,25) (see Figure 23.14). Does this make it more plausible that sunspots can affect climate?

FIGURE 23.14
Correlation between midwestern droughts and Zürich sunspot number. (S. Schneider, *The Genesis Strategy,* copyright 1975 by Plenum Press)

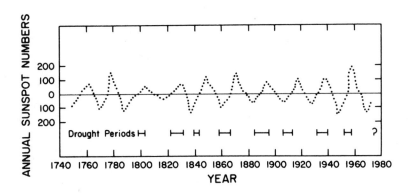

24

CLIMATE CHANGE AND HUMAN ACTIVITY

Knowing the basic non-human-generated causes of climate change from the last chapter, we consider here the effects of the "human volcano." People have historically affected local conditions, contributing to desertification in some regions, to the long-term fertility of soils in other regions, and establishing and perpetuating prairielands in temperate regions.

The industrial revolution led to the burning of fossil fuels in growing yearly volumes. In Chapter 22, we discussed the effects of "pollutants" in the effluent from burning processes. Another component of the effluent—considered harmless until only a few years ago—is carbon dioxide. Humanity has embarked unknowingly upon a long-term experiment in climatic stability. We examine some possible consequences of the experiment in this chapter.

KEY TERMS   *desiccation*
savanna
greenhouse effect
general circulation model
biological pump
glacial surge

HISTORICAL EVIDENCE FOR ANTHROPOGENIC CLIMATE CHANGE

We examine in this chapter the relationship between climate change and human activity (we call human-caused change *anthropogenic* change). In the previous chapter, we discussed weather and climate. We saw that not long ago, the globe experienced a mild ice age that had profound consequences for people living at the higher latitudes. The current mild weather is decidedly not the norm. The Earth is in an ice age configuration (1,2) of 80,000 to 100,000 year glacials alternating with 10,000 to 15,000 year interglacials (2). While we all tend to take our interglacial climate as the way things should be, we are enjoying climatic beneficence, and the stability is extremely fragile (as we saw in considering pack-ice cover in the preceding chapter). It is sobering that the world's ice age climate differed so little from that of today (3,4), while the effect was so striking.

Relatively minor changes could spark the advent of new ice ages. Human activity seems quite capable of supplying minor changes—indeed, we are now shaping the future climate. Perhaps the most important ways we can cause climatic change are by changing the albedo (or amount of light reflected), by changing the amount of dust in the atmosphere, by causing a redistribution of heat, and by increasing or decreasing the amounts of certain gases in the atmosphere (3).

Albedo, or reflectivity, can readily be changed by human intervention. At relatively minor cost, dust or soot could be dumped on the North Pole ice cap, which would cause it to absorb more sunlight and melt (5). But even normal activity can result in albedo change. As Chapter 17 showed, many peoples practice swidden (slash and burn) agriculture. Substantial areas can be put to the torch (Figure 24.1 [6]). It is well known that much of the Great American Prairie was maintained by Indian-set fires (7,8). The prairie reflects much more sunlight than forest does. Evidence for burning's altering the climate was also found in Australia (9).

Overgrazing also alters albedo. Desert expansion, such as has been seen in the recent advance of the Sahara into the Sahel, can be stimulated by increasing the albedo (5,10). An example of the effect of overgrazing in the spread of deserts was seen in 1974 satellite photos of the Sahel showing desert surrounding a fenced-in green area (11). The green area was a ranch with natural local grass; it was operated to produce cattle by a commercial establishment that allowed the land to regenerate after grazing. In neighboring regions, cattle belonging to individuals had eaten every blade, altering the microclimate enough to prevent grass from regrowing.

Sagan, Toon, and Pollack (7) argue that people have influenced the climate by alteration of abundance and distribution of vegetation beginning as early as 20,000 years ago. The name "Tierra del Fuego" (land of fire) was given by Magellan to the southernmost tip of South America because of the many fires set by the natives, burning off all the local bushes and trees. Hanno referred to annual burning south of the Sahara, which was followed by desiccation. Deforestation had destroyed 60 percent of the central European forests since the eleventh century. Traces of Rome's public road building from 171 B.C. caused changes in pollen species seen in lake sediments in Italy, as Italy lost forest cover and started to suffer from floods (12). Italy suffered "an *ecological catastrophe* during ... Antiquity" evidenced in sediments and caused by people (13).

Table 24.1 (7) lists changes in albedo due to desertification, salinization, deforestation, and urbanization. The Rājasthān Desert, now a million km² in area, was once the site of the Indus valley civilization. The desert appears to have been

FIGURE 24.1
Seasonal burns, 1972 and 1973 dry season, in equatorial Africa. (J. Hidore, copyright 1978 by Society for Geographical Education)

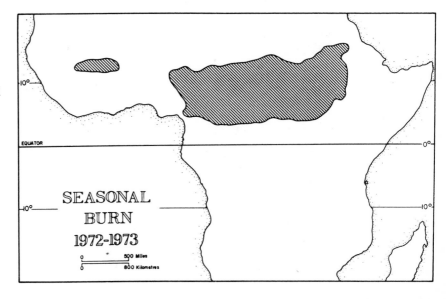

TABLE 24.1
Anthropogenic albedo changes.

Albedo Process	Surface Albedo	Global change	
		Last Few Million Yr	Last 25 Yr
Desertification			
(savanna→desert)	0.16–0.36	0.004	0.0006
Salinization			
(open field→saltflat)	0.1–(0.25–0.50)	0.00015–0.0001	0.00025–0.0004
Temperate deforestation			
(forest→grassland field)			
Summer	0.12–0.15	0.00025	Small
Winter	0.25–0.60	0.0004	
Tropical deforestation			
(forest→field, savanna)	0.07–0.16	0.001	0.00035
Urbanization	0.17–0.15	− 0.0025	− 0.001
Totals		0.006	0.001

SOURCE: C. Sagan, O. B. Toon, and J. B. Pollack, Reference 7. Copyright 1979 by AAAS.

caused by overgrazing, since grass grows rapidly when it is protected from overgrazing. Ancient farming villages are buried in the Sahara (7), and North Africa was the breadbasket of the Roman Empire; ecological change was hastened by human abuse and importation of goats by Arab invaders. Central Anatolia, now semidesert, was once the breadbasket of Byzantium before goat husbandry destroyed the delicate ecological balances and helped cause desertification. All major deserts are subject to a radiation deficit; they reflect so much that they radiate more heat than they absorb (14,15). The Sahara Desert loses about 35 W/m² in radiation; the subsequent sinking of air and its heating supplies this energy (14).

Iraq, which was once Mesopotamia, home of several mighty civilizations, has had 20 to 30 percent of its farmland destroyed by salt (16). The Aswan Dam, in preventing the annual flooding of Egyptian fields, has led to increased salinization there (11).

Irrigation in the American West, a dry region, increases the leaching of salts upward as the water carries the salt away. Irrigation water has caused American rivers to become salt-loaded; the dams built to regulate water flow for irrigation and to supply electricity have also contributed to the salt

problem. Salty water from upstream can harm farmers downstream who try to use the water to irrigate. The United States and Mexico signed a treaty in 1944 to try to regulate the salinity of the Colorado River; a desalting plant has been built under the treaty terms (17).

Deforestation has turned 40 percent of the African equatorial forest into savanna over the last several thousand years (7). The deforestation of the Amazon is occurring quickly (18). The British energy crisis in the mid-sixteenth century was caused by a shortage of wood, and in turning the British to coal, may have sparked the Industrial Revolution (19). Deforestation has a dramatic effect on winter albedo in snow areas: Snow-covered farmland has double the albedo of snow-covered trees (7); over deciduous forests, visible and near-infrared albedos are almost equal; over coniferous forests, near-infrared albedo is greater than that in the visible; and over scrubland and snow-covered fields, the visible albedo is greater than that in the near infrared (20).

Albedo is extremely important in assessing climate change because, as implied by climate models discussed subsequently, a change in the 0.3 albedo by 0.01 can cause a change in mean annual temperature of 2°C (7). Sagan et al. conclude that it is "likely that the human species has made a substantial and continuing impact on climate since the invention of fire."

The burden of human-generated dust now in the atmosphere is comparable to that emitted by volcanic action (21,22). Reid Bryson has spoken of the "human volcano." An article by *New York Times* science writer Walter Sullivan that appeared over two decades ago discussed human-caused desertification and salinization; its sardonic title: "Is there Intelligent Life on Earth?"

CARBON DIOXIDE AND THE GREENHOUSE EFFECT

Perhaps the three most important trace gases in the atmosphere are ozone, water vapor, and carbon dioxide. Ozone is important because it absorbs ultraviolet radiation from sunlight. High levels of ultraviolet light could lead to widespread eye damage and increases in incidence of skin cancer (see Chapter 1). Both water vapor (which is concentrated near the Earth's surface) and carbon dioxide (which is distributed throughout the atmosphere) act to absorb infrared radiation, which raises the Earth's temperature. Both carbon dioxide and water vapor absorb the long wavelength (infrared) radiation emitted by the Earth, and re-emit it in all directions. As a result, the ambient average temperature is not 254 K but 287 K, about 30°C higher. Water vapor is extremely important in this process because it typically appears in the lower atmosphere at a concentration of 10,000 to 50,000 parts per million (ppm). Of course, water vapor also forms clouds, which help cool Earth by reflecting sunlight. Carbon dioxide is relatively uniformly distributed with height, at an average concentration of about 350 ppm, but varies from a low in the Antarctic to a high in the Arctic (23). (See problem 14 at the end of this chapter.)

The Greenhouse Effect

Water vapor and carbon dioxide act as two layers of a multilayer window to surround the Earth and warm it. A somewhat similar thing happens

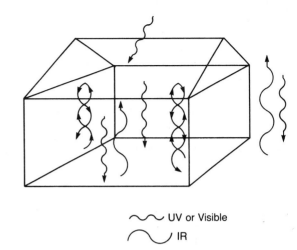

~~~ UV or Visible

⌒ IR

FIGURE 24.2
In a greenhouse, the glass transmits visible electromagnetic radiation, but reflects infrared. As a result, the temperature of the quiescent air rises.

in a closed-up car during the spring and summer. The glass allows light in, but reflects the infrared radiation coming from the interior of the car, ultimately heating up the car. A greenhouse works in just this way (Figure 24.2). Of course, having the car or greenhouse closed so that the air is quiescent is important for them but irrelevant for the atmosphere. Because of the functional similarity between the glass in the greenhouse and the CO_2 and water vapor in the atmosphere in trapping radiation, this trapping is known as the *greenhouse effect*. Venus, with a high carbon dioxide concentration (95 percent) and a surface temperature of 470°C, is an example of runaway greenhouse effect (24).

Establishing the Increase in Carbon Dioxide

Carbon dioxide concentrations are increasing partly because of the burning of 3 gigatonnes (Gtonnes) of coal, 3 Gtonnes of oil, and 1.6 Gm³ of natural gas each year (25). As a result, some 5 Gtonnes of carbon are released each year from fossil fuels alone, four times the rate of release in 1950, and ten times the rate in 1900 (25).

Other contributions than fossil fuels have added to the carbon dioxide burden of the air. Deforestation contributes between 0.5 and 5 Gtonnes to the carbon balance each year (26). People burning trees, brush (6,8,19,25), and fossil fuels have managed to increase the carbon dioxide concentration in the air from about 260 ppm in the Arctic (27) and the Antarctic (28) during a long period prior to the mid-nineteenth century to around 350 ppm now (5,25,27–29). The recent trends—including seasonal variations, with less carbon dioxide during summer—are shown in Figure 24.3 (30), which shows measurements made at Mauna Loa, Hawaii: A continuous increase of about one part per million per year (30,31) is measured. Mauna Loa is a very high mountain, and Hawaiian air is just about as clean as air can be. Hence the CO_2 levels from Mauna Loa must reflect the average atmospheric CO_2 concentration in the Northern Hemisphere.

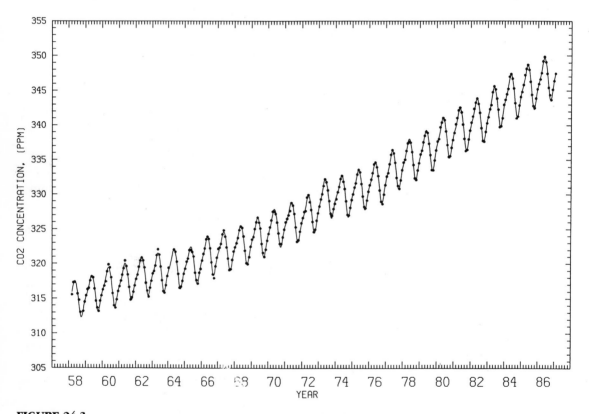

FIGURE 24.3

The concentration (in parts per million) of atmospheric carbon dioxide measured at Mauna Loa Observatory, Hawaii, which has very clean air. The gas analyzer samples northern hemisphere CO_2 values. The dots indicate monthly average concentrations. Note that, in the so-called Keeling plot, summer values are lower and winter values are higher (as one might expect). (Data are from C. D. Keeling and represent a 1987 update of C. D. Keeling, R. B. Bacastow, and T. P. Whorf in *Carbon Dioxide Review: 1982* and Ref. 30 in C. D. Keeling, D. J. Moss, and T. P. Whorf, "Measurements of the Concentrations of Atmospheric Carbon Dioxide at Mauna Loa Observatory, Hawaii 1958–1986." The measurements were obtained in a cooperative program between NOAA and the Scripps Institute of Oceanography.)

Some people take comfort from the fact that, despite this measured increase, the mean annual temperature, which increased by 0.6°C from 1880 to 1940 (32), decreased 0.3°C from 1940 to about 1970 (15,21,32), as can be seen from Figure 24.4 (25,33). We explained in Chapter 23 how the relative increase of volcanic activity in the 1960s as well as other albedo-increasing factors had decreased the atmosphere's transparency. The drop was evanescent, and has been overwhelmed by the long-term increase.

An analysis over a span of eight hundred years showed the effects of CO_2 to be very strong, the effects due to volcanic signal to be weak, and the effects due to solar fluctuations to be uncertain (34). Without the anthropogenic CO_2 increase, the 1980s would have seen CO_2 levels similar to those in the Little Ice Age (34). A different analysis using historical archives and more modern data from land and oceans to determine temperature between 1861 and 1984 shows a clear warming trend (35). The years 1980, 1981, and 1983 were the warmest on record in the world (35) until 1987, the warmest year ever.

The estimate of a pre-industrial CO_2 level of about 260 ppm depended on the analysis of air bubbles trapped in polar ice (27,28). The ice is crushed in an evacuated vessel, and the strength of molecular CO_2 response is examined to determine the proportion of CO_2. Extrapolation of the data of Figure 24.3 backward in time gives a best estimate of the pre-industrial level of carbon dioxide of 290 ppm (24,25,33,36). Analysis from tree rings using the $^{13}C/^{12}C$ ratio yields values between 260 and 276 ppm for this parameter (37). If the smaller values of CO_2 concentration are correct, there must be a biotic source of carbon (i.e., carbon contributed to the atmosphere by living matter) (38).

The atmosphere of Earth was probably originally nitrogen (N_2), methane (CH_4), and carbon dioxide. Because life existed, it was able to increase the concentration of O_2 to its current 20 percent value and decrease the concentration of CO_2 and methane. The amount of oxygen is in a long-term steady state. The amount stored in sed-

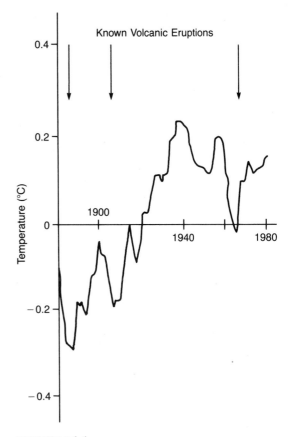

FIGURE 24.4
The course of the global yearly mean temperature from 1880 to 1980. Note that known volcanic eruptions correlate with times of lower-than-normal mean temperature. (Adapted from Refs. 25 and 33, copyright 1981 by AAAS)

iments each year is just about what we get from the weathering of now-exposed ancient sediments each year (39,40). Carbon dioxide is present in such small concentrations that it is not in equilibrium in the atmosphere, and its concentration has fluctuated rapidly in response to short-term changes in conditions. It has a residence time of only ten years in the atmosphere (40).

Historically, the concentration of carbon dioxide has been within a narrow range. If it were less than a third of its present value, photosynthesis would cease (39). During the last ice age (about 20,000 years ago), it was at about 200 ppm (25,36,39,41), and the warm period following it (about 5000 years ago) had concentrations of about 400 ppm (25). Recent data indicate that the changes in concentration of CO_2 in the atmosphere were 50 to 80 ppm within a century (42,43). The records indicate abrupt changes, almost as if there were a switch from one equilibrium situation to another over a short time. Such changes led the Energy Committee of the German Physical Society to warn that (25):

By the time that a direct causal link between CO_2 and temperature rise is established, it will be too late to apply effective preventive measures. This constricted time horizon makes the CO_2 problem especially urgent. [Author's translation]

Somewhat more sedately, but just as compellingly, the U.S. Committee for the Global Atmospheric Research Program said (36):

Our vulnerability to climatic change is seen to be all the more serious when we recognize that our present climate is in fact highly *abnormal,* and that we may already be producing climatic changes as a result of our own activities.

The Carbon Budget

We now turn to a look at the carbon budget and the models that led to the concern expressed in the last two quotations. The oceans contain about 50 petatonnes of sedimentary carbonates, 38 teratonnes of dissolved inorganic carbon, and about 20 petatonnes of organic carbon (44). The atmosphere contains around 700 gigatonnes (Gtonne) of carbon (12,44,45). Terrestrial vegetation contains from 500 to 900 Gtonne of carbon, and the soil contains about twice that much (46,47). Forests contain about ten to twenty times more carbon than cropland (44). Each year, about 5 Gtonne of carbon is released by burning of fossil fuels (25), and the cumulative release since 1850 is in the neighborhood of 120 Gtonne (47). Figure 24.5 (48) indicates how this has risen since 1860. These almost unimaginable figures correspond to a release in the 1980s of about 1 tonne of CO_2 for each person on Earth (37), and the release rate is increasing (as Figure 24.3 shows).

We have seen that carbon dioxide acts as a trap for infrared radiation. Hence, increasing the carbon dioxide content of the atmosphere warms

FIGURE 24.5
Estimated global emissions of carbon dioxide from 1860 to 1974. (C. F. Baes et al., copyright 1977 by Society of the Sigma Xi)

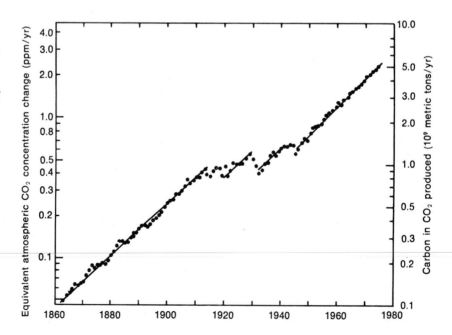

the Earth; this means more water vapor in the air (the warmer the air, the more water it can hold, as discussed in Chapter 18), which warms the atmosphere still more. Also, as the temperature is raised, liquid water can hold less carbon dioxide gas in solution, further increasing the carbon dioxide concentration in the atmosphere. Other feedback loops in the ocean have to do with the fact that current surface water is more acidic than the water in the ocean deeps. As this water circulates, it will cause calcium carbonate to dissolve, releasing more carbon dioxide into the seawater; much of this will be released to the atmosphere over a several hundred year time period (43).

The atmosphere–ocean feedback loops are complex and not entirely understood. Nevertheless, the most complex and realistic models of the atmosphere account for them as best they can. So-called general circulation models of the atmosphere, which also incorporate the ocean, typified by the analysis of Manabe and Wetherald (49) in 1967, indicated that a doubling of carbon dioxide would cause an average temperature increase of 2 to 3°C overall, with a temperature rise of 8 to 10°C near the poles. More recent calculations give reasonably similar results. An atmospheric profile typical of such models is shown in Figure 24.6 (49) and Figure 24.7 (50).

The sea takes up carbon dioxide in a surface layer. Some carbon dioxide is used in marine biomass (total mass about 3 Gtonnes); the CO_2 is then cycled into sediments or into the carbon reservoir of the deeper ocean at the rate of 2–3 Gtonnes per year (51). Most carbon in the ocean is in the form of ions of carbonate or bicarbonate, not as dissolved CO_2. The oceans react to an increase in CO_2 in the air by changing by only about a ninth as much (44,52). In the oceans, a "biological pump" operates (44,53–55) to take CO_2 from the atmosphere into the ocean: The plants and animals taking up the carbon convert it to skeletal material, live out their life cycles, and die. Their skeletons sink to the ocean depths, where the carbonate is partly deposited in sediment and partially dissolves but cannot mix due

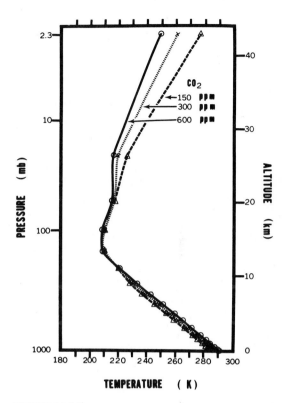

FIGURE 24.6

The temperature of the atmosphere varies with altitude. The effects of carbon dioxide increases may be calculated in general circulation models of the atmosphere. Effects depend on altitude. (Manabe and Wetherald, copyright 1967 by American Meteorological Society)

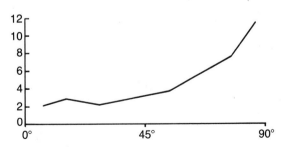

FIGURE 24.7

Estimated effect, by latitude, on surface temperatures as a consequence of doubling atmospheric carbon dioxide levels. (Manabe and Wetherald, copyright 1975 by American Meteorological Society)

MILANKOVITCH CYCLES AND CO$_2$

Recent information on the abruptness of the CO$_2$ concentration changes (33,55) and the wide range of the swing in the concentrations (25,39,41) led to consideration of CO$_2$ as the agent amplifying the effects of the Milankovitch cycles (53,55). Polar ice core records indicate that orbital changes precede changes in CO$_2$ concentrations, which precede ice sheet growth or melting. The 100,000 year cycle appears to be enhanced by this mechanism in ways that still are not clear. It may be that there are two (or more) stable modes for the climate; an "ice age" one, with reduced deep water production in the North Atlantic, and an interglacial, in which water is pumped from North Atlantic to North Pacific. The climate could "snap" from one to the other (54) the way a bicycle wheel, tied to the ceiling by rubber bands fastened at one particular place on the rim, is stable in two positions (Figure 24.8).

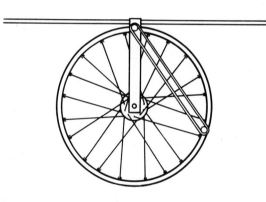

FIGURE 24.8
Climate can "snap" from one stable equilibrium to another, much as a bicycle wheel hung from the ceiling and also attached to a point on the ceiling by a rubber band can be stable in two positions.

to temperature and salinity differences. On the shallow continental shelves, some carbon is cycled into sediment along with nitrogen and oxygen (12).

The oceans and the biota clearly must be involved in the carbon cycle, because the CO$_2$ content of the atmosphere has risen only about half as much as if all CO$_2$ had been released from fossil-fuel production (26,52). Also, there is less carbon monoxide in the air than would be expected. Carbon monoxide is increasing rapidly, at a rate of about 2 percent per year (56). The ocean deeps, which hold 35,000 billion tonnes of carbon dioxide at present, have a virtually unlimited capacity for absorbing carbon (57). The problem is that the transfer of carbon between the surface and the abysses is slow (transfer takes

hundreds of years in the Atlantic and thousands of years in the Pacific) and thus cannot account for the relative dearth of carbon dioxide in the atmosphere. The rates of transfer to the abyssal regions are well known from studies with tritium from oceanic nuclear weapons tests and carbon-14 (radioactive carbon). The dilution of carbon-14 in the atmosphere, called the *Suess effect,* has been extensively studied. The global carbon cycle is, at the moment, still shrouded in mystery.

Global Warming

To try to understand what is happening, consider the somewhat more sophisticated geochemical carbon dioxide model illustrated in Figure 24.9 (36). Hansen et al. (33) provided the definitive

FIGURE 24.9

Scheme of interaction among components in the climate system. Open arrows indicate internal processes in the atmosphere. (National Academy of Sciences)

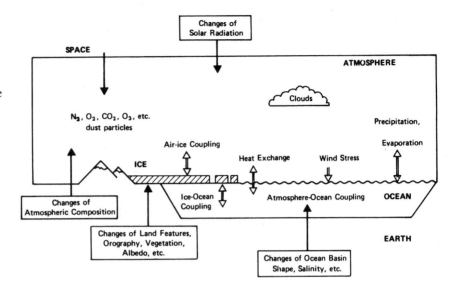

calculation of the most recent models. Their results are shown in Table 24.2. Overall they determined that a rise in carbon dioxide from 300 to 600 ppm would lead to a temperature increase of 3°C–4.5°C, including the effects of trace gases. They conclude that (33):

the general agreement between modeled and observed temperature trends strongly suggests that CO_2 and volcanic aerosols are responsible for much of the global temperature variation of the past century.

The NRC panel on climatic variations concurred that the best estimate of long-term global warming due to doubling of CO_2 is 3°C ± 1.5°C (57). Of course, world temperature will continue to increase as long as CO_2 and other "greenhouse gases" continue to increase.

Geochemical models seem to reproduce temperature profiles, as shown in Figure 24.10 (33). In such models, the biota constitute a sink of carbon (38). Ecological models such as those in References 12, 31, and 45 have the biota as a net

TABLE 24.2

Changes in the atmosphere and their effects in a one-dimensional model of the atmosphere.

Change	Temperature Change (°C)
CO_2 (300–600 ppm)	2.8°
Solar luminosity (+ 1%)	1.6°
Stratospheric aerosols (Optical density change 0.2)	−1.9°
Tropospheric aerosols (optical density change 0.02)	−1.2°
Land albedo (+0.05, implying global change 0.015)	−1.3°
Low clouds (2% of globe)	−1.4°
Middle clouds (2% of globe)	−0.4°
High clouds (2% of globe)	0.9°
Nitrous oxide, N_2O (0.28–0.56 ppm)	0.6°
Methane, CH_4 (1.6–3.2 ppm)	0.2°
$C Cl_2F_2$, $C Cl_3F_3$ (0–2 ppb)	0.5°
Ozone, O_3 (−25%)	−0.5°

SOURCE: J. Hansen et al., Reference 33. Copyright 1981 by AAAS.

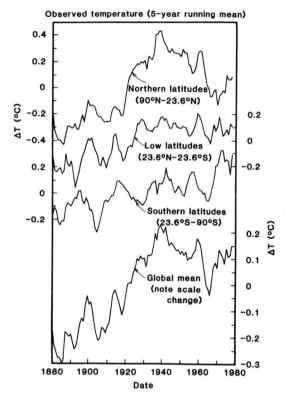

FIGURE 24.10

Observed temperature variation, by latitude, from 1880 to 1980. (Ref. 33, copyright 1981 by AAAS)

source of CO_2. For example, Woodwell et al. (47) estimate that the biota release between 1.8 and 4.7 Gtonne of CO_2 per year. Typical results of the calculations are shown in Figure 24.11 (38). The geochemical models use the biota "box" to balance their flux of carbon, as the ecological models use the oceans. Perhaps an intermediate model will someday succeed. Meanwhile, the geochemists can use the known rates of ocean carbon transfer to pooh-pooh the ecologists' models, while the ecologists can point to the huge releases from burning wood (1.5 Gtonne CO_2 per year, or ≈30 percent of fossil fuel release [45]) and from deforestation, and ask how the geochemists can see the biota as a carbon sink.

Human activity has increased rates of carbonate sedimentation by large factors. The expla-

nation of the missing sink of carbon may lie in increased sedimentation rates. Walsh (12) estimates this at a net storage of about 1.8 Gtonne of carbon per year; this could be the missing "sink."

Trace Gases Trace gases, such as nitrous oxide, methane, and chlorofluorocarbons, seem to be increasing at an extremely rapid rate. It was first proposed in 1976 (58) that trace gases could cause warming. A World Meterological Organization panel concluded that (see Table 24.2) the combined effects of changes in the trace gases are as large as that due to CO_2 (59). It is for this reason that an EPA study on prevention of greenhouse warming found that little could be done to reduce the impact of the warming by changes in energy policy now (60).

There is no doubt that the chlorofluorocarbons CFC-11 and CFC-12 will have a pronounced

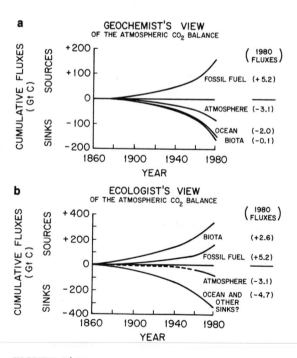

FIGURE 24.11

Two different interpretations of the sources and sinks in the carbon budget between 1860 and 1980. (J. Hobbie et al., courtesy BioScience)

heating effect. These gases absorb infrared in a window in the water vapor spectrum, and they are extremely strong absorbers (29,59). Absorption of CFC-11 in the infrared is eight times as strong as that of one of CO_2's principal bands (59). Also, because the gases are *extremely* dilute (unlike CO_2), the increase in absorption is directly proportional to the increase in concentration, currently about 6 percent per year (29).

The trace gas N_2O, increasing at 0.2 to 0.4 percent per year, gets into the atmosphere, even in the absence of human intervention, from lightning and volcanic gases. However, there is no doubt that nitrogen oxides from the burning of fossil fuel, application of nitrogen fertilizer, and discharge of sewage have a greater effect in causing the rise in the world average temperature (59). A sensitive indicator of this warming will be the growth or continued melting of the lower part of valley glaciers (61).

It has been variously guessed that the increased methane has been coming from flatulence in cattle (since their numbers are increasing), from the digestive tracts of termites (termite numbers increase due to increases in forest clearing), or from rice paddies (whose area

worldwide is on the increase) (59,62). Methane has increased substantially; analysis of air bubbles in glacial ice indicates that the atmospheric concentration was a steady 0.7 ppm from 27,000 years ago up to 500 years ago (63). By 25 years ago, it had risen to 1.25 ppm (63), and today the concentration of methane is 1.6 ppm (47); methane has increased by 1.1 ± 0.2 percent per year for the last 30 years (64). The fact that methane levels decreased during the 1982–1983 El Niño, by which mechanism it is not understood, illustrates our ignorance of how methane enters and leaves the atmosphere (65).

Effects of Temperature Changes

We have seen the historical record of decreases of 1°C in temperature: Food production drops because of a decrease of two weeks in the growing season (66). The response to temperature increases of 1°C is not so clear, but the record does show that the climate was wetter in North Africa and India in such times (67). It also shows that a 1°C warming caused a 10 percent decrease in precipitation and reduced the Colorado River flow by 25 percent (68). A schematic representation of the changes is shown in Figure 24.12

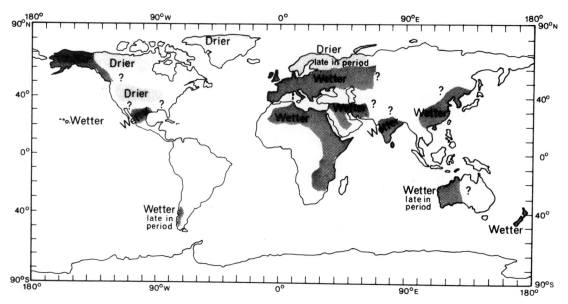

FIGURE 24.12
Estimated rainfall patterns, 4000 to 8000 years before the present, when the annual
mean temperature was several degrees warmer than now. Such a pattern could
well develop in the near future as the world mean temperature rises by several
degrees. (W. Kellogg, in *Climate Change,* copyright 1978 by Cambridge University
Press; reprinted with permission.)

(Kellogg, Reference 67). The USSR might do
somewhat better and the United States somewhat
worse under the inevitable changes. If CO_2 doubles, what are now the fruitful areas of the Mediterranean, the United States, and the USSR will
probably be tropical deserts; northern Canada
and Siberia will warm up and become wetter
(25). The boundary between permafrost (or tundra) and normal soil moves north by 100 to 200
km for each degree Celsius of temperature increase (69). Much more of Canada should be able
to grow wheat, and wheat production in the
southern Great Plains will decrease (70). In the
warm Alithermal era, from about 5000 to 10,000
years ago, rain decreased by 10 to 25 percent
over what is now the U.S. Midwest (71) while
temperature rose 0.5°C to 2.0°C. Africa was much
wetter; Lake Chad had ten times its present area
(67). The climate models (33) also predict a larger
increase in rain and runoff at high latitudes. Be-

tween latitudes of 35° and 45°, there is predicted
to be less rain and more evaporation (19,44).

It is ironic that, if such a change does come
to pass, the United States would have been mostly
responsible for its own predicament. The United
States burns a quarter of all the coal burned in
the world, and the USSR and Canada together
burn another quarter (76). In view of the possible
advantages to be gained, perhaps the USSR and
Canada would want to encourage continued reliance on fossil fuels, but the same could not be
said for the United States.

**Changes Stemming from Increased CO_2
Levels**   The species of plants grown would
change as well. After a glacial episode when climate changes, vegetation does shift. It has recently been found, however, that plant
communities persist long after the climate change
has wiped out favorable habitats—about two mil-

lenia separate climate changes from the consequent changes in vegetation (77). Farmers, one would hope, would change their crops and methods or, more likely, just move north. A recent study predicts a shift in the U.S. corn belt of 175 km SSW or NNE for each 1°C change in the growing season mean daily temperature (78). Projections indicate yield decreases in the major crops of 2 to 12 percent: The South Dakota yield is predicted to decrease 11 percent, while the Kansas yield decrease is predicted at only 5 percent (68). Soil moisture in the Great Plains, western Europe, northern Canada, and Siberia would decrease (79). Historically speaking, changes in grain available for world trade of about 1 percent have sparked price increases (see Chapter 17) (80). An increase of 1°C increases the rate of forest respiration (the decay of biomass, with release of carbon dioxide), as extrapolated from measurements made on Long Island, by 3 to 25 percent (26). This is another feedback loop accelerating the warning.

There may be favorable consequences as well—for example, increased CO_2 concentrations close stomata in plants, reducing water loss (68). Several groups (81) have argued that there is evidence for this effect, but the studies are too restricted to be definitive (82). Waggoner and others have estimated that crop yields may increase 3 to 12 percent by this mechanism (83), although tests have shown that, while many plants grow faster in increased CO_2 environments, many also grow more slowly (68). On the whole, however, photosynthesis appears to proceed best at CO_2 concentrations of around 600 ppm, which bodes well for future agriculture.

In the drier Midwest, the growing season will be shortened by ten days, droughts will become more frequent, and irrigated farms will be in trouble, because they depend for supply of irrigation water on the small difference between rains and evaporation (68,83). Fossil water reservoirs there are already depleted. There will certainly be squabbles over Colorado River water. The Colorado River Compact was developed in the 1930s to allocate water among the various

states. In the early twentieth century, there was unprecedented water flow available (80). More water from the Colorado River has been allocated than actually flows in most parts. We have already remarked that a 1°C rise in temperature would reduce river flow by 25 percent; a 2°C rise would cause a 40 percent decrease in the flow (83). The changes would demand radical measures, such as the end of irrigated agriculture in the Colorado basin (44,83). In the 1930s, the dust bowl brought in its wake pests such as the jack rabbit and grasshoppers, whose populations exploded (68). Such "shifty pests" may throw all projections of crop production awry (83).

Of course, the warming would permit growing in more northern latitudes, but these soils are generally poorer than the ones that would be covered by the water. Furthermore, as we have previously mentioned, warm periods have often resulted in dust bowl conditions in central continents (1,5,21,23,25,44,72,84).

Sea Level Rise   Sea levels have risen 140 mm over the last century, with a 230 mm rise in the last fifty years (85). Part of the rise (30 to 60 mm) may be due to thermal expansion (85) from the 0.5°C rise in temperature since 1880. So much dam building has occurred during the last half-century that less rise has been seen than there would have been without storage of 3000 to 6000 km^3 of water (86,87). We have stored about 125 km^3/yr in reservoirs, but we would need a rate of 500 km^3/yr to stop all the rise in sea level (86).

Another effect should follow as a corollary to the greater temperature rise predicted near the poles. The North Pole ice cap is only meters thick and rests on seawater. While the melting of the ice sheet over the North Pole will not contribute to a rise in water level (just as a glass of water filled to the brim with water with ice cubes floating on it does not overflow when the ice melts), an open Arctic Ocean would be warmer by 10°C at the edge in winter. This could affect the climate in unpredictable ways. One predictable way it would affect climate is that such a melting would strengthen the climatic trend by decreasing Earth's albedo (20), since ice reflects sunlight more effectively than open water or ground.

In the Antarctic, part of the ice sheet (the Ross ice shelf) lies over seawater, but is anchored to the ice on the continent itself, although the sheet is much thicker than the north polar ice. These sheets are currently at temperatures that average -4 or $-5°C$. The ice will begin melting when the average temperature exceeds 0°C by some small amount. The result could well be the melting of these ice sheets over only a few decades (23), with a concurrent rise in sea level (due to melting of the Ross ice shelf) of 6 meters or so. Such a rise would drown most world seaports as well as flood some of the world's most fertile soils. A total of 10 million square kilometers of land would disappear, 70 percent in the Northern Hemisphere, if the sea rose 10 m (87). The United States would lose only about 2 percent of its land area, but one-quarter of Florida and Louisiana would be flooded, as would one-sixth of Delaware and the District of Columbia, one-eighth of Maryland, one-tenth of New Jersey (mostly heavily developed land near New York City), and one-twelfth of North Carolina (88). About 12 million people would have to be relocated, and $300 million in property would be flooded (88). Corpus Christi and Galveston in Texas, New Orleans in Louisiana, Savannah in Georgia, Charleston in South Carolina, most of Washington, D.C., Wilmington in Delaware, Atlantic City in New Jersey, and Boston, Massachusetts would be submerged (88).

The historical record of sea level shows that changes in the volume of glacial ice are correlated with climate change and sea level (89). The most dangerous effect of the polar temperature rise could be due to "glacier surge." Glaciers have been observed to move very rapidly for a few months to several years, then slow down again. This results from melting at the bottom of the glacier, allowing semi-frictionless sliding. The motion generates heat, which feeds back to prolong the process. If this were to happen to the Ross ice sheet, the 6 m rise in sea level would come almost all at once, with disastrous results (67). Sea level has changed rapidly at the beginnings and endings of glacial episodes (90). About 95,000 years ago, sea levels rose 15 to 20 m within a century (86). Rapid sea level change is associated with climatic change.

If the heating is prolonged enough, the Greenland and Antarctic ice sheets (about 8 percent and 90 percent, respectively, of the total ice in the world) would melt over a period of centuries, bringing the sea level up another 60 m or so (25). This would cause an even larger shift, but since the change would take place slowly, it would probably not be socially wrenching. The shorelines of the continents would be *very* different if all the ice were to melt.

It may be too late to prevent a significant rise in average world temperature. Even a drastic reduction in dependence on fossil fuels to half of today's by a century from now is not likely to prevent the CO_2 concentration from climbing to about 450 ppm (25,60). The most effective way to change the rate of increase of CO_2 is to reduce the dependence of developed countries on fossil fuels—for example, by wider use of nuclear energy or by development of viable solar energy (25). Conservation and prevention of waste of energy (see Chapters 18, 19, and 20) should be pursued with vigor. The adjustments to a warmer world will have to be made, since prevention of the consequences of warming is impossible.

W. W. Kellogg (67), in looking at the future "warmer Earth" sees a balance: Some countries will be better off, others worse. Earth, he believes, will be better able to support its increased population. Despite such optimism, he still sees the grim reaper of starvation as part of a change to a "warmer Earth." We can only hope that we are farsighted enough to prepare to meet the consequences that we were apparently not farsighted enough to prevent. If we manage to prepare, Kellogg's nightmare need not become reality.

SUMMARY

Climate has changed over the course of millenia. Climate changes for various physical reasons, but also because human beings can promote desertification by cutting too many trees, or by causing overgrazing, or change the albedo in other ways.

Volcanic activity injects aerosols into the upper atmosphere. These aerosols lead to a lower surface temperature. Increased volcanic activity in the mid-twentieth century has decreased the effects of the carbon dioxide increase to an extent.

The least expected agents of climate change have been found in the gases that are normally present in small amounts in the atmosphere. Carbon dioxide and other "trace gases" can cause the Earth's mean temperature to rise through the greenhouse effect. While the carbon cycle seems to be reasonably well understood in its overall characteristics, its details are still not understood.

There is a sink of carbon somewhere—but where? At any rate, the CO_2 that does stay in the atmosphere will raise Earth's mean temperature, with effects on humanity that are still the subject of speculation, although the effects of CO_2 increase on climate are relatively certain: a lot of warming at the poles, and a very small amount of warming at the equator. The United States will be worse off overall, with a smaller area available for agriculture. Other countries, now in the Third World, will make gains. Sea level will rise 1 to 6 meters during the next century, causing social dislocation. We may hope that we will do better at managing the change than we have done in preventing it.

REFERENCES

1. W. W. Kellogg and S. H. Schneider, *Science* 186 (1974): 1163.
2. C. B. Beaty, *Am. Sci.* 66 (1978): 452.
3. A. L. Hammond, *Science* 191 (1976): 455; and J. W. Sawyer, *Environment* 20, no. 2 (1978): 25.
4. W. L. Gates, *Science* 191 (1976): 1138.
5. S. H. Schneider, *The Genesis Strategy* (New York: Plenum, 1976).
6. J. J. Hidore, *J. Geog.* 77 (1978): 214.
7. C. Sagan, O. B. Toon, and J. B. Pollack, *Science* 206 (1979): 1363.
8. L. I. Wilder, *Little House on the Prairie* (New York: Harper & Row, 1970).
9. A. P. Kershaw, *Nature* 322 (1986): 47.
10. J. Charney, P. H. Stone, and W. J. Quirk, *Science* 190 (1974): 741.
11. N. Wade, *Science* 185 (1974): 234.
12. J. J. Walsh, *BioScience* 34 (1984): 499.
13. H. Brückner, *Geo. J.* 13 (1986): 7.
14. J. P. Peixoto and A. H. Oort, *Rev. Mod. Phys.* 56 (1984): 365; W. L. Gates, in *Projecting the Climatic Effects of Increasing Carbon Dioxide* ed. M. C. MacCracken and F. M. Luther (Washington, D.C.: Department of Energy, DOE/ER-0237, 1986), 57ff; and M. E. Schlesinger, also in *Projecting the Climatic Effects of Increasing Carbon Dioxide,* 281ff.
15. E. Raschke and J. Schmetz, *Reports of the Deutsche Forschunggemeinshaft* 1 (1984): 10.
16. W. S. Smith, *Nat. Hist.* 92, no. 11 (1983): 36; and W. Sullivan, *New York Times,* 31 Dec. 1985.
17. R. Caputo, *National Geographic* 167 (1985): 577; and A. F. Pillsbury, *Sci. Am.* 244, no. 1 (1981): 54.
18. S. Postel, *Futurist* 18, no. 4 (1984): 39.
19. J. U. Nef, *Sci. Am.* 237, no. 5 (1977): 140.
20. D. A. Robinson and G. Kukla, *J. Clim. Appl. Met.* 23 (1984): 1626.
21. J. M. Mitchell, in *Global Effects of Environmental Pollution,* ed. F. Singer (Dodrecht, Neth.: D. Reidel Publ. Co., 1970); and W.H. Matthews, *Intern. J. Environmental Studies* 4 (1973): 283.

22. H. H. Lamb, *Ecologist* 4 (1974): 10.

23. R. H. Gammon, E. T. Sundquist, and P. J. Fraser, in *Carbon Dioxide and the Global Carbon Cycle* ed. J. R. Trabalka (Washington, D.C.: Department of Energy, DOE/ER-0239, 1986), 25ff.

24. U. von Zahn, *Alexander von Humboldt-Stiftung Mitteilungen* 39, no. 9 (1981): 15.

25. Energy Committee of the German Physical Society, *Phys. Bl.* 39 (1983): 320.

26. G. M. Woodwell, *Oceanus* 29, no. 4 (1986): 71.

27. U. Siegenthaler and H. Oeschger, *Science* 199 (1978): 388; U. Siegenthaler and H. Oeschger, *Ann. Glaciology* 5 (1984): 153; and G. I. Pearman et al., *Nature* 320 (1986): 248.

28. A. Neftel et al., *Nature* 315 (1985): 45; D. Raymond and J. M. Barnala, *Nature* 315 (1985): 309; and H. Friedli et al., *Nature* 324 (1987): 237.

29. R. E. Dickinson and R. J. Cicerone, *Nature* 319 (1986): 109.

30. R. B. Barcastow, C. D. Keeling, and T. P. Whorf, *J. Geophys. Res.* 90 (1985): 10529; C. D. Keeling et al., *J. Geophys. Res.* 90 (1985): 10511; C. D. Keeling, private communication.

31. G. M. Woodwell, *Sci. Am.* 238, no. 1 (1978): 34; G. M. Woodwell et al., *Science* 199 (1978): 141.

32. P. V. Hobbs, H. Harrison, and E. Robinson, *Science* 189 (1974): 909.

33. J. Hansen et al., *Science* 213 (1981): 957.

34. C.-D. Schonwiese, *Arch. Met. Geophy. Biocl.* B35 (1984): 155.

35. P. D. Jones, T. M. L. Wigley, and P. B. Wright, *Nature* 322 (1986): 430.

36. Panel on climatic variations, U.S. Committee for the Global Atmospheric Research Program, National Research Council, *Understanding Climatic Change* (Detroit: Grand River Books, 1980).

37. R. A. Kerr, *Science* 222 (1983): 1107.

38. J. Hobbie et al., *BioScience* 34 (1984): 492.

39. R. M. Garrels, A. Lerman, and F. T. Mackenzie, *Am. Sci.* 64 (1976): 306; and J. C. G. Walker, *BioScience* 34 (1984): 486.

40. H. D. Holland, B. Lazar, and M. McCaffrey, *Nature* 320 (1986): 27; and M. O. Andreae, *Oceanus* 29, no. 4 (1986): 27.

41. N. J. Shackleton et al., *Nature* 306 (1983): 319; see also J. M. Palais, *Oceanus* 29, no. 4 (1986): 55.

42. H. Oeschger and B. Stauffer, *Science* 222 (1983): 1107; and B. Stauffer et al., *Ann. Glaciology* 5 (1984): 160.

43. J. J. McCarthy, P. G. Brewer, and G. Feldman, *Oceanus* 20, no. 4 (1986): 16.

44. R. Revelle, *Sci. Am.* 247, no. 2 (1982): 35; and B. Moore III and B. Bolin, *Oceanus* 29, no. 4 (1986): 9.

45. C. S. Wong, *Science* 200 (1978): 197.

46. D. B. Botkin et al., *BioScience* 34 (1984): 508; and R. A. Houghton et al., in *Carbon Dioxide and the Global Carbon Cycle,* ed. J. R. Trabalka (Washington, D.C.: Department of Energy, DOE/ER-0239, 1986), 113ff.

47. G. M. Woodwell et al., *Science* 222 (1983): 1081.

48. C. F. Baes et al., *Am. Sci.* 65 (1977): 310.

49. S. Manabe and R. T. Wetherald, *J. Atmos. Sci.* 24 (1967): 241; and S. Manabe, in *Man's Impact on the Climate,* ed. W. H. Matthews, W. W. Kellogg, and G. D. Robinson (Cambridge, Mass.: MIT Press, 1971).

50. S. Manabe and R. T. Wetherald, *J. Atmos. Sci.* 32 (1975): 3.

51. C. F. Baes, A. Björkström, and P. J. Mulholland, in *Carbon Dioxide and the Global Carbon Cycle,* ed. J. R. Trabalka (Washington, D.C.: Department of Energy, DOE/ER-0239, 1986), 81ff.
52. W. S. Broecker et al., *Science* 206 (1979): 409.
53. R. A. Kerr, *Science* 223 (1984): 1053.
54. W. S. Bracken, S. M. Peteet, and D. Rind, *Nature* 315 (1985): 21; T. P. Barnett, in *Detecting the Climatic Effects of Increasing Carbon Dioxide,* ed. M. C. MacCracken and F. M. Luther (Washington, D.C.: Department of Energy,DOE/ER-0235, 1986), 149ff; and W. S. Broecher, *Nature* 328 (1987): 123.
55. N. G. Pisias and J. Imbrie, *Oceanus* 29, no. 4 (1986): 43.
56. C. P. Rinsland and J. S. Levine, *Nature* 318 (1985): 250.
57. Panel on climatic variations, National Research Council, *Carbon Dioxide and Climate: A Second Assessment* (Washington, D.C.: National Academy Press, 1982).
58. W. C. Wang et al., *Science* 194 (1976): 685.
59. R. A. Kerr, *Science* 220 (1983): 1364; G. Marland and R. M. Rotty, *J.A.P.C.A.* 35 (1985): 1033.
60. S. Seidel and D. Keyes, *Can We Delay a Greenhouse Warming?* (Washington, D.C.: Office of Policy and Resources Management, EPA, 1983).
61. J. Oerlemans, *Nature* 320 (1986): 607.
62. D. H. Ehhalt, *Environment* 30, no. 10 (1985): 6.
63. B. Stauffer et al., *Science* 229 (1985): 1386. See also R. A. Kerr, *Science* 226 (1984): 954.
64. C. P. Rinsland, J. S. Levine, and T. Miles, *Nature* 318 (1985): 245; and R. A. Rasmussen and M. A. K. Khalil, *Science* 232 (1986): 1623.
65. M. A. K. Khalil and R. A. Rasmussen, *Science* 232 (1986): 56.
66. L. M. Thompson, *Science* 188 (1975): 535; D. Pimentel and J. Krummel, *Ecologist* 7 (1977): 254; R. A. Bryson, *Science* 184 (1974): 753; and S. W. Matthews, *National Geographic* 150 (1976): 576.
67. W. W. Kellogg, in *Climate Change,* ed. J. Gribben (New York: Cambridge Univ. Press, 1978). See also M. E. Schlesinger and J. F. B. Mitchell, in *Projecting the Climatic Effects of Increasing Carbon Dioxide,* ed. M. C. MacCracken and F. M. Luther (Washington, D.C.: Department of Energy, DOE/ER-0237, 1986), 81ff.
68. P. E. Waggoner, *Am. Sci.* 72 (1984): 179.
69. R. G. Barry, in *Detecting the Climatic Effects of Increasing Carbon Dioxide* ed. M. C. MacCracken and F. M. Luther (Washington, D.C.: Department of Energy, DOE/ER-0235, 1986), 109ff.
70. C. Rosenzweig, *Climate Change* 7 (1985): 367.
71. T. P. Bennett, in *Detecting the Climatic Effects of Increasing Carbon Dioxide* ed. M. C. MacCracken and F. M. Luther (Washington, D.C.: Department of Energy, DOE/ER-0235, 1986), 149ff.
72. R. A. Kerr, *Science* 226 (1984): 326.
73. D. M. McLean, *Science* 201 (1978): 401.
74. D. A. Russell, *Sci. Am.* 248, no. 1 (1983): 58; and L. W. Alvarez, *Phys. Today* 40, no. 7 (1987): 24.
75. D. R. Prothero, *Science* 229 (1985): 550.
76. R. A. Bryson and B. M. Goodman, *Science* 207 (1980): 1041.
77. R. Lewin, *Science* 228 (1985): 165.

78. J. E. Newman, *Biometeorology* 7 (1980): 128. See also W. L. Dicker, V. Jones, and R. Achutuni, in *Characterization of Information Requirements for Studies of Carbon Dioxide Effects: Water Resources, Agriculture, Fisheries, Forests, and Human Health,* ed. M. R. White (Washington, D.C.: Department of Energy, DOE/ER-0236, 1986), 69ff.

79. S. Manabe and R. T. Wetherald, *Science* 232 (1986): 626.

80. S. H. Schneider and R. L. Temkin, in *Climate Change,* ed. J. Gribben (New York: Cambridge Univ. Press, 1978).

81. A. R. Aston, *J. Hydrol* 67 (1984): 273; V. C. LaMarche et al., *Science* 225 (1984): 1019; and S. B. Idso and J. Brazel, *Nature* 312 (1984): 51.

82. T. M. L. Wigley, K. R. Briffa, and P. D. Jones, *Nature* 312 (1984): 102; B. Acock and L. H. Allen, in *Direct Effects of Increasing Carbon Dioxide on Vegetation,* ed. B. R. Strain and J. D. Cure (Washington, D.C.: Department of Energy, DOE/ER-0238, 1986), 53ff; B. A. Kimball, also in *Direct Effects of Increasing Carbon Dioxide on Vegetation,* 185ff.

83. P. E. Waggoner, "Agriculture and a Climate Changed by More Carbon Dioxide," NRC paper; N. J. Rosenberg, in *Carbon Dioxide Review,* ed. W. C. Clark (New York: Oxford Univ. Press, 1982); W. C. Clark, *Climatic Change* 7 (1985): 5; and J. M. Callaway and J. W. Currie, in *Characterization of Information Requirements for Studies of Carbon Dioxide Effects: Water Resources, Agriculture, Fisheries, Forests, and Human Health,* ed. M. R. White (Washington, D.C.: Department of Energy, DOE/ER-0236, 1986) 23ff.

84. J. Norwine, *Environment* 19, no. 8 (1977): 7.

85. J. E. Hansen, *Nature* 313 (1985): 349.

86. W. S. Newman and R. W. Fairbridge, *Nature* 320 (1986): 319.

87. A. Henderson-Sellers and K. McGuffie, *New Scientist* 110, no. 1512 (1986): 24.

88. S. H. Schneider and R. S. Chen, *Ann. Rev. Energy* 5 (1980): 107.

89. B. U. Hag, J. Hardenbol, and P. R. Vail, *Science* 235 (1987): 1156; see also R. A. Kerr, *Science* 235 (1987): 1141.

90. F. P. Bretherton, *Oceanus* 29, no. 4 (1986): 3.

91. G. J. MacDonald, *Bull. Am. Phys. Soc.* 24 (1979): 31.

PROBLEMS AND QUESTIONS

True or False

1. There appears to be more carbon dioxide in the atmosphere than there should be on the basis of known production rates.

2. Animals can have an effect on climate.

3. Gases in equilibrium in the atmosphere are not subject to rapid (year-to-year) fluctuations.

4. Most carbon dioxide in the ocean is present as dissolved gas.

5. The mean Earth temperature is expected to rise by about 3°C by the mid–twenty-first century.

6. Glacial ages have been associated with changes in the numbers of species of plankton in the oceans.

7. The concentration of carbon dioxide has decreased steadily in the past several years.

8. A sea level rise of a few meters will have a negligible effect on people.

9. Changes in mean temperature of only one or two degrees could have a large effect on agricultural production.

10. There is disagreement as to the exact levels of pre-industrial carbon dioxide.

Multiple Choice

11. Records of glacial ages in past times may be found in

    a. the amount of calcium car-
bonate deposited on ocean
floors.

    b. the ratio of ^{18}O to ^{16}O in
plankton in the Pacific Ocean.

    c. accumulation of wind-blown
soil containing cold-resistant
species.

    d. pollen found in ancient lake
bottoms.

    e. all of the above alternatives.

12. The amount of carbon dioxide in the atmosphere is

    a. increasing rapidly (about
10%/year).

    b. increasing slightly (about
½%/year).

    c. remaining the same.

    d. decreasing slightly (about
½%/year).

    e. decreasing rapidly (about
10%/year).

13. Earth's present climate may best be described as

    a. glacial.

    b. marginally glacial.

    c. marginally interglacial.

    d. interglacial.

    e. tropical.

14. Figure 24.13 shows Earth's "breathing." Which statement below based on the
figure is *false*?

    a. There is a gradual increase in
CO_2 at high southern latitudes
(near edge).

    b. Carbon dioxide levels drop
in summer because of growing plants in the Northern
Hemisphere.

    c. There must be more land in
the temperate Northern
Hemisphere than in the
Southern Hemisphere.

    d. The CO_2 levels in the North-
ern Hemisphere are not ris-
ing.

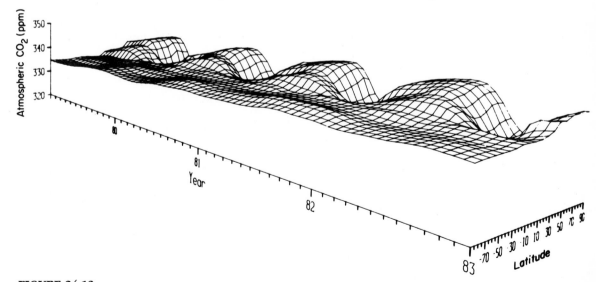

FIGURE 24.13
The Earth appears to "breathe," that is, the mean annual level fluctuates about the
mean rise due to seasonal changes. (R. Gammon, Ref. 23)

15. The greenhouse effect is responsible for the fact that the Earth is slightly warmer than it would otherwise have been. This occurs because of trapping of infrared radiation by absorption by gases. There is fear that people are causing conditions that could lead to a substantial change in the Earth's average temperature. Which of the following is (are) implicated in possible changes in average temperature?
    a. Increased amounts of carbon dioxide in the atmosphere
    b. Increases in particulate matter in the atmosphere
    c. Increases in the amounts of "trace gases," such as methane, in the atmosphere
    d. Changes in the amount of water vapor (clouds) in the atmosphere
    e. All of the above are implicated in possible temperature changes.

16. Which of the following atmospheric "trace gases" may be partly responsible for the increase in mean Earth temperature expected by the middle of the next century?
    a. CH_4
    b. N_2O
    c. CFC-11
    d. CO_2
    e. All of the above

17. The greenhouse effect can occur in the atmosphere because the carbon dioxide absorbs
    a. microwaves.
    b. infrared radiation.
    c. visible light.
    d. ultraviolet radiation.
    e. all light from the sun.

18. Scientists working on climate are generally persuaded that increased carbon dioxide levels will cause increased temperatures because
    a. their laboratory experiments have verified the effect.
    b. historical data on increased carbon dioxide levels show ambient temperature increases.
    c. they are certain that high carbon dioxide levels in the atmosphere during the era of the dinosaurs caused that era to be very warm.
    d. they have confidence in the general circulation models of the atmosphere, which predict effects that have been observed in addition to predicting the warming.
    e. they believe what they are told.

19. Which of the following could cause Earth's albedo to change?
    a. The clearing of tropical forests
    b. Dusting of the Greenland ice cap with carbon black
    c. Increase in the amount of ice cover in polar regions
    d. Replacement of temperate forest by savanna
    e. All of the above could change the albedo.

20. The water vapor in the atmosphere could affect the Earth's mean temperature by
    a. causing increased numbers of clouds to form; the clouds reflect incident sunlight, causing the temperature to fall.
    b. absorbing infrared radiation, causing the temperature to rise.
    c. both methods referred to in the first two possible answers.
    d. neither of the methods referred to in the first two possible answers.

Discussion Questions

21. Human activity is causing release of 7.2 gigatonnes of carbon into the air yearly (5.7 from burning of fossil fuels, 1.5 from the biota). Coal contributes *more* carbon per kWh_e than either oil or gas (if coal is used instead of oil, 25 percent more carbon is emitted; if coal is used instead of natural gas, 72 percent more carbon is emitted). Yet the United States is planning to emphasize the burning of coal in order to escape dependence on foreign suppliers of oil, and because of the shortage of natural gas. What effect might this decision have on the environment?

22. There are some indications that some natural gas (containing methane) is not of fossil origin. Lake Kivu in the Great Rift Valley contains 2 Tft^3 of dissolved natural gas. In this region, all evidence of organic activity has been wiped out by the active volcanism (91). The natural gas is hypothesized to come from Earth's outgassing. If this were true, there might be a good deal more gas than had been thought. Would it be a good idea in that case to encourage use of natural gas rather than coal or oil? Why or why not?

23. What human activities in addition to the ones discussed in this chapter could influence the environment?

24. Can you think of any places for the missing carbon monoxide and carbon dioxide to go? For example, the soil contains a lot of carbon in dead organic matter (humus, peaty soils). The amount of carbon stored here is known only to within a factor of 2 (1000 to 2000 Gtonnes). Are there other such possible sinks for carbon?

25

SAFETY AND NUCLEAR ENERGY

Chapters 21, 22, and 24 should have convinced you that energy from fossil fuels can be hazardous to human health. If solar energy, conservation, and recycling cannot replace necessary electricity in the short run, and if the hazards of fossil fuels are recognized, where shall we get our electricity?

In this chapter, we examine the nuclear option. Two events of recent times have made the public increasingly chary of nuclear energy: the Three Mile Island accident in 1979 and the Chernobyl disaster in 1986. We examine the risks—what could go wrong in the reactors involved in nuclear energy generation; in addition, we consider the health effects of exposure to ionizing radiation.

The nuclear industry will have to convince the populace that nuclear wastes may be stored safely and that Chernobyl-type disasters are not possible before more people will again accept the nuclear alternative. The chapter compares realistic risks among various methods of generating energy, with results that may surprise some readers.

China syndrome
event tree
fault tree
voluntary risk
involuntary risk
source terms
exposure
dose
curie
becquerel
LET
rad
rem
tailing
working level
linear dose theory
threshold
absolute risk model
capacity factor
water migration
deep burial

KEY TERMS   worst-case assumptions
ECCS
LOCA (core melt)
transients

The nuclear age, properly speaking, dates to the late 1930s, when a series of experiments, culminating in the observation of fission by Otto Hahn and Lise Meitner, demonstrated the potential for controlled nuclear fission to those physicists working in the exciting new field. It was but a short time until Enrico Fermi and his co-workers had shown controlled fission in a squash court under Stagg Field at the University of Chicago. Even as the newly christened "nuclear scientists" worked on nuclear weapons for the Manhattan Project, many were already thinking of the blessings they expected to be conferred upon society by the cheap nuclear energy they would make available. Nuclear energy would be too cheap to bother metering. These scientists recognized early on that safety would be an issue, and they were convinced that they could make nuclear energy absolutely safe. Their insistence that nuclear energy would be absolutely safe has haunted the nuclear industry up to the present and is partly responsible for public disenchantment with nuclear energy.

It was the prospect of peaceful uses for nuclear energy that led to the formation of the Atomic Energy Commission (AEC) after World War II; the agency had the responsibility to promote and to regulate nuclear energy in the United States. History of this time, which describes the interplay of personality and politics, is available in very readable form in Jungck's book *Brighter than a Thousand Suns* (1).

In 1974, Congress split the AEC into two parts: the Nuclear Regulatory Commission (NRC) and the Energy and Resources Development Administration (ERDA). In 1977, ERDA was absorbed into the newly created Department of Energy. The reason for the split is that, over the years, the tension between the AEC's "watchdog" role and its mandate to boost nuclear energy had become more and more pronounced. There was a public feeling that perhaps safety and prudence could have been compromised by the close association between the regulators and the regulated. When people hear news of the leakage of 115,000 gallons of radioactive waste from the Hanford holding tanks in one accident, and then learn that about 425,000 gallons had been spilled since 1958 and worse (2), they can be forgiven for worrying about what the AEC was doing about nuclear wastes. The revelation that some high Washington AEC officials admitted "censoring"

reports to the AEC itself also had a disquieting effect (3,4). It was hoped that the creation of the separate NRC to do the regulating would change situations such as that in which, of 3333 violations uncovered by AEC inspectors between 1 July 1973 and 30 June 1974 (98 of which were serious), only 8 led to the imposition of any form of punishment (5). The 4 tonnes of enriched uranium missing from one plant, and the news that the gaseous diffusion plant at Oak Ridge was missing material that "runs into tons" (6), raised questions in the public mind about AEC management of the nuclear program.

The 1979 accident at Three Mile Island in the reactor itself propelled the question of nuclear reactor safety from the position of being an issue for a minority to a serious concern to the majority. In such a state of uncertainty as to the hazards of nuclear energy, it is no wonder that citizens have signed petitions, demonstrated at construction sites, or voted to prohibit state licensing of nuclear reactors in some states. California enacted a law that effectively put a stop to further plant construction there. Sweden suspended all reactor construction until a government commission satisfactorily addresses the problem of waste disposal (as of 1985, the results were still not satisfactory). The people of Sweden voted to retire all reactors as they became obsolete and to replace the energy with that from other sources. Voters in Austria have kept a completed reactor from opening. Since the Three Mile Island accident, there has been no new nuclear plant ordered in the United States. The serious accident at Chernobyl in the Ukraine has further hardened the opposition to nuclear energy.

There are many issues to address in any examination of the safety of nuclear plants: proliferation of nuclear weapons, catastrophic accidents at nuclear reactors, chronic health-endangering emissions from reactors, health-related concerns from other parts of the nuclear fuel cycle (see Chapter 14), and the ultimate disposal of nuclear wastes. All these issues are discussed here, but I shall examine in the most detail questions of catastrophic accidents and of relative and absolute health dangers of nuclear energy as compared to other forms of energy generation.

CATASTROPHIC ACCIDENT

We have had an incident at Three Mile Island (TMI) and a disaster at Chernobyl, but there has never been a truly catastrophic accident, one that released a large proportion of the core's radioactivity, with a nuclear reactor. While the lack of catastrophe is desirable, it means that it is hardly possible to know the consequences of potential catastrophes. This lack of experience then means that other approaches must be adopted if anything at all is to be learned about the possibility of accidents. These alternative approaches are only as good as the assumptions are reasonable. Several quantitative studies of BWRs (boiling water reactors) and PWRs (pressurized water reactors) have been undertaken (7–10) of the hypothesized consequences of large reactor accidents.

The earliest report, known as WASH-740 (7), assumed that an accident would release 50 percent of all fission products in the reactor core to the environment, and also assumed the worst possible weather conditions. Such worst-case assumptions had to be made at this time because of the lack of any experience and because of the desire to make any estimates conservative. The result of all these pessimistic assumptions was that this accident in a 167 MW_e nuclear plant could result in 3400 deaths, cause 43,000 acute illnesses, contaminate up to 150,000 square miles, and do $30 billion damage (1984 dollars). The 1964 revision to WASH-740 was suppressed by the AEC until 1973, when portions were released under the Freedom of Information Act (3). The results of the revised study were even more pessimistic than WASH-740: Fatalities could be as high as 45,000, and an area "the size of the State of Pennsylvania" could be contaminated. The probability of this happening is so extremely low that it is not possible to produce reasonable estimates of it.

These reports were useful because they forced the AEC to formulate a safety philosophy and because they caused the AEC and the reactor

operators to take steps that had not been—but could be—taken to improve safety. The NRC (as did the AEC) demands three levels of safety (8):

  i maximum safety in normal operations,
  ii backup safety systems to minimize the effects of possible accidents, and
  iii additional safety systems in case the safety systems in (ii) fail when an accident occurs.

In backup ii, for example, is the provision for onsite electric power systems to run pumps in case of loss of offsite power, and for SCRAM (self-controlled remote automatic) systems (for quickly shutting the reactor down). In backup iii, an emergency core cooling system (ECCS) is designed to deal with a loss-of-coolant accident (LOCA). One of the most telling criticisms of the early reports was that no adequate tests of the emergency core cooling system were available, and that no adequate idea of the probability of an accident is known. For this purpose, the AEC commissioned WASH-1400 (9), known as the Rasmussen report. The purpose of the study was to estimate risk to the public from potential accidents in commercial nuclear power plants and to compare that risk to other risks. Most of the serious accidents it examined followed from LOCAs or from *transients* (short-lived events). It also identified valve malfunction, human error, and loss of offsite power as major contributors to these accidents' causes.

In a LOCA, an instantaneous break occurs in the reactor coolant system. Immediately after shutdown (SCRAM), the fuel is still generating about 7 percent of the heat generated under normal operating conditions. Heat generated decreases rapidly after shutdown, and the reactor would cool if the heat could be carried away. Since in a LOCA the cooling system is inoperative, the heat stays in the reactor core, where it could cause the core to melt. The molten core would generate steam as the molten fuel drops into the water in the reactor pressure vessel, and cause chemical reactions between hot fuel and structural material and steam. The reactions could generate substantial amounts of hydrogen, which could cause an explosion and the consequent

RISK VERSUS BENEFIT

The risk one is willing to take seems to increase faster than the ensuing benefit ($R \approx B^3$). Note, however, that people seem to tolerate risks that are voluntary at a level about 10,000 times greater than risks that are involuntary, for the same benefit. The disease risk (about 10^{-6} in Figure 25.6) seems to provide the benchmark for voluntary toleration of risk (13). There are also indications that acceptable risk decreases in response to involvement of larger numbers of people—for example, as indicated previously in the comparison of risk of auto and air accidents. Accidents threatening society itself (very large mortality over very short time) are not acceptable until the risk is *much* lower than indicated in Figure 25.6. Other characteristics influencing risk assessment include familiarity, level of knowledge about risk, and considerations of equity (14).

In order to illustrate these ideas, Starr and his associates considered the case of a fire in the 2 million barrels of oil stored for a 1000 MW_e oil-fired plant as compared to an equivalent catastrophic release from a nuclear reactor (13). They found the fire to be substantially greater as a real hazard than the nuclear accident, although it seems that the nuclear accident was more frightening. This is similar to the conclusions reached by the Rasmussen report, and serves to demonstrate that the report is not *totally* unrealistic.

Society is not risk-free, and it must decide how to allocate rare resources to manage risks. The refusal on the part of some—those who refuse to recognize TANSTAAFL (Chapter 1)—to accept that there is no absolute safety may lead the great mass of people astray. To illustrate this, consider dams. Dams generate hydroelectricity and are widely perceived as benign. However, dams have failed at a rate of 1/5000 per dam per year historically, and a UCLA study indicated that 250,000 deaths could result from failure of the largest California dam (15). The researchers also found that the failure rate was probably closer to 0.01 per dam per year. In the San Fernando earthquake of 1971, the Van Norman Dam almost failed because of soil liquefaction; had it gone, the water from the dam would have caused 50,000 to 100,000 deaths among the people living in suburban communities just below it (15). Every alternative has its drawbacks.

Many dam failures have caused minimal loss of life (for example, the Baldwin Hills, California, dam, the Teton dam in Idaho, the Malpasset dam, near Frejus, France), but the failure of the Vaiont dam in northern Italy took 1925 lives (16). Thus, the risk of failure can be great if population density is high near the dam.

The most common cause of simple gravity dam failure is sliding along the foundation, but dams are also subject to flood failure, earthquake, human misjudgment, and so on (16). Gravity dam failure probabilities are typically estimated at $\approx 10^{-5}$ to 10^{-4} per dam per year, roughly the same as the historical dam failure rate (15,17).

escape of fission products into the surroundings (8,9).

Transient events could cause the reactor's power to increase beyond the heat removal system capacity or cause the heat removal capacity to drop below the core's rate of heat generation. Such transient events could result from operator error, malfunction, or failure of equipment.

Of course, no matter how it is caused, a core meltdown would take some time to occur, during which the emergency core cooling system would be presumed to be operating. This time delay could allow for evacuation of the area immediately adjacent to the power plant. (We learned during the TMI accident that, despite the Rasmussen report's recommendation that evacuation plans be prepared, this had not been done by the time of the TMI event.)

Despite these efforts to cool the core, there may be a *core melt,* in which the core would melt straight through the containment vessel. The fear that it would melt its way through the Earth to the other side characterizes the "China syndrome." This cannot happen; the core is relatively safe in the ground unless it encounters large quantities of ground water (11). It would melt downward until its rate of heat output is the rate at which heat is conducted away from the molten rock around the core; these rates are equal at a depth of about 3 m (12). As the core cools, the rock would fuse into a glassy surface, which would prevent immediate water contamination. Ground water would carry heat away; after the core cooled, however, there would be a danger of slow release of radioactivity through leaching action of ground water on rock. If the glassy material cracked, then there could be a rapid release of radioactivity, but this is not thought likely. The main danger in such a core melt is formation of a blowhole through the top from which vaporized core material could escape. This *would* release considerable radioactive material and cause health hazards, in a manner similar to what happened at Chernobyl.

The Rasmussen Report

The Rasmussen report analyzed risks for plants such as are now in operation in the United States. To do this, they chose two designs for modeling: one each for boiling water reactors (BWRs) and pressurized water reactors (PWRs). (See figures 25.1 and 25.2.) Because of the wide variation in design, some features of the modeled reactors

FIGURE 25.1
Histogram of PWR radioactive release probabilities. (U.S. Nuclear Regulatory Commission)

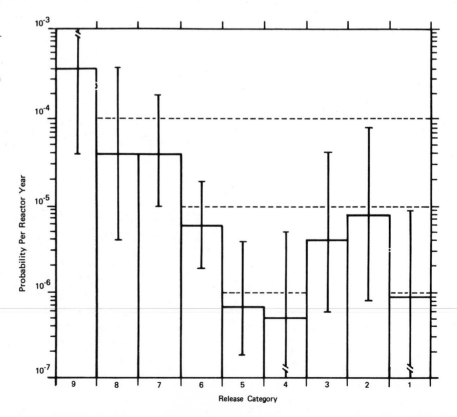

FIGURE 25.2
Histogram of BWR radioactive release probabilities. (U.S. Nuclear Regulatory Commission)

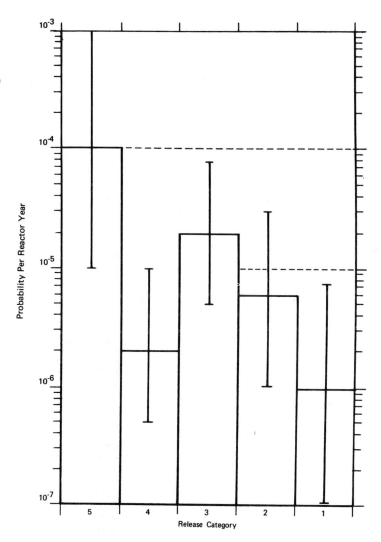

are not found in all reactors in use. The study relied heavily on the use of *fault tree* and *event tree* analysis. An event tree begins with an event and then traces all paths stemming from an event (within reason) to see if an accident results. A fault tree begins with an accident and attempts to trace backward in time to discover the circumstances leading to the accident in question. Illogical sequences are eliminated by computer, and sequences characterized by small consequences are eliminated by an analyst.

The analysis described resulted in identification of 78 accident sequences of interest (from a possible 130,000) falling into fourteen groups

of accidents. These were studied in great detail. The results of the analysis are summarized in Table 25.1, and the health effects—immediate and delayed consequences—as given in the Rasmussen report are given in Tables 25.2, 25.3, and 25.4. The categories of release in Table 25.1 correspond roughly (for each reactor type) to increasing probability of occurrence (see Figures 25.1 and 25.2) as well as to decreasing severity. Roughly, the worse the consequences, the less probable the accident.

The study was supposed to compare the risks of nuclear energy to other risks. Some of the most controversial results are presented in Fig-

TABLE 25.1
Summary of accidents involving core.

Release Category	Probability per Reactor-Yr	Time of Release (h)	Duration of Release (h)	Warning Time for Evacuation (h)	Elevation of Release (meters)	Containment Energy Release (10^6 Btu/h)	Fraction of Core Inventory Released[a]							
							Xe-Kr	Org I	I	Cs-Rb	Te-Sb	Ba-Sr	Ru[b]	La[c]
PWR 1	9×10^{-7}	2.5	0.5	1.0	25	520[d]	0.9	6×10^{-3}	0.7	0.4	0.4	0.05	0.4	3×10^{-3}
PWR 2	8×10^{-6}	2.5	0.5	1.0	0	170	0.9	7×10^{-3}	0.7	0.5	0.3	0.06	0.02	4×10^{-3}
PWR 3	4×10^{-6}	5.0	1.5	2.0	0	6	0.8	6×10^{-3}	0.2	0.2	0.3	0.02	0.03	3×10^{-3}
PWR 4	5×10^{-7}	2.0	3.0	2.0	0	1	0.6	2×10^{-3}	0.09	0.04	0.03	5×10^{-3}	3×10^{-3}	4×10^{-4}
PWR 5	7×10^{-7}	2.0	4.0	1.0	0	0.3	0.3	2×10^{-3}	0.03	9×10^{-3}	5×10^{-3}	1×10^{-3}	6×10^{-4}	7×10^{-5}
PWR 6	6×10^{-6}	12.0	10.0	1.0	0	NA	0.3	2×10^{-3}	8×10^{-4}	8×10^{-4}	1×10^{-3}	9×10^{-5}	7×10^{-5}	1×10^{-5}
PWR 7	4×10^{-5}	10.0	10.0	1.0	0	NA	6×10^{-3}	2×10^{-5}	2×10^{-5}	1×10^{-5}	2×10^{-5}	1×10^{-6}	1×10^{-6}	2×10^{-7}
PWR 8	4×10^{-5}	0.5	0.5	NA	0	NA	2×10^{-3}	5×10^{-6}	1×10^{-4}	5×10^{-4}	1×10^{-6}	1×10^{-8}	0	0
PWR 9	4×10^{-4}	0.5	0.5	NA	0	NA	3×10^{-6}	7×10^{-9}	1×10^{-7}	6×10^{-7}	1×10^{-9}	1×10^{-11}	0	0
BWR 1	1×10^{-6}	2.0	2.0	1.5	25	130	1.0	7×10^{-3}	0.40	0.40	0.70	0.05	0.5	5×10^{-3}
BWR 2	6×10^{-6}	30.0	3.0	2.0	0	30	1.0	7×10^{-3}	0.90	0.50	0.30	0.10	0.03	4×10^{-3}
BWR 3	2×10^{-5}	30.0	3.0	2.0	25	20	1.0	7×10^{-3}	0.10	0.10	0.30	0.01	0.02	3×10^{-3}
BWR 4	2×10^{-6}	5.0	2.0	2.0	25	NA	0.6	7×10^{-4}	8×10^{-4}	5×10^{-3}	4×10^{-3}	6×10^{-4}	6×10^{-4}	1×10^{-4}
BWR 5	1×10^{-4}	3.5	5.0	NA	150	NA	5×10^{-4}	2×10^{-9}	6×10^{-11}	4×10^{-9}	8×10^{-12}	8×10^{-14}	0	0

[a] A discussion of the isotopes used in the study is found in Appendix VI of Reference 9. Background on the isotope groups and release mechanisms is found in Appendix VII of Reference 9.

[b] Includes Mo, Rh, Tc, Co.

[c] Includes Nd, Y, Ce, Pr, La, Nb, Am, Cm Pu, Np, Zr.

[d] A lower energy release rate than this value applies to part of the period over which the radioactivity is being released. The effect of lower energy release rates on consequences is found in Appendix VI of Reference 9.

TABLE 25.2
Consequences of reactor accidents for various probabilities for one reactor.

Chance per Reactor-Year	Consequences				
	Early Fatalities	Early Illness	Total Property Damage 10^9	Decontamination Area ~ Square Miles	Relocation Area Square Miles
One in 20,000[a]	<1.0	<1.0	<0.1	<0.1	<0.1
One in 1,000,000	<1.0	300	0.9	2000	130
One in 10,000,000	110	3000	3	3200	250
One in 100,000,000	900	14,000	8	—	290
One in 1,000,000,000	3300	45,000	14	—	—

[a]This is the predicted chance of core melt per reactor year.

TABLE 25.3
Consequences of reactor accidents for various probabilities for one reactor.

Chance Per Reactor-Year	Consequences		
	Latent Cancer[b] Fatalities (per year)	Thyroid Nodules[b] (per year)	Genetic Effects[c] (per year)
One in 20,000[a]	<1.0	<1.0	<1.0
One in 1,000,000	170	1400	25
One in 10,000,000	460	3500	60
One in 100,000,000	860	6000	110
One in 1,000,000,000	1500	8000	170
Normal incidence	17,000	8000	8000

[a]This is the predicted chance of core melt per reactor year.
[b]This rate would occur approximately in the 10 to 40 year period following a potential accident.
[c]This rate would apply to the first generation born after a potential accident. Subsequent generations would experience effects at a lower rate.

TABLE 25.4
Consequences of reactor accidents for various probabilities for 100 reactors.

Chance Per Year	Consequences				
	Early Fatalities	Early Illness	Total Property Damage 10^9	Decontamination Area Square Miles	Relocation Area Square Miles
One in 200[a]	<1.0	<1.0	<1.0	<1.0	<1.0
One in 10,000	<1.0	300	0.9	2000	130
One in 100,000	110	300	3	3200	250
One in 1,000,000	900	14,000	8	b	290
One in 10,000,000	3300	45,000	14	b	b

[a]This is the predicted chance per year of core melt considering 100 reactors.
[b]No change from previously listed values.

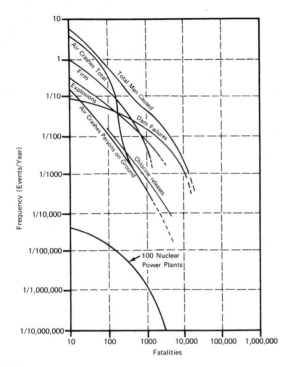

FIGURE 25.3
Frequency of man-caused events involving fatalities.
(U.S. Nuclear Regulatory Commission)

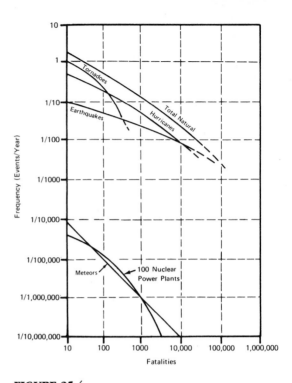

FIGURE 25.4
Frequency of natural events involving fatalities. (U.S.
Nuclear Regulatory Commission)

ures 25.3, 25.4, and 25.5. The comparison shown
in Figure 25.4, in particular, between fatalities
caused by being struck by meteorites and fatalities caused by accidents at a hundred nuclear
power plants, was overstressed in press reports.
However, the graphs do serve to express the main
conclusion of the Rasmussen report—that the
danger of nuclear energy pales in comparison to
most other hazards. As a result, the study has
been hailed by the nuclear industry as a vindication of nuclear power as is.

Many believe that the report's flaws do not
overwhelm the reasonableness of its conclusions.
The Birkhofer report (10) prepared for the West
German government reaches conclusions similar
to those of the Rasmussen report: 104,000 eventual fatalities over thirty years for the worst case,
which had a probability of 5×10^{-7} per reactor-
year.

Risk–Benefit Analyses

Let us suppose for the sake of argument that the
report's conclusions are correct insofar as the
probability of catastrophic hazard is small. This
might not be as reassuring as it sounds at first.
Large accidents cause more impact (even if rare)
than many smaller accidents of similar or even
larger total mortality. Many people who have no
qualms about driving a car (and in the United
States alone there are about 50,000 deaths from
auto accidents per year) are unwilling to travel
by air (typically 2000 deaths per year); deaths
from auto accidents occur by ones and twos, while
those from air disasters occur by the hundreds
(see Table 25.5). Some of these considerations
have been investigated by Chauncey Starr and his
coworkers (13). Starr observed that risk (R) and
benefit (B) seem to be related for both voluntary
and involuntary activities (see Figure 25.6).

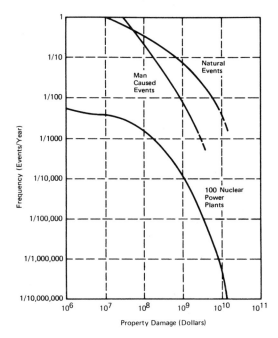

FIGURE 25.5
Frequency of accidents involving property damage.
(U.S. Nuclear Regulatory Commission)

The Lewis Report

The Rasmussen report stimulated a great deal of controversy after its release in draft form, some of which subsided with release of the final version in 1975 (9). The NRC was impelled by the furor to set up an independent review board, which produced what is now known as the "Lewis report" (18). Professor Lewis was the chairman of the American Physical Society (APS) study group on light water reactor safety, which issued a critique of the Rasmussen report (19). The conclusions of the Lewis report have led the NRC to recognize that the task assigned the Rasmussen group must be continually repeated and updated, and the NRC has also backed away from total acceptance of the report as is.

The Lewis report considers the Rasmussen report to be a substantial advance over previous attempts to quantify the results of accidents. It found the fault tree/event tree method useful,

and praised the report for demonstrating the importance of other than early fatalities, and for identifying small LOCAs, transient events, and human error as important to overall risk (contrary to previous belief). The Lewis report recommends that these newly discovered weak points should determine priorities of the NRC in the areas of safety research programs and in enforcement of regulations.

The APS study group (19) raised some questions that remain unanswered, and the Lewis report has made some very strong criticisms of the Rasmussen report. Among the most important are:

- both studies question the weather (dispersion) model used in WASH-1400;
- the numbers characterizing risk garnered from multiplications of probabilities (each uncertain) "should not be used uncritically" (17), even though *most* are probably correct to within a factor of 10; it was recommended that the numbers be presented as a range of probabilities rather than as one number (the square-root-bounding-model probability, which is the square root of the product of the high and low probabilities);
- the radiation risk model used by WASH-1400 may not be applicable in all cases (see subsequent discussion of the absolute risk model); if the other model is adopted, fatalities would increase by a factor of 1.5 over a thirty year plateau period or by a factor of 4 for a lifetime at risk;
- the assumed evacuation of the affected population was characterized as more pious hope than reality; lack of evacuation would increase early fatalities by a factor of 3;
- the use of a dose-reduction factor from the linear dose model was unjustified;
- the estimated numbers of fatalities are probably good only to within a factor of 5 to 10;
- the neglect of fires, earthquakes, and human error in determining overall risk is probably unjustified; and
- the next report should address the problem of sabotage if possible.

TABLE 25.5
Individual risk of early fatality by various causes (U.S. population average, 1969).

Accident Type	Total Number for 1969	Approximate Individual Risk Early Fatality Probability/yr[a]
Motor vehicle	55,791	3×10^{-4}
Falls	17,827	9×10^{-5}
Fires and hot substance	7451	4×10^{-5}
Drowning	6181	3×10^{-5}
Poison	4516	2×10^{-5}
Firearms	2309	1×10^{-5}
Machinery (1968)	2054	1×10^{-5}
Water transport	1743	9×10^{-6}
Air travel	1778	9×10^{-6}
Falling objects	1271	6×10^{-6}
Electrocution	1148	6×10^{-6}
Railway	884	4×10^{-6}
Lightning	160	5×10^{-7}
Tornadoes	118[b]	4×10^{-7}
Hurricanes	90[c]	4×10^{-7}
All others	8695	4×10^{-5}
All accidents	115,000	6×10^{-4}
Nuclear accidents (100 reactors)	—	2×10^{-10}[d]

[a]Based on total U.S. population, except as noted.
[b]1953–1971 average.
[c]1901–1972 average.
[d]Based on a population at risk of 15×10^{6}.

FIGURE 25.6
Model of risk versus perceived benefit for various activities. (C. Starr, copyright 1969 by AAAS)

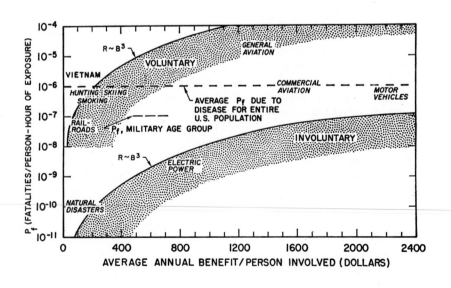

According to these responsible criticisms, then, the Rasmussen report (while representing a substantial advance over previous attempts) should not be taken uncritically, and particularly, the use of numbers for probabilities is tantamount to misrepresentation. We should be extremely skeptical of results presented in the manner of Figures 25.3, 25.4, and 25.5. Perhaps the most compelling reason for treating the Rasmussen report with caution is the occurrence of the accident at unit 2 on Three Mile Island near Harrisburg, Pennsylvania. There have been no modeling studies of graphite-steam reactors such as the one at Chernobyl, and so, despite the accident there, one can make no conclusions about the value of modeling from it.

THE THREE MILE ISLAND ACCIDENT

On 28 March 1979, a serious accident involving a partial core melt took place at the three-month-old Three Mile Island pressurized water reactor (PWR) owned by the Metropolitan Edison Company, a subsidiary of General Public Utilities Corporation. Middletown, Pennsylvania, became the focus of unprecedented public attention; public opinion polls after the accident showed that only 33 percent of those polled were willing to have a nuclear plant in their communities, down from the 56 percent of the previous poll (20). The subsequent events at Chernobyl (see below) had an even stronger negative public impact.

It now appears that valve malfunction, pump malfunction, faulty design, and human error were responsible for the TMI accident. Ironically, it was just such causes that were implicated as major factors by the Rasmussen report (9). The course of events in the accident (21,22) actually began two weeks before, when two valves were mistakenly left closed after a routine maintenance check. The accident itself began when a pump stopped. One of the operators had been working near the pump just before it failed. The pump that failed should have been recirculating water from the condenser back to the reactor vessel. Three backup feedwater pumps should have be-

gun to operate when the pump failed—to take off the excess heat generated—but did not, because the closing of the valves two weeks previously had deprived them of their water supply. Since heat removal was inadequate, the temperature rose, increasing the pressure inside the reactor vessel. An automatic reactor shutdown accompanied by a shutting off of the turbine occurred because of the high pressure.

When the pressure rose unnoticed through about 153 times atmospheric pressure (2250 psi), a relief valve opened as it had been designed to do. This reduced the pressure. As the pressure dropped back through normal operating pressure, 150 times atmospheric pressure (2200 psi), the valve failed to close as it should have, and the control board incorrectly showed that it had. The pressure therefore continued to drop. When the pressure hit 109 times atmospheric (1600 psi), the high-pressure coolant injection system, part of the emergency core cooling system, automatically turned on (21).

The operators turned off one pump for this system two minutes after the system turned on, and turned off a second pump ten minutes thereafter. This was done because a water indicator gave a faulty reading that indicated the water level was high enough (it was built for steady operation, and was unreliable during the rapid changes that were occurring). This act guaranteed that the core would be uncovered. The operators also turned off one of the primary circulation pumps an hour and a quarter into the accident and turned off the other pump after 1.8 hours, because the pumps were vibrating. This course of action again caused heat buildup and exposed even more of the core (22).

At the worst point in the accident, 15 to 30 percent of the core was uncovered and apparently remained so for 13 ½ hours (21). The uncovered part of the core became very hot because there were steam voids, and so there was no water to conduct the heat away as would have happened if water had been there. As a result, an estimated 45 percent of the core melted and as much as 70 percent was damaged. Fuel assemblies broke open and dumped rubble into the lower part of the reactor (23). The damage was known just after the accident to have been severe because of the presence of a "hot" spot in the core of the cooled reactor. However, decontamination revealed that about 20 tonnes of core material was sitting at the bottom of the reactor vessel, a much more severe consequence than expected. This discovery raised the question of how the vessel was able to maintain its structural integrity during the accident (23).

When the pumps were finally restarted, it was found that a large gas bubble had developed in the core that contained about 2 percent hydrogen. At the time, this development raised fears of an explosion. The presence of hydrogen actually helps oxygen to reform water, and these fears appear baseless. Incidentally, the production of hydrogen gas by a hot reactor core was anticipated and discussed in the Rasmussen report (Reference 9, Sec. 3.3, main report).

The accident became more severe when the liquid released from the reactor vessel by the stuck valve caused a seal to rupture in a holding tank, ultimately spilling some 400,000 gallons of radioactive water onto the floor of the containment building.

One of the problems contributing to the accident was a lack of skilled operators and an overly casual attitude about the work. Another problem was the engineering design of the control room: The TMI control room was an intimidating place, with alarms going off occasionally. The instrumentation was designed for ease of fabrication, not for communication of information (24). Human error cannot totally be eliminated, but faulty design and lack of training are reparable.

Decontamination of the reactor has taken more than eight years. The major release from the plant was radioactive iodine, which was produced when the fuel rods cracked; much less escaped than would have been anticipated. No radioactive krypton or xenon, released at the same time, got out of the plant. The total dose (see the explanation of dose in the section on health effects

immediately following) to the population adjacent to the plant (21,25) appears to be impossible to determine exactly but is thought to have been low.

Estimates of the effects on health using the standard linear dose model (see discussion later in this chapter) indicate less than one extra death among the 2 million people exposed. The severest critic, using a model enhancing the effect of low doses, made a prediction that about fifty extra deaths would result over the course of thirty years (compared to the 110,000 normal cancer deaths over this time). It will be seen that it will be extremely difficult to find any effect of the accident on health. The statistical fluctuation on the estimate of 110,000 is the square root of 110,000, about 330 deaths, over six times the number predicted to die in this worst case.

Another cause for concern brought to light by the accident is the misrepresentation of the facts by the utility, Metropolitan Edison, before the NRC became involved. One observer has characterized their description of the reactor as "stable" as roughly equivalent to the description of an out-of-control train as stable because its speed is not changing. This sort of misrepresentation is unfortunate and contributes to the public mistrust of nuclear energy and of the utilities. It also weakens the voice of the responsible commentators in the nuclear energy debate.

Further, it lends credence to the suspicion that all utility executives are venal, as they are characterized in the film *China Syndrome,* a fictional account of a nuclear accident.

The TMI accident was almost repeated at the Davis-Besse nuclear plant in Ohio in June, 1985 (26). The same set of pumps failed; the valve did not reclose; the operator pushed the wrong buttons in response to system failure; and fourteen separate pieces of equipment failed. This time, however, the problem was recognized in time to prevent damage.

One of the most useful results of the TMI accident was the reanalysis it spurred of the release of radioactivity (so-called source terms). It had been expected that a lot of iodine-131 would be released. There were 64 megacuries (see the definition of the curie in the section on health effects) of radioactive iodine in the containment vessel, but only 17 curies was released, one four-millionth of what had been expected. Three reviews were done: the American Nuclear Society (ANS); the industry degraded-core rule making program (IDCOR); and the American Physical Society (APS). All studies agreed that source terms are lower than calculated in WASH-1400, so that many of those accident sequences have lower releases (12,27,28).

The Wilson report (28), the APS study, was more cautious than the others, suggesting that it is impossible to say that the "calculated source term for any accident sequence involving any reactor plant would always be a small fraction of the fission product inventory at reactor shutdown" (28). The studies agree that the iodine forms cesium iodide (less volatile than iodine) and that tellurium, another radioactive element, forms compounds with zirconium and stainless steel. The Wilson report presents its caution with a scenario for release of material from containment: With the power off and primary coolant lost, the ECCS is activated. The core eventually melts through the containment vessel and gets to the floor of the containment building. The core boils off the water surrounding it, melts again, and interacts with the concrete of the structure to generate hydrogen and carbon monoxide (12,27,28). The pressure could cause confinement to be breached, emitting aerosols to the atmosphere. Such a sequence of events could cause contamination of soil and bodily harm.

CHERNOBYL

On April 25, 1986, the worst known nuclear accident of the nuclear age came to pass at the Ukrainian generating facility at Chernobyl. A fire combined with a breach of the reactor's containment structure to spray radioactive contamination over the local area as well as over much of eastern and western Europe. According to the report written by the USSR State Committee on Utilization of Atomic Energy for the International Atomic Energy Agency (29), the Chernobyl accident occurred mainly because of human error. The operators committed at least six serious violations of the operations protocol, including disabling all technical protection systems. "The designers of the reactor facility did not provide for protective safety systems capable of preventing (such) an accident . . . since they considered such a conjunction of events to be impossible" (29).

The conditions that led to the accident arose because of the operators' concern about what would happen if there were a failure in the offsite electrical supply. All nuclear generating stations draw their operating electricity from offsite, and all have local backup generators to provide onsite energy in case of failure of offsite supply. The concern was justified, since an offsite electricity loss was actually experienced at the Kursk nuclear station in 1980 (30,31). For the Russian reactors of the type RBMK, it is especially important to protect against loss of offsite electricity because of the necessity for sufficient water circulation (30) and because of the need for computer control of the reactor at all times, since its design entailed a "positive void coefficient," which made it susceptible to runaway reactions (see Chapter 14).

The nuclear engineers in the USSR decided to use the kinetic energy of the turbines at the stations to supply electricity to the pumps, computers, etc., during the time required to engage the diesel backups (fifteen to sixty seconds). This method had already been subject to testing at various stations, including Chernobyl (29). During the afternoon of 25 April 1986, the operators of the graphite-steam nuclear reactor unit 4 at Chernobyl were engaged in such a "turbine inertia" test.

The test began at 1 o'clock, as operators reduced the power output to half (1600 MW_t) over a twelve hour period. At 13:05 o'clock, one of the turbines was switched off (in accord with the test protocol), and at 14 o'clock, the emergency cooling system was disconnected. At this point, the shutdown was stopped because the electricity from the reactor was needed in the distribution system (29,30). Stopping a shutdown is a violation of experimental and operating protocol. At 23:10 o'clock, the shutdown resumed, and the test (to take place in the vicinity of 700 to 1000 MW_t) continued. However, as the reactor was being slowly shut down, neutron-absorbing xenon had been building up. As a result, when the operators shut off the local automatic regulating system according to the operating rules, the power output plunged to 30 MW_t.

The operators pulled all the manual control rods to raise the power output, and attained 200 MW_t by 1 o'clock on the 26th (30). This placed the reactor in a precariously stable condition; despite this, it was decided to conduct the experiment. Two additional pumps were started in order to join the six operational pumps, so that four could be shut down during the test. This caused the flow rate to jump and the reactor steam pressure to drop toward the emergency trip level. This was a violation of operating instructions; when too much water flows through the pipes, regions of voids, or bubbles, build up and cause vibrations (29,30).

The operators cut off information to prevent the automatic trip, and ignored a printout requiring immediate shutdown. Because of the drop in steam production, all the automatic control rods withdrew (30). At 1:23 o'clock the operators blocked the closing of the emergency regulating valves, so that the test could be repeated if necessary, again in violation of all operating procedure and test protocol.

At 1:23:40 o'clock, the shift foreman realized that something had gone dreadfully wrong and ordered emergency SCRAM. The control rods began to engage, but then halted with a shock. The Russian computer analysis indicated that within three seconds of SCRAM, the power rose above 530 MW_t for some seconds. The increase in heat output probably caused some pressure tubes to rupture (29,32,33). As the pressure tubes ruptured, water reacted with zirconium from the fuel

rod cladding to produce hydrogen, and steam and graphite reacted endothermically to produce hydrogen and carbon monoxide (29,30). Presumably, the high pressure breached the containment around the many pipes that penetrate the containment vessel, allowing oxygen inside the containment. It is probably at this point that the 1000 tonne cover plate lifted (34). This led to immediate ignition of the hydrogen at the high temperature inside the core.

At 1:24 o'clock on 26 April 1986, there was a loud bang, followed seconds later by a fireball and two explosions. The explosions pulverized fuel rods, and the rising plume from the explosion carried the debris upwards about 500 meters (35–37), protecting the region nearby from immediate contamination. Flames rose over 30 meters into the air as the graphite caught fire.

The high flames carried most radiation upward, so that relatively few deaths occurred (all thirty-one of the dead were plant workers or firefighters) (33,34), and the evidence of severe irradiation of firefighters and workers did not appear until the next day in most cases. An additional fireman and several workers developed acute radiation sickness. The fire itself was put out by about 5 o'clock that morning, although smoldering continued until the core became buried by the lead and boron dropped by helicopter.

Local authorities apparently thought they had a "normal" fire, not a nuclear accident, according to a report by *New York Times* correspondent Harrison Salisbury, an expert on the Soviet Union (38). As a result, no one in authority knew what the Swedes were talking about when they detected the radioactive cloud and demanded an explanation on 28 April (38,39). This misconception on the part of the local authorities may explain why the unit 3 reactor was not shut off until 3½ hours after the start of the accident; why units 1 and 2 were not shut down until a day *after* the accident; and why local residents were not evacuated until over 36 hours after the start of the accident. The Russian delegates at the IAEA meeting in Vienna in August, 1986, were reportedly still incredulous at the circumstances surrounding the accident (30). The report (29) itself said that

the prime cause of the accident was an extremely improbable combination of violations of instructions and operating rules committed by the staff of the unit. The accident assumed catastrophic proportions because the reactor was taken by the staff into a non-regulation state in which the positive void coefficient of reactivity was able substantially to enhance the power excursion.

J. M. Hendrie, former Nuclear Regulatory Commission chairman, agreed with the Russian analysis of the actions of the operators at Chernobyl (29): "What they did sounds like a very poorly considered invitation to disaster."

The command system itself should not be exonerated of responsibility. The RBMK reactor (see Chapter 14) was developed from a military reactor designed for plutonium production and converted to civilian use. It is for this reason that the reactor is designed for refueling while running. This leads to the presence of the thousands of pipes cutting through the upper surface of the containment vessel, which is a rectangular cavity more susceptible to breach than the curved containment vessels used elsewhere. It is known from the isotopes observed in the fallout that the Chernobyl reactor itself was not used for plutonium production (33).

The reactor design itself was faulty because of the positive void coefficient, and the instability problems plaguing the reactor were avoidable. The bureaucratic mentality was involved as well; nuclear energy expert R. Wilson says that "(t)hey specified a set of operating rules to be rigidly followed. But they forgot that rules that are not understood are often not complied with ... " (34).

The accident released about 2.5 percent of the radioactivity in the core into the environment (39). That makes it by far the largest nonweapon release of radiation publicly known. The radiation, in addition to being detected in Europe (36,39,40), was seen clearly on Japanese monitoring devices (41). Deposition of radioactivity was extremely nonuniform; where rain fell, much was deposited. Debris from the accident was carried so far because it was carried upward by the hot gas plume from the graphite fire. Levels of radioactivity were higher in countries bordering the USSR.

There had been an accidental release from a graphite reactor in 1957, at Windscale (now Sellafield) in England. The reactor produced plutonium for weapons. Because of operator error, it caught fire during a routine procedure to release "Wigner energy," an energy of dislocation caused by collisions of neutrons with the carbon atoms. The fire burned for about four days and released a great amount of radioactivity locally (32,42). That accident released only one-150th of what was released at Chernobyl. Consequences of this release and that of Chernobyl are discussed later in the chapter after some definition of terms.

HEALTH EFFECTS: THE PROBLEM OF CHRONIC EMISSIONS

In order to be able to discuss health effects in nuclear pollution, it is necessary to discuss radioactive *exposure* and *dose*. Radioactive decay involves emission of one to several particles from a nucleus. These particles may have high or low energy. The biological effects ensue due to energy loss along the path of the particle through the body. For example, a particle in the soft tissues of the body can ionize water, which reacts destructively with other cell units. For this reason, this radiation is known as ionizing radiation.

Exposure and Dose

An amount of radioactive nuclei or particles can be characterized by the number of nuclear disintegrations per unit of time: the *curie* (Ci) is 3.7×10^{10} disintegrations per second. The curie represents the activity of one gram of radium. It is being replaced by the *becquerel,* which is one disintegration per second. There are many dif-

NONIONIZING RADIATION

It may be helpful to contrast ionizing radiation, i.e., high-energy x rays or gamma rays, to nonionizing radiation such as microwaves. The energy of ionizing radiation, which has frequencies above 10^{15} Hz, is enough to expel electrons from atoms. For microwaves, with frequencies of 30 MHz to 300 GHz, this is not possible. The frequency used in microwave ovens, 2.45 GHz, corresponds to a photon energy of 10^{-4} eV (44). It takes several eV to ionize an atom—the ionization energy of hydrogen, for example, is 13.6 eV.

High intensities of gamma rays produce many ions in body cells. The ions disrupt cell function and may even lead to cell death. The difference between high- and low-intensity gamma rays is in the number of ions left behind. Exposure to high intensities of microwave radiation causes the polar molecules in tissues (mainly water) to rotate, and the rotating molecule collides with tissue, generating heat. Exposure to 1 kW/m² of microwave radiation will result in cooking the tissue.

Animal studies have shown that microwaves cause transient changes in the number of blood cells, changes in the structure of the cell nucleus, and changes in the way cells divide. Protracted exposure may affect animals' immune systems (44). Exposure of the testes causes temporary sterility.

Microwaves do not penetrate deeply in the body. Absorption is in the outer 20 mm of tissue (44). Thus, people do not get as much exposure as, say, mice or guinea pigs. Mice absorb thirty times more energy per unit mass than humans at 2.50 GHz.

---

ferent decay products in disintegrations, each having a different effect on health because of differing biological activity. For example, soon after the discovery of radioactivity, it was noticed that at least three different kinds of radioactivity were present: the so-called alpha rays, beta rays, and gamma (or x) rays. The alpha particles (helium nuclei—the bound states of two protons plus two neutrons) travel only a few tens of millimeters in air and may be stopped by cloth or cardboard; the beta particles (electrons from the nucleus) travel up to a few meters in air and are stopped by varying thicknesses of aluminum, iron, and lead; and the gamma rays (electromagnetic energy from the nucleus) are stopped only by substantial thicknesses of lead. Although alpha particles are easily stopped, if they do penetrate tissue, they are over ten times as dangerous to health as gamma rays.

The amount of energy deposited per unit length, known as *linear energy transfer* (LET), is high for alpha particles, lower for beta particles, and lowest for gamma rays. This means high-LET, or charged particle, radiation is more damaging than low-LET radiation (22,43).

To discuss the relative hazards of radiation, then, requires a unit that incorporates both energy loss and biological effectiveness. It is easy to do this for gamma rays, since they penetrate everything. The *röntgen* or *roentgen* (abbreviated R), named for the discoverer of radioactivity, measures the ionization of gamma radiation: This is a unit of exposure from gamma radiation producing one unit of charge per cubic centimeter in dry air at 0°C and atmospheric pressure (corresponding to an energy loss of 0.0877 J/kg air). Since we wish to discuss the effects on humans, a better unit of measurement is the *gray,* which

TABLE 25.6

Somatic effects of radiation dose.

Dose (rem)	Short-Term Effect
0–25	None detectable
25–50	Slight, temporary blood change
100–200	Nausea, fatigue
300–500	Half of exposed people die (LD_{50})
10,000	Most tissue destroyed

is a measure of absorbed energy: It is the dose from gamma radiation losing 1 J/kg of tissue. At this time, the *rad*—the dose from gamma radiation losing 0.01 J/kg of tissue—is more commonly used. The rad (0.01 gray) represents deposition from exposure to 1 röntgen in soft body tissue (22).

While all doses could now be measured in grays or rads, the biological effectiveness is different for the different types—protons, neutrons, and alpha particles are many times more dangerous than electrons or gamma rays. The biological effectiveness of alpha particles increases as their energy increases, in contrast to the case for gamma rays and electrons. It is thus convenient to define a unit based on the gray or the rad which also takes this difference into account. This unit is the *sievert,* which is equivalent in biological effect to one gray of gamma rays. The more common unit in use at present is the *rem* (*r*oentgen *e*quivalent, *m*an). For electrons and gamma or x rays, 1 rem ≈ 1 rad. For alpha particles at fission energies, 1 rem ≈ 1/20 rad. For

protons and neutrons, 1 rem ≈ 1/10 rad. The same relationship prevails between the sievert and the gray.

One effect we have not considered here is that doses from some sources concentrate in specific body organs. For example, iodine goes directly to the thyroid. The effects of various doses on people are given in Table 25.6.

Anyone who lives anywhere in the world is exposed to radiation. Radiation comes from cosmic rays, from bricks, from weapons fallout, from diagnostic x rays, and so on. In Table 25.7, we present the exposures in mSv (10^{-3} Sv) per person (45) and in total dose per year to the U.S. population (46). Some differences would exist in Table 25.7 for newer exposures, but they are not major. For example, as of 1984, the estimate for medical uses in the United States averaged 75 mrem/yr (47), and in West Germany, 50 mrem/yr (48). (The mean bone marrow dose from x rays varies from 52 mrem/yr for young adults and adolescents to 151 mrem/yr for persons over 65; a chest x ray gives a 10 mrem dose when properly administered, and a barium enema, 875 mrem [47].) Also, new consumer products such as smoke detectors using americium-241 contribute a dose estimated at 1 to 4 mrem/yr (47).

The population has been exposed to relatively high levels of ^{90}Sr and ^{137}Cs from fallout from weapons tests in the 1950s and 1960s. The nuclear test ban treaty in 1963 had clear effects on the amount of radioactivity in milk (Figure 25.7) (48). The dose, even in 1963—the year of maximum exposure—is about 3 percent of the natural back-

TABLE 25.7

Radiation exposures.

	U.S. Average Personal Dose (mSv)	Total Dose (10^3 person-Sv/yr)
Natural background	1.26	200
Medical arts	0.75 (1984)	170 (1978)
Technological (tailings, wastes)	0.02	10
Weapons development	——	1.65×10^{-3}
Nuclear fallout	0.04 (1963)	10–16 (1978)
Nuclear energy	0.002	0.56

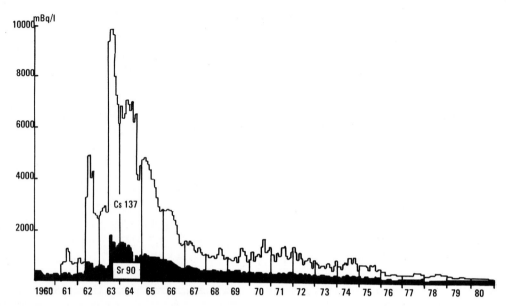

FIGURE 25.7

Radioactivity in milk in West Germany due to fallout from weapons testing. (West German Interior Ministry)

ground exposure. People engage voluntarily in activities, such as travel by airplane, that cause an increase in yearly dose by amounts larger than this 3 percent.

The total dose a person gets depends strongly on location. For example, the annual background dose in Denver from cosmic rays is some 0.60 mSv higher than at sea level. At some locations in the world, radioactivity from subsurface uranium deposits is an order of magnitude higher than at Denver. There have also been large exposures to some individuals because of inappropriate use of mine refuse (see accompanying box). With the existence of human exposure to radiation, whether or not there are nuclear weapons tests or nuclear power plants, it is now incumbent upon advocates of nuclear energy to estimate the health cost (remember, TANSTAAFL) and to attempt to compare these costs to the costs of competing energy sources.

The Three Mile Island accident resulted in the exposure of nearby residents to some radioac-

tivity. The accident released 630 GBq (17 curies) (28) of radiation into the environment. Estimates of dose range from 16 person-sievert to 53 person-sievert within 80 kilometers of the plant (with about 90 percent of the dose concentrated within 15 kilometers); an average release estimate of 33 person-sievert has been adopted. The highest conceivable dose—that to a person who standing at the plant boundary throughout the course of the accident—is 1 mSv (100 mrem). People who were within 1.5 kilometers of the plant the entire time probably received a dose of about 0.5 mSv, and those continuously within 3 to 5 kilometers of the plant probably received about 0.3 mSv (21,25).

The Chernobyl accident released about 1.85 EBq (or 50 MCi) into the environment, as totaled on May 6, 1986 (29). Because of the decay in activity of many of the products released during the time between the start of the accident and May 6th, the actual *integrated* exposure is to about 2.5 EBq (70 MCi). Because the plume carried the activity up so high, the exposed population at risk totals about 400 million. Levels of radioactivity measured outside the USSR ranged from 130 to 750 times background levels in Poland; 1 to 40 times background in Sweden; 1 to 30 times background in Finland, Switzerland, and West Germany; 1 to 10 times background in Austria; and 1 to 2 times background in Britain (36,39–41).

In West Germany, activity of 8–12 kBq/m³ of air were measured after Chernobyl; for comparison, weapons test fallout over West Germany in the 1960s produced measurements of 0.8 kBq/m³ (36). Thus, peak exposures in western Europe from Chernobyl were ten to fifteen times that from fallout. Of course, the exposure from fallout lasted much longer than that from Chernobyl. Simulations of the Chernobyl accident by Lawrence Livermore Lab (32) and the United Kingdom Atomic Energy Agency (49) imply that the dose from ^{131}I was greater than or equal to 10 Sv within a relatively small area running along a line from Kiev to Stockholm. The area within which the dose was likely to be above 1 Sv ranges up between the Norwegian border and western Finland, stretches from Danzig through Minsk to Voronezh and down through Prague to the Austrian border. The area over which a dose of 0.10 Sv is absorbed is bounded in the USSR by Odessa and Tula, and in the west, blankets Dresden, Munich, Graz, and Vienna.

The heaviest exposures at Chernobyl were suffered by the 444 technicians and firefighters on the plant site; 30 of these people died of radiation, and one is probably buried within the plant itself. The 45,000 people of the town of Pripyat were evacuated on April 27th, 1½ days after the accident. By early morning of that day, the exposure in the streets nearest the power plant was 0.18–0.6 R/h (50), and it was increasing. The IAEA report estimates their probable dose from gamma rays at 1.5–5 rem and the exposure from beta radiation to the skin at 10–20 rads (50). Somewhat later, an additional 90,000 people within a 30 km radius were evacuated. The total amount of radioactivity within this zone was estimated to be 0.7 EBq (51).

Residents of Pripyat received a total average dose of 30 mSv, and the 24,000 people living in the zone between 3 and 15 kilometers radius received a total average dose of 430 mSv (34). The report estimates a total collective dose of 16,000 person-sieverts (50) for all evacuees.

The IAEA report claims that there are no direct consequences for people outside the 30 km zone; they will be exposed for a short while to contaminated milk and will continue to suffer exposure from contaminated plant matter until at least 2036. It estimates a collective dose to the 75 million people at risk inside the Soviet Union is 86,000 person-sievert for 1986 and will total 290,000 person-sievert over fifty years (50). The figure for collective dose for 1986 is just less than the dose from natural background radiation (94,500 person-sievert) and somewhat larger than the dose from medical irradiation (56,000 person-sievert). Over fifty years, the exposure is about 6 percent of background exposure, 4.7 M person-Sv.

LINEAR DOSE THEORY

Comparison of health costs of competing energy sources is very difficult indeed, as well as controversial. We do have *some* evidence of radiation effect at very high doses, but what happens at low doses is difficult to determine (refer to the discussion of the Three Mile Island accident). However, there must be ways to estimate the effects, so that something can be said about low doses. The problem is that of making a *responsible* estimate.

The problem here is similar to that of determining if some chemical compound, say saccharin, is carcinogenic. Investigators could feed doses of saccharin to rats and observe the resulting number of cancers. However, there would probably be some cancers even if no rat had ingested saccharin. The investigator would then have to find the difference in the numbers of

INDUSTRIAL WASTES, MINE TAILINGS, AND EXPOSURE

There are many parts of the nuclear fuel cycle that generate radioactive waste. One that had attracted little notice until 1966 was that of mine tailings. Tailings are the ground rock left over after the uranium-bearing rock is processed to recover the uranium. Frequently, tailings were used as landfill and as building material. They are dangerous because of the radium and the radon gas produced when radium decays. The gas goes into the air, and the radium itself can be leached into the ground water (52). The ground rock is left in huge piles in the open, where wind can pick up the dust and carry it for miles (53). In fact, left as is, it is predicted that the tailings themselves will constitute a hazard greater than that of the reprocessed waste generated by the nuclear plants that used the uranium (52,54). This danger could be almost eliminated rather easily (see Appendix D of Reference 52) simply by putting a 150 mm (6 inch) clay cap on the pile. However, in many cases, the decline in demand for uranium has put the uranium mills out of business (employment fell from 22,000 in 1978 to 2000 in 1985) (55). This leaves the cleanup to be done by the federal government, at an estimated cost of $2 billion to $4 billion (55).

The rather casual attitude toward exposure to radioactivity in the past has led to discovery of abandoned tailings piles in downtown Salt Lake City (52) and in Denver (which also has some twenty sites contaminated by radium from the watch-dial industry) (56).

An even worse problem is that the tailings were used as landfill under building sites. The city of Grand Junction, Colorado, has hundreds of buildings in which permissible exposure limits have been exceeded many times. Many other towns in Colorado, as well as Salt Lake City, Utah, have similar, if smaller, problems.

Some Navajo Indians were exposed to high levels of radiation as a result of using rock left over from uranium mining for home building (57). Background exposure is 12–14 μR/h in the Navajo reservation, and inside exposures ranged from 11.3 to 33.6 μR/h (57).

Inadvertent exposure has also occurred in Tennessee, where local shale was used to make cement block. (Shale was mentioned as a source of reactor fuel in

cancers between two identical groups—one fed saccharin, one not. If we suppose that the risk of cancer is relatively low (for example, one in a million per year per rat), then for a sample size of two thousand rats, one would have to wait a thousand years to see a difference of one in the number of cancers. Unfortunately for the test results, most rats will have died of other causes before the end of this time! The investigator could use 2 million rats, and then at the end of the year see a difference of one cancer. But rats have

cancer at some rate—say, one in ten thousand. Then there would be 100 natural cancers, and 101 in the group treated with saccharin, which is within the statistical error of 10 in this example. It is clear that the investigator has caused a boom in the rat supply market. Not much else is clear.

If the investigator could feed (and not feed) a reasonable number of rats a very large amount of saccharin, an effect would probably be quite noticeable. If the investigator then knew how low doses of saccharin are related to high doses of

Chapter 14.) Doses as high as 0.7 Sv/yr could be received by inhabitants of houses made of such block (58,59). The discovery of low levels of radiation in some 100,000 homes and schools built from brick made with gravel from Florida phosphate mines (60) would come as no surprise to health physicists (58).

A similar discovery was made in Berks County, Pennsylvania. Stanley J. Watras, a worker helping to construct a nuclear generating plant, brought in enough radiation to trigger alarms on site (61). It was discovered that the levels of exposure in the Watras living room were so high that lung cancer was a certain result. Even a few years' exposure would have produced a high level of risk. A subsequent survey revealed that 40 percent of Berks County houses had unsafe radon levels (61). The problem here was that the so-called Reading Prong (running from Reading, Pennsylvania, through Peekskill, New York) has high uranium levels. Over 100,000 people live over the Reading Prong, most in the Morristown area of New Jersey (61). The affected states will have to deal with this problem (62).

Radon is part of a chain of decays descending from uranium. Radon itself is not dangerous; the decay products (daughters) are hazardous. Since the daughters of radon are positively charged, they are attracted to dust, which can then be inhaled. This internal exposure to alpha particles is very dangerous, so it is common to consider the dose to be that of all alpha-emitting radon daughters together, partly because the half-lives are so short. The total alpha energy emissions are then given in terms of working levels (WL); an exposure to 1.3×10^5 MeV per liter of air (or about 3700 disintegrations per liter of air) is called a WL (63,64). Typical background levels are 5×10^{-3} WL, and EPA recommends a maximum of 0.02 WL in a home.

The Watras living room registered 16 WL (61); the maximum exposure of the Navajo was 0.04 WL, much lower but still twice as high as the maximum recommended (57).

Many people around the world get very high doses from living on outcrops containing high levels of uranium and thorium. In the United States, in addition to high levels in Arizona, Colorado, and Pennsylvania, the Gassoway member of the Chattanooga shale has higher levels than background. In Brazil, a few people live in an area of monazite sands and get over ten times the normal dose, as do roughly 100,000 people in Kerala, India (47). There are no obvious health effects on these people.

saccharin, then something will have been learned. What the investigator does for saccharin dose is to assume a *linear dose theory* for dose-response. By this is meant that a dose 1/10 as large as that given would produce 1/10 as many cancers, or a dose 10 times larger would cause 10 times as many cancers (65). It was just such a procedure that led to the Food and Drug Administration's proposed ban on saccharin as a probable carcinogen. The linear dose theory specifically rules out the presence of thresholds below which there is no significant effect as a result because of cell repair or some other mechanism. There is some evidence for thresholds (66), but there is no proof either way.

There are problems with using mice and rats. Toxicity and carcinogenicity vary among species, as well as by the method of administration of the chemical (66). Cancer rates should be greater in humans than in mice, say, because humans have so many more cells, and thus more places for things to go wrong. Old mice should have much

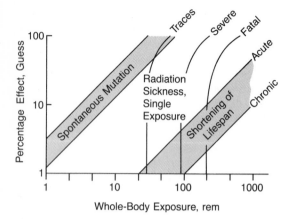

FIGURE 25.8

Effect of whole-body exposure to ionizing radiation. (U.S. Atomic Energy Commission)

lower rates of cancer than old humans; however, this is not found, which suggests that mice are thirty thousand to a billion times more susceptible than humans (66). Additionally, there are mouse cancers of organs that do not exist in human beings (67), which complicates study interpretations.

The investigator who concentrates on the effects of nuclear doses, and who faces similar problems, adopts an approach identical to the approach discussed for exposure to chemicals.

That is, the linear theory of dose-response is used to estimate the effect of low doses of radiation. There is the possibility that some dose would produce *no* response, of course—in this case, there is said to be a threshold dose. In the absence of proof that there are thresholds, the conservative thing to do is to choose the linear dose theory. Then any error made will result in an overestimate of the effect, an error on the side of caution. The linear dose theory is an expression of scientific ignorance. In the linear dose theory, we assume:

- there is no recovery from any dose;
- doses add no matter how much time has elapsed between them; and
- the graph of hazard versus dose is a straight line.

The theory is not without experimental support. Figure 25.8 (58) shows some indication of such straight-line behavior. A study of cancer risk to children due to prenatal x rays gives an apparent straight line (Figure 25.9 [58]). Another support comes from experiments in induction of mutations in fruit flies (Figure 25.10 [45]). Note that the mutation rate is not zero, even when no dose is applied; this is the mutation rate due to background radiation (and in fruit flies accounts for only about 1 percent of spontaneous mutations).

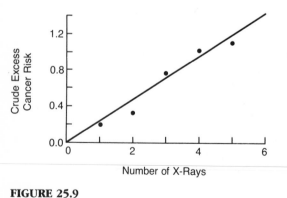

FIGURE 25.9

The risk of cancer grows as the number of X rays to which one is exposed increases. (U.S. Atomic Energy Commission)

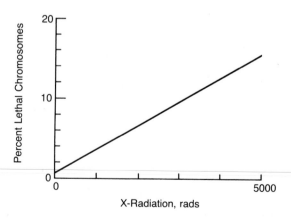

FIGURE 25.10

The genetic effect of exposure to X radiation.

Other support for linearity comes from studies of the leukemia rate in Hiroshima and Nagasaki after World War II.

The estimates of effects of nuclear doses are critically dependent on the experience of the survivors of the Hiroshima and Nagasaki bombings. For this reason, the recent controversy over the actual dose at Hiroshima is very important (68). Two groups, one at Lawrence Livermore Lab and another at Oak Ridge National Lab, agree that accepted figures for the neutron (high-LET) dose were grossly overstated—for example, they calculate that the neutron radiation at 1.18 km from the center was previously overestimated by a factor of 6 to 10 (68–70). If this recalculation is accepted, the Hiroshima and Nagasaki data can be combined, and this results in an increase in our estimate of the biological effectiveness of lower doses by a factor of 2–3 (68,69).

The effect of some doses is shown in Table 25.8 (71). The frequency of chromosomal aberrations is 0.1 per cell per sievert overall, and there are $4–18 \times 10^{-4}$ excess cancers/year/sievert ($2–10 \times 10^{-4}$ of which are fatal) from two to ten years after irradiation (43). The lifetime excess cancers are $14–100 \times 10^{-3}$/sievert, of which $7–50 \times 10^{-3}$/sievert are fatal (43). Overall, 1 to 3 percent of all cancers in the general population are attributable to natural background.

There has recently been some controversy over the appropriateness of the linear dose theory, with a few investigators claiming that risk *increases* at low dose (72), but the great majority of experts are now arguing among themselves over whether there are or are not thresholds (71). A strong majority of these experts is still of the opinion that the linear dose-response theory is the best method available for estimating risk from ionizing radiation. All the risks commonly quoted use this method. Thresholds are more likely for chemical mutagens, according to the latest ideas.

One of the problems with the acceptance of the Rasmussen report was the use of a dose-reduction factor for low dose. That is, the report assumes that low dose entails less risk than would be present if the linear hypothesis were accepted. The BEIR (Biological Effects of Ionizing Radiation) reports (73,74) have set no precedent for this, and the linear standard is the generally accepted one. In the opinion of the experts, the Rasmussen dose-reduction factor seems unjustified. Another problem with the report is the use of an absolute risk model (18,73). Again, the procedure is to try to adopt a conservative model of cancer risk in light of the lack of information on risk and the latency period for cancers (as much as four decades). The *absolute risk* model assumes that the radiation-induced cancer rate is

TABLE 25.8
Health effects of radiation.

Cancer Type	Maximum Chance of Early Death/rem/ Million people	Average Shortening of Life (days per rem per million people)
Leukemia	23	0.2
Lung	25	0.64
Stomach	12	0.065
Alimentary Canal	4	0.0073
Pancreas	4	0.0073
Breast	29	0.062
Bone	5	0.029
Thyroid	10	0.0062
Other	20	0.11

SOURCE: Reprinted with permission from *Health Physics* 35, 563 by R. L. Gotchy, Copyright 1978, Pergamon Journals, Ltd.

an extra additive risk independent of the rate of incidence of the particular type of cancer; the *relative risk* model assumes that the risk is some multiple of the extant rate for that type of cancer. The latter model gives higher numbers than the former because the cancer rate increases with age. Unfortunately, there is no way to decide the correct risk model—which model is appropriate may depend on the cancer type (18). However, the relative risk model satisfies the criterion of conservatism and therefore should have been used in the Rasmussen report.

Reactor Safety Standards

Now that we have some idea of the risk of radiation dose, it is possible to discuss reactor safety standards. The history of these standards shows a continual lowering of occupational and allowable public exposures, except in the area of medical technology (58,75). The permissible population exposure limits in the United States have most recently been lowered from 170 mrem/yr to 5 mrem/yr for emissions from nuclear power plants. The limits were lowered in response to the BEIR reports (73,74) and because of the previous use of too low a cancer/leukemia ratio by the International Committee on Radiation Practices. The studies on which the standards had been based had been conducted over too short a term to measure the rates adequately, because of the long delay time for the occurrence of some types of cancer (75,76). A related problem has to do with biological concentration mechanisms (77). Concentration travels up the food chain; it can be dangerous to eat fish because of their high radiation levels, even though they are taken from water that meets the safety standards for emission from nuclear power plants. (Biological concentration is discussed in Chapters 11 and 22.)

Estimating Health Effects

The estimate of the cancer death rate from the latest BEIR revision for which a consensus existed (73) is 7 to 35.3 fatal cancers per thousand per-son-sievert per year. This is about 30 to 140 person-sievert per fatal cancer.

Current workers disagree on risk estimates, but generally accept higher values (lower risk) than in BEIR: The DOE has adopted 23 cancers per thousand person-sievert (43 person-sievert/cancer) (78); the EPA uses 20 cancers per thousand person-sievert (50 person-sievert/cancer) (78); Wilson and Crouch give 12.5 cancers per thousand person-sievert (80 person-sievert/cancer) (79); and the 1977 United Nations benchmark is 10 cancers per thousand person-seivert (100 person-sievert/cancer) (78). To be most conservative, we overestimate the risk by continuing to adopt the BEIR values.

If the average allowable dose to a person from a particular nuclear plant is 5 mrem, we may use the linear theory (and BEIR values) and assume that *at worst* the entire U.S. population gets the average dose (although this is not likely, it would be allowed). Then, with a population of 240 million, the total yearly dose to the population would be 12,000 person-sievert (or 1.2 million person-rem). The plant could cause 84 to 422 extra fatal cancers per year under present standards.

The normal cancer death rate in the United States is about 360,000 per year, of which 80 to 90 percent are due to environmental effects (80). To make the estimate another way, leukemia occurs naturally at one to two cases of leukemia per million person-rem per year. For the allowed nuclear plant dose, 12,000 person-sievert, we could have as many as 1.2 to 2.4 extra leukemia deaths. Using an estimate of twenty eventual cancers per leukemia produced (54) leads to twenty-four to forty-eight extra cancers per year from nuclear plants. Notice that this is in rough agreement with the numbers of the BEIR report. It also gives a flavor of the sort of accuracy with which the estimates are made.

The dose from Three Mile Island, according to the BEIR guidelines, will cause 0.1 to 1.9 extra cancers per thousand years. In the same way, the Windscale accident, which released 300 kilocuries of ^{133}Xe, 20 kilocuries of ^{131}I, and 12 kilocuries of ^{132}Te (36,42), is estimated to have caused an extra thirty-nine fatalities (81).

CARELESSNESS AND THE NUCLEAR INDUSTRY

The nuclear industry has been plagued by a greater amount of both carelessness and publicity than some of the other health-threatening industries in America (see Chapter 21, 22, and 24). As we have seen, all three major reactor accidents were caused by human error. Even small accidents have attracted wide attention. Many mistakes have been made because the industry was learning to use nuclear energy at the same time it was operating the plants. There was a similar learning period, filled with both small and very large accidents, for the coal- and oil-fired plants when the utilities were learning to operate these facilities.

The industry has been criticized for poor engineering, carelessness, and failure to live up to the promises some boosters have made. A few examples are worth noting.

Valves have been a problem from the infancy of the industry. Indeed, at one point, defective valves killed workers in one plant. When the AEC asked utility companies whether they were using this defective type of valve, some utilities did not know that their plants used valves at all (85)!

Fuel rod specifications for reactors were not changed when the utilities began building PWRs. Because the pressures in the newer reactors were so much higher, the fuel pellets suffered a malady called densification (86). The fuel rods containing such densified pellets could, and some did, easily collapse. When the rods collapsed, radioactivity leaked into the water in the core.

Cracks were discovered in piping. The cracks had escaped discovery during the routine ultrasonic testing procedures.

Impurities were discovered in welds in some reactor vessels. Such impurities cause the weld to become more brittle with age, and thence susceptible to thermal shock (86). The operators of the Rancho Seco power plant in Sacramento, California, subjected that reactor to thermal shock by overcooling and overpressurizing the system in a way that would most stress brittle steel in such welds (87); luckily for everyone, there were no unfortunate consequences from the incident.

Many early supporters of nuclear energy had thought that the reactors would operate 80 to 90 percent of the time. This would help lower the cost of nuclear as opposed to fossil-fuel generation. Some critics have claimed that nuclear reactors are operating far below capacity, thus raising costs (88). Most analyses find that capacity factors for coal and nuclear are similar (89). The busbar costs (that is, the costs of the electricity as it leaves the facility) for older nuclear plants are consistently lower for nuclear than for other sources of energy (89,90). A promising incentive for raising the capacity factor is the effort of some states to allow the utilities to charge higher rates for higher reliability (91).

Waste treatment facilities have suffered from shoddy safety practices and poor engineering. A plant in Erwin, Tennessee, exposed more than 50 workers to radiation levels far above those permissible (5). The now-decommissioned Nuclear Fuel Services facility in West Valley, New York, exceeded allowed emissions and violated numerous safety regulations, and will cost up to 1.1 billion dollars to clean up (92). The Kerr-McGee involvement in the plutonium poisoning of Karen Silkwood, and possibly in her mysterious death while on her way to meet a *New York Times* reporter to discuss the company's safety violations, was the subject of a trial that led to a huge award to the Silkwood estate. It also reflected badly on the entire industry, and was publicized in a 1983 movie.

The Chernobyl disaster resulted in two fatalities during the fire as well as twenty-nine deaths from among the total of 203 people diagnosed as suffering from acute radiation sickness. The long-term followup data from the Chernobyl survivors will be of immense value to the rest of the world in determination of the effects of low doses of radiation. Elaborate plans for the followup should assure its success (50).

Using the BEIR numbers and the linear dose theory, the 135,000 evacuated people are expected to have about 17,000 spontaneous cancer deaths. The report (50) says that their natural death rate from cancer will be increased less than 2 percent as a result of Chernobyl for external exposures, i.e., less than 2700 extra deaths. The rest of the USSR is expected to suffer a comparable number of deaths.

The Poles'ye soils of Byelorussia and the Ukraine are humus-poor, and cesium will be absorbed into plants at rates one to two orders of magnitude greater than for other soil types. Hence, radioactive cesium will be a more pernicious influence on agricultural production there. The report estimates a collective internal dose for the region's population of 2.1 million person-sievert, which will cause an increase over seventy years of about 0.4 percent above natural mortality, causing an estimated extra 40,000 deaths (82).

The major internal dose will come from ^{131}I in milk, but this is rather easily controlled by a program of monitoring. Few additional deaths should result, except for those already exposed unwittingly to unmonitored contaminated milk in the early days after the accident (some to very high levels). The report estimates about 1500 additional deaths because of this (50,82). All in all, the number of additional deaths in the USSR should be less than fifty thousand if the basis of the calculations is correct.

These estimates are disputed by Gofman (83), who estimates a *minimum* of 330,000 fatalities on the basis of known dose effects, 145,000 of these in the USSR alone. Cochran estimates about 100,000 deaths in the USSR (82), in good agreement with Gofman. Other scientists believe the Russians may even have *overestimated* mortality by a factor of about 2 to 5 (82).

Estimates of Hohenemser et al. (36) for the effect on the city of Konstanz, West Germany, estimate that the 100,000 people will suffer 1.6 to 2.9 cancers per year from the effects of the extra exposure. A group this size would see 250 people die of cancer during a given year without any exposure above background. Such an effect is undetectable. For comparison, weapons test fallout of the early 1960s is presumed to be responsible for 1.5 to 1.6 extra cancers per year in Konstanz (36). The average dose in West Germany after one year of exposure is estimated by the German government to be 0.7 to 1.6 mSv for children and 0.5 to 1.1 mSv for adults (84). This would imply an extra mortality for all of West Germany of one to two thousand. By Gofman's estimates, on the other hand, West Germany will suffer an extra 27,000 cancer deaths due to Chernobyl (the only country to have a greater number of predicted fatalities is Rumania, with 46,000; the next worst effect is in Poland, for which 25,000 excess fatalities are predicted by Gofman) (83).

Over a thirty-year plateau period, the worst allowable extra burden of a nuclear power plant is thirty to seventy extra leukemia cases and 2300 to 12,000 extra cancer deaths out of a total exposed population of 10 million. This assumes that the plant operators are careful and the techniques of nuclear energy perfected. This is a somewhat unrealistic assumption, perhaps, as suggested in the discussion of the Chernobyl and TMI accidents.

We can go further and estimate the health cost to people in various professions. In the United States, radiologists die 5.2 years sooner than other doctors. If we assume a 10 sievert lifetime dose, taking delayed effects into account by dividing by 5, the linear dose theory tells us that a 1 sievert dose shortens life by about a day. Thus, we can see the effects on medical x ray technicians, who get an average whole-body dose of 3.0 to 3.5 sievert/yr; dental x ray technicians, who get an average dose of 0.5 to 1.25 sievert/yr; medical technicians handling radionuclides, getting a dose

FALLOUT AND NUCLEAR ACCIDENTS

Fallout from nuclear testing in the atmosphere increased in the 1950s and then decreased through the 1960s as a result of the nuclear test ban treaty. The U.S. government has been accused by "atomic veterans" of being less than candid about the risks of exposure to nuclear blasts when it sent troops into the Nevada desert during nuclear weapons tests. The tests also exposed unwitting civilians as well. Utah was in the weather path from the test site. As a result, residents of sparsely settled Washington County, Utah, received ten times the global average dose (94). Residents of Salt Lake City were exposed on those occasions when the cloud moved further north than expected. They consequently received four times the global average dose (94). The 133,000 people in high-fallout counties were exposed to 0.86 ± 0.14 R, while the 556,000 people in lower fallout areas were exposed to 1.3 ± 0.3 R (94).

Fallout will be important for accidents as well. The fallout is, of course, much less than that from a nuclear weapon. The effects of a worst-case reactor accident, in which the core melts and the containment vessel ruptures, and a thermonuclear detonation (95) were compared. The comparison reveals that the lethal zone, defined as the area that would expose people to a dose in excess of 4 kSv/day, is less than 1 square mile for the reactor and about 400 square miles for the 1 megaton bomb. The Hiroshima and Nagasaki bombs had a lethal zone of about 29 square miles. The area of contamination for which the dose would exceed 20 Sv/yr for over a month was 1800 mi² for the reactor accident, and 20,000 mi² for the bomb (95). Of course, dropping a bomb on a nuclear plant is a wonderful way to lay waste to large areas. An attack on a nuclear reactor at the confluence of the Rhine and Neckar Rivers in West Germany would render a third of the country uninhabitable (95).

of 3.6 to 5.4 sievert/yr; and workers in the civilian nuclear power industry, getting 6 to 8 sievert/yr (73). Over a forty year working life, people in these professions should lose about 14, 5, 22, and 32 days of life, respectively, for practicing their professions. For comparison, we give up, on average, half a year of life because of our desire to drive cars, given the probability of fatal accidents. The dose equivalent to smoking a cigarette is 0.07 Sv (7 mrem) (74), so each cigarette shortens life by about two hours.

In addition to direct hazards of ionizing radiation, there is a *genetic* hazard for humanity. Figure 25.10 (45) showed how lethal chromosomes increase in fruit flies with dose; so it must be also with human beings. Radiation alters genes and chromosomes. The nonlethal mutations are of greatest long-term worry, because the offspring will survive and perhaps pass on damaged germ cells in turn. Also, the increase in the mutation rate eventually means a corresponding increase in the genetic death rate.

Everyone has deleterious genetic material (to some extent), and this is passed on to the next generation. What is deleterious to some may be advantageous to others in other circumstances. This helps give humans the *genetic* ability to adjust to change.

Nobel Prize winner Joshua Lederberg (93) has estimated that at least 25 percent of our health care burden is of genetic origin. By acting in a humanitarian way, we can help add to our racial genetic load. There is no excuse for gratuitous additions to the genetic burden; we should be as careful as we possibly can. This is not always the course that is taken. In Western society, men wear pants, which heats the gonads, increasing the mutation rate (54). If males were to wear kilts, it would reduce the rate of mutations, but I've not seen any mass movement toward kilt wearing!

Waste from Nuclear Energy

No discussion of commercial nuclear energy can be complete without considering the problem of disposal of the waste from power plants. The current waste recovery system leaves much to be desired. For a slight cost increase, estimated at 0.02 cents per kWh_e (90,96), much of the waste could be recycled to recover 99 to 99.9 percent of all useful materials, rather than throwing much of it away. With reprocessing, the million year waste problem becomes a 700 year problem (96). The problem with reprocessing is that the plutonium recovered in the reprocessing itself can be used by nations having the reactors to build bombs, even with imposition of inspections (as with India). It takes but a small amount of even dilute plutonium to make a bomb—in contrast to the case for a uranium bomb (97). The difficulty is that, with expensive oil and the recent U.S. tendency to reduce its role as uranium supplier, more nations will decide to "go nuclear" and reprocess their own waste (98). Various proposals to avoid such developments include construction of breeder reactors, each of which would supply fuel for three to four LWRs (99) (recall that the United States has suspended its breeder research); the development of so-called advanced converter reactors (ACRs) using the uranium–thorium cycle with quick removal of spent fuel (100); spread of the newly developed CIVEX ion exchange process in place of the more risky old PUREX ion exchange process could help thwart diversion (101). There is practically no U.S. research on ACRs, either, though breeder and ACR research are continuing in France and Russia. The breeder has the burden of producing a lot of plutonium, and is also more prone to core disassembly (a euphemism for nuclear runaway) (102). Also, if one considers nuclear energy to be a short-term solution to the energy supply problem, the economic argument for the breeder is perhaps somewhat questionable.

Our nuclear wastes are being temporarily stored until a decision is made on permanent disposal. Congress tried to encourage states to join compacts to set up nuclear waste disposal facilities for the 140,000 m^3 of low-level wastes being generated in the United States each year (103). After January 1986, the regional compacts were supposed to be able to exclude wastes from nonmember states, but movement in this direction is glacial. It looks as if the waste will ultimately, through a political decision, be stored in Nevada.

The danger of the waste to health is best seen in Figure 25.11 (54), which shows the consequences of distributing the waste as widely as possible. The problem is to find some way to eliminate this danger. Some critics of the waste disposal policy claim that the most investigated option, that of storage in salt formations, is a *fait accompli*. They feel that the NRC has a don't-bother-me-my-mind's-already-made-up attitude.

Salt formations were originally favored because the heat and pressure would cause the salt to melt and fill up any cracks or cavities. However, the salt is impermeable only when no water is present. Investigations of the AEC in Lyons, Kansas, and the AEC/NRC in southeast New Mexico have shown that salt formations are not so isolated from water as had previously been thought (54,104,105). A pocket of brine in the salt formation is only about 200 m from the proposed storage site at the Waste Isolation Pilot Project (WIPP) (105). Water would migrate toward the waste, which would cause severe corrosion problems for any container (106). This migration of water takes place because solubility increases as temperature increases. The part of a brine pocket close to a hot waste container would take in salt from the formation, while the salt in the cooler

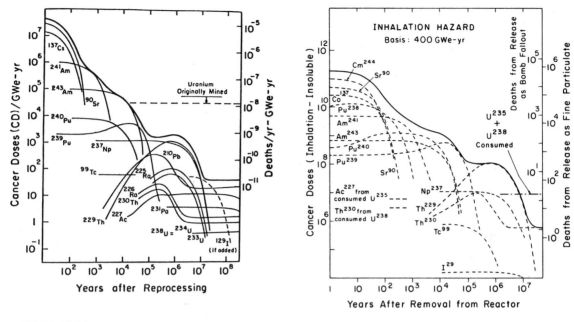

FIGURE 25.11

Consequences of nuclear industry waste generation (B. Cohen, Ref. 51). (a) Number of cancer-generating doses in wastes from 400 GWyr of electricity generated in nuclear fission reactors, if it were eaten in digestible form. (b) Number of cancer-generating doses in wastes from 400 GWyr of electricity generated in nuclear fission reactors, if it were inhaled as fine, insoluble particles.

brine further away would precipitate out. The salt water in contact with the metal will eat it away in short order. The temperature will be high initially—a typical canister of one-year-old waste generates 25 kW_t, and after about ten years, the heat flux drops to 3.5 kW_t (107,108). Even glass is susceptible to high leaching rates at the high temperatures expected in the brine (109). Ceramic containers hold out the best prospect of attaining impermeability.

Tests with heated canisters in holes drilled in rock were run for over a year and showed that the theory of rock movement was still not understood. The rock moved less than predicted (108). In the same experiment, the hydraulic conductivity was measured to be about 10^{-11}m/s (103). Water movement through the formations adjacent to the WIPP facility is also slow (107). Rock appears to be a better medium for disposal than salt.

Most other disposal schemes envision deep burial: for example, in the sea bottom (105), or in tunnels dug in Nevada for nuclear tests (110), or in shallow craters in the Nevada desert caused by underground nuclear tests (111), or at depths below 600 meters (54). The last proposal relies on the protection afforded by the geology and by the slow movement of waste atoms through the ground water system. It may take more than a thousand years for water from 600 m depth to reach the surface (54), or as much as 2 Myr if the water transit experiment in a Swedish mine is accurate (108). If the containers themselves fail, this method may not be as safe as had originally been believed. Some elements travel relatively quickly through the overburden, even in the most secure geological formations (112), although the radioactive elements at Oklo (see Chapter 14) did not travel much at all over a very long timespan.

TABLE 25.9
Dumping of low-level radioactive-wastes.

Year	Amount in Tonnes	Alpha (actinides) (curies)	Beta-emitters (curies)
1967	10,840	250	7600
1969	9180	500	22,000
1971	3970	630	11,200
1972	4130	680	21,600
1973	4350	740	12,600
1974	2270	420	100,000
1975	4460	780	60,500
1976	6770	880	53,500
1977	5605	958	74,450
1978	8000	1100	19,000
1979	5400	1400	83,000

SOURCE: Council on Environmental Quality, Reference 114.

TABLE 25.10A
Risks to individuals.

Activity	Persons Involved (Number)	Cost ($)	Annual Cost per Person ($)
Uranium mining and milling	620	463,200	747.09
Manufacturing	33,724	1,519,300	45.00
Reactor operation	1290	81,890	63.00
Reprocessing	800	91,720	115.00
Public near reactor	33,841,000	19,410	0.0004
Total U.S.	200,000,000	2,175,520	0.10

TABLE 25.10B
Health effects of civilian nuclear power, per 1000 MW$_e$ plant-year.

Activity	Accidents (not radiation-related)	Fatalities — Radiation-related (cancers and genetic)	Fatalities — Total	Injuries (days off)
Uranium mining and milling	0.173	0.001	0.174	330.5
Fuel processing and reprocessing	0.048	0.040	0.088	5.6
Design and manufacture of reactors, instruments, etc.	0.040		0.040	24.4
Reactor operation and maintenance	0.037	0.107	0.144	158
Waste disposal		0.0003	0.0003	
Transport of nuclear fuel	0.036	0.010	0.046	
Totals	0.334	0.158	0.492	518

The International Council of Scientific Unions studied the problem of high-level waste (113). They made many recommendations for further study, including development of underground laboratories to test behavior in actual conditions for underground storage, study of disposal in subduction regions, study of motion of ions through the seafloor, study of adsorption of radionuclides by sediment systems and the interface between sediment and seawater, and the need for data on concentration in marine organisms. These last data are sorely needed, since the nuclear nations are dumping large quantities of radioactive wastes into the world's oceans. Table 25.9 (114) shows some recent history of ocean dumping.

RISKS RELATIVE TO OTHER METHODS OF GENERATING ENERGY

As previously stated, the most important point in any decision to choose or to eschew nuclear energy is that there be a fair comparison of the real costs of nuclear energy and of the alternatives to it. It is reasonable to compare the nuclear industry to the coal industry. Two estimates of risk,

Table 25.10 (13), and a summary table, Table 25.11 (96,115), make this comparison. Tables 25.12 and 25.13 (8) show the costs of coal mining itself. Figures 25.12 (8) and 25.13 (116) show cost comparisons of various energy strategies. It can be seen in both figures that nuclear energy appears to be one of the least dangerous ways to generate energy. As of the 1970s, the numbers of deaths and injuries associated with coal mining and processing had decreased from those shown in Tables 25.12 and 25.13 (117): Coal mining claimed 0.33 deaths per Mtonne and coal processing 0.019 deaths per Mtonne; the number of disabling injuries is 25/Mtonne in mining and 1.2/Mtonne in processing. B. L. Cohen, a nuclear energy expert, points out that a single 1000 MW_e coal-fired plant causes twenty-five fatalities, sixty thousand cases of respiratory disease, and $12 million in property damage, as well as emitting an amount of NO_x equivalent to twenty thousand cars per year (54). It also produces ashes and sludge. In Cohen's view, nuclear energy is a means of cleansing the Earth, since net activity declines.

Another sort of comparison as shown in Figure 25.13 (116) purports to show that solar energy generating facilities are almost as dangerous as

TABLE 25.11
Summary of nuclear health risks.

	Fatalities	Days Lost
Sagan	0.390	1022
Rose	0.492	513
Hub et al.	0.932	373
AEC	0.161–0.364	

TABLE 25.12
Occupational fatalities due to accidents[a] (fatalities per 1000 MW_e plant year[b]).

	Fossil		
	Coal	Oil	Gas
Mining/pumping	0.96	0.06	0.02
Fuel processing	0.02	0.04	—
Transportation	0.05	0.03	0.02
Power plant operation	0.03	0.03	0.03
Totals	1.1	0.16	0.07

[a]From 1965–1970 injury rates and production data.
[b]Basis: 1000 MW_e power plant generating 6.6 billion kilowatt hours of electricity at 75 percent capacity factor.

TABLE 25.13
Occupational nonfatal injuries due to accidents[a] (injuries per 1000 MW$_e$ plant year[b]).

| | Nuclear | |
	PWR	BWR
Mining	0.09	0.09
Milling	0.003	0.003
Conversion	0.0003	0.0002
Enrichment	0.001	0.001
Fabrication	0.0004	0.0005
Reprocessing	0.0001	0.0001
Transportation	0.002	0.002
Power plant operation	0.01	0.01
Totals	0.1	0.1

[a]From 1965–1970 injury rates and production data.

[b]Basis: 1000 MW$_e$ power plant generating 6.6 billion kilowatt hours of electricity at 75 percent capacity factor.

FIGURE 25.12
Person-days lost per MWyr over the lifetime of various energy generating systems. (U.S. Atomic Energy Commission)

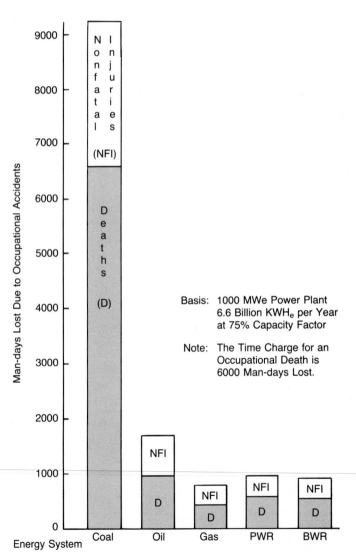

Basis: 1000 MWe Power Plant 6.6 Billion KWH$_e$ per Year at 75% Capacity Factor

Note: The Time Charge for an Occupational Death is 6000 Man-days Lost.

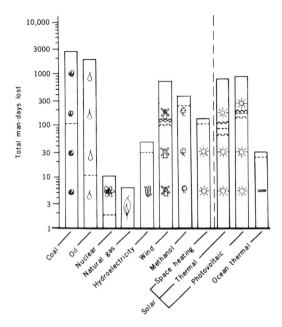

FIGURE 25.13
Cost of several strategies for generating electricity.
(H. Inhaber, copyright 1979 by AAAS)

coal- or oil-fired generating facilities. Much of the cost of the unconventional sources comes from the cost of the backup systems that are necessary. In the study, the backup was assumed to be a coal power plant, which is the most hazardous of the alternatives. This contributes much of the hazard risk shown.

To close the discussion, we note that fossil fuel contains radioisotopes (58). An analysis shows that coal-fired plants can release substantial radioactivity (118). By extrapolating the data in Reference 100 on dose from coal plants, I find a 1000 MW_e plant can produce 36 mrem/yr. The coal effluents are especially dangerous for bone dose (119), and emissions are higher in total than allowed for nuclear reactors, even assuming only 1 percent coal ash in the smoke (10 percent is more typical) (119). In terms of chronic emissions, a nuclear utility is probably safer than a coal-fired plant. As discussed in the last chapters, coal effluents are produced in substantially greater

volume than nuclear effluents, and the scrubber residue in particular presents a long-term disposal hazard. Overall, it appears to be safer to generate nuclear electrical energy than coal electrical energy.

SUMMARY

The nuclear industry was expected by its advocates to be safe and to generate energy so cheaply that it would not have to be metered. An extensive research program quantified the hazards of radioactivity, and nuclear plants were designed with safeguards in mind. Because so much more was known about nuclear hazards, and because of assurances that nuclear energy would be absolutely safe, people set standards for nuclear generation that are much more stringent than those for fossil-fuel generation. In addition, expensive changes were required in already-operating reactors. These costs, together with the extra costs caused because most American nuclear plants are one of a kind, have made nuclear energy very expensive indeed.

Reactors are designed to foreclose the possibility of catastrophic accident, although it is impossible to ensure that an accident will not occur. Reactors cannot cause a nuclear explosion because the fuel is too dilute. They are housed in reinforced concrete containment vessels. Heat is transferred by heat exchangers to isolate the radioactivity.

The Rasmussen report analyzed the probability of catastrophic accident and concluded that chances are quite small. The German Birkhofer report reached similar conclusions, that chances of serious accident are on the order of one per million reactor years.

The accident at Three Mile Island crystallized opposition to nuclear energy in the United States. No new reactors have been ordered in the States since, and many previously on order were cancelled. Since most of the problems at TMI involved the operators, the NRC has made changes in licensing requirements. The small release of activity led advisory groups to the conclusion that risks, as quantified before TMI, had probably been overstated.

The accident at Chernobyl was a nightmare come true. People everywhere felt threatened by nuclear energy in a way they never had before. The ubiquity of human error in accidents reinforced conclusions reached in the wake of TMI. The long-term consequences of the Chernobyl accident include 50,000 to 330,000 premature deaths, as well as a greater awareness of the fragility of the technology of nuclear energy.

The unit of exposure is the röntgen. Units of dose are the gray and the sievert; at present, the rad and the rem are more commonly used. Natural background is about 1 sievert. Exposure from nuclear energy is less than 10 mSv. The linear dose theory is used to extrapolate effects from high dose, where the effects can be measured, to low dose, where they cannot. The health consequences of bomb blasts for people living in Hiroshima and Nagasaki are extremely important in determining risk; at present, there is concern that the exposure data in use are incorrect, and that people are at higher risk than previously estimated.

Wastes from nuclear energy will have to be dealt with. Current thinking is that they should be buried in salt deposits. Others propose deep-rock burial. Waste volumes from the civilian nuclear industry are much smaller than those from military uses. No completely satisfactory solution is at hand.

In comparison to other forms of energy generation, nuclear energy is relatively safe. The public should probably be made more aware of the risks inherent in other common energy generation choices.

It is appropriate to have stringent standards for nuclear energy generation. These protection standards should be applied to non-nuclear energy generation as well. Remember TANSTAAFL! Energy benefit has its cost. Those enjoying the benefits should be prepared to pay the costs; these costs should not be borne disproportionately.

REFERENCES

1. R. Jungck, *Brighter than a Thousand Suns* (New York: Harcourt Brace Jovanovitch, 1958).
2. R. Gillette, *Science* 181 (1973): 728; D. Burnham, *New York Times,* 21 Oct. 1976; *New York Times,* 15 April 1979; D. Burnham, *New York Times,* 6 May 1979; T. Lash, *Amicus J.* 1, no. 2 (1979): 24; and E. Marshall, *Science* 236 (1987): 1616.
3. D. Burnham, *New York Times,* 10 Nov. 1974.
4. R. Gillette, *Science* 177 (1972): 771, 867, 970, and 1080.
5. D. Burnham, *New York Times,* 25 August 1974.
6. D. Burnham, *New York Times,* 29 Dec. 1974.
7. United States Atomic Energy Commission, *Theoretical Possibilities and Consequences of Major Accidents in Large Nuclear Power Plants,* WASH-740, March 1957.
8. United States Atomic Energy Commission, *The Safety of Nuclear Power Reactors and Related Facilities,* WASH-1250, July 1973.
9. United States Nuclear Regulatory Commission, *Reactor Safety Study* (Rasmussen report), WASH-1400 (NUREG-75/014), Oct. 1975.
10. *Die Deutsche Universitäts-Zeitung* 33 (1979): 530 describes the Birkhofer report to the German Ministry for Research and Technology.

11. R. P. Hammond, *Am. Sci.* 62 (1979): 155.
12. C. Norman, *Science* 228 (1985): 31.
13. C. Starr, *Science* 165 (1969): 1232; C. Starr, M. A. Greenfield, and D. F. Hansknecht, *Nuclear News,* Oct. 1972; and C. Starr and C. Whipple, *Science* 208 (1980): 1114.
14. P. Slovic, *Science* 236 (1987): 280.
15. D. Okrent, *Science* 208 (1980): 372.
16. Anonymous, *Water, Power, and Dam Const.* 37, no. 11 (1985): 33. This is a report on the International Workshop on Dam Failures from 1985.
17. K. V. Bury and H. Kreuzer, *Water, Power, and Dam Const.* 37, no. 11 (1985): 45.
18. United States Nuclear Regulatory Commission, *Risk Assessment Review Group Report to the United States Nuclear Regulatory Commission,* NUREG/CR-0400, Sept. 1978.
19. American Physical Society Study Group on Light Water Reactor Safety, *Rev. Mod. Phys.* 47, Supplement 1 (1975): 1.
20. E. Marshall, *Science* 204 (1979): 152; and C. R. Herron and D. Lewis, *New York Times,* 15 April 1979.
21. E. Marshall, *Science* 204 (1979): 280, 594; and B. G. Levi, *Phys. Today* 32, no. 6 (1979): 77.
22. S. Glasstone and W. H. Jordan, *Nuclear Power and Its Environmental Effects* (LaGrange Park, Ill.: American Nuclear Society, 1980).
23. W. Booth, *Science* 238 (1987): 1342.
24. J. W. Sanders, *Psychology Today* 13, no. 11 (1980): 52.
25. A. O. Sulzberger, Jr., *New York Times,* 22 April 1979; and C. Mohr, *New York Times,* 13 May 1979.
26. M. L. Wald, *New York Times,* 23 June 1985.
27. B. G. Levi, *Phys. Today* 38, no. 5 (1985): 67.
28. R. Wilson et al., *Rev. Mod. Phys.* 57, Supplement 1 (1985): 1.
29. A. A. Abagyan et al., *The Accident at the Chernobyl' Nuclear Power Plant and Its Consequences* Commission on the Causes of the Accident at the Fourth Unit of the Chernobyl' Nuclear Power Plant, appointed by the USSR State Committee on the Utilization of Atomic Energy, 1986.
30. J. F. Ahearne, *Science* 236 (1987): 673.
31. Annex 2, Reference 29.
32. B. G. Levi, *Phys. Today* 39, no. 7 (1986): 17; F. von Hippel and T. B. Cochran, *Bull. At. Sci.* 43, no. 1 (1986): 18.
33. C. Norman, *Science* 232 (1986): 1331; C. Norman and D. Dickson, *Science* 233 (1986): 1141; E. Marshall, *Science* 233 (1986): 1375. See also S. Diamond, *New York Times,* 16 August 1986.
34. R. Wilson, *Science* 236 (1987): 1636.
35. S. Diamond, *New York Times,* 18 August 1986.
36. C. Hohenemser et al., *Environment* 28, no. 5 (1986): 6.
37. S. Schmeman, *New York Times,* 6 May 1986; and S. Diamond, *New York Times,* 13 May 1986.
38. H. Salisbury, *New York Times Magazine,* 27 July 1986, p. 18.
39. L. Devell et al., *Nature* 321 (1986): 192.

40. F. A. Fry, R. H. Clarke, and M. C. O'Riordan, *Nature* 321 (1986): 569; C. R. Hill et al., *Nature* 321 (1986): 655; C. Hohenemser et al., *Nature* 321 (1986): 817; A. J. Thomas and J. M. Martin, *Nature* 321 (1986): 818; B. Holliday, K. C. Simmonds, and B. T. Wilkins, *Nature* 321 (1986): 821; F. B. Smith and M. J. Clark, *Nature* 327 (1986): 690; I. R. Falconer, *Nature* 322 (1986): 692; N. G. Alexandropoulos et al., *Nature* 322 (1986): 779; and K. Bangert et al., *Nature* 73 (1986): 495.

41. M. Aoyama et al., *Nature* 321 (1986): 819.

42. R. H. Clarke, *Ann. Nucl. Sci. Eng.* 1 (1974): 73.

43. A. C. Upton, *Sci. Am.* 246, no. 2 (1982): 41.

44. S. F. Cleary, *BioScience* 33 (1983): 269. See also D. I. McRee, *J.A.P.C.A.* 24 (1974): 122.

45. I. Asimov and T. Dobzhansky, *The Genetic Effects of Radiation* (Washington, D.C.: AEC, 1966).

46. Report of the 1978 Interagency Task Force on Ionizing Radiation, as quoted in J. L. Marx, *Science* 204 (1979): 162.

47. M. Eisenbud, *Environment* 26, no. 10 (1984): 6.

48. Bundesministerium des Innern, *Umweltradioactivität und Strahlenbelastung* (West Germany, 1983).

49. P. Taylor, *New Scientist* 110, no. 1508 (1986): 24.

50. Annex 7, Reference 29.

51. Annex 4, Reference 29.

52. American Physical Society Study Group on Nuclear Fuel Cycles and Waste Management, *Rev. Mod. Phys.* 50, Supplement 1 (1978): 1.

53. L. J. Carter, *Science* 202 (1978): 191; and G. Lichtenstein, *New York Times,* 22 May 1976.

54. B. L. Cohen, *Phys. Today* 29, no. 1 (1976): 9; *Rev. Mod. Phys.* 49 (1977): 1; *Sci. Am.* 236, no. 6 (1977): 21; *Phys. Teacher* 18 (1978): 526; and *Am. J. Phys.* 54 (1986): 38.

55. M. Crawford, *Science* 229 (1985): 537.

56. Anonymous, *New York Times,* 9 March 1979.

57. H. W. Tso, *AAPT Announcer* 15, no. 2 (1985): 45.

58. K. Z. Morgan, in *The Environmental and Ecological Forum, 1970–1971,* ed. A. B. Kline, Jr. (Washington, D.C.: AEC, 1972).

59. A. F. Gabrysh, and F. J. Davis, *Civil Eng.* 839 (December 1974): 89; *Nucleonics* 13, no. 1 (1955): 50; as quoted in Reference 36.

60. H. Raines, *New York Times,* 16 March 1979; and D. Henry, *New York Times,* 13 May 1979.

61. P. Shabecoff, *New York Times,* 19 May 1985; A. P. Dispatch, *New York Times,* 7 July 1985; and W. Greer, *New York Times,* 28 Oct. 1985. See also P. Shabecoff, *New York Times,* 14 July 1985; radon at dangerous levels may be contaminating a million homes and causing 5000 to 20,000 cases of lung cancer per year; R. Hanley, *New York Times,* 3 April 1986 and 20 April 1986; A. A. Narvaez, *New York Times,* 13 August 1986.

62. Anonymous, *New York Times,* 16 August 1986.

63. For a further discussion of radon hazards, see the review article by W. J. Fisk et al., *Indoor Air Quality Control Techniques: A Critical Review,* LBL-16493 (1983).

64. D. J. Crawford and R. W. Leggett, *Am. Sci.* 68 (1980): 526.

65. T. H. Maugh II, *Science* 202 (1978): 37.

66. G. B. Gori, *Science* 208 (1980): 256.

67. L. B. Lave, *Science* 236 (1987): 291.

68. E. Marshall, *Science* 212 (1981): 900; C. Haberman, *New York Times,* 4 August 1985; L. Roberts, *Science* 238 (1987): 1649.

69. K. Z. Morgan, in *Nuclear Power: Both Sides,* ed. M. Kaku and J. Trainer (New York: W. W. Norton & Co., 1982), 35ff.

70. C. Haberman, *New York Times,* 4 August 1985.

71. A. Brodsky, in *Nuclear Power: Both Sides,* ed. M. Kaku and J. Trainer (New York: W. W. Norton & Co., 1982), 46 ff. The data are based on R. L. Gotchy, *Health Physics* 35 (1978): 563.

72. C. Holden, *Science* 204 (1979): 155; and J. L. Marx, *Science* 204 (1979): 160.

73. E. Marshall, *Science* 204 (1979): 711; M. E. Jacobs, *Phys. Today* 32, no. 7 (1979): 78.

74. BEIR report, *The Effects on Populations of Exposure to Low Levels of Ionizing Radiation* (Washington, D.C.: National Academy of Sciences–National Research Council, 1972).

75. K. Z. Morgan, *Environment* 13, no. 1 (1971): 28.

76. J. Gofman and J. R. Tamplin, *Poisoned Power* (New York: Signet, 1974), Appendix I; J. Gofman, in *The Environmental and Ecological Forum, 1970–1971,* ed. A. B. Kline (Washington, D.C.: AEC, 1972).

77. F. G. Laoman, T. R. Rice, and F. A. Richards, in *Radioactivity in the Marine Environment* (Washington, D.C.: National Academy of Sciences–National Research Council, 1971), Ch. 7. M. E. Eisenbud, *Environment* 20, no. 8 (1978): 6.

78. E. Marshall, *Science* 236 (1987): 658.

79. R. Wilson and E. A. C. Crouch, *Science* 236 (1987): 267.

80. D. P. Burkitt, *Intern. J. Environmental Studies* 1 (1971): 275; and R. J. C. Harris, *Intern. J. Environmental Studies* 1 (1971): 59.

81. I. Barringer, *New York Times,* 20 July 1986.

82. C. Norman and D. Dickson, *Science* 233 (1986): 1141.

83. J. W. Gofman, *Assessing Chernobyl's Cancer Consequences,* talk delivered to the American Chemical Society meeting, Sept. 1986.

84. Der Bundesminister für Umwelt, Naturschutz, und Reaktorsicherheit, *Bericht über den Reaktorunfall in Tschernobyl, seine Auswirkungen und die getroffenen bzw. zu treffenden Vorkehrungen* (West Germany, 1986).

85. R. Gillette, *Science* 179 (1973): 360.

86. R. Gillette, *Science* 177 (1972): 330; AEC, *Annual Report to Congress, 1973.*

87. E. Marshall, *Science* 215 (1982): 1596.

88. S. Novick, *Environment* 16, no. 10 (1974): 6; D. Burnham, *New York Times,* 9 March 1975.

89. D. B. Myers, *The Nuclear Power Debate* (New York: Praeger Pubs., 1977); A. D. Rosser and T. A. Rieck, *Science* 201 (1978): 582; J. M. Fowler, R. L. Goble, and C. Hohenemser, *Environment* 20, no. 3 (1978): 25; R. L. Goble and C. Hohenemser, *Environment* 21, no. 8 (1979): 32.

90. M. R. Copulos, *National Review* 31 (1979): 156; A. J. Parisi, *New York Times,* 8 April 1979.

91. M. L. Wald, *New York Times,* 8 Dec. 1985.

92. S. R. Weisman, *New York Times,* 21 March 1979; Anonymous, *New York Times,* 22 Sept. 1974.

93. J. Lederberg, *Washington Post,* 19 July 1970.

94. H. L. Beck and P. W. Krey, *Science* 220 (1983): 18.

95. S. A. Fetter and K. Tsipis, *Sci. Am.* 244, no. 4 (1981): 41.

96. A. S. Kubo and D. J. Rose, *Science* 182 (1973): 1205; and D. J. Rose, *Science* 184 (1974): 351.

97. E. J. Moniz and T. L. Neff, *Phys. Today* 31, no. 4 (1978): 42; and T. B. Taylor, *Ann. Rev. Nucl. Sci.* 25 (1975): 407.

98. D. J. Rose and R. K. Lester, *Sci. Am.* 238, no. 4 (1978): 46.

99. C. L. Rickard and R. C. Dahlberg, *Science* 202 (1978): 581.

100. H. A. Ferverson, F. von Hippel, and R. H. Williams, *Science* 203 (1979): 330.

101. V. K. McElheney, *New York Times,* 28 Feb. 1978.

102. S. Rattner, *New York Times,* 16 March 1978.

103. C. Norman, *Science* 223 (1984): 258; and *Science* 227 (1985): 448.

104. L. J. Carter, *Science* 200 (1978): 1135.

105. R. A. Kerr, *Science* 204 (1979): 603.

106. M. Sun, *Science* 215 (1982): 1483; L. J. Carter, *Science* 222 (1983): 1104.

107. F. Donath, in *Nuclear Power: Both Sides,* ed. M. Kaku and T. Trainer (New York: W. W. Norton & Co., 1982), 115ff.

108. P. A. Witherspoon, N. G. W. Cook, and J. E. Gale, *Science* 211 (1981): 894.

109. R. A. Kerr, *Science* 204 (1979): 289; and R. O. Pohl, in *Nuclear Power: Both Sides,* ed. M. Kaku and J. Trainer (New York: W. W. Norton & Co., 1982), 123ff.

110. R. P. Hammond, *Am. Sci.* 67 (1979): 146.

111. I. J. Winograd, *Science* 212 (1981): 1457.

112. A. Barbreu et al., *Science* 197 (1977): 519.

113. J. M. Harrison, *Science* 226 (1984): 11.

114. Council on Environmental Quality, *Environmental Quality,* 11th ed. (Washington, D.C.: GPO, 1980).

115. L. A. Sagan, *Science* 177 (1972): 487.

116. H. Inhaber, *Science* 203 (1979): 718.

117. H. W. Lorber and A. Ford, *Environment* 22, no. 4 (1980): 25.

118. J. E. Martin, E. D. Harward, and D. T. Oakley, in *Power Generation and Environmental Change,* ed. D.A. Berkowitz and A.M. Squires (Cambridge, Mass.: MIT Press, 1971).

119. J. P. McBride et al., *Science* 202 (1978): 1045.

PROBLEMS AND QUESTIONS

True or False

1. If the low probabilities of the Rasmussen report are to be believed, there will never be a reactor accident.

2. The *rem* is a unit of dose that has taken the biological effect of radiation into account.

3. Average levels of manmade radiation are of the same order of magnitude as natural background radiation levels.

4. Genetic mutations would cease if all *radiation* were eliminated.

5. The Rasmussen report evacuation extrapolation is reasonable.

6. People voluntarily incur risks up to a thousand times greater than those they would incur involuntarily.

7. A coal plant is safer (in terms of cost to people) to supply than a nuclear plant (assuming no catastrophic accidents, etc.).

8. The accident at Three Mile Island was one of a kind.
9. There is no disposal problem for nuclear waste.
10. The inert gas radon is a health hazard because it decays radioactively into alpha-emitters, which can be breathed.

Multiple Choice

11. If we denote the hazard of a 1 rem dose by 1, what is the expected hazard of a 100 millirem dose?
    a. 1
    b. 0.1
    c. 0.01
    d. None of the above

12. Which of the units in the following list refers to exposure to radioactivity?
    a. roentgen
    b. becquerel
    c. rad
    d. rem
    e. gray

13. What is the maximum safe number of working levels in a home?
    a. 5×10^{-3} WL
    b. 0.02 WL
    c. 1 WL
    d. 16 WL
    e. 1.3×10^5 WL

14. What is (are) the cause(s) of the accident at Three Mile Island?
    a. Human error
    b. Pump malfunction
    c. Valve malfunction
    d. All of the above were causes of the accident.
    e. None of the above was a cause of the accident.

15. What is (are) the cause(s) of the accident at Chernobyl?
    a. Human error
    b. Pump malfunction
    c. Valve malfunction
    d. All of the above were causes of the accident.
    e. None of the above was a cause of the accident.

16. One of the most important consequences of the accident at Three Mile Island is that
    a. the reactor did not explode.
    b. the hydrogen bubble did not ignite.
    c. many cancers are expected in its wake.
    d. the amounts of radioactivity actually released are much smaller than had been estimated before the accident.
    e. the fuel rods fractured, releasing radioactive iodine, xenon, and krypton.

17. Which factor(s) listed below help explain why the public is uncertain about the acceptability of nuclear energy?
    a. Nuclear energy is portrayed as magic in films.
    b. Referenda get people upset.
    c. TV stars are opposed to nuclear energy.
    d. The federal government has lied about the drawbacks of nuclear energy.
    e. The public was led to expect riskless generation of energy by nuclear fission.

18. For release of radioactivity *inside* the body, which of the following is most dangerous to health?
    a. Alpha emission
    b. Beta emission
    c. Gamma emission
    d. Emission of neutrons
    e. Emission of neutrinos

19. For release of radioactivity *outside* the body, which of the following is most dangerous to health?
    a. Alpha emission
    b. Beta emission
    c. Gamma emission
    d. Emission of neutrons
    e. Emission of neutrinos

20. Mine tailings are dangerous because
    a. people use the materials for building supplies.
    b. radioactive materials can blow off tailings piles and be carried in the wind.
    c. radium in the tailings can be leached into ground water.
    d. All of the above are reasons mine tailings are dangerous.
    e. None of the above is a reason mine tailings are dangerous.

21. Which of the following methods of generating electricity is the most dangerous in terms of deaths and days lost to injury?
    a. Coal
    b. Oil
    c. Gas
    d. BWR
    e. PWR

Discussion Questions

22. Enumerate the issues involved in nuclear power plant adoption.
23. What do you think of the risk–benefit analysis as an assessment tool?
24. Is another Chernobyl-type accident possible? Why or why not?
25. What would you characterize as the most important outcome of the accident at Chernobyl?
26. How is the U.S. nuclear industry going to be able to survive? In your answer, consider public perception as well as the changes you might recommend.
27. How does the linear dose model express "conservatism"? In other words, describe how something "worse" could be "better."
28. Argue both sides of the question as to which type of electricity generation—coal-fired or nuclear—is most hazardous to the human race in the long run.
29. Explain how there could be a disagreement about the effects of dose on health. What evidence or lack of evidence can you identify?
30. Explain why we would multiply numbers for the additional cancers per year by a factor of 10 or so if the exposure were from alpha particles rather than from gamma rays.

26
TOCSIN

The alarm has been sounded. What we human beings choose to do now will determine our collective future. Since change will occur, let us agree to make it change for the better!

KEY TERMS
Spaceship Earth
dust bowl
Sahel
albedo
savanna
desiccation
"green line"
allelopathy
Lotka–Volterra equation
K-selected species
r-selected species
tocsin
O'Neill habitats
technological fix
appropriate technology
hard energy path
soft energy path

No man is an island, entire of itself,
every man is a piece of the continent,
  a part of the main;
if a clod be washed away by the sea,
  Europe is the less;
as well as it a promontory were,
as well as it a manor of thy friends
  or of thine own were.
Any man's death diminishes me,
  because I am involved in mankind;
and therefore never send to know for whom the
  bell tolls;
it tolls for thee.

John Donne, *Devotions* XVII (1632)

INTERCONNECTIONS

Society has reached the state described in the last chapters because, often, people who make the decisions leading to these consequences are not aware of those consequences, or of the interrelationship of resource use, energy use, population pressure, and the like. People sometimes turn a blind eye when personal gain is involved. John Stuart Mill (1) characterized Americans as people "devoted to dollar-hunting." Perhaps business people, not only American ones, ignore information that could be used to persuade them that they were wrong. I would prefer to think that we are not so crass in our dishonesties. In any case, it is never easy to visualize the action of a complex system when some part of it is changed, even when one tries with the purest of intentions.

So that we can try to understand how action on one matter might cause some unexpected reaction on some completely different matter, we must try to picture the entire network of relationships in some way. Figure 26.1 is my attempt to indicate the most important reticulations in the closed system making up "Spaceship Earth." Notice that everything affects practically everything else. Of course, the categories are grossly simplified in Figure 26.1, but it does give a hint of the *complexity* of the system. The complexity of the system in some sense guarantees the stability of the system in the face of perturbations. Since I am concerned in this book with investigating the interaction of people with everything else, let me briefly present (or remind you of, as the case may be) a few of the interconnections in Figure 26.1.

People and the Land

Rene Dubos (2) has emphasized that the world around us is very much a manmade world, even to our natural surroundings. In Eurasia, with historically high population densities over millennia, the landscape has been shaped by people. (This is less true of Africa south of the Sahara and of the Americas.) The very geography has been in some way domesticated.

The Bad News   In recent times, we all have come to appreciate some of the changes people can wreak on the unprotected land. During 1934 alone, a total of 700 million tons of topsoil is estimated to have been blown off the North American continent (3) during the notorious "dust bowl" era. Perhaps these numbers seem horribly large. Brace yourself! The Soil Conservation Service of the Agriculture Department estimated that, in 1977, 5 billion tons of topsoil either washed away (40 percent of it) or blew away (60 percent of it), a record high (4). In many states, topsoil totaling over 10 tons per acre of cropland is being

FIGURE 26.1
Interconnectivity of the support systems of the Earth.

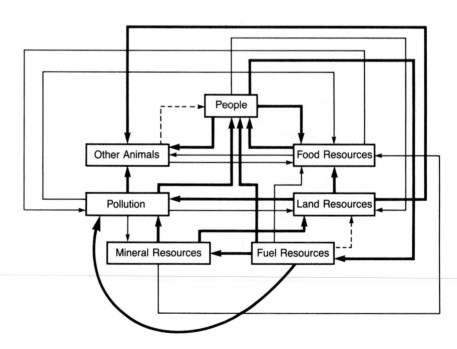

lost each year. The reasons: Marginal cropland is being used; farmers have dropped soil-protecting crop rotation practices in favor of chemical fertilizers and herbicides; and big farm machines are obliterating the terraces built to conserve soil in the days of smaller machines. This profligacy is leading to the silting up of rivers and lakes— silt is being carried at the rate of 400 tons per minute to the Mississippi's mouth (4).

Disease of the land has hit worldwide. The appalling spread of the Sahara desert during the Sahel drought—it *had* been advancing at about 1 km/year (5) before the drought, and during the drought advanced at 50 km/year—appears to be a result of a combination of resource mismanagement and slow climatic changes. Mismanagement has occurred because the Sahelian pastoralist views his herd, not the herd plus its food (grass), as the resource base (6). This leads to an increase in herd size as a hedge against dry years. The vegetation is reduced by the consequent overgrazing, which increases the albedo and leads to decreased rainfall (7), a classic example of "tragedy of the commons."

Most of the deserts in the Mideast and Asia were probably caused at least partly by overgrazing (2,8), or overcropping (9). We know of several relatively modern examples. The goat is characterized by an ability to find food in practically any organic material (although not in the proverbial tin cans). The goat was first domesticated in a region of very dry climate just because of this ability. The Seljuk and Othmani (Ottoman) Turks were keepers of goats in their homeland in central Asia on the fringes of the Gobi desert. Under the twin urgings of Genghis Khan and the gradual desiccation of their land, they moved west, where they were converted to Islam and became the spearhead of Islamic penetration of the Anatolian flank of the Byzantine Empire. This region of the Empire was its breadbasket: Wheat and other crops grew in abundance, and the region exported grain and imperial levies. As the Turkish tide gradually overwhelmed Byzantine resistance, the goats they brought with them re-

duced the area to the semidesert state of today. The pressure of goat husbandry on North Africa has contributed to desertification in a region so rich that Carthage almost had the resources to defeat Rome. After the close of the Punic wars, North Africa became the breadbasket of Imperial Rome.

The Good News   As we have seen in Chapter 24, desertification and deforestation are not the only illustrations of human effects on climate. Most of the world's savanna is the result of human intervention: burning off the forest (2,8). The creation of prairie grassland has produced some of the richest soils in the world. However, the results of human intervention have not always been so serendipitous.

Europe has suffered extreme deforestation over the ages. Plato blamed the barrenness of his native Greece on ruthless deforestation (2). The water storage capacity of denuded hills is much smaller than that of the same hills when they were covered with forest. Soil erosion consequently increases, as does the frequency and severity of floods. The story of more, and more serious, floods in Florence over the last half millennium seems to rest with the denudation of the once timber-covered hills (10). In Algeria, the rainy season now washes some 80,000 acres of land cover out to sea (11).

The deforestation is continuing apace. Only in Europe and North America has the tide turned in favor of increasing forest cover. Today, about 25 percent of the land area of Europe is in forest. Firewood is the chief fuel in poor countries, and major cities there are surrounded by treeless belts. Firewood, once cheap in Katmandu, Nepal, is a major expense for the people living there now. According to Revelle (12):

It is no coincidence that the forests of all countries with major crop failures in recent years due to droughts or floods—Bangladesh, Ethiopia, India, Pakistan, and the Sahel countries—had been razed to the ground.

Forests were once heavy in Asia, but now constitute only 11 percent of the land area (13). The tree *Leucaena leucocephala,* which can grow to a height of 20 m in six to eight years, holds some promise for stemming the loss of Asian woodland if authorities can get the people to grow them in community woodlots (13). The trees produce 10 to 45 tonnes of wood per hectare per year, which means the trees are 0.3 to 1.4 percent efficient in their use of solar energy (13).

Desert land can be reclaimed. Planting trees can serve to hold back the desert and help change the local microclimate (14) by cooling the air to a height seven times the tree height. The Algerians are planting a great wall of trees, in some places as wide as 24 km (11,15). The trees are being planted between the 200 mm and the 300 mm rainfall line, in regions with thirty-seven to eighty days of rainfall per year (14,16). Desertification is occurring between the lines of 100 mm and 200 mm rainfall. The trees also help prevent wind from blowing at such great speeds, protecting farmland from erosion and preventing the spread of the desert (14,16,17). The trees are mostly Aleppo pine, with a mixture of cedars, poplars, and Arizona cypress (14,16,17). The "green line" is now about 80,000 hectares in area (14). Israel has had success in increasing precipitation by planting millions of trees since 1948. The desert has been pushed back from near Tel Aviv almost to Beersheva (18). China embarked on a massive reforestation effort that has met with mixed success. Korea, whose forests were devastated in the Korean War (1949—1953), embarked on a large reforestation program in 1973. Despite pessimism because of the severity of the problem, Korea succeeded in restoring forest to two-thirds of total land area by 1985 (19).

People and Plants

It is seldom realized that North American agriculture is imported agriculture (20). Potatoes come from South America, as do sweet potatoes. Corn originated in Central America, wheat in the Middle East, rye in northern Europe. The United States of the late nineteenth century actually had a government agency whose sole purpose was to introduce new plant and animal species here.

In 1916 alone, 2294 species were introduced into the country for the first time (21).

The Bad News   Some imports of plants have been less than successful. Consider milfoil (*Myriophyllum* spp.). This aquatic plant was imported to be put into aquariums and ornamental ponds (22). It managed to escape and has become a nationwide nuisance as it clogs water supplies and chokes out fish. It is virtually impossible to chop weeds fast enough to stay ahead of the growth of the plant. A similar situation exists in the southeast, especially Florida, where the water hyacinth is busily strangling lakes.

Another interesting case involves Johnson grass (*Sorghum halepense*) (23). Introduced here in 1880 as a forage grass, this adaptable grass could not be kept out of cultivated fields. It simply crowded out the planted crops. It thrives by producing its own herbicide that kills competing plants (a phenomenon called *allelopathy* [24]). Since it grows to a height of 8 feet, it also kills by putting competing plants in the shade. It carries viruses that are transmitted by aphids or leaf hoppers and can infect corn. Even though it is a warm weather plant, it has now been encountered as far north as Syracuse, New York (23).

Manmade pollutants such as ragweed pollen, filbert pollen, and grass pollen have made the Willamette Valley of Oregon the hay fever capital of the United States. Such pollutants are manmade in the sense that people create the surroundings in which these plants flourish, whether on purpose or inadvertently (as with ragweed).

The American chestnut has become nearly extinct because of the chestnut blight fungus introduced from Europe in the 1920s. There may be some hope for the chestnut because a viral infection of the blight fungus appeared spontaneously in Italy in the mid-1970s (25). The elm may be less fortunate. A shipload of infected wood brought Dutch elm disease from the Netherlands during the 1920s. The disease is slowly spreading westward, killing the elms as it goes.

These problems with new species introduction and the alteration of old species habitat are serious enough, but they pale into insignificance in comparison to the truly awesome conundrum we are creating for ourselves. A mere fifteen plants account for over 75 percent of human dietary calories, as uniformity in agriculture is spread in the wake of the Green Revolution (26). Our reliance on such a narrow food base is fraught with risks. Our prehistoric ancestors had a repertoire of about five thousand food plants to support them. Many of these species are truly cultivated plants, dependent on people for their survival. As uniformity spreads, a whole species may be lost as seed is used for food. Genetic material is stored only in living things, and when a species disappears, the loss is irreversible.

The Good News   Over millenia of interaction with humanity, plants evolve to depend on care—ground preparation, weeding, sowing of seed, protection during growth, and collection of seed. Over these many years, plants have changed into a variable, integrated, adapted group of plants resembling one another but containing the diversity not found in industrial agriculture. These groups are called *land races*.

The land races contain genetic material that can be used to breed new varieties of our agriculturally important plants. A land race of rice was extremely important in the development of the rice of the "Green Revolution," because it had extremely desirable traits that could be used to keep the rice plants erect. The traits hidden among the land races may be needed for future crops. If the land races were to vanish, a mutated rust, mold, fungus, or virus might be able to eliminate totally a cultivated species. It is the genetic diversity available to a species that can act to allow it to adapt to changing situations. Without the land races, the new varieties of our major crops that we must produce every four or five years to preclude crop failures such as the 1970 corn blight would simply not be available (27).

Of the world's ten thousand endangered plant species, many are or may be of future value. The land races of some food species are in this list, including such plants as *Amaranthus* (South America), a cereal rich in protein and the amino acid lysine (one not ordinarily found in plants); the wax gourd (Asia), which produces three crops of fruit per year and can be stored without refrigeration for extended periods; the mangosteen (Southeast Asia), thought by some to be the world's best-tasting fruit; and the tamaringo, a forage crop that grows in abundance in the Atacama desert, sometimes through a meter of salt (28). Other plants of possible importance include guayule (discussed in Chapter 16), a possible source of natural rubber, and plants that can survive extremes of heat and cold. It is hoped that researchers can discover just how these plants manage to thrive in such hostile environments. For example, the plant *Tidestromia oblongifolia* grows at temperatures of 50°C, the *Alocasia macrorrhia* grows in perpetually dark regions, and *Atriplex glabriuscula* grows at 10°C (29).

More familiar plants may also yield unexpected benefits. Alfalfa contains a substance, triacontanol, that can cause spectacular increases in plant growth rates (30). A common African plant, *Ajugara remota,* seems to have an "anti-feedant" substance in its leaves—it stops insects from feeding on it.

Private seed banks exist, and some governments have been working to gather, categorize, and save this genetic heritage. The decade of the 1980s has seen very positive developments in awareness of the fragility of the genetic variability of useful plants.

Insects, Other Animals, and People

The Bad News   In addition to an imported agriculture, we have imported many animals, such as starlings, sparrows, and carp. Some imports have been successful; some have not. For example, importation of the brown trout has proved beneficial, while the carp is destroying aquatic habitats and spreading disease (21). The huge numbers of animals brought in are truly mind-boggling. In 1968, we imported 120,000 mammals, and in 1972, 100 million fish were brought into the country.

INTERSPECIES COMPETITION

The adaptation of species to their environments depends on the carrying capacity of the environment and the number of competing species, as well as the growth rates of the various species. The most common way to describe the interspecies competition and its effect on the number of individuals in a particular species—denoted N_i for species i—is called the Lotka–Volterra equation (31):

$$\frac{dN_i}{dt} = \frac{N_i r_i}{K_i} \times (K_i - N_i - \sum_{\substack{j=1 \\ j \neq i}}^{n} \alpha_{ij} N_j).$$

Here, dN_i/dt is the rate of change in the number of individuals in species i, K_i is the carrying capacity or maximum number of individuals of species i the environment can support, r_i is the net rate of increase (birth rate minus death rate), and α_{ij} describes the effect of an extra individual of species j upon the ability of the environment to support individuals of species i. For simplicity, if we neglect interspecies competition for resources, the equation reads

$$dN/dt = (Nr/K)(K-N),$$

which has the solution (for N as a function of time)

$$N(t) = N_0\, e^{rt}/[1 + (N_0/K)(e^{rt} - 1)].$$

If N_0/K is very small, the number of individuals in the species grows exponentially. Even if N_0/K is not small, for small times t (compared to the time $1/r$), the number grows approximately exponentially. Thus, a species exploiting a niche will at first increase exponentially. Then as t becomes much larger than $1/r$, the number approaches the carrying capacity of the niche, K. This behavior is illustrated for K 10 N_0 in Figure 26.2, and is essentially the logistic curve discussed in several early chapters.

The introduction of a single pair of African snails into the Miami area in 1967 led to an infestation within three years, one that cost $500,000 to eradicate (32). The introduction of the giant snail *Achatina fulica* into Hawaii in 1936 caused introduction of sixteen *other* species (seven beetles, two flies, and seven snails) in the attempt to control *A. fulica* (21). It cost $11 million to eradicate the Mediterranean fruit fly (medfly) from Florida in the late 1960s (33). It cost far more to eliminate the medfly from California in 1981 (34). The medfly has continued to appear occasionally in Florida during the 1980s.

Perhaps the worst insect pest ever released in the United States is the fire ant, which made its way into the country through Mobile, Alabama, in 1918. The stings of the ant can cause blisters and (occasionally) allergic reactions. In order to try to control the ant, the federal government supported application of the insecticide Mirex to the tune of $148 million. The program was cancelled because the insecticide has been implicated in fish kills and may cause cancer (35).

The Good News   Many insect and animal pests have natural predators they may have left behind. Judicious use of these natural enemies (sometimes imported) can control the pests or ameliorate the consequences of their presence. Understanding which virus attacks the gypsy moth

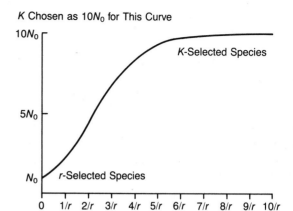

FIGURE 26.2
K-selected species are on the right-hand end and *r*-selected species on the left hand of this logistic curve. The parameter *K* is $10N_0$ for the curve shown.

In a given environment, a species may be close to its carrying capacity. The population is stable, and the rate of increase is low, as is mortality. Such species are called *K*-selected. Predator species are usually *K*-selected.

If members of a species have short lives, brief spans between generations, and explosive ability to increase ($1/r$ small), they can take advantage of transient environments. Such species are called *r*-selected. Bacteria and viruses are *r*-selected species.

K-selected species are especially vulnerable to disturbance of their environments, so it comes as no surprise that most endangered species are *K*-selected (31). The *r*-selected species include pests such as rats, which are able to "make hay while the sun shines"—they are population explosions waiting for the opportunity to happen.

can help lead to control of the pest. The deer population of many states, out of balance with the environment because of the eradication of wolves, could be controlled by judicious reimportation of wolves to reestablish the balance.

Many plants produce natural insecticides. Over ten thousand compounds protecting plants from insects are known (24). This promises hope for the future.

Toxic Waste versus People

The Bad News The story of industrial wastes has qualities reminiscent of a nightmare. Disease, miscarriage, and birth defects were found among residents in the suburban tracts built over an old waste dump near the Love Canal in Niagara Falls, New York (36). The publicity served to remind us all that over 250 million tonnes of hazardous industrial waste must be disposed of each year (37). The gruesome stories roll on of toxic waste pollution: Lewiston, New York (38); New Jersey, with fourteen thousand factories using chemicals (39); Connecticut, with six thousand sources of industrial waste chemicals (40); Love Canal and Hooker Chemical (41); Chester, Pennsylvania, with a site next to a liquid natural gas storage facility on which over ten thousand 55-gallon drums burst into flame (42); Bayou Sorrel, Louisiana, where a truck driver died from fumes from a waste pit

(43); over 100 million liters of waste in the String-fellow dump in California, some of which overflowed into a flood control channel (44); a legal dump in Charles City, Iowa, which is contaminating well water with arsenic and other chemical compounds (45); and a Swartz Creek, Michigan, dump, where several million liters of PCB-contaminated oil, hydrochloric acid, and cyanide were contaminating ground water (46). In Times Beach, Missouri, the government had to step in and buy the city from its inhabitants because a waste hauler had dumped oil laced with deadly dioxin in the area (46,47); the city of Joliet, Illinois, will have to spend $30 billion over ten years to clean up illegally disposed waste (48).

Europe has not fared much better. In Seveso, Italy, a factory explosion contaminated an entire town with dioxin. In 1983, residue dioxin from the disaster made legal and illegal moves all over Europe (49,50). The U.S. Army is suing Shell Oil for $1.9 G for contamination of the Rocky Mountain Arsenal (51). A chemical fire in Basel, Switz-erland, led to the pollution of the entire Rhine River to its mouth, and to massive fish kills.

There are over 7.5 million chemical compounds catalogued, seventy thousand chemical products on the market, and over a thousand new commercial chemicals appearing each year (48). These numbers are astounding. The waste disposal business gets bigger each year. It is now a $10 billion to $12 billion business annually; hazardous waste disposal alone is a billion-dollar-a-year business (52). There are 546 hazardous waste sites identified for action under Superfund (Figure 26.3) (53), which was created by a tax on chemical production as part of a bill enacted by Congress in 1980 to allow cleanup of contaminated disposal sites in the aftermath of Love Canal. Superfund has proved insufficient for the task, causing Congress to pass a further $10 billion program in 1985, funded by a tax of $3.1 billion on petroleum companies, $2 billion on chemical companies, and $2 billion on products of toxic wastes (54). The Germans are now worrying about

FIGURE 26.3
Hazardous waste sites as of August, 1983. The total number of sites in the U. S. is 546. These are to be cleaned up under the Comprehensive Environmental Response Compensation and Liability Act of 1980, the "Superfund," which was renewed in 1987. (U.S. Environmental Protection Agency)

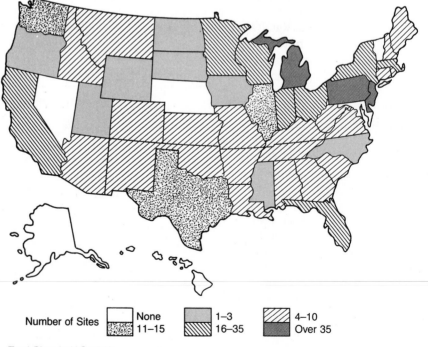

Number of Sites | None | 1–3 | 4–10
11–15 | 16–35 | Over 35

Total Sites in U.S.: 546

PCBs and other chemicals (55), and are contemplating setting up a Superfund of their own (56).

The Good News Among the good news is that Love Canal contamination does not appear to be spreading (57). Prevention of conditions that could lead to new Love Canals is the highest priority. Research is proceeding on methods for fixing the wastes chemically in a matrix (46,58), on new techniques for incineration—shipboard incineration and incineration on a bed of molten sodium carbonate (59)—and on recycling of chemical wastes to save money (60). Dow Chemical burns most of its wastes as fuel (48); but companies such as 3M, Allied Chemical, and Monsanto are getting more involved in recycling (60). By 1984, the Chemical Manufacturers Association claimed that a third of all chemicals produced were being recycled. Much of this reflects the rise in price for disposal of a barrel of waste from $24 in the late 1970s to about $100 in the mid-1980s (60). If all this waste were recycled and its venom made innocuous, we would be fulfilling our moral obligation to prevent harm to others, now and in the future.

Analysis of ground water has revealed significant contamination by toxic chemicals in 30 percent of the systems supplying water to U.S. municipalities of more than ten thousand people (61). As a result, Congress is considering new legislation that would supersede the Safe Drinking Water Act of 1974, and affect sixty thousand water supply companies. This legislation would not have been possible without great advances in water cleaning techniques, especially filtration with activated charcoal. It may even be possible to clean ground water in place.

In addition to contamination by toxic chemicals from dumps and industrial processes, it has been estimated that a hundred thousand underground gasoline storage tanks are leaking into ground water (62). A pioneering water cleaning plant was developed in Provincetown, Massachusetts, to deal with a small gasoline leak that had contaminated local ground water. It consists of four stages between recovery and ground discharge: separation, in which gasoline floats to the surface for removal; air stripping, in which gasoline evaporates from the water spray; sand and carbon filtration; and a final filtration with activated charcoal (62).

In this work, humanity may get help from Mother Nature. There are organisms that eat oil spilled on the oceans and on land (General Electric has patents on the bacteria that do it) (46). A white rot fungus (*Phanerochaete chrysosporium*) appears to be able to digest the insecticides lindane and DDT, as well as other, less well-known ones, to render them harmless (63).

PROSPECTS FOR THE FUTURE

It must seem that I am terribly pessimistic about the world and its future from the almost endless list of difficulties I have recounted through four of the last five chapters. I have proposed few solutions, and outlined few courses of action, because an important purpose of this book is to sound the alarm, the *tocsin,* and to arm the reader with the facts about energy and the environment necessary to deal rationally with the myriad of interrelated problems facing the human race. In this, the end of the book, I want to indulge my exhibitionist tendencies, and share with you, the reader, the reasons that I feel guardedly optimistic about the prospects for the survival and prosperity of the human race.

I do not believe that I am being Panglossian in my vision of the future. Why not? My father always said to me that things I would call problems (whatever they happened to be), were not *problems*; they were *opportunities*. In some sense, all the problems I have listed here are opportunities waiting for someone to take advantage of them. For example, I referred in the introduction to this chapter to "Spaceship Earth." John Stuart Mill (as early as 1857) eloquently defended the need of the closed world to move toward a stationary-state economy and a limited population (1):

If the Earth must lose that great portion of its pleasantness which it owes to things that the unlimited increase of wealth and population would extirpate from it, for the mere purpose of enabling it to support a larger, but not a happier or a better population, I sincerely hope, for the sake of posterity, that they will be content to be stationary, long before necessity compels them to it.

The stationary-state economy has been discussed at some length in a modern detailed analysis (64, 65). All such ideas are based on the closed aspect of the Earth system. But the "Spaceship Earth" analogy is at least inaccurate, if not downright misleading. After all, how did the Earth come to possess supplies of fossil fuel? The fossils were living plants or animals that lived by feeding on plants or by feeding on other animals who fed on plants, once long ago. The plants, the ultimate food source, grew because the sun shone, supplying the energy necessary for photosynthesis. All our fossil fuels are simply stored solar energy. We are not self-sufficient in this sense. It is solar energy that supplies energy sufficient to allow a decrease in the entropy here on Earth (although clearly the entropy of the universe increases in the process). The Earth is simply not a closed system.

It might be more accurate if we were to call it "Spaceship Solar System" instead of "Spaceship Earth." If we think of the solar system as a closed system, then some of our problems become opportunities. There are a lot of minerals available in the solar system—those in the asteroid belt are especially easy to send to Earth because there is no energy penalty for climbing out of a gravitational well. Of course, we do not know the mineral assay of the asteroids at present, but we should have one fairly soon (by which I mean the next twenty-five to fifty years). There must be sufficient minerals in the asteroid belt to supply mankind for eons. By suitably adjusting the speed of an asteroid at an appropriate point in its orbit, we ought to be able to make the asteroid pass through the Earth's orbit. We might also process asteroids onsite and ship the refined materials directly to Earth. Even for refined minerals, the asteroids have a great advantage over the

other planets as a source. This is so because we do not have to spend energy fighting a planet's gravitational pull as we would if we were transporting something from, say, Mars. We can use much less energy to transport the same mass of minerals from the asteroid belt.

In fact, if we were to establish factories in orbit, perhaps at the Lagrange points á la O'Neill's proposal to build habitats there (66), we could run industrial processes in space. It would be impossible to pollute the Earth, then, in the processing of materials. Of course, it would be difficult to run industrial processes as they are run today, because in space there are no free goods: air or water to be used at will. Industrial design in space would have to be sensible and conservative (in the best sense). This is possible only because of the huge amounts of solar energy available to orbiting industrial establishments. Gossamer-thin reflective plastic could easily be shaped into an extremely large parabolic mirror that could concentrate solar energy for use by the orbiting factories. This should be economically viable because only finished products would have to be sent down to the ground. Finished products are worth much more per unit mass than raw ore. This is important because the energy cost of bringing down material depends on the mass involved. A down-to-earth example is the ratio of cost of a computer chip made of silicon and germanium to that of the raw materials—a very large number!

This solution is essentially what is known in the trade as a "technological fix." There seems of late a great animosity toward anything hinting of technology. I think that there is nothing intrinsically wrong with using technology to improve human life, and to repair some of the mistakes we have made in applying the fruits of science through technology. As an example of technology applied to problems, electrons from the decay of ^{60}Co can be used to cleanse sewage sludge, and near Munich, West Germany, a gamma ray facility is doing a similar job (67). The irony of the antitechnology stance is that we just have no choice but to embrace technology. I am unwilling to "turn back the clock," and I have no intention of making my children turn it back either. Alvin Toffler, in *Future Shock* (68), writes

To turn our back on technology would not only be stupid but immoral. Given that a majority of men still figuratively live in the twelfth century, who are we even to contemplate throwing away the key to economic advance? Those who prate antitechnological nonsense in the name of some vague "human values" need to be asked "which humans?" To deliberately turn back the clock would be to condemn billions to enforced and permanent misery at precisely the moment in history when their liberation is becoming possible. We clearly need not less but more technology.

I do not mean to imply here that only large-scale technological endeavor is important or even that only huge technological projects are useful. The writings of E. F. Schumacher (69) emphasize the application of technology appropriate to the necessary end. Should tractors, say, be exported to Chad to help the country become agriculturally self-sufficient in lieu of encouraging the country toward a quasi-1890s technological base (reapers, improved harnesses, etc.) grown indigenously? This level of technology teaches the user and accustoms people to the use of comprehensible technology. Grafting the complete 1980s technology onto the unsophisticated, unsuspecting, and semifeudal Chadian economic base would not work—the graft would be rejected by a national "immune response system."

Lovins (70) has emphasized the desirability of the "soft energy path," one that is dispersed and technically simple, in contrast to the "hard energy path" of centralized large-scale energy generation. Lovins sees the two paths as mutually exclusive, nay, contradictory; I am sure that synthesis is possible. Each has its own appropriate place in an organized economy. A recent social trend of the sort to illustrate this complementarity can be observed in the phenomenon of the small, cottage industry business. No large concern would consider running a business that merely breaks even, providing employees' salaries and covering costs. Grocery chains routinely close profitable stores because their rates of return are not as

high as the chain's management desires. Thus, the advent of the small handcraft or exclusive service business providing employment but not gigantic profit gives balance to the economy and provides a measure of self-respect to the people involved, people who otherwise might have been reduced to bitter parasitism. The point is that in diversity lies strength—for energy sources, for plant species, for economic networks, and for application of technology. I have already discussed how plant monoculture can easily be destroyed by small changes in conditions, while natural plant systems recover from even rather large perturbations; how specialized predators can easily be wiped out when specific prey vanish. The same applies to applied technology. There will be no utopia resulting from utilizing only the soft or only the hard path, from using only small-scale or only large-scale technology. (Thank goodness. I have always thought that utopias were especially boring places.) Orbiting factories will not reconstitute the Garden of Eden (71); they will just serve to increase the resilience of the system. No matter which way we turn, no matter how different the conditions, people will be people in the same old familiar endearingly human, frustratingly human ways.

Returning to the problem of vanishing resources, I can see an opportunity for ingenuity closer to home: the mining of garbage dumps for the mineral resources we need (72). This is not being done to a great extent (Figure 26.4), even though it is generally much cheaper energetically to do so (Chapter 20). Our modern midden heaps are an atrocity in conception and in realization.

Of course, even with enhanced recycling of materials, there will still be resources "falling through the cracks." In the computer scenarios of the Club of Rome report (73), the recycling of materials merely staved off disaster for some short time interval. We shall have to continue exploiting resources whether here on Earth or elsewhere. It is very clear that recycling and reuse is a moral necessity. John Donne's words from the beginning of the chapter remind us: "No man

FIGURE 26.4

Old scrap reclaimed in the United States, 1976. (U.S. Bureau of Mines)

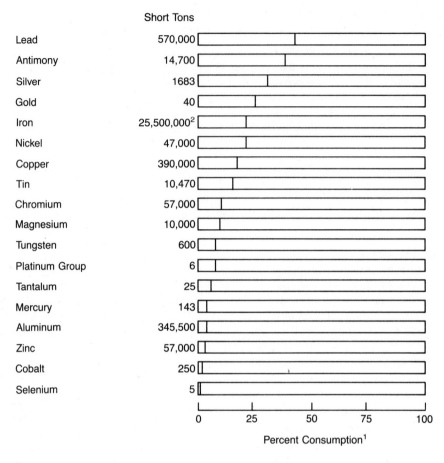

	Short Tons	
Lead	570,000	
Antimony	14,700	
Silver	1683	
Gold	40	
Iron	25,500,000[2]	
Nickel	47,000	
Copper	390,000	
Tin	10,470	
Chromium	57,000	
Magnesium	10,000	
Tungsten	600	
Platinum Group	6	
Tantalum	25	
Mercury	143	
Aluminum	345,500	
Zinc	57,000	
Cobalt	250	
Selenium	5	

Percent Consumption[1]

[1] Apparent Consumption = U.S. Primary + Secondary Production + Imports − Exports + Adjustments for Government and Industry Stock Changes.

[2] Includes Exports

is an island, entire of itself." We are obligated not to sacrifice our children's future to prolong our ease right now.

The developed nations of the world consume most of the world's nonrenewable resources to support their living standards. If it is moral to wish everyone to have a standard of living comparable to our own, we must either be prepared to lower our own standard of living to help achieve the balance, or we must find new sources of resources so that there is enough for all to raise standards in the rest of the world. This is the nub of the mistake made by those who argue that it is more immoral for Americans to have children than, say, Indians, because of the much greater impact of Americans on the environment. What is desirable is for all to live in plenty (or at least free of want); the ultimate impact of an Indian child is the same as that of an American child. Ultimately, both persons, or their descendants, will have to be supported in a similar style of life.

We may try to export population to other planets or to O'Neill's space colonies to raise living standards for those who remain. Europe chose this sort of solution. Over 50 million left to take up new lives in the Americas, out of a population varying between 250 million (1850) and 450 million (1950). Despite this emigration, despite the loss of millions killed in each of the two world wars, the population had doubled over the century. It is unlikely that emigration will, in the short run, ease the pressure of demand for resources.

The idea of lowering my personal living standard is unpleasant but not intolerable. However, I would much prefer to hope that unbridled imagination will lead to a world in which standards of living will level off upwards. Humanity has many talents, and many opportunities for exercising them now that the mass of people has begun to awaken to an awareness of the seriousness of the current energy and pollution situation. Will we use our talents and imagination to end the folly of our behavior? The conundrum is expressed by the historian Ortega y Gasset (74):

> To modern man is happening what was said of the Regent during the minority of Louis XV: he had all the talents except the talent to make use of them.

SUMMARY

Everything is connected to everything else. It is a sometimes sorry history—Earth's exploitation by its inhabitants meddling in ecosystems with disastrous consequences such as desertification or epidemics of plant diseases; disappearance of species from the Earth; throwing away of valuable resources; and careless handling of toxic wastes. It would be easy to succumb to despair.

Nevertheless, where there is life, there is hope. The solar system has resources in plenty to use to make life pleasant for all Earthlings. The sun is an energy source for the next 4 billion years or so. The asteroids contain mineral resources in easily collectible form. Two other planets in our solar system can most likely be made habitable. All this is possible.

Technology misuse has caused problems, but at this stage technology (applied science) is humanity's hope for the future. *Ad astra per aspera* ("Through work to the stars").

REFERENCES

1. J. S. Mill, *Principles of Political Economy* (Toronto: Univ. of Toronto Press, 1965).
2. R. Dubos, *The God Within* (New York: Charles Scribner's Sons, 1972).
3. G. T. T. Molitor, *The Futurist* 8, (1974): 169; P. Sears, in *Patient Earth,* ed. J. Harte, R. E. Socolow (New York: Holt, Rinehart & Winston, 1971).
4. S. S. King, *New York Times,* 10 Dec. 1978.
5. I. Campbell, *Ecologist* 4 (1974): 164.
6. R. Baker, *Ecologist* 4 (1974): 170.
7. J. Charney, P. H. Stone, and W. J. Quirk, *Science* 190 (1974): 741.
8. W. H. Matthews, *Intern. J. Environmental Studies* 4 (1973): 283; W. W. Kellogg and S. H. Schneider, *Science* 186 (1974): 1163; and C. Sagan, O. B. Toon, and J. B. Pollack, *Science* 206 (1979): 1363.
9. E. P. Eckholm, *Losing Ground* (New York: W. W. Norton, 1975).
10. R. M. Klein, in *Ants, Indians, and Little Dinosaurs,* ed. A. Ternes (New York: Charles Scribner's Sons, 1975).
11. E. Furness, *New York Times,* 26 Jan. 1975.
12. R. Revelle, *Science* 209 (1980): 164.
13. Quote from United Nations journal *Ceres,* as given in Reference 9, pg. 22.
14. B. Kadik, *Algerian Technical Bulletin* No. 8, 4/1982.
15. W. Campbell-Purdie, *Ecologist* 4 (1974): 300.
16. Algerian Ministry of Water, Environment, and Forests, *The Battle Against Desertification in Algeria: Experience of the Green Line,* 1984.
17. B. Kadik, *Algerian Technical Bulletin,* No. 7, 1/1982.
18. P. J. Vesilind, *Nat. Geog.* 168 (1985): 2.
19. C. Haberman, *New York Times,* 7 July 1985.

20. J. R. Harlan, *Science* 188 (1975): 618.

21. W. R. Courtenay, Jr. and C. R. Robins, *BioScience* 25 (1975): 306.

22. D. S. Brody, *Christian Science Monitor,* 14 May 1975.

23. H. Faber, *New York Times,* 20 April 1975.

24. S. H. Wittwer, *Science* 188 (1975): 579; J. E. Brody, *New York Times,* 26 May 1987.

25. W. Sullivan, *New York Times,* 10 Jan. 1978.

26. G. Wilkes, *Ecologist* 7 (1977): 312.

27. A. Anderson, Jr., *New York Times Magazine,* 27 April 1975, p. 15; D. W. Ehrenfeld, *Am. Sci.* 64 (1976): 648; M. Reisner, *NRDC Newsletter* 6, no. 1, (1977): 1; and E. Eckholm, *Futurist* 12 (1978): 289.

28. B. Webster, *New York Times,* 28 March 1976.

29. S. Seagrave, *BioScience* 26 (1976): 153.

30. B. Rensberger, *New York Times,* 14 April 1977; anonymous item, *New York Times,* 14 Jan. 1979.

31. T. W. Schoener, *Science* 185 (1974): 27; R. S. Miller and D. B. Botkin, *Am. Sci.* 62 (1974): 172.

32. W. B. Ennis, Jr., W. M. Dowler, and W. Klassen, *Science* 188 (1975): 593.

33. M. Beroza, *Am. Sci.* 59 (1971): 320.

34. E. Marshall, *Science,* 213 (1981): 417, 849.

35. J. E. Brody, *New York Times,* 20 April 1975.

36. J. Crossland, *Environment* 19, no. 5 (1977): 6; P. G. Hollie, *New York Times,* 22 March 1978; M. H. Brown, *New York Times Magazine,* 21 Jan. 1979, p. 23.

37. P. Shabecoff, *New York Times,* 18 Nov. 1984; and C. Norman, *Science* 220 (1983): 34.

38. R. Smothers, *New York Times,* 17 Dec. 1978.

39. M. Waldron, *New York Times,* 17 Dec. 1978.

40. M. L. Wald, *New York Times,* 17 Dec. 1978.

41. J. Miller, *New York Times,* 27 April 1979.

42. B. A. Franklin, *New York Times,* 27 April 1979.

43. F. F. Marcus, *New York Times,* 27 April 1979.

44. G. Hill, *New York Times,* 27 April 1979.

45. W. Robbins, *New York Times,* 27 April 1979.

46. A. A. Boraiko, *Nat. Geog.* 167 (1985): 318.

47. See, for example, M. Sun, *Science* 219 (1983): 468.

48. W. Williams, *New York Times,* 13 March 1983.

49. D. Dickson, *Science* 220 (1983): 1362.

50. M. Dowling, *Environment* 27, no. 3 (1985): 18.

51. J. Peterson, *New York Times,* 11 Dec. 1983.

52. R. Blumenthal, *New York Times,* 25 Nov. 1984.

53. Department of Commerce, *Statistical Abstract of the United States, 1984* (Washington, D.C.: GPO, 1984), Figure 7.2.

54. P. Shabecoff, *New York Times,* 11 Dec. 1985.

55. W. Gehrmann, *die Zeit,* 26 July 1985.

56. I. Meyer-List, *die Zeit,* 5 July 1985.

57. T. H. Maugh II, *Science,* 215 (1982): 490; R. J. Smith, *Science* 217 (1982): 714; and R. J. Smith, *Science* 217 (1982): 808.

58. T. H. Maugh II, *Science,* 204 (1979): 1295.

59. T. H. Maugh II, *Science,* 204 (1979): 1188.

60. S. J. Marcus, *New York Times,* 8 Jan. 1984.

61. L. A. Daniels, *New York Times,* 3 Nov. 1985.

62. Anonymous, *New York Times,* 8 Dec. 1985.

63. J. A. Bumpus et al., *Science* 228 (1985): 1434.

64. H. E. Daly, in *Patient Earth,* ed. J. Harte and R. E. Socolow (New York: Holt, Rinehart & Winston, 1971).

65. D. L. Meadows, ed., *Alternatives to Growth—I* (Cambridge, Mass: Ballinger, 1977).

66. G. K. O'Neill, *New York Times Magazine,* 18 Jan. 1976, p. 10.; G. K. O'Neill, *The High Frontier* (New York: William Morrow & Co., 1977).

67. J. G. Trump, *Am. Sci.* 69 (1981): 276.

68. A. Toffler, *Future Shock* (New York: Random House, 1970).

69. E. F. Schumacher, *Small is Beautiful* (New York: Perennial Library, 1975).

70. A. Lovins, *Foreign Affairs* 55 (1976): 65; and A. Lovins, *Soft Energy Paths* (Cambridge, Mass.: Friends of the Earth—Ballinger, 1977).

71. P. Csonka, *Futurist* 11 (1977): 285.

72. R. R. Grinstead, *Environment* 14, no. 3 (1972): 2 and 14, no. 4 (1972): 34; J. G. Abert, H. Alter, and J. F. Bernheisel, *Science* 183 (1974): 1052; J. McCaull, *Environment* 16, no. 1 (1974): 6; L. D. Orr, *Environment* 18, no. 10 (1976): 33; P. K. DeJoie, *Environment* 19, no. 7 (1977): 32; L. Hastings, *Environment* 19, no. 7 (1977): 38; A. Van Dam, *New Ecologist* 1 (1978): 20; J. P. Sterba, *New York Times,* 16 May 1978; and R. Smothers, *New York Times,* 14 Jan. 1979.

73. D. H. Meadows et al., *The Limits to Growth* (New York: Signet, 1972); M. Mesarovic and E. Pestel, *Mankind at the Turning Point* (New York: E.P. Dutton, 1974).

74. J. Ortega y Gasset, *The Revolt of the Masses* (New York: W. W. Norton & Co., 1974).

PROBLEMS AND QUESTIONS

1. Can you think of other dangerous dabbling by the human race?

2. How does the removal of cover affect flat land? hilly land? semi-arid land?

3. It is thought that the ancestors of most Europeans originally shared the Mongolian plateau with the Mongolian peoples of Asia. It is known that the Hittites, the Scythians, and the various Aryan invaders of India came from that region over quite a long timespan. This area is now the Gobi desert. Advance scenarios for the historical migrations of these various peoples.

4. In Roman times, it took a man on a horse three days' ride into Libya to reach the desert. Now the desert is everywhere within 50 km of the coast. What could have happened? Are the people entirely at fault?

5. Discuss the advantages and disadvantages of using only a few types of plants to feed humanity.

6. Ehrenfeld (27) argues that non-resources should be conserved. By this he means that conservation cannot and should not rely on economic or ecological justifications. He thinks that each species on the Earth, evolved over millions of years, is natural art, and that a species lost is permanently gone. He says: "What sort of change in the world view would favor the conservation of non-resources? Nothing less than a rejection of the heroic, Western ethic with its implicit denial of man's biological roots and evolved structure."
Comment on Ehrenfeld's ideas.

7. Fifty years ago, 80 percent of the wheat grown in Greece was of the native strains; today, almost all native strains seem to have disappeared. Is this desirable?

8. Michael Brown (36) reports that much toxic material is illegally dumped. Violators have loosened tank-truck valves and dumped contaminants along the road (a New York company spilled out polychlorinated biphenyls—PCBs— from a truck onto 270 miles of North Carolina highway). What are the attitudes of company officials that could tolerate this sort of atrocity? Is indiscriminate dumping a national pastime? Where do you dispose of trash and garbage on long car trips?

9. Discuss the issues in the controversy over whether one should follow a "hard" or "soft" energy path.

10. E. F. Schumacher has discussed the use of technology that fits the people's level of sophistication. How can this best be done? Why would Schumacher be against use of Western technology in developing countries?

11. Toxic wastes have caused problems for municipal water supplies. What can be done to clean up water? How could this problem be prevented?

12. How is the concept "Spaceship Earth" an inaccurate analogy?

13. How is the introduction of foreign flora and fauna into new areas advantageous? Disadvantageous?

APPENDIX 1: SCIENTIFIC NOTATION

KEY TERMS   *scientific notation*
              multiplication
              commutativity
              division
              equation
              powers of ten
              eka-
              peta-
              tera-
              giga-
              mega-
              kilo-
              milli-
              micro-
              nano-
              pico-
              femto-
              atto-
              centi-
              hecto-
              order of magnitude

The language of science is mathematics. Most scientific ideas are presented most succinctly as equations. In order to talk about science, we have to recall some mathematical ideas and agree on a way of talking that will reduce confusion. In this section, we briefly review multiplication, division, the meaning of an equation, and the manipulation of equations. We then continue to the main business at hand, scientific notation.

MULTIPLICATION

When we take two numbers and multiply them together—for example 3×6—we are forming the number that results from taking the second number and adding it to itself the first number of times (in the example, $6 + 6 + 6 = 18$). Alternatively, we take the first number and add it to itself the second number of times (in the example, $3 + 3 + 3 + 3 + 3 + 3 = 18$). The operation is more complex when the numbers are not integer, but the concept itself is unchanged. Note that it is a property of the real numbers (all the infinite number of numbers running in the continuum from $-\infty$ to $+\infty$) that they are commutative; this means that the positions of the first and second and so on numbers do not matter to the result of the operation. Specifically, the operation of multiplication is commutative.

DIVISION

When we divide one number by another, we are asking how many times the denominator (the number on the bottom) fits into the numerator (the number on the top). For example, $18/3$ is a shorthand for the number that results when 18 is broken up in units of 3; the result is 6. The operation might work this way: we begin with 18. Now, $18 - 3$ is 15, and 3 has fit once so far. Next, $15 - 3$ is 12, and 3 fits another time; that is, twice so far. Continuing, $12 - 3$ is 9, and 3 has so far fit $2 + 1 = 3$ times. In a similar way, 9 $- 3$ is 6, with 4 threes so far; $6 - 3$ is 3. At last, we find that 4 plus 2 is 6, the total number of 3s in 18.

Although a division such as $10.8/2.42$ is mechanically more complicated, the procedure is the same; we ask how many times the number 2.42 fits into the number 10.8 or, how big 10.8 is when it is broken into units of size 2.42.

There is no restriction whatsoever on the numbers allowed in either position. For example, 2 divided by 5 $(2/5)$ asks how big 2 is in units 5 big. This is, of course, a number smaller than one, and we have a way of expressing it—as a decimal. In the example, $2/5$ is 0.4. The extra zero in front of the decimal point does not matter. It is simply put in that position for clarity in the written expression.

MEANING AND MANIPULATION OF EQUATIONS

An equation is a statement. It says that one collection of numbers (or symbols) has the same value as another collection of numbers (or symbols). An example of an equation containing numbers is $18/3 = 6$. This equation states that the numerical value on the left-hand side of the equal sign is the same as the numerical value on the right-hand side of the equal sign.

An equation is invariably such a statement of numerical equivalence. However, it may appear to be more subtle in some cases because of unfamiliarity with the content of the equation, when that equation contains symbols rather than numbers. An example of a simple equation containing both symbols and numbers is $2.5z = 10$. In this example, the numerical value of the left-hand side of the equation must be the same as the numerical value of the right-hand side of the equation, just as before. In this particular example, the numerical value of z is fixed; only one number when substituted for z will leave the numerical value of the right-hand side the same as that of the left-hand side. This number is 4, because 4×2.5 is $2.5 + 2.5 + 2.5 + 2.5$, or 10.

There is another way to achieve the same result. Notice that if we divide both sides of the equation $2.5z = 10$ by the same number, the resulting statement is still an equation. That is, the number on the right-hand side and the number on the left-hand side have the same numerical value. The number will in general be different from the number that was previously the equivalent on both sides of the equal sign. Suppose we divide both sides of $2.5z = 10$ by the number 2.5. On the left-hand side, $2.5z/2.5$ asks how many times 2.5 fits into $2.5z$. By our definitions above, this is simply z. On the right-hand side, $10/2.5$ is just 4 by our previous discussion of division. We may then restate the equation $2.5z = 10$ as $z = 4$.

When an equation has mixed numbers and symbols on both sides, it is a general statement of relation. For example, the equation $v^2 = 2gx$ states that the product of v with itself is the same as 2 times g times x. If we are given that g is some number (say, 10), we can simplify the right-hand side: $2gx = 2(10)x = 20x$. Here we have said that, if g is equal to 10, the equation is the same whether the symbol or the number is used. Our example equation now takes the form $v^2 = 20x$. Since we do not know a specific value for v or for x, we cannot specifically solve the equation numerically—that is, we cannot state the numerical values of v and x the way we did the value for z in the previous example above. What we *can* say is that there is a relation between the *possible* values of v and the *possible* values of x.

If we knew x were 20, for example, we would know v to be 20; if x were zero, v would be zero; and so on. There are an infinite number of possible values of x for the infinite possible values of v, two for each value of x. There are two v terms for each x because the sign of v is not determined.

The most concise way to present this infinite amount of information is in a graph, which can show some sample of the behavior of v for different values of x. The graph is shown in Figure A.1.1. Note that both positive and negative values of v give the same value of x.

Equations in physics are often shorthand ways of presenting just such an infinitude of information. If x is the symbol for the height in meters above the ground, and v is the symbol for speed in meters/second, then the equation $v^2 = 2(10$ m/s$^2)x$ describes their relation to one another. The relation is more important than the numerical values because the equation is a completely general statement about any body's speed when that body is any arbitrary distance x above the ground.

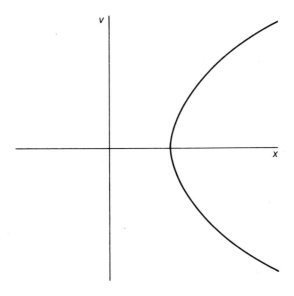

FIGURE A1.1
The graphic relation between v and x following from the equation $v^2 = (\text{constant})x$.

We also may solve equations that are general relations to isolate the dependence of one of the unknown quantities in terms of the others. For example, we may have an equation $ab = cd/e$. Suppose we wish to solve for c in terms of the other unknown quantities. This involves manipulations of the equation given in ways that do not change its identity as an equation. We multiply both sides of the equation by e: $abe = cd$. Then we divide both sides of the equation by d: $abe/d = c$. By this process, the desired unknown, c, has been isolated. We often go through such manipulations in this book.

POWERS OF TEN

Numbers are expressible as powers of other numbers. By the power of a number we mean the number of times it multiplies itself; 10 to the power 2 (10^2) means 10 times 10 ($10^2 = 10 \times 10$). An interesting property of such numbers is that two numbers expressed this way are easily multiplied. As an example, $10^2 \times 10^3$ is a shorthand for $[(10 \times 10) \times (10 \times 10 \times 10)]$, which is otherwise known as 10^5. That is, we may multiply two numbers of this form by simply adding their exponents. The word *exponent* is a synonym for the power of a number; in 10^A, A is the power, or exponent, of 10. Although we have so far restricted ourselves to integer exponents, there is nothing to prevent noninteger exponents. These are discussed in Chapter 8 and Appendix 3. In general, $10^A \times 10^B$ is $10^{A + B}$.

$$\underbrace{(10 \times 10 \times \ldots \times 10)}_{A \text{ tens}} \times \underbrace{(10 \times \ldots \times 10)}_{B \text{ tens}}$$

$$= \underbrace{10 \times 10 \times \ldots \times 10}_{A + B \text{ tens}}$$

These numbers, 10 and its powers, arise naturally in our number system. Because we have ten fingers, and our ancestors used them to keep track of numbers of things, we have ten unit numbers, ten units of tens of units, ten units of hundreds of units, and so forth. Our numbers are implicitly expressed in terms of powers of ten.

Imagine drawing a line across the page and dividing it into columns under the heading units (1), tens (10), hundreds (10^2), thousands (10^3), and so on. The year of the first notice of the energy crises by the general public, 1973, is, in terms of entries under this column, 1 in the thousands column, 9 in the hundreds column, 7 in the decades column, and 3 in the units column

10^5	10^4	10^3	10^2	10	1
		1	9	7	3

Note that this just reproduces the number 1973 as the column entries. In fact, the number 1973 is just the column-entry shorthand for one thousand, nine hundreds, seven decades, and three units. We could also write 1973 as $1 \times 10^3 + 9 \times 10^2 + 7 \times 10 + 3$. Any number in our number system can be expressed in one of these alternative ways.

So far we have discussed only multiplication of these numbers. Division is also simple. Consider the division of 10^3 by 10, which we will write as the multiplication $10^3 \times (1/10)$. The result with this rewriting is easily obtained

$$(10 \times 10 \times 10) \times (\tfrac{1}{10}) = (10 \times 10) \times (10 \times \tfrac{1}{10})$$
$$= (10 \times 10) \times 1 = 10^2.$$

We may follow this pattern for any number of the form 10^A

$$10^A/10 = 10^A \times \tfrac{1}{10} = 10^{A-1+1} \times \tfrac{1}{10} =$$
$$10^{A-1} \times 10 \times \tfrac{1}{10} = 10^{A-1}.$$

It is clear that if we wish to follow the pattern $10^A \times 10^B = 10^{A+B}$, we must write 1/10 as 10^{-1}. It follows that $1/100 = 1/(10 \times 10)$ is 10^{-2}, $1/10^3 = 10^{-3}$, etc.

Now suppose we were to multiply 100 by 1/100. The result is of course 1. If we use our notation, though, $10^2 \times 10^{-2} = 10^{2-2} = 10^0$. In order to continue to use the rule, we must choose $10^0 = 1$. With this completion, we can write our column headings more fully

$$\ldots \quad 10^5 \quad 10^4 \quad 10^3 \quad 10^2 \quad 10^1 \quad 10^0 \quad 10^{-1} \quad 10^{-2} \quad \ldots$$

Suppose we filled in a number in our columns

10^1	10^0	10^{-1}	10^{-2}
1	5	2	5

$\ldots$ before 10^1 and after 10^{-2} $\ldots$

Had we filled in only the two columns under 10^1 and 10^0, we would know how to write the number: 15. We cannot write the number 1525, because that would mean 1×10^3 plus $\ldots$ instead of 1×10^1 plus $\ldots$. In the scheme we have adopted, position matters. To supplement the original scheme and allow translation of the column headings (to eliminate some of the clumsiness), we add a positional indicator: the decimal point. The decimal point is the indicator of the shift from numbers whose size is greater than 1, to numbers whose size is smaller than 1. In the case of the example given above, the number is written 15.25. The decimal point thus marks the switch from powers greater than zero to powers less than zero.

SCIENTIFIC NOTATION

In an attempt to preserve clarity in the transmission of scientific information, scientists have made an agreement to use a special way of writing numbers. A number such as the speed of light, 300,000,000 m/s, is so large it boggles the mind. It would be very easy to add or lose a zero in writing such a number. To eliminate such possible confusion as could result, all numbers of interest are written in a notation $M \times 10^N$, where M is a number between 1.0 and $9.999 \ldots$, and N is an integer exponent.

The speed of light in this notation is 3.0×10^8 m/s. There is a greatly reduced possibility of error in this form of expression.

UNIT PREFIXES

In ordinary speech, we have a shorthand for large numbers. They are grouped and given names such as thousands, billions, etc. As in common speech, it is usual to give special names to factors 10^3 apart in magnitude. Large international conventions of scientists agree on the names to be given to multipliers, and assign symbols to stand for the size. We shall here illustrate the most common names by their use as prefixes for lengths. The meter is the standard of length for every country in the world, and we shall explore its prefixes:

10^{18} m	Em	exameter
10^{15} m	Pm	petameter
10^{12} m	Tm	terameter
10^9 m	Gm	gigameter
10^6 m	Mm	megameter
10^3 m	km	kilometer
10^0 m	m	meter
10^{-3} m	mm	millimeter
10^{-6} m	μm	micrometer
10^{-9} m	nm	nanometer
10^{-12} m	pm	picometer
10^{-15} m	fm	femtometer
10^{-18} m	am	attometer

There are two other prefixes in common use. Centi- means one one-hundredth, and is commonly used with meter: a centimeter is slightly less than ½ inch. Hect(o)- means one hundred. Areas are commonly given in hectares (10^4 m^2). A hectare is slightly less than 2½ acres (2.471 acres). Occasionally, volumes are given in hectoliters.

ORDER OF MAGNITUDE

An *order of magnitude* is a factor of 10^1.

The important unit names are separated by three orders of magnitude; that is, by a factor of $(10^1)^3 = 10^3$.

A human ovum may be about 1 μm in diameter. The human adult has a typical height of 1½ to 2 m. Thus, we say that the size of an adult human is about 6 orders of magnitude larger than the ovum from which it developed. A newborn child is, say, about 50 cm (20 in.) long. Thus an adult is only a factor of 3 or 4 longer than an infant. Since this is not a factor of 10, we may say that the adult and the baby have lengths of about the same order of magnitude.

A child at birth typically weighs 5 to 9 lb and an adult, 100 to 200 lb (this depends on the adult's sex). Since the ratio of weights is 10 to 40, we may say that an adult's weight is about an order of magnitude greater than a child's.

Beware the often-found usage of order of magnitude for a factor of 2.

**PROBLEMS
AND QUESTIONS**

True or False
1. A meter is an order of magnitude longer than a centimeter.
2. A gallon of milk contains an order of magnitude more milk than a pint.
3. The number 13×10^{-7} is expressed in scientific notation.
4. Multiplication is just repeated addition.

Multiple Choice
5. The population of Mauritius is 900,000. If infant mortality is 70 per 1000 infants born, and the birthrate is 31 per 1000 population, the yearly number of dead infants is
   a. 1.95×10^2.
   b. 2.03×10^7.
   c. 3.99×10^4.
   d. 4.15×10^8.
   e. none of the above.

Problems
6. $10^2 \times 10^5 =$
7. $10^{17} \times 10^{-24} =$

8. $10^{-5} \times 10^{-10} =$

9. $10^4/10^{-5} =$

10. $10^6 \times 10^{-6} =$

11. The Republic of South Africa had an estimated 1970 population of 20,100,000. Its people belong to the following "racial" groups: blacks, 67 percent; whites, 20 percent; "colored," 10 percent; "Asians," 3 percent. In addition, 60 percent of the whites speak Afrikaans and the rest speak English. Express all these numbers in scientific notation and calculate the number of people in each subgroup (expressed in scientific notation).

12. Write in scientific notation
    a. 230,000,000 (U.S. population)
    b. 1/10,000
    c. 3,617,204 (U.S. area, in mi²)
    d. 96,981 (Oregon area, in mi²)
    e. 668,700 (number of American Indians in United States in 1970)

13. A person can run 100 meters in about 10 seconds. A cheetah can run about 300 m; a jet plane can go about 2.7 kilometers; and an *Apollo* spacecraft can go 20 kilometers in about the same time. Express the respective speeds in meters/second in scientific notation. Express the speed of light in the same units in scientific notation. Calculate the ratios of the above speeds to the speed of light and express in scientific notation. The speed of light is 300,000,000 m/s.

14. Solve the equation $F = qvB$ for v.

15. Solve the following equation for v_1:

$$mgh_1 + \tfrac{1}{2} mv_1^2 = mgh_2$$

Discussion Questions

16. For what reasons do you think it would be useful to develop scientific notation? Need it have been in powers of ten?

17. Of what utility is the idea of "order of magnitude"?

18. Use the ideas introduced in the section on division to show that 1 over a fraction is the reciprocal of the fraction: $1/(a/b) = b/a$.

APPENDIX 2: LOGARITHMS

When noninteger real numbers appear as powers of ten, the resulting number can be *any* number. As we vary the exponent of 10 between 0 and 1, all numbers between 1 and 10 result. There is a one-to-one correspondence between the numbers in the interval 0 to 1 and the numbers in the interval 1 to 10.

Such a table is clearly useful if only because of the difficulties of translation of numbers such as 10^{ct}. In fact, such tables were constructed long ago: They are called tables of logarithms. The logarithm (log for short) of a number is the power of some base number which corresponds one-to-one with that number. The number ct is the log to the base 10 of 10^{ct} (we write $ct = \log_{10} 10^{ct}$). Since $2^3 = 8$, we could say that the log of 8 to the base 2 is 3. Any number whatsoever can be used as a base. The most commonly used bases are 10 and $e = 2.718 \ldots$. The former base we have considered in some detail; the latter base is noninteger, and logarithms expressed to this base are known as natural logarithms. In this natural base, equations of the form $\Delta N = aN\Delta t$ have the simple solution $N(t) = N_0 e^{at}$.

A table of logarithms to the base 10 may be found at the end of this appendix. In Table A2.1, we present selected integer logarithms. The log tables contain only logs of numbers between 1 and 10. It is easy to understand why these are all

TABLE A2.1
Selected integer logarithms.

$10^{-4} = 1/10{,}000$	$\log_{10} 10^{-4} = -4$
$10^{-3} = 1/1000$	$\log_{10} 10^{-3} = -3$
$10^{-2} = 1/100$	$\log_{10} 10^{-2} = -2$
$10^{-1} = 1/10$	$\log_{10} 10^{-1} = -1$
$10^0 = 1$	$\log_{10} 10^0 = 0$
$10^1 = 10$	$\log_{10} 10^1 = 1$
$10^2 = 100$	$\log_{10} 10^2 = 2$
$10^3 = 1000$	$\log_{10} 10^3 = 3$

that are necessary. Consider the multiplication of two numbers of the form 10^A and 10^B. The resultant number, R, is $10^A \times 10^B = 10^{A+B}$. By our definition of $\log_{10}$, then

$$\log_{10} R = \log_{10} 10^A + \log_{10} 10^B = A + B$$

This is simply a re-expression of the rule just discussed in Appendix 1: The log of the product of two numbers is the sum of their respective logs. Thus, the $\log_{10}$ of a number such as 62 is

$$\log_{10} 62 = \log_{10} (6.2 \times 10) = \log_{10} 6.2 + \log_{10} 10$$
$$= \log_{10} 6.2 + 1.$$

We need only look up the number representing 6.2 in the log table. To find the $\log_{10}$ of a number such as 0.037, we rewrite 0.037 as 3.7×10^{-2}

531

EXPONENTIAL GROWTH

How is it possible that $N(t) = N_0 10^{ct}$ is a solution to $N = aN\Delta t$? Let us suppose that it is a solution, and see if we can calculate ΔN. We write

$$\Delta N = N(t + \Delta t) - N(t)$$
$$= N_0 10^{c}(t + \Delta t) - N_0 10^{ct}.$$

Using the fact that $10^{ct + c\Delta t} = 10^{ct} \times 10^{c\Delta t}$, we gather common factors

$$\Delta N = N_0 10^{ct}(10^{c\Delta t} - 1).$$

If Δt were $\approx$ zero (i.e., the time interval is small), the factor in parentheses would be near zero because $10^0 = 1$. So $\Delta N = 0$ if $\Delta t = 0$. In fact, for small Δt, ΔN is proportional to Δt, as seems plausible from our argument: $\Delta N = N_0 10^{ct} (c \log_e 10 \, \Delta t)$. Since $N_0 10^{ct}$ is just $N(t)$,

$$\Delta N = (c \log_e 10)N(t) \, \Delta t.$$

Therefore, $c = a/\log_e 10$, and

$$N(t) = N_0 e^{at}.$$

Here e represents a special base in which the solution to $\Delta N = aN\Delta t$ is particularly simple.

EXPONENTIAL DECAY—RADIOACTIVE DECAY

Radioactive decay in nuclei is a random occurrence; whether a particular nucleus will undergo decay is an unanswerable question. However, if a substantial number of the same sort of radioactive nuclei are gathered together, we can predict that a certain percentage will decay spontaneously in a given time. That is, ΔN, the number of nuclei that have decayed in time Δt, is proportional to N. Again, we get a geometrical (or exponential) change. Such a situation is described by $N(t) = N_0 e^{-at}$; because of the minus sign, N decreases with time instead of growing, as in the examples in the text. The time required for one-half the nuclei in a group to decay is called the half-life of the nucleus. A free neutron, for example, has a half-life of about ten minutes. If we begin with 1 million free neutrons when the clock starts ticking, we will have only 500,000 ten minutes later. In a further ten minutes, this is halved to 250,000; and in the next ten minutes, it is halved again to 125,000. This will continue until there are too few neutrons left for us to predict decay with any certainty.

Many nuclei have half-lives less than ten minutes, and many others have very long half-lives. Particles involved in so-called strong nuclear interactions may have lifetimes as short as 10^{-21}s. The proton has been predicted by Grand Unified Theories to have a lifetime in excess of 10^{31} years, and recent experiments have shown that it must actually be greater than 10^{32} years.

EXAMPLES OF CALCULATIONS WITH LOGARITHMS

1. $\log_{10}(10^{-3} \times 10^{-4}) = \log_{10} 10^{-3} + \log_{10} 10^{-4}$
$$= (-3) + (-4) = -7$$

2. $32 \times 98 = ?$

$\log_{10}(32 \times 98) = \log_{10} 32 + \log_{10} 98 = \log_{10}(3.2 \times 10^1) + \log_{10}(9.8 \times 10^1)$
$$= \log_{10} 3.2 + 1 + \log_{10} 9.8 + 1.$$

Now read off the logs from Table A2.2:

$\log_{10} 3.2 = 0.5051.$
$\log_{10} 9.8 = 0.9912.$

Then, substituting

$\log_{10}(32 \times 98) = 0.5051 + 1 + 0.9912 + 1 = 3.4963.$

Thus the answer to 32×98 is $10^{3.4963} = 10^3 \times 10^{0.4963} = 1000 \times 10^{0.4963}$. Looking up 0.4963 in the table of logarithms, Table A2.2, we find t is between 3.13 and 3.14, closer to 3.14. For most purposes, this answer, 3.140 is good enough. Note that the answer resulting from multiplication is 3.136. We could actually approximate this rather well by interpolation: 0.4963 is 8/14 of the way from the log of 3.13 to the log of 3.14; $8/14 = 0.57 \approx 0.6$. Thus, the answer is approximately 3.136.

3. $32/98 = ?$

Since division is the inverse of multiplication (see Appendix 1), the rule is that the log of one number dividing another is the respective difference of the logs

$\log_{10}(32/98) = \log_{10} 32 - \log_{10} 98 = 1.5051 - 1.9912 = -0.4861,$

from the previous example. It is not clear at first how to handle this number until we recognize that, as with the last example, we eventually wish to have the number in the form of scientific notation before we can decipher it. Thus we write $10^{-0.4861}$ as $10^{-1 + 0.5139} = 10^{-1} \times 10^{0.5139}$. In this form, we see that the number having the log of 0.5139 is between 3.26 and 3.27, closer to 3.27 ($7/13 = 0.54$ of the way). Hence, we get 3.265×10^{-1}. By long division we get $0.32653 \ldots$.

$\log_{10} 0.037 = \log_{10}(3.7 \times 10^{-2})$
$$= \log_{10} 3.7 + \log_{10} 10^{-2}$$
$$= \log_{10} 3.7 + (-2) = \log_{10} 3.7 - 2.$$

Some further examples are presented in the box.

Logarithms are useful because they provide a way for us to graph exponential, or geometrical, growth easily. Since the logarithm of a number changes from 0 to 1 as the number changes from 1 to 10, we may choose to make an axis of our graph paper proportional on a linear scale to the log of the number. This provides a nonlinear, logarithmic, scale for the number itself.

TABLE A2.2

Table of logarithms for numbers from 1.00 to 9.99.

1.0	0	1	2	3	4	5	6	7	8	9
.0	.000	.004	.009	.013	.017	.021	.025	.029	.033	.037
.1	.041	.045	.049	.053	.057	.061	.064	.068	.072	.076
.2	.079	.083	.086	.090	.093	.097	.100	.104	.107	.111
.3	.114	.117	.121	.124	.127	.130	.134	.137	.140	.143
.4	.146	.149	.152	.155	.158	.161	.164	.167	.170	.173
.5	.176	.179	.182	.185	.188	.190	.193	.196	.199	.201
.6	.204	.207	.210	.212	.215	.217	.220	.223	.225	.228
.7	.230	.233	.236	.238	.241	.243	.246	.248	.250	.253
.8	.255	.258	.260	.262	.265	.267	.270	.272	.274	.276
.9	.279	.281	.283	.286	.288	.290	.292	.294	.297	.299

2.0	0	1	2	3	4	5	6	7	8	9
.0	.301	.303	.305	.307	.310	.312	.314	.316	.318	.320
.1	.322	.324	.326	.328	.330	.332	.334	.336	.338	.340
.2	.342	.344	.346	.348	.350	.352	.354	.356	.358	.360
.3	.362	.364	.365	.367	.369	.371	.373	.375	.377	.378
.4	.380	.382	.384	.386	.387	.389	.391	.393	.394	.396
.5	.398	.400	.401	.403	.405	.407	.408	.410	.412	.413
.6	.415	.417	.418	.420	.422	.423	.425	.427	.428	.430
.7	.431	.433	.435	.436	.438	.439	.441	.442	.444	.446
.8	.447	.449	.450	.452	.453	.455	.456	.458	.459	.461
.9	.462	.464	.465	.467	.468	.470	.471	.473	.474	.476

3.0	0	1	2	3	4	5	6	7	8	9
.0	.477	.479	.480	.481	.483	.484	.486	.487	.489	.490
.1	.491	.493	.494	.496	.497	.498	.500	.501	.502	.504
.2	.505	.507	.508	.509	.511	.512	.513	.515	.516	.517
.3	.519	.520	.521	.522	.524	.525	.526	.528	.529	.530
.4	.531	.533	.534	.535	.537	.538	.539	.540	.542	.543
.5	.544	.545	.547	.548	.549	.550	.551	.553	.554	.555
.6	.556	.558	.559	.560	.561	.562	.563	.565	.566	.567
.7	.568	.569	.571	.572	.573	.574	.575	.576	.577	.579
.8	.580	.581	.582	.583	.584	.585	.587	.588	.589	.590
.9	.591	.592	.593	.594	.595	.597	.598	.599	.600	.601

4.0	0	1	2	3	4	5	6	7	8	9
.0	.602	.603	.604	.605	.606	.607	.609	.610	.611	.612
.1	.613	.614	.615	.616	.617	.618	.619	.620	.621	.622
.2	.623	.624	.625	.626	.627	.628	.629	.630	.631	.632
.3	.633	.634	.635	.636	.637	.638	.639	.640	.641	.642
.4	.643	.644	.645	.646	.647	.648	.649	.650	.651	.652
.5	.6532	.6542	.6551	.6561	.6571	.6580	.6590	.6599	.6609	.6618
.6	.6628	.6637	.6646	.6656	.6665	.6675	.6684	.6693	.6702	.6712
.7	.6721	.6730	.6739	.6749	.6758	.6767	.6776	.6785	.6794	.6803
.8	.6812	.6821	.6830	.6839	.6848	.6857	.6866	.6875	.6885	.6893
.9	.6902	.6911	.6920	.6928	.6937	.6946	.6955	.6964	.6972	.6981

TABLE A2.2 continued

5.0	0	1	2	3	4	5	6	7	8	9
.0	.6990	.6998	.7007	.7016	.7024	.7033	.7042	.7050	.7059	.7067
.1	.7076	.7084	.7093	.7101	.7110	.7118	.7126	.7135	.7143	.7152
.2	.7160	.7168	.7177	.7185	.7193	.7202	.7210	.7218	.7226	.7235
.3	.7243	.7251	.7259	.7267	.7275	.7284	.7292	.7300	.7308	.7316
.4	.7324	.7332	.7340	.7355	.7356	.7364	.7372	.7380	.7388	.7396
.5	.7404	.7412	.7419	.7427	.7435	.7443	.7451	.7459	.7466	.7474
.6	.7482	.7490	.7497	.7505	.7513	.7520	.7528	.7536	.7543	.7551
.7	.7559	.7566	.7574	.7582	.7589	.7597	.7604	.7612	.7619	.7627
.8	.7634	.7642	.7649	.7657	.7664	.7672	.7679	.7686	.7694	.7701
.9	.7709	.7716	.7723	.7731	.7738	.7745	.7752	.7760	.7767	.7774

6.0	0	1	2	3	4	5	6	7	8	9
.0	.7782	.7789	.7796	.7803	.7810	.7818	.7825	.7832	.7839	.7846
.1	.7853	.7860	.7868	.7875	.7882	.7889	.7896	.7903	.7910	.7917
.2	.7924	.7931	.7938	.7945	.7952	.7959	.7966	.7973	.7980	.7987
.3	.7993	.8000	.8007	.8014	.8021	.8028	.8035	.8041	.8048	.8055
.4	.8062	.8069	.8075	.8082	.8089	.8096	.8102	.8109	.8116	.8122
.5	.8129	.8136	.8142	.8149	.8156	.8162	.8169	.8176	.8182	.8189
.6	.8195	.8202	.8209	.8215	.8222	.8228	.8235	.8241	.8248	.8254
.7	.8261	.8267	.8274	.8274	.8287	.8293	.8299	.8306	.8312	.8319
.8	.8325	.8331	.8338	.8344	.8351	.8357	.8363	.8370	.8376	.8382
.9	.8388	.8395	.8401	.8407	.8414	.8420	.8426	.8432	.8439	.8445

7.0	0	1	2	3	4	5	6	7	8	9
.0	.8451	.8457	.8463	.8470	.8476	.8482	.8488	.8494	.8500	.8506
.1	.8513	.8519	.8525	.8531	.8537	.8543	.8549	.8555	.8561	.8567
.2	.8573	.8579	.8585	.8591	.8597	.8603	.8609	.8615	.8621	.8627
.3	.8633	.8639	.8645	.8651	.8657	.8663	.8669	.8675	.8681	.8686
.4	.8692	.8698	.8704	.8710	.8716	.8722	.8727	.8733	.8739	.8745
.5	.8751	.8766	.8762	.8768	.8774	.8779	.8785	.8791	.8797	.8802
.6	.8808	.8814	.8820	.8825	.8831	.8837	.8842	.8848	.8854	.8859
.7	.8865	.8871	.8876	.8882	.8887	.8893	.8899	.8904	.8910	.8915
.8	.8921	.8932	.8932	.8938	.8943	.8949	.8954	.8960	.8965	.8971
.9	.8976	.8982	.8987	.8993	.8998	.9004	.9009	.9015	.9020	.9025

8.0	0	1	2	3	4	5	6	7	8	9
.0	.9031	.9036	.9042	.9047	.9053	.9058	.9063	.9069	.9074	.9079
.1	.9085	.9090	.9096	.9101	.9106	.9112	.9117	.9122	.9128	.9133
.2	.9138	.9143	.9149	.9154	.9159	.9165	.9170	.9175	.9180	.9186
.3	.9191	.9196	.9201	.9206	.9212	.9217	.9222	.9227	.9232	.9238
.4	.9243	.9248	.9253	.9258	.9263	.9269	.9274	.9279	.9284	.9289
.5	.9294	.9299	.9304	.9309	.9315	.9320	.9325	.9330	.9335	.9340
.6	.9345	.9350	.9355	.9360	.9365	.9370	.9375	.9380	.9385	.9390
.7	.9395	.9400	.9405	.9410	.9415	.9420	.9425	.9430	.9435	.9440
.8	.9445	.9450	.9455	.9460	.9465	.9369	.9375	.9479	.9484	.9489
.9	.9494	.9499	.9504	.9509	.9513	.9518	.9523	.9528	.9533	.9538

TABLE A2.2 continued

9.0	0	1	2	3	4	5	6	7	8	9
.0	.9542	.9547	.9552	.9557	.9562	.9566	.9571	.9576	.9581	.9586
.1	.9590	.9595	.9600	.9605	.9609	.9614	.9619	.9624	.9628	.9633
.2	.9638	.9643	.9647	.9652	.9657	.9661	.9666	.9671	.9675	.9680
.3	.9685	.9689	.9694	.9699	.9703	.9708	.9713	.9717	.9722	.9727
.4	.9731	.9736	.9741	.9745	.9750	.9754	.9759	.9763	.9768	.9773
.5	.9777	.9782	.9786	.9791	.9795	.9800	.9805	.9809	.9814	.9818
.6	.9823	.9827	.9832	.9836	.9841	.9845	.9850	.9854	.9859	.9863
.7	.9868	.9872	.9877	.9881	.9886	.9890	.9894	.9899	.9903	.9908
.8	.9912	.9917	.9921	.9926	.9930	.9934	.9939	.9943	.9948	.9952
.9	.9956	.9961	.9965	.9969	.9974	.9978	.9983	.9987	.9991	.9996

PROBLEMS AND QUESTIONS

Multiple Choice

1. The logarithm of 4.2 is 0.6232. The logarithm of 3 is 0.4771. The logarithm of $420 \times (3 \times 10^{-4})$ is

   a. -0.8997
   b. 0.1260
   c. 0.7148
   d. 3.1003
   e. 7.1003

Problems

(Use the logarithm table, Table A2.2, to solve.)

2. $12 \times 20 =$
3. $(6 \times 10^8) \times (1.6 \times 10^{27}) =$
4. $(7 \times 10^{-8}) \times (7 \times 10^{20}) =$
5. $(9 \times 10^{-17}) \times (8 \times 10^{21}) =$
6. (800 million) $\times$ (0.1%) $=$

APPENDIX 3: UNDERSTANDING TABULAR DATA

Often the information a person needs is in tables full of numbers; this can seem so overwhelming that the search for the information is given up in disgust. Organizing data into tabular form is certainly one way to try to put information into a form suitable for interpretation. Such attempts usually are successful only when the table is sufficiently short, or when only modest correlations are presented.

It is often possible to get relevant information from government sources in tabular form. In this section, we tackle the question of how to interpret and use a barrage of tables to examine questions of public interest.

Tables A3.1–A3.6 are Tables 20, 21, 29, 30, 31, and 32 from the government-commissioned SRI report *Patterns of Energy Consumption in the United States* (1). These tables indicate the extent of saturation (percent of American households owning a particular item) and of energy use, for the years 1960 to 1968. The amount of information in these tables is overwhelming. How can any sense be made of them?

Perhaps the first thing to do is to try to understand the trend. We can move our eyes down the columns one at a time. Looking at Table 20, for example, we see that the number of households increased steadily between 1960 and 1968.

So too did the number of gas and electric water heaters. By looking at the saturation column, we discover that both gas and electric water heating use grew faster than the number of households. In addition, we see that gas water heaters are growing in use more rapidly than electric water heaters. Rapid scanning of the other tables gives much the same trend.

You now have some information. Is there more information in the tables? The answer is yes. If you were an appliance manufacturer, these tables could tell you which items enjoy the biggest possibilities for growth—namely, those with the smallest market saturation. You would compare and then make a decision on the basis of that comparison. Such work is tedious indeed with the number of tables to be considered even here. In a real-life situation, more alternatives would probably be considered. The problem is one of *organization* and *presentation*.

By constructing comparison tables from the original tables, we can make the requisite comparisons, but this is a tedious task, and the comparisons are still difficult to make and interpret.

The simplest solution is to use graphs to organize the data in the tables. We might plot the saturation of many items on one sheet of graph paper. This allows the observers to use their eyes

TABLE A3.1
(TABLE 20)
Saturation of water heaters in residences, 1960–68.

	Number of Households* (millions)	Percentage Saturation**		Number of Water Heaters in Use (millions)	
		Gas	Electric	Gas	Electric
1960	53.0	54%	20%	28.6	10.6
1961	53.3	56	21	29.8	11.2
1962	54.7	58	22	31.7	12.0
1963	55.2	59	22	32.7	12.2
1964	56.0	61	23	34.2	12.9
1965	57.3	63	23	36.1	13.2
1966	58.1	65	23	37.8	13.7
1967	58.8	66	24	38.8	14.1
1968	60.4	68	24	41.2	14.5

Statistical Abstract of the United States, various years; household count as of March of years shown; mobile homes are included.

**Merchandising Week;* 1960 Census, Gas Appliance Manufacturers Association, and SRI estimates.

TABLE A3.2 (TABLE 21)
Residential energy consumption for water heating.

	Electricity				Gas			
			Total Consumption				Total Consumption†	Total Energy Consumption
Year	Units (millions)	Consumption* (kWh)	(billions of kWh)	(trillions of Btu)	Units (millions)	Unit Consumption** (millions of Btu)	(trillions of Btu)	(trillions of Btu)
1960	10.6	4272	45.3	429	28.6	25.5	730	1159
1961	11.2	4272	47.8	444	29.8	25.5	761	1205
1962	12.0	4290	51.6	474	31.7	25.5	808	1282
1963	12.2	4300	51.7	480	32.7	25.8	846	1326
1964	12.9	4320	55.6	512	34.2	26.0	889	1401
1965	13.2	4390	57.8	532	36.1	26.4	952	1484
1966	13.7	4400	60.1	563	37.8	26.8	1013	1576
1967	14.1	4420	62.4	579	38.8	27.0	1049	1628
1968	14.5	4490	65.4	613	41.2	27.2	1125	1738
		12,816 (1960)				7508 (1960)		

*Source is Edison Electric Institute with SRI estimates.

**Source is American Gas Association with SRI estimates.

†Both natural and liquefied petroleum gas.

TABLE A3.3
(TABLE 29)
Saturation of television sets in residences (thousands of units).

	Sales*		Replacements**		Sets in Use	
	Monochrome	Color	Monochrome	Color	Monochrome	Color
Base†					44,924	1389
1960	5605	117	2242	47	48,287	1459
1961	6047	144	2419	58	51,915	1545
1962	6460	438	2584	175	55,791	1808
1963	7141	749	2856	300	60,076	2257
1964	8542	1541	3417	616	65,201	3182
1965	8954	2827	3582	1131	70,573	4878
1966	7904	5549	3162	2220	75,315	8207
1967	5384	6496	2154	2598	78,545	12,105
1968	6296	7865	2518	3146	82,323	16,824
1969	7270	6607	2908	2643	86,685	20,788

*Source is Current Industrial Reports, Bureau of the Census.

**Assumed at 40% of sales.

†Source is National Survey of Television Sets in U.S. households, June 1967, Advertising Research Foundation.

TABLE A3.4 (TABLE 30)
Residential energy consumption for television.

	Monochrome Sets			Color Sets			Total Energy Consumption	
Year	Number of Units (millions)	Unit Consumption* (kWh)	Total Consumption (billions of kWh)	Number of Units (millions)	Unit Consumption (kWh)	Total Consumption (billions of kWh)	(billions of kWh)	(trillions of Btu)
1960	48.2	345	16.6	1.5	450	0.7	17.3	163
1961	51.9	345	17.9	1.5	450	0.7	18.6	172
1962	55.8	346	19.3	1.8	455	0.8	20.1	186
1963	60.1	349	21.0	2.3	460	1.1	22.1	203
1964	65.2	350	22.8	3.2	465	1.5	24.3	225
1965	70.6	352	24.9	4.9	470	2.3	27.2	251
1966	75.3	356	26.8	8.2	475	3.9	30.7	288
1967	78.5	359	28.2	12.1	482	5.8	34.0	315
1968	82.3	360	29.6	16.8	490	8.2	37.8	352

*Source is Edison Electric Institute with SRI estimates.

TABLE A3.5 (TABLE 31)

Saturation of air-conditioning units in residences, (thousands of units) 1960–68.

	Electric Room Air Conditioners				Electric Central Air Conditioners				Gas Central Air Conditioners	
Base††	Sales*	Replacement Units (50%)	Incremental Units (50%)	Units in Use	Shipments**	Replacement Units (30%)	Incremental Units (70%)	Units in Use	Sales†	Units in Use
				7126				996		
1960	1402	701	701	7827	312	94	218	1214		
1961	1327	663	664	8491	366	110	256	1470		
1962	1445	722	723	9214	467	140	327	1797		15***
1963	1868	934	934	10,148	580	174	406	2203	11	26
1964	2565	1282	1283	11,431	701	210	491	2694	17	43
1965	2755	1377	1378	12,809	828	248	580	3274	20	63
1966	3101	1550	1551	14,360	960	288	672	3946	29	92
1967	3839	1919	1920	16,280	1047	314	733	4679	29	121
1968	3747	1873	1874	18,154	1165	350	815	5494	39	160

*Source is Current Industrial Reports, Bureau of the Census.
**Source is Current Statistical Review, Metal Products Manufacturing.
†Source is H. R. Linden, "Current Trends in U.S. Gas Demand and Supply," Public Utilities Fortnightly.
††Source is 1960 Census of Housing.
***SRI estimate.

TABLE A3.6 (TABLE 32)

Residential energy consumption for air conditioning.

	Electric								Gas			
	Room Air Conditioning			Central Air Conditioning			Total Consumption		Central Air Conditioning			Total Energy Consumption (trillions of Btu)
Year	Units (millions)	Unit Consumption* (kWh)	Total Consumption (billions of kWh)	Units (thousands)	Unit Consumption** (kWh)	Total Consumption (billions of kWh)	(billions of kWh)	(trillions of Btu)	Units (thousands)	Unit Consumption (millions of Btu)	Total Consumption (trillions of Btu)	
1960	7.8	1250	10	1214	3600	4	14	133	—	—	—	133
1961	8.5	1265	11	1470	3600	5	16	150	—	—	—	150
1962	9.2	1280	12	1797	3600	6	18	167	15	20	—	167
1963	10.1	1295	13	2203	3600	8	21	195	26	20	1	196
1964	11.4	1310	15	2694	3600	10	25	230	43	20	1	231
1965	12.8	1325	17	3274	3600	12	29	272	63	20	1	273
1966	14.4	1310	19	3916	3600	14	33	310	92	20	2	312
1967	16.3	1360	22	4679	3600	17	39	361	121	20	2	363
1968	18.2	1375	25	5494	3600	20	45	423	160	20	3	426

*Source is Edison Electric Institute with SRI estimates.
**SRI estimate.

to make visual comparisons more easily than if reading the same data from tables. With this method we can compare saturation and observe the trends in saturation with time.

It is clear from Figure A3.1 that there is really no possibility of great sales of refrigerators, ranges, or washing machines. You sell only to new families or to replace old units. Color TV, air conditioners, electric dryers, and dishwashers offered (as of 1968) considerable scope for increasing market penetration. There are great possibilities for sales to families not owning these items. Notice that many households have two black and white TVs or one color and one black and white TV.

If you sell electricity, the saturation curves in Figure A3.1 are interesting but insufficient. In order to supply electricity, we also must know how much energy is used per item as well as how fast the saturation changes. Note the energy use curves in Figure A3.2, in which all energies are given in kWh$_e$. This is partly due to population growth, but it also depends on efficiency.

We can see about how population increases by looking at the cooking or washing curves for energy consumption, since from Figure A3.1 these are about constant in saturation. The refrigeration curve is growing faster since people are replacing older refrigerators with newer frost-free models that are less efficient. Another way to see the same result is to present the *per household* energy consumption, as in Figure A3.3.

These three curves give us a fairly complete idea of all the information contained in the tables.

FIGURE A3.1
Saturation index for selected items.

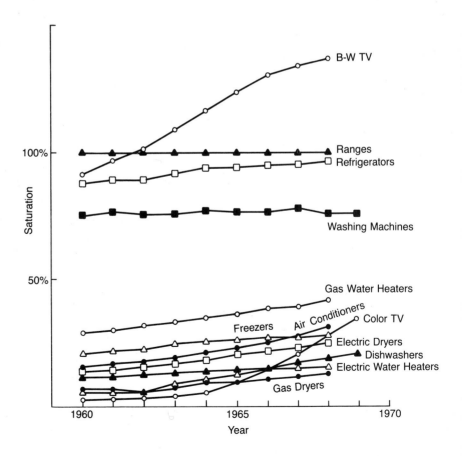

The information is more readily used in graphical form and the comparisons are much easier. The graph is one of the most important tools of the informed person.

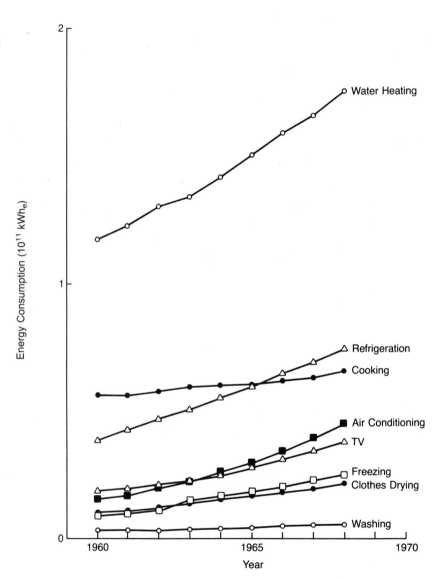

FIGURE A3.2
Energy use for specific purposes.

FIGURE A3.3
Per household energy
consumption.

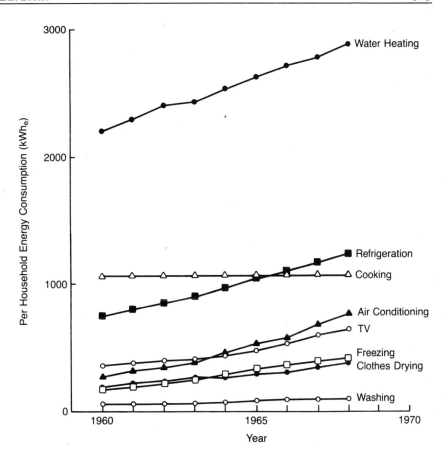

REFERENCES

1. Stanford Research Institute, *Patterns of Energy Consumption in the United States* (Washington, D.C.: GPO, 1972).

INDEX

Note: **Boldface** numbers indicate definition.

544